Lecture Notes in Computer Science 12911

Advanced Research in Computing and Software Science

Subline of Lecture Notes in Computer Science

More information about this subseries at http://www.springer.com/series/7407

Organization

Program Committee

Akanksha Agrawal	Indian Institute of Technology Madras, India
Therese Biedl	University of Waterloo, Canada
Sergio Cabello	University of Ljubljana, Slovenia
Steven Chaplick	Maastricht University, The Netherlands
Vincent Cohen-Addad	Google Zürich, Switzerland
Jiří Fiala	Charles University, Czech Republic
Robert Ganian	Technische Universität Wien, Austria
Gwenaël Joret	Université Libre de Bruxelles, Belgium
Eun Jung Kim	Université Paris-Dauphine, France
Sándor Kisfaludi-Bak	Max Planck Institute for Informatics, Saarbrücken, Germany
Łukasz Kowalik (Co-chair)	University of Warsaw, Poland
Marvin Künnemann	Max Planck Institute for Informatics, Saarbrücken, Germany
O-joung Kwon	Incheon National University, South Korea
Erik Jan van Leeuwen	Utrecht University, The Netherlands
Martin Milanič	University of Primorska, Slovenia
Michał Pilipczuk (Co-chair)	University of Warsaw, Poland
Paweł Rzążewski (Co-chair)	Warsaw University of Technology and University of Warsaw, Poland
Sophie Spirkl	University of Waterloo, Canada
Kavitha Telikepalli	Tata Institute of Fundamental Research, Mumbai, India
Przemysław Uznański	University of Wrocław, Poland
Birgit Vogtenhuber	Graz University of Technology, Austria

Organizing Committee

Łukasz Kowalik	University of Warsaw, Poland
Konrad Majewski	University of Warsaw, Poland
Karolina Okrasa	University of Warsaw & Warsaw University of Technology, Poland
Michał Pilipczuk	University of Warsaw, Poland
Marcin Pilipczuk	University of Warsaw, Poland
Paweł Rzążewski	Warsaw University of Technology & University of Warsaw, Poland

Additional Reviewers

Jungho Ahn
Aditya Anand
Guillaume Aubian
Benjamin Bergougnoux
Ahmad Biniaz
Henning Bruhn-Fujimoto
Binh-Minh Bui-Xuan
Yixin Cao
Jérémie Chalopin
Guilherme de Castro Mendes Gomes
Vaggos Chatziafratis
Pratibha Choudhary
Brian Cloteaux
Sabine Cornelsen
Logan Crew
Ágnes Cseh
Radu Curticapean
Clément Dallard
Sanjoy Dasgupta
Dariusz Dereniowski
Michał, Dębski
Cemil Dibek
Stephan Dominique Andres
Feodor Dragan
Pavel Dvořák
Eduard Eiben
David Eppstein
Günş Erdoğan
Thomas Erlebach
Bruno Escoffier
Andreas Emil Feldmann
Stefan Felsner
Laurent Feuilloley
Celina Figueiredo
Arnold Filtser
Till Fluschnik
Esther Galby
Cyril Gavoille
Yuval Gitlitz
Petr Golovach
Daniel Gonçalves
Fabrizio Grosso
Bruno Guillon
Gregory Gutin
Sepehr Hajebi
Thekla Hamm
Tesshu Hanaka
Meike Hatzel
Klaus Heeger
Michael Hoffmann
Svein Høgemo
Patrick Hompe
Shenwei Huang
Chien-Chung Huang
Vesna Iršič
Lars Jaffke
Pallavi Jain
Xinrui Jia
Carl Johan Casselgren
Lawqueen Kanesh
Ringi Kim
Fabian Klute
Dušan Knop
Petr Kolman
Dariusz Kowalski
Laszlo Kozma
Murali Krishna Enduri
Jari J. H. de Kroon
Peter L. Erdos
Bundit Laekhanukit
Abhiruk Lahiri
Juho Lauri
Ray Li
Yanjia Li
Jason Li
Paloma de Lima
Patrick Lin
Maarten Löffler
Zbigniew Lonc
Jayson Lynch
Jayakrishnan Madathil
Konrad Majewski
Diptapriyo Majumdar
Frederik Mallmann-Trenn
Mathieu Mari
Tomáš Masařík

Simon Mauras
Neeldhara Misra
Pranabendu Misra
Matthias Mnich
Wojciech Nadara
Meghana Nasre
André Nichterlein
Jana Novotná
André Nusser
Karolina Okrasa
Krzysztof Onak
Tim Ophelders
Deryk Osthus
Kenta Ozeki
Peter Palfrader
Pan Peng
Marta Piecyk
Théo Pierron
Mikaël Rabie
M. S. Ramanujan
Felix Reidl
Marc Roth
M. S. Ramanujan
R. B. Sandeep
Manfred Scheucher
Šimon Schierreich
Ildikó Schlotter
Mordechai Shalom
Sebastian Siebertz
Giannos Stamoulis
Raphael Steiner
Kenny Štorgel
Prafullkumar Tale
Till Tantau
Oleg Verbitsky
Mathilde Vernet
Roland Vincze
Magnus Wahlström
Zhiyu Wang
Sebastian Wiederrecht
Marcin Wrochna
Mingxian Zhong
Anna Zych

The Long Tradition of WG

WG 1975	U. Pape – Berlin, Germany
WG 1976	H. Noltemeier – Göttingen, Germany
WG 1977	J. Mühlbacher – Linz, Austria
WG 1978	M. Nagl, H. J. Schneider – Burg Feuerstein, near Erlangen, Germany
WG 1979	U. Pape – Berlin, Germany
WG 1980	H. Noltemeier – Bad Honnef, Germany
WG 1981	J. Mühlbacher – Linz, Austria
WG 1982	H. J. Schneider, H. Göttler – Neuenkirchen, near Erlangen, Germany
WG 1983	M. Nagl, J. Perl – Haus Ohrbeck near Onasbrück, Germany
WG 1984	U. Pape – Berlin, Germany
WG 1985	H. Noltemeier – Schloß Schwanberg near Würzburg, Germany
WG 1986	G. Tinhofer, G. Schmidt – Stift Bernried near Munich, Germany
WG 1987	H. Göttler, H. J. Schneider – Kloster Banz near Bamberg, Germany
WG 1988	J. van Leeuwen – Amsterdam, The Netherlands
WG 1989	M. Nagl – Castle Rolduc, The Netherlands
WG 1990	R. H. Möhring – Johannesstift Berlin, Germany
WG 1991	G. Schmidt, R. Berghammer – Fischbachau near Munich, Germany
WG 1992	E. W Mayr – Wilhelm-Kempf-Haus, Wiesbaden-Naurod, Germany
WG 1993	J. van Leeuwen – Utrecht, The Netherlands
WG 1994	G. Tinhofer, E. W. Mayr, G. Schmidt – Munich, Germany

WG 1995	M. Nagl – Haus Eich, Aachen, Germany
WG 1996	G. Ausiello, A. Marchetti-Spaccamela – Cadenabbia near Como, Italy
WG 1997	R. H. Möhring – Bildungszentrum am Müggelsee, Berlin, Germany
WG 1998	J. Hromkovič, O. Sýkora – Smolenice Castle, Slovakia
WG 1999	P. Widmayer – Monte Verità, Ascona, Switzerland
WG 2000	D. Wagner – Waldhaus Jakob, Konstanz, Germany
WG 2001	A. Brandstädt, Boltenhagen near Rostock, Germany
WG 2002	L. Kučera – Český Krumlov, Czech Republic
WG 2003	H. L. Bodlaender – Elspeet, The Netherlands
WG 2004	J. Hromkovič, M. Nagl – Bad Honnef, Germany
WG 2005	D. Kratsch – Île du Saulcy, Metz, France
WG 2006	F. V. Fomin – Sotra near Bergen, Norway
WG 2007	A. Brandstädt, D. Kratsch, H. Müller – Jena, Germany
WG 2008	H. Broersma, T. Erlebach – Durham, UK
WG 2009	C. Paul, M. Habib – Montpellier, France
WG 2010	D. M. Thilikos – Zarós, Crete, Greece
WG 2011	J. Kratochvíl – Teplá Monastery, West Bohemia, Czech Republic
WG 2012	M. C. Golumbic, G. Morgenstern, M. Stern, A. Levy – Israel
WG 2013	A. Brandstädt, K. Jansen, R. Reischuk – Lübeck, Germany
WG 2014	D. Kratsch, I. Todinca – Le Domaine de Chalès, Orléans, France
WG 2015	E. W. Mayr – Garching near Munich, Germany
WG 2016	P. Heggernes – Rumeli Hisarüstü, Istanbul, Turkey
WG 2017	H. L. Bodlaender, G. J. Woeginger – Eindhoven, The Netherlands
WG 2018	A. Brandstädt, E. Köhler, K. Meer – Cottbus, Germany
WG 2019	I. Sau, D. M. Thilikos – Vall de Núria, Catalunya, Spain
WG 2020	I. Adler, H. Müller – Leeds, UK (virtual)
WG 2021	Ł. Kowalik, M. Pilipczuk, P. Rzążewski – Warsaw, Poland (virtual)

Contents

Preprocessing to Reduce the Search Space: Antler Structures for Feedback Vertex Set

Huib Donkers(✉) and Bart M. P. Jansen

Eindhoven University of Technology, Eindhoven, The Netherlands
{h.t.donkers,b.m.p.jansen}@tue.nl

Abstract. The goal of this paper is to open up a new research direction aimed at understanding the power of preprocessing in speeding up algorithms that solve NP-hard problems exactly. We explore this direction for the classic Feedback Vertex Set problem on undirected graphs, leading to a new type of graph structure called *antler decomposition*, which identifies vertices that belong to an optimal solution. It is an analogue of the celebrated *crown decomposition* which has been used for Vertex Cover. We develop the graph structure theory around such decompositions and develop fixed-parameter tractable algorithms to find them, parameterized by the number of vertices for which they witness presence in an optimal solution. This reduces the search space of fixed-parameter tractable algorithms parameterized by the solution size that solve Feedback Vertex Set.

Keywords: Kernelization · Preprocessing · Feedback vertex set · Graph decomposition

1 Introduction

The goal of this paper is to open up a new research direction aimed at understanding the power of preprocessing in speeding up algorithms that solve NP-hard problems exactly [25,30]. In a nutshell, this new direction can be summarized as: how can an algorithm identify part of an optimal solution in an efficient preprocessing phase? We explore this direction for the classic [36] Feedback Vertex Set problem on undirected graphs, leading to a new graph structure called *antler* which reveals vertices belonging to an optimal feedback vertex set.

We start by motivating the need for a new direction in the theoretical analysis of preprocessing. The use of preprocessing, often via the repeated application

This project has received funding from the European Research Council (ERC) under the European Union's Horizon 2020 research and innovation programme (grant agreement No. 803421, ReduceSearch).

The original version of this chapter was previously published non-open access. A Correction to this chapter is available at https://doi.org/10.1007/978-3-030-86838-3_32

Ł. Kowalik et al. (Eds.): WG 2021, LNCS 12911, pp. 1–14, 2021.
https://doi.org/10.1007/978-3-030-86838-3_1

of reduction rules, has long been known [3,4,41] to speed up the solution of algorithmic tasks in practice. The introduction of the framework of parameterized complexity [20] in the 1990s made it possible to also analyze the power of preprocessing *theoretically*, through the notion of kernelization. It applies to *parameterized decision problems* $\Pi \subseteq \Sigma^* \times \mathbb{N}$, in which every instance $x \in \Sigma^*$ has an associated integer parameter k which captures one dimension of its complexity. For FEEDBACK VERTEX SET, typical choices for the parameter include the size of the desired solution or structural measures of the complexity of the input graph. A kernelization for a parameterized problem Π is then a polynomial-time algorithm that reduces any instance with parameter value k to an equivalent instance, of the same problem, whose total size is bounded by $f(k)$ for some computable function f of the parameter alone. The function f is the *size* of the kernelization.

A substantial theoretical framework has been built around the definition of kernelization [16,21,26,28,30]. It includes deep techniques for obtaining kernelization algorithms [11,27,37,40], as well as tools for ruling out the existence of small kernelizations [12,18,22,29,31] under complexity-theoretic hypotheses. This body of work gives a good theoretical understanding of polynomial-time data compression for NP-hard problems.

However, we argue that these results on kernelization *do not* explain the often exponential speed-ups (e.g. [3], [5, Table 6]) caused by applying effective preprocessing steps to non-trivial algorithms. Why not? A kernelization algorithm guarantees that the input *size* is reduced to a function of the parameter k; but the running time of modern parameterized algorithms for NP-hard problems is not exponential in the total input size. Instead, fixed-parameter tractable (FPT) algorithms have a running time that scales polynomially with the input size, and which only depends exponentially on a problem parameter such as the solution size or treewidth. Hence an exponential speed-up of such algorithms cannot be explained by merely a decrease in input size, but only by a decrease in the *parameter*!

We therefore propose the following novel research direction: to investigate how preprocessing algorithms can decrease the parameter value (and hence search space) of FPT algorithms, in a theoretically sound way. It is nontrivial to phrase meaningful formal questions in this direction. To illustrate this difficulty, note that strengthening the definition of kernelization to "a preprocessing algorithm that is guaranteed to always output an equivalent instance of the same problem with a strictly smaller parameter" is useless. Under minor technical assumptions, such an algorithm would allow the problem to be solved in polynomial time by repeatedly reducing the parameter, and solving the problem using an FPT or XP algorithm once the parameter value becomes constant. Hence NP-hard problems do not admit such parameter-decreasing algorithms. To formalize a meaningful line of inquiry, we take our inspiration from the VERTEX COVER problem, the fruit fly of parameterized algorithms.

A rich body of theoretical and applied algorithmic research has been devoted to the exact solution of the VERTEX COVER problem [5,23,32,33]. A standard

2-way branching algorithm can test whether a graph G has a vertex cover of size k in time $\mathcal{O}(2^k(n+m))$, which can be improved by more sophisticated techniques [14]. The running time of the algorithm scales linearly with the input size, and exponentially with the size k of the desired solution. This running time suggests that to speed up the algorithm by a factor 1000, one either has to decrease the input size by a factor 1000, or decrease k by $\log_2(1000) \approx 10$.

It turns out that state-of-the-art preprocessing strategies for VERTEX COVER indeed often *succeed* in decreasing the size of the solution that the follow-up algorithm has to find, by means of crown-reduction [2,15,24], or the intimately related Nemhauser-Trotter reduction based on the linear-programming relaxation [39]. Recall that a vertex cover in a graph G is a set $S \subseteq V(G)$ such that each edge has at least one endpoint in S. Observe that if $H \subseteq V(G)$ is a set of vertices with the property that there exists a minimum vertex cover of G containing all of H, then G has a vertex cover of size k if and only if $G - H$ has a vertex cover of size $k - |H|$. Therefore, if a preprocessing algorithm can identify a set of vertices H which are guaranteed to belong to an optimal solution, then it can effectively reduce the parameter of the problem by restricting to a search for a solution of size $k - |H|$ in $G - S$.

A *crown decomposition* (cf. [1,15,24], [16, §2.3], [28, §4]) of a graph G serves exactly this purpose. It consists of two disjoint vertex sets $(\mathsf{head}, \mathsf{crown})$, such that crown is a non-empty independent set whose neighborhood is contained in head, and such that the graph $G[\mathsf{head} \cup \mathsf{crown}]$ has a matching M of size $|\mathsf{head}|$. As crown is an independent set, the matching M assigns to each vertex of head a private neighbor in crown. It certifies that any vertex cover in G contains at least $|\mathsf{head}|$ vertices from $\mathsf{head} \cup \mathsf{crown}$, and as crown is an independent set with $N_G(\mathsf{crown}) \subseteq \mathsf{head}$, a simple exchange argument shows there is indeed an optimal vertex cover in G containing all of head and none of crown. Since there is a polynomial-time algorithm to find a crown decomposition if one exists [2, Thm. 11–12], this yields the following preprocessing guarantee for VERTEX COVER: if the input instance (G, k) has a crown decomposition $(\mathsf{head}, \mathsf{crown})$, then a polynomial-time algorithm can reduce the problem to an equivalent one with parameter at most $k - |\mathsf{head}|$, thereby giving a formal guarantee on reduction in the parameter based on the structure of the input.[1]

As the first step of our proposed research program into parameter reduction (and thereby, search space reduction) by a preprocessing phase, we present a graph decomposition for FEEDBACK VERTEX SET which can identify vertices S that belong to an optimal solution; and which therefore facilitate a reduction from finding a solution of size k in graph G, to finding a solution of size $k - |S|$

[1] The effect of the crown reduction rule can also be theoretically explained by the fact that interleaving basic 2-way branching with exhaustive crown reduction yields an algorithm whose running time is only exponential in the *gap* between the size of a minimum vertex cover and the cost of an optimal solution to its linear-programming relaxation [38]. However, this type of result cannot be generalized to FEEDBACK VERTEX SET since it is already NP-complete to determine whether there is a feedback vertex set whose size matches the cost of the linear-programming relaxation (see the full version [19]).

in $G - S$. While there has been a significant amount of work on kernelization for FEEDBACK VERTEX SET [10,13,34,35,42], the corresponding preprocessing algorithms do not succeed in finding vertices that belong to an optimal solution, other than those for which there is a self-loop or those which form the center a flower (consisting of $k + 1$ otherwise vertex-disjoint cycles [10,13,42], or a technical relaxation of this notion [34]). In particular, apart from the trivial self-loop rule, earlier preprocessing algorithms can only conclude a vertex v belongs to all optimal solutions (of a size k which must be given in advance) if they find a suitable packing of cycles witnessing that solutions without v must have size larger than k. In contrast, our argumentation will be based on *local* exchange arguments, which can be applied independently of the global solution size k.

We therefore introduce a new graph decomposition for preprocessing FEEDBACK VERTEX SET. To motivate it, we distill the essential features of a crown decomposition. Effectively, a crown decomposition of G certifies that G has a minimum vertex cover containing all of head, because (i) any vertex cover has to pick at least $|\mathsf{head}|$ vertices from $\mathsf{head} \cup \mathsf{crown}$, as the matching M certifies that $\text{VC}(G[\mathsf{head} \cup \mathsf{crown}]) \geq |\mathsf{head}|$, while (ii) any minimum vertex cover S' in $G - (\mathsf{head} \cup \mathsf{crown})$ yields a minimum vertex cover $S' \cup \mathsf{head}$ in G, since $N_G(\mathsf{crown}) \subseteq \mathsf{head}$ and crown is an independent set. To obtain similar guarantees for FEEDBACK VERTEX SET, we need a decomposition to supply disjoint vertex sets $(\mathsf{head}, \mathsf{antler})$ such that (i) any minimum feedback vertex set contains at least $|\mathsf{head}|$ vertices from $\mathsf{head} \cup \mathsf{antler}$, and (ii) any minimum feedback vertex set S' in $G - (\mathsf{head} \cup \mathsf{antler})$ yields a minimum feedback vertex set $S' \cup \mathsf{head}$ in G. To achieve (i), it suffices for $G[\mathsf{head} \cup \mathsf{antler}]$ to contain a set of $|\mathsf{head}|$ vertex-disjoint cycles; to achieve (ii), it suffices for $G[\mathsf{antler}]$ to be acyclic, with each tree T of the forest $G[\mathsf{antler}]$ connected to the remainder $V(G) \setminus (\mathsf{head} \cup \mathsf{antler})$ by at most one edge (implying that all cycles through antler intersect head). We call such a decomposition a 1-antler. Here *antler* refers to the shape of the forest $G[\mathsf{antler}]$, which no longer consists of isolated spikes of a crown (see Fig. 1 in the full version [19]). The prefix 1 indicates it is the simplest type of antler; we present a generalization later. An antler is *non-empty* if $\mathsf{head} \cup \mathsf{antler} \neq \emptyset$, and the *width* of the antler is defined to be $|\mathsf{head}|$.

Unfortunately, assuming $\mathsf{P} \neq \mathsf{NP}$ there is no *polynomial-time* algorithm that always outputs a non-empty 1-antler if one exists. We prove this in the full version [19]. However, for the purpose of making a preprocessing algorithm that reduces the search space, we can allow FPT time in a parameter such as $|\mathsf{head}|$ to find a decomposition. Each fixed choice of $|\mathsf{head}|$ would then correspond to a reduction rule which identifies a small ($|\mathsf{head}|$-sized) part of an optimal feedback vertex set, for which there is a simple certificate for it being part of an optimal solution. Such a reduction rule can then be iterated in the preprocessing phase, thereby potentially decreasing the target solution size (and search space) by an arbitrarily large amount. Hence we consider the parameterized complexity of testing whether a graph admits a non-empty 1-antler with $|\mathsf{head}| \leq k$, parameterized by k. On the one hand, we show this problem to be W[1]-hard in the full version [19]. This hardness turns out to be a technicality based on the forced bound on $|\mathsf{head}|$, though. We provide the following FPT algorithm which yields a search-space reducing preprocessing step.

Theorem 1. *There is an algorithm that runs in $2^{\mathcal{O}(k^5)} \cdot n^{\mathcal{O}(1)}$ time that, given an undirected multigraph G on n vertices and integer k, either correctly determines that G does not admit a non-empty 1-antler of width at most k, or outputs a set S of at least k vertices such that there exists an optimal feedback vertex set in G containing all vertices of S.*

Hence if the input graph admits a non-empty 1-antler of width at most k, the algorithm is guaranteed to find at least k vertices that belong to an optimal feedback vertex set, thereby reducing the search space.

Based on this positive result, we go further and generalize our approach beyond 1-antlers. For a 1-antler $(\mathsf{head}, \mathsf{antler})$ in G, the set of $|\mathsf{head}|$ vertex-disjoint cycles in $G[\mathsf{head} \cup \mathsf{antler}]$ forms a very simple certificate that any feedback vertex set of G contains at least $|\mathsf{head}|$ vertices from $\mathsf{head} \cup \mathsf{antler}$. We can generalize our approach to identify part of an optimal solution, by allowing more complex certificates of optimality. The following interpretation of a 1-antler is the basis of the generalization: for a 1-antler $(\mathsf{head}, \mathsf{antler})$ in G, there is a subgraph G' of $G[\mathsf{head} \cup \mathsf{antler}]$ (formed by the $|\mathsf{head}|$ vertex-disjoint cycles) such that $V(G') \supseteq \mathsf{head}$ and head is an optimal feedback vertex set of G'; and furthermore this subgraph G' is simple because all its connected components, being cycles, have a feedback vertex set of size 1. For an arbitrary integer z, we therefore define a z-antler in an undirected multigraph graph G as a pair of disjoint vertex sets $(\mathsf{head}, \mathsf{antler})$ such that (i) any minimum feedback vertex set in G contains at least $|\mathsf{head}|$ vertices from $\mathsf{head} \cup \mathsf{antler}$, as witnessed by the fact that $G[\mathsf{head} \cup \mathsf{antler}]$ has a subgraph G' for which head is an optimal feedback vertex set and with each component of G' having a feedback vertex set of size at most z; and (ii) the graph $G[\mathsf{antler}]$ is acyclic, with each tree T of the forest $G[\mathsf{antler}]$ connected to the remainder $V(G) \setminus (\mathsf{head} \cup \mathsf{antler})$ by at most one edge. (So condition (ii) is not changed compared to a 1-antler.) Our main result is the following.

Theorem 2. *There is an algorithm that runs in $2^{\mathcal{O}(k^5 z^2)} \cdot n^{\mathcal{O}(z)}$ time that, given an undirected multigraph G on n vertices and integers $k \geq z \geq 0$, either correctly determines that G does not admit a non-empty z-antler of width at most k, or outputs a set S of at least k vertices such that there exists an optimal feedback vertex set in G containing all vertices of S.*

In fact, we prove a slightly stronger statement. If a graph G can be reduced to a graph G' by iteratively removing z-antlers, each of width at most k, and the sum of the widths of this sequence of antlers is t, then we can find in time $f(k, z) \cdot n^{\mathcal{O}(z)}$ a subset of at least t vertices of G that belong to an optimal feedback vertex set. This implies that if a complete solution to FEEDBACK VERTEX SET can be assembled by iteratively combining $\mathcal{O}(1)$-antlers of width at most k, then the entire solution can be found in time $f'(k) \cdot n^{\mathcal{O}(1)}$. Hence our work uncovers a new parameterization in terms of the complexity of the solution structure, rather than its size, in which FEEDBACK VERTEX SET is fixed-parameter tractable.

Our algorithmic results are based on a combination of graph reduction and color coding. We use reduction steps inspired by the kernelization algorithms [10,42] for FEEDBACK VERTEX SET to bound the size of antler in the size of head. After such reduction steps, we use color coding [6] to help identify antler structures. A significant amount of effort goes into proving that the reduction steps preserve antler structures and the optimal solution size.

2 Preliminaries

Due to space restrictions, proofs of statements marked ★ have been deferred to the appendix. For any family of sets $X_1, \ldots, X_\ell$ indexed by $\{1, \ldots, \ell\}$ we define for all $1 \leq i \leq \ell$ the following $X_{<i} := \bigcup_{1 \leq j < i} X_j$, $X_{>i} := \bigcup_{i < j \leq \ell} X_j$, $X_{\leq i} := X_i \cup X_{<i}$ and $X_{\geq i} := X_i \cup X_{>i}$. For a function $f\colon A \to B$, let $f^{-1}\colon B \to 2^A$ denote the *preimage function of* f, that is $f^{-1}(a) = \{b \in B \mid f(b) = a\}$.

All graphs considered in this paper are undirected multigraphs, which may have loops. Based on the incidence representation of multigraphs we represent a multigraph G by a vertex set $V(G)$, an edge set $E(G)$, and a function $\iota\colon E(G) \to 2^{V(G)}$ where $\iota(e)$ is the set of one or two vertices incident to e for all $e \in E(G)$. In the context of an algorithm with input graph G we use $n = |V(G)|$ and $m = |E(G)|$. We assume we can retrieve and update number of edges between two vertices in constant time, hence we can ensure in $\mathcal{O}(n^2)$ time that there are at most two edges between any to vertices, meaning $m \in \mathcal{O}(n^2)$. For a vertex set $X \subseteq V(G)$ let $G[X]$ denote the subgraph of G induced by X. For a set of vertices and edges $Y \subseteq V(G) \cup E(G)$ let $G - Y$ denote the graph obtained from $G[V(G) \setminus Y]$ by removing all edges in Y. For a singleton set $\{v\}$ we write $G - v$ instead of $G - \{v\}$. For two graphs G and H the graph $G \cap H$ is the graph on vertex set $V(G) \cap V(H)$ and edge set $E(G) \cap E(H)$. For $v \in V(G)$ the open neighborhood of v in G is $N_G(v) := \{u \in V(G) \mid \exists e \in E(G)\colon \{u, v\} = \iota(e)\}$. For $X \subseteq V(G)$ let $N_G(X) := \bigcup_{v \in X} N_G(v) \setminus X$. The degree $\deg_G(v)$ of a vertex v in G is the number of edge-endpoints incident to v, where a self-loop contributes two endpoints. For two disjoint vertex sets $X, Y \subseteq V(G)$ the number of edges in G with one endpoint in X and another in Y is denoted by $e(X, Y)$. To simplify the presentation, in expressions involving $N_G(..)$ and $e(.., ..)$ we may use a subgraph H as argument instead of the set $V(H)$ that is formally needed.

3 Feedback Vertex Cuts and Antlers

In this section we present properties of antlers and related structures. A Feedback Vertex Set (FVS) in a graph G is a vertex set $X \subseteq V(G)$ such that $G - S$ is acyclic. The feedback vertex number of a graph G, denoted by $\textsc{fvs}(G)$, is the minimum size of a FVS in G. A *Feedback Vertex Cut* (FVC) in a graph G is a pair of disjoint vertex sets (C, F) such that $C, F \subseteq V(G)$, $G[F]$ is a forest, and for each tree T in $G[F]$ we have $e(T, G - (C \cup F)) \leq 1$. The *width* of a FVC (C, F) is $|C|$, and (C, F) is *empty* if $|C \cup F| = 0$. The set C intersects any cycle that contains a vertex from F, explaining the name Feedback Vertex Cut.

Observation 1. *If (C, F) is a FVC in G then any cycle in G containing a vertex from F also contains a vertex from C. The set C is a FVS in $G[C \cup F]$, hence $|C| \geq \text{FVS}(G[C \cup F])$.*

Observation 2. *If (C, F) is a FVC in G then for any $X \subseteq V(G)$ we have that $(C \setminus X, F \setminus X)$ is a FVC in $G - X$. Additionally, for any $Y \subseteq E(G)$ we have that (C, F) is a FVC in $G - Y$.*

We now present one of the main concepts for this work. An *antler* in a graph G is a FVC (C, F) in G such that $|C| \leq \text{FVS}(G[C \cup F])$. Then by Observation 1 the set C is a minimum FVS in $G[C \cup F]$ and no cycle in $G - C$ contains a vertex from F. We observe:

Observation 3. *If (C, F) is an antler in G, then $\text{FVS}(G) = |C| + \text{FVS}(G - (C \cup F))$.*

For a graph G and vertex set $C \subseteq V(G)$, a *C-certificate* is a subgraph H of G such that C is a minimum FVS in H. We say a C-certificate has *order z* if for each component H' of H we have $\text{FVS}(H') = |C \cap V(H')| \leq z$. For an integer $z \geq 0$, a z-antler in G is an antler (C, F) in G such that $G[C \cup F]$ contains a C-certificate of order z. Note that a 0-antler has width 0.

Observation 4. *If (C, F) is a z-antler in G for some $z \geq 0$, then for any $X \subseteq C$, we have that $(C \setminus X, F)$ is a z-antler in $G - X$.*

While antlers may intersect in non-trivial ways, the following proposition relates the sizes of the cross-intersections.

Proposition 1 (★). *If (C_1, F_1) and (C_2, F_2) are antlers in G, then $|C_1 \cap F_2| = |C_2 \cap F_1|$.*

Lemma 1 shows that what remains of a z-antler (C_1, F_1) when removing a different antler (C_2, F_2), again forms a smaller z-antler. We will rely on this lemma repeatedly to ensure that after having found and removed an incomplete fragment of a width-k z-antler, the remainder of that antler persists as a z-antler to be found later.

Lemma 1 (★). *For any integer $z \geq 0$, if a graph G has a z-antler (C_1, F_1) and another antler (C_2, F_2), then $(C_1 \setminus (C_2 \cup F_2), F_1 \setminus (C_2 \cup F_2))$ is a z-antler in $G - (C_2 \cup F_2)$.*

Lemma 2 shows that we can consider consecutive removal of two z-antlers as the removal of a single z-antler.

Lemma 2 (★). *For any integer $z \geq 0$, if a graph G has a z-antler (C_1, F_1) and $G - (C_1 \cup F_1)$ has a z-antler (C_2, F_2) then $(C_1 \cup C_2, F_1 \cup F_2)$ is a z-antler in G.*

The last structural property of antlers, given in Lemma 3, derives a bound on the number of trees of a forest $G[F]$ needed to witness that C is an optimal FVS of $G[C \cup F]$.

Lemma 3 (★). *Let (C, F) be a z-antler in a graph G for some $z \geq 0$. There exists an $F' \subseteq F$ such that (C, F') is a z-antler in G and $G[F']$ has at most $\frac{|C|}{2}(z^2 + 2z - 1)$ trees.*

4 Finding Feedback Vertex Cuts

As described in Sect. 1, our algorithm to identify vertices in antlers uses color coding. To allow a relatively small family of colorings to identify an entire antler structure (C, F) with $|C| \leq k$, we need to bound $|F|$ in terms of k as well. We therefore use several graph reduction steps. In this section, we show that if there is a width-k antler whose forest F is significantly larger than k, then we can identify a reducible structure in the graph. To identify a reducible structure we will also use color coding. In Sect. 5 we show how to reduce such a structure while preserving antlers and optimal feedback vertex sets.

Define the function $f_r : \mathbb{N} \to \mathbb{N}$ as $f_r(x) = 2x^3 + 3x^2 - x$. We say a FVC (C, F) is *reducible* if $|F| > f_r(|C|)$, and (C, F) is a *single-tree* FVC if $G[F]$ is connected.

Definition 1. *A FVC (C, F) is* simple *if $|F| \leq 2f_r(|C|)$ and one of the following holds: (a) $G[F]$ is connected, or (b) all trees in $G[F]$ have a common neighbor v and there exists a single-tree FVC (C, F_2) with $v \in F_2 \setminus F$ and $F \subseteq F_2$.*

In the full version [19] we show that if a graph G contains a single-tree reducible FVC (C, F), then G contains a simple reducible FVC (C, F'). In turn, such a simple reducible FVC can be found using color coding. A vertex coloring of G is a function $\chi : V(G) \to \{\dot{\mathsf{C}}, \dot{\mathsf{F}}\}$. We say a simple FVC (C, F) is *properly colored* by a coloring χ if $F \subseteq \chi^{-1}(\dot{\mathsf{F}})$ and $C \cup N_G(F) \subseteq \chi^{-1}(\dot{\mathsf{C}})$.

Lemma 4 (★). *Given a graph G and coloring χ of G that properly colors a simple reducible FVC (C, F), a reducible FVC (C', F') can be found in $\mathcal{O}(n^3)$ time.*

It can be shown that whether a FVC of width k is properly colored is determined by at most $1 + k + 2f_r(k) = \mathcal{O}(k^3)$ relevant vertices. By creating an $(n, \mathcal{O}(k^3))$-universal set for $V(G)$ using [16, Theorem 5.20], we can obtain in $2^{\mathcal{O}(k^3)} \cdot n \log n$ time a set of $2^{\mathcal{O}(k^3)} \cdot \log n$ colorings that contains a coloring for each possible assignment of colors for these relevant vertices. By applying Lemma 4 for each coloring we obtain the following lemma:

Lemma 5 (★). *There exists an algorithm that, given a graph G and an integer k, outputs a (possibly empty) FVC (C, F) in G. If G contains a reducible single-tree FVC of width at most k then (C, F) is reducible. The algorithm runs in time $2^{\mathcal{O}(k^3)} \cdot n^3 \log n$.*

5 Reducing Feedback Vertex Cuts

We apply reduction operations inspired by [10,42] on the subgraph $G[C \cup F]$ for a FVC (C, F) in G. We give 5 reduction operations and show at least one is applicable if $|F| > f_r(|C|)$. The operations reduce the number of vertices $v \in F$ with $\deg_G(v) < 3$ or reduce $e(C, F)$. The following lemma shows that this is sufficient to reduce the size of F.

Lemma 6 (★). *Let G be a multigraph with minimum degree at least 3 and let (C, F) be a FVC in G. We have $|F| \leq e(C, F)$.*

Next, we give the reduction operations. These operations apply to a graph G and yield a new graph G' and vertex set $S \subseteq V(G) \setminus V(G')$. An operation with output G' and S is *FVS-safe* if for any minimum feedback vertex set S' of G', the set $S \cup S'$ is a minimum feedback vertex set of G. An operation is *antler-safe* if for all $z \geq 0$ and any z-antler (C, F) in G, there exists a z-antler (C', F') in G' with $C' \cup F' = (C \cup F) \cap V(G')$ and $|C'| = |C| - |(C \cup F) \cap S|$.

Operation 1. *If $u, v \in V(G)$ are connected by more than two edges, remove all but two of these edges to obtain G' and take $S := \emptyset$.*

Operation 2. *If $v \in V(G)$ has degree exactly 2 and no self-loop, obtain G' by removing v from G and adding an edge e with $\iota(e) = N_G(v)$. Take $S := \emptyset$.*

Operations 1 and 2 are well established and FVS-safe. Additionally Operation 1 can easily be seen to be antler-safe. To see that Operation 2 is antler-safe, consider a z-antler (C, F) in G for some $z \geq 0$. If $v \notin C$ it is easily verified that $(C, F \setminus \{v\})$ is a z-antler in G'. If $v \in C$ pick a vertex $u \in N_G(v) \cap F$ and observe that $(\{u\} \cup C \setminus \{v\}, F \setminus \{u\})$ is a z-antler in G'.

Operation 3. *If (C, F) is an antler in G, then $G' := G - (C \cup F)$ and $S := C$.*

Operation 4. *If (C, F) is a FVC in G and for some $v \in C$ the graph $G[F \cup \{v\}]$ contains a v-flower of order $|C| + 1$, then $G' := G - v$ and $S := \{v\}$.*

Operation 5. *If (C, F) is a FVC in G, $v \in C$, and $X \subseteq F$ such that $G[F \cup \{v\}] - X$ is acyclic, and if T is a tree in $G[F] - X$ containing a vertex $w \in N_G(v)$ such that for each $u \in N_G(T) \setminus \{v\}$ there are more than $|C|$ other trees $T' \neq T$ in $G[F] - X$ for which $\{u, v\} \subseteq N_G(T')$, then take $S := \emptyset$ and obtain G' by removing the unique edge between v and w, and adding double-edges between v and u for all $u \in N_G(V(T)) \setminus \{v\}$.*

In the full version [19] we prove that the last three operations are both FVS-safe and antler-safe. Finally we show that when we are given a reducible FVC (C, F) in G, then we can find and apply an operation in $\mathcal{O}(n^2)$ time. With a more careful analysis better running time bounds can be shown, but this does not affect the final running time of the main algorithm.

Lemma 7 (★). *Given a graph G and a reducible FVC (C, F) in G, we can find and apply an operation in $\mathcal{O}(n^2)$ time.*

6 Finding and Removing Antlers

We will find antlers using color coding, using coloring functions of the form $\chi\colon V(G) \cup E(G) \to \{\dot{\mathsf{F}}, \dot{\mathsf{C}}, \dot{\mathsf{R}}\}$. For all $c \in \{\dot{\mathsf{F}}, \dot{\mathsf{C}}, \dot{\mathsf{R}}\}$ let $\chi_V^{-1}(c) = \chi^{-1}(c) \cap V(G)$. For any integer $z \geq 0$, a z-antler (C, F) in a graph G is *z-properly colored* by a coloring χ if all of the following hold: (i) $F \subseteq \chi_V^{-1}(\dot{\mathsf{F}})$, (ii) $C \subseteq \chi_V^{-1}(\dot{\mathsf{C}})$, (iii) $N_G(F) \setminus C \subseteq \chi_V^{-1}(\dot{\mathsf{R}})$, and (iv) $G[C \cup F] - \chi^{-1}(\dot{\mathsf{R}})$ is a C-certificate of order z. Recall that $\chi^{-1}(\dot{\mathsf{R}})$ can contain edges as well as vertices so for any subgraph H of G the graph $H - \chi^{-1}(\dot{\mathsf{R}})$ is obtained from H by removing both vertices and edges. It can be seen that if (C, F) is a z-antler, then there exists a coloring that z-properly colors it. Consider for example a coloring where a vertex v is colored $\dot{\mathsf{C}}$ (resp. $\dot{\mathsf{F}}$) if $v \in C$ (resp. $v \in F$), all other vertices are colored $\dot{\mathsf{R}}$, and for some C-certificate H of order z in $G[C \cup F]$ all edges in H have color $\dot{\mathsf{F}}$ and all other edges have color $\dot{\mathsf{R}}$.

Lemma 8 (★). *A $n^{\mathcal{O}(z)}$ time algorithm exists taking as input an integer $z \geq 0$, a graph G, and a coloring χ and producing as output a z-antler (C, F) in G, such that for any z-antler $(\hat{C}, \hat{F})$ that is z-properly colored by χ we have $\hat{C} \subseteq C$ and $\hat{F} \subseteq F$.*

If a graph G contains a reducible single-tree FVC of width at most k then we can find and apply an operation by Lemma 5 and 7. If G does not contain such a FVC, but G does contain a non-empty z-antler (C, F) of width at most k, then using Lemma 3 we can prove that whether (C, F) is z-properly colored is determined by the color of at most $26k^5z^2$ relevant vertices and edges. Using two $(n + m, 26k^5z^2)$-universal sets, we can create a set of colorings that is guaranteed to contain a coloring that z-properly colors (C, F). Using Lemma 8 we find a non-empty z-antler and apply Operation 3. We obtain the following:

Lemma 9 (★). *Given a graph G and integers $k \geq z \geq 0$. If G contains a non-empty z-antler of width at most k we can find and apply an operation in $2^{\mathcal{O}(k^5z^2)} \cdot n^{\mathcal{O}(z)}$ time.*

Note that applying an operation reduces the number of vertices or increases the number of double-edges. Hence by repeatedly using Lemma 9 to apply an operation we obtain, after at most $\mathcal{O}(n^2)$ iterations, a graph in which no operation applies. By Lemma 9 this graph does not contain a non-empty z-antler of width at most k. Further analysis shows that this method reduces the solution size at least as much as iteratively removing z-antlers of width at most k. This is described in Theorem 3. By taking $t = 1$ we obtain Theorem 2.

Theorem 3 (★). *Given as input a graph G and integers $k \geq z \geq 0$ we can find in $2^{\mathcal{O}(k^5z^2)} \cdot n^{\mathcal{O}(z)}$ time a vertex set $S \subseteq V(G)$ such that*

1. *there is a minimum FVS in G containing all vertices of S, and*
2. *if $C_1, F_1, \ldots, C_t, F_t$ is a sequence of disjoint vertex sets with for all $1 \leq i \leq t$ the pair (C_i, F_i) is a z-antler of width at most k in $G - (C_{<i} \cup F_{<i})$, then $|S| \geq |C_{\leq t}|$.*

As a corollary to this theorem, we obtain a new type of parameterized-tractability result for Feedback Vertex Set. For an integer z, let the z-antler complexity of FVS on G be the minimum number k for which there exists a (potentially long) sequence $C_1, F_1, \ldots, C_t, F_t$ of disjoint vertex sets such that for all $1 \leq i \leq t$, the pair (C_i, F_i) is a z-antler of width at most k in $G - (C_{<i} \cup F_{<i})$, and such that $G - (C_{\leq t} \cup F_{\leq t})$ is acyclic (implying that $C_{\leq t}$ is a feedback vertex set in G). If no such sequence exists, the z-antler complexity of G is $+\infty$.

Intuitively, Corollary 1 states that optimal solutions can be found efficiently when they are composed out of small pieces, each of which has a low-complexity certificate for belonging to some optimal solution.

Corollary 1 (★). *There is an algorithm that, given a graph G, returns an optimal feedback vertex set in time $f(k^*) \cdot n^{\mathcal{O}(z^*)}$, where (k^*, z^*) is any pair of integers such that the z^*-antler complexity of G is at most k^*.*

To conclude, we reflect on the running time of Corollary 1 compared to running times of the form $2^{\mathcal{O}(\text{FVS}(G))} \cdot n^{\mathcal{O}(1)}$ obtained by FPT algorithms for the parameterization by solution size. If we exhaustively apply Lemma 7 with the FVC $(C, V(G) \setminus C)$, where C is obtained from a 2-approximation algorithm [9], then this gives an *antler-safe* kernelization: it reduces the graph as long as the graph is larger than $f_r(|C|)$. This opening step reduces the instance size to $\mathcal{O}(\text{FVS}(G)^3)$ without increasing the antler complexity. As observed before, after applying $\mathcal{O}(n^2)$ operations we obtain a graph in which no operations can be applied. This leads to a running time of $\mathcal{O}(n^4)$ of the kernelization. Running Theorem 3 to solve the reduced instance yields a total running time of $2^{\mathcal{O}(k^5 z^2)} \text{FVS}(G)^{\mathcal{O}(z)} + \mathcal{O}(n^4)$. This is asymptotically faster than $2^{\mathcal{O}(\text{FVS}(G))}$ when $z \leq k = o(\sqrt[7]{\text{FVS}(G)})$ and $\text{FVS}(G) = \omega(\log n)$, which captures the intuitive idea sketched above that our algorithmic approach has an advantage when there is an optimal solution that is large but composed of small pieces for which there are low-complexity certificates.

7 Conclusion

We have taken the first steps of a research program to investigate how and when a preprocessing phase can guarantee to identify parts of an optimal solution to an NP-hard problem, thereby reducing the search space of the follow-up algorithm. Aside from the technical results concerning antler structures for Feedback Vertex Set and their algorithmic properties, we consider the conceptual message of this research program an important contribution of our theoretical work on understanding the power of preprocessing and the structure of solutions to NP-hard problems.

This line of investigation opens up a host of opportunities for future research. For combinatorial problems such as Vertex Planarization, Odd Cycle Transversal, and Directed Feedback Vertex Set, which kinds of substructures in inputs allow parts of an optimal solution to be identified by an efficient preprocessing phase? Is it possible to give preprocessing guarantees not in terms of the size of an optimal solution, but in terms of measures of the stability [7,8,17] of optimal solutions under small perturbations? Some questions also remain open concerning the concrete technical results in the paper. Can the running time of Theorem 2 be improved to $f(k) \cdot n^{\mathcal{O}(1)}$? We conjecture that it cannot, but have not been able to prove this.

References

1. Abu-Khzam, F.N., Collins, R.L., Fellows, M.R., Langston, M.A., Suters, W.H., Symons, C.T.: Kernelization algorithms for the vertex cover problem: theory and experiments. In: Proceedings of 6th ALENEX/ANALC, pp. 62–69 (2004)
2. Abu-Khzam, F.N., Fellows, M.R., Langston, M.A., Suters, W.H.: Crown structures for vertex cover kernelization. Theory Comput. Syst. **41**(3), 411–430 (2007). https://doi.org/10.1007/s00224-007-1328-0
3. Achterberg, T., Bixby, R.E., Gu, Z., Rothberg, E., Weninger, D.: Presolve reductions in mixed integer programming. Technical report 16-44, ZIB, Takustr.7, 14195 Berlin (2016). http://nbn-resolving.de/urn:nbn:de:0297-zib-60370
4. Achterberg, T., Wunderling, R.: Mixed integer programming: analyzing 12 years of progress. In: Jünger, M., Reinelt, G. (eds.) Facets of Combinatorial Optimization, pp. 449–481. Springer, Heidelberg (2013). https://doi.org/10.1007/978-3-642-38189-8_18
5. Akiba, T., Iwata, Y.: Branch-and-reduce exponential/FPT algorithms in practice: a case study of vertex cover. Theor. Comput. Sci. **609**, 211–225 (2016). https://doi.org/10.1016/j.tcs.2015.09.023
6. Alon, N., Yuster, R., Zwick, U.: Color-coding. J. ACM **42**(4), 844–856 (1995)
7. Angelidakis, H., Awasthi, P., Blum, A., Chatziafratis, V., Dan, C.: Bilu-Linial stability, certified algorithms and the independent set problem. In: Bender, M.A., Svensson, O., Herman, G. (eds.) Proceedings of 27th ESA. LIPIcs, vol. 144, pp. 7:1–7:16. Schloss Dagstuhl - Leibniz-Zentrum für Informatik (2019). https://doi.org/10.4230/LIPIcs.ESA.2019.7
8. Awasthi, P., Blum, A., Sheffet, O.: Center-based clustering under perturbation stability. Inf. Process. Lett. **112**(1–2), 49–54 (2012). https://doi.org/10.1016/j.ipl.2011.10.006
9. Bafna, V., Berman, P., Fujito, T.: A 2-approximation algorithm for the undirected feedback vertex set problem. SIAM J. Discrete Math. **12**(3), 289–297 (1999). https://doi.org/10.1137/S0895480196305124
10. Bodlaender, H.L., van Dijk, T.C.: A cubic kernel for feedback vertex set and loop cutset. Theory Comput. Syst. **46**(3), 566–597 (2010). https://doi.org/10.1007/s00224-009-9234-2
11. Bodlaender, H.L., Fomin, F.V., Lokshtanov, D., Penninkx, E., Saurabh, S., Thilikos, D.M.: (meta) Kernelization. J. ACM **63**(5), 44:1–44:69 (2016). https://doi.org/10.1145/2973749
12. Bodlaender, H.L., Jansen, B.M.P., Kratsch, S.: Kernelization lower bounds by cross-composition. SIAM J. Discrete Math. **28**(1), 277–305 (2014). https://doi.org/10.1137/120880240
13. Burrage, K., Estivill-Castro, V., Fellows, M.R., Langston, M.A., Mac, S., Rosamond, F.A.: The undirected feedback vertex set problem has a poly(k) kernel. In: Bodlaender, H.L., Langston, M.A. (eds.) IWPEC 2006. LNCS, vol. 4169, pp. 192–202. Springer, Heidelberg (2006). https://doi.org/10.1007/11847250_18

14. Chen, J., Kanj, I.A., Xia, G.: Improved upper bounds for vertex cover. Theor. Comput. Sci. **411**(40–42), 3736–3756 (2010). https://doi.org/10.1016/j.tcs.2010.06.026
15. Chor, B., Fellows, M., Juedes, D.W.: Linear kernels in linear time, or how to save k colors in $O(n^2)$ steps. In: Hromkovič, J., Nagl, M., Westfechtel, B. (eds.) WG 2004. LNCS, vol. 3353, pp. 257–269 (2004). https://doi.org/10.1007/978-3-540-30559-0_22
16. Cygan, M., et al.: Parameterized Algorithms. Springer, Cham (2015). https://doi.org/10.1007/978-3-319-21275-3
17. Daniely, A., Linial, N., Saks, M.E.: Clustering is difficult only when it does not matter. CoRR abs/1205.4891 (2012). http://arxiv.org/abs/1205.4891
18. Dell, H., van Melkebeek, D.: Satisfiability allows no nontrivial sparsification unless the polynomial-time hierarchy collapses. J. ACM **61**(4), 23:1–23:27 (2014). https://doi.org/10.1145/2629620
19. Donkers, H., Jansen, B.M.P.: Preprocessing to reduce the search space: antler structures for feedback vertex set. CoRR abs/2106.11675 (2021). https://arxiv.org/abs/2106.11675
20. Downey, R., Fellows, M.R.: Parameterized Complexity. Monographs in Computer Science. Springer, New York (1999). https://doi.org/10.1007/978-1-4612-0515-9
21. Downey, R.G., Fellows, M.R.: Fundamentals of Parameterized Complexity. TCS. Springer, London (2013). https://doi.org/10.1007/978-1-4471-5559-1
22. Drucker, A.: New limits to classical and quantum instance compression. SIAM J. Comput. **44**(5), 1443–1479 (2015). https://doi.org/10.1137/130927115
23. Dzulfikar, M.A., Fichte, J.K., Hecher, M.: The PACE 2019 parameterized algorithms and computational experiments challenge: the fourth iteration (invited paper). In: Jansen, B.M.P., Telle, J.A. (eds.) Proceedings of 14th IPEC. LIPIcs, vol. 148, pp. 25:1–25:23. Schloss Dagstuhl - Leibniz-Zentrum für Informatik (2019). https://doi.org/10.4230/LIPIcs.IPEC.2019.25
24. Fellows, M.R.: Blow-ups, win/win's, and crown rules: some new directions in FPT. In: Bodlaender, H.L. (ed.) WG 2003. LNCS, vol. 2880, pp. 1–12. Springer, Heidelberg (2003). https://doi.org/10.1007/978-3-540-39890-5_1
25. Fellows, M.R.: The lost continent of polynomial time: preprocessing and kernelization. In: Bodlaender, H.L., Langston, M.A. (eds.) IWPEC 2006. LNCS, vol. 4169, pp. 276–277. Springer, Heidelberg (2006). https://doi.org/10.1007/11847250_25
26. Flum, J., Grohe, M.: Parameterized Complexity Theory. TTCSAES, Springer, Heidelberg (2006). https://doi.org/10.1007/3-540-29953-X
27. Fomin, F.V., Lokshtanov, D., Misra, N., Saurabh, S.: Planar $\mathcal{F}$-deletion: approximation, kernelization and optimal FPT algorithms. In: Proceedings of 53rd FOCS, pp. 470–479 (2012). https://doi.org/10.1109/FOCS.2012.62
28. Fomin, F.V., Lokshtanov, D., Saurabh, S., Zehavi, M.: Kernelization: Theory of Parameterized Preprocessing. Cambridge University Press, Cambridge (2019). https://doi.org/10.1017/9781107415157
29. Fortnow, L., Santhanam, R.: Infeasibility of instance compression and succinct PCPs for NP. J. Comput. Syst. Sci. **77**(1), 91–106 (2011). https://doi.org/10.1016/j.jcss.2010.06.007
30. Guo, J., Niedermeier, R.: Invitation to data reduction and problem kernelization. SIGACT News **38**(1), 31–45 (2007). https://doi.org/10.1145/1233481.1233493
31. Hermelin, D., Kratsch, S., Soltys, K., Wahlström, M., Wu, X.: A completeness theory for polynomial (turing) kernelization. Algorithmica **71**(3), 702–730 (2015). https://doi.org/10.1007/s00453-014-9910-8

32. Hespe, D., Lamm, S., Schulz, C., Strash, D.: Wegotyoucovered: the winning solver from the PACE 2019 implementation challenge, vertex cover track. CoRR abs/1908.06795 (2019)
33. Hespe, D., Schulz, C., Strash, D.: Scalable kernelization for maximum independent sets. ACM J. Exp. Algorithm. **24**(1), 1.16:1–1.16:22 (2019). https://doi.org/10.1145/3355502
34. Iwata, Y.: Linear-time kernelization for feedback vertex set. In: Proceedings of 44th ICALP. LIPIcs, vol. 80, pp. 68:1–68:14 (2017). https://doi.org/10.4230/LIPIcs.ICALP.2017.68
35. Jansen, B.M.P., Raman, V., Vatshelle, M.: Parameter ecology for feedback vertex set. Tsinghua Sci. Technol. **19**(4), 387–409 (2014). https://doi.org/10.1109/TST.2014.6867520
36. Karp, R.M.: Reducibility among combinatorial problems. In: Jünger, M., et al. (eds.) 50 Years of Integer Programming 1958–2008. TTCSAES, pp. 219–241. Springer, Heidelberg (2010). https://doi.org/10.1007/978-3-540-68279-0_8
37. Kratsch, S., Wahlström, M.: Representative sets and irrelevant vertices: new tools for kernelization. In: Proceedings of 53rd FOCS, pp. 450–459 (2012). https://doi.org/10.1109/FOCS.2012.46
38. Lokshtanov, D., Narayanaswamy, N.S., Raman, V., Ramanujan, M.S., Saurabh, S.: Faster parameterized algorithms using linear programming. ACM Trans. Algorithms **11**(2), 15:1–15:31 (2014). https://doi.org/10.1145/2566616
39. Nemhauser, G., Trotter, L.: Vertex packings: structural properties and algorithms. Math. Program. **8**, 232–248 (1975). https://doi.org/10.1007/BF01580444
40. Pilipczuk, M., Pilipczuk, M., Sankowski, P., van Leeuwen, E.J.: Network sparsification for steiner problems on planar and bounded-genus graphs. ACM Trans. Algorithms **14**(4), 53:1–53:73 (2018). https://doi.org/10.1145/3239560
41. Quine, W.V.: The problem of simplifying truth functions. Am. Math. Mon. **59**(8), 521–531 (1952)
42. Thomassé, S.: A $4k^2$ kernel for feedback vertex set. ACM Trans. Algorithms **6**(2) (2010). https://doi.org/10.1145/1721837.1721848

Parameterized Complexity of BANDWIDTH of Caterpillars and WEIGHTED PATH EMULATION

Hans L. Bodlaender(✉)

Department of Information and Computing Sciences, Utrecht University, P.O. Box 80.089, 3508 TB Utrecht, The Netherlands
h.l.bodlaender@uu.nl

Abstract. In this paper, we show that BANDWIDTH is hard for the complexity class $W[t]$ for all $t \in \mathbf{N}$, even for caterpillars with hair length at most three. As intermediate problem, we introduce the WEIGHTED PATH EMULATION problem: given a vertex-weighted path P_N and integer M, decide if there exists a mapping of the vertices of P_N to a path P_M, such that adjacent vertices are mapped to adjacent or equal vertices, and such that the total weight of the pre-image of a vertex from P_M equals an integer c. We show that WEIGHTED PATH EMULATION, with c as parameter, is hard for $W[t]$ for all $t \in \mathbf{N}$, and is strongly NP-complete. We also show that DIRECTED BANDWIDTH is hard for $W[t]$ for all $t \in \mathbf{N}$, for directed acyclic graphs whose underlying undirected graph is a caterpillar.

Keywords: Bandwidth · Parameterized complexity · Weighted path emulation · W-hierarchy · Caterpillars

1 Introduction

The BANDWIDTH problem is one of the classic problems from algorithmic graph theory. In this problem, we are given an undirected graph $G = (V, E)$ and integer k, and want to find a bijection from V to $\{1, 2, \ldots, n\}$, with $n = |V|$, such that for each edge $\{v, w\} \in E$: $|f(v) - f(w)| \leq k$. The problem was proved to be NP-complete in 1976 by Papadimitriou [21]. Later, several special cases were proven to be NP-complete. In 1986, Monien [19] showed that BANDWIDTH stays NP-complete when the input is restricted to caterpillars with hair length at most three. A caterpillar is a tree where all vertices of degree at least three are on the same path; the hairs are the paths attached to this central path, and have here at most three vertices.

In this paper, we consider the parameterized complexity of this problem. We consider the standard parameterization, i.e., we ask for the complexity of BANDWIDTH as a function of n and k. This problem is long known to belong to XP: already in 1980, Saxe [22] showed that BANDWIDTH can be solved in time $f(k) \cdot n^{k+1}$ for some function f; this was later improved to $f(k) \cdot n^k$ [17].

L. Kowalik et al. (Eds.): WG 2021, LNCS 12911, pp. 15–27, 2021.
https://doi.org/10.1007/978-3-030-86838-3_2

In 1994, Bodlaender et al. [5] reported that BANDWIDTH is $W[t]$-hard for all positive integers t, even when we restrict the input to trees. However, the proof of this fact was so far never published. In the current paper, we give the proof of a somewhat stronger result: BANDWIDTH is $W[t]$-hard for all positive integers t, even when we restrict the input to caterpillars with maximum hair length three. A sketch of a proof that BANDWIDTH is $W[t]$-hard for all positive integers t for general graphs appears in the monograph by Downey and Fellows [10]. In recent years, Dregi and Lokshtanov [12] gave a proof that BANDWIDTH is $W[1]$-hard for trees of pathwidth at most two, and showed that there does not exist an algorithm for BANDWIDTH on such trees with running time of the form $f(k)n^{o(k)}$ assuming that the Exponential Time Hypothesis holds.

Our proof uses techniques from the NP-hardness proof for BANDWIDTH on caterpillars by Monien [19]. In particular, one gadget in the proof is identical to a gadget from Moniens proof. Also, the proof is inspired by ideas behind the proof of the result reported in [5], and a proof for $W[t]$-hardness of a scheduling problem for chains of jobs with delays, which was obtained by Bodlaender and van der Wegen [7].

To obtain our main result, we obtain an intermediate result that is also interesting on itself. We consider a variation of the notion of uniform emulation. The notion of emulation was introduced by Fishburn and Finkel [15], to describe the simulation of processor networks on smaller processor networks. An emulation of a graph $G = (V, E)$ on a graph $H = (W, F)$ is a function $f : V \rightarrow W$, such that for each edge $\{v, w\} \in E$, $f(v) = f(w)$ or $\{f(v), f(w)\} \in F$, i.e., neighboring vertices are mapped to the same or neighboring vertices. An emulation is uniform when each vertex in H has the same number of vertices mapped to it, i.e., there is a constant c, called the *emulation factor*, such that for all $w \in W$: $|f^{-1}(w)| = c$. An analysis of the complexity to decide whether for given G and H, there exists a uniform emulation was made by Bodlaender and van Leeuwen [8], and Bodlaender [2]. In particular, in [2], the complexity of deciding if there is a uniform emulation on a path or cycle was studied. It was shown that UNIFORM EMULATION ON A PATH belongs to XP, parameterized by the emulation factor c, belongs to XP for connected graphs and is NP-complete, even for $c = 4$, when we allow that G is not connected. Bodlaender et al. [5] claimed that UNIFORM EMULATION ON A PATH is hard for $W[t]$ for all positive integers t. In this paper, we show a variation of this result, where the input is a *weighted path*. We name the problem of finding an uniform emulation of a weighted path to a path WEIGHTED PATH EMULATION. It is straightforward to modify the algorithm from [2] to weighted graphs. This shows that WEIGHTED PATH EMULATION belongs to XP, with the emulation factor as parameter.

There is a sharp distinction between the complexity of the BANDWIDTH problem for caterpillars with hairs of length at most two, and caterpillars with hairs of length three (or larger). Assmann et al. [1] give a characterization of the bandwidth for caterpillars whose hair length is at most two, and show that one can compute a layout of optimal width in $O(n \log n)$ time. This contrasts with the NP-hardness and fixed parameter intractability for caterpillars with hairs of length three, by Monien [19] and this paper. For related results, see [18,20,23].

Very recently, the results in the paper were strengthened, and it was shown that BANDWIDTH for caterpillars of hair length at most three, and WEIGHTED PATH EMULATION are complete for a class of parameterized problems called XNLP. For a brief discussion of these results, see Sect. 6.

This paper is organized as follows. In Sect. 2, we give a number of definitions. In Sect. 3, we discuss hardness for the WEIGHTED PATH EMULATION problem. Section 4 gives the main result: hardness for the bandwidth of caterpillars with hairs of length at most three. In Sect. 5, we discuss a variation of the proof, to obtain that DIRECTED BANDWIDTH is hard for $W[t]$ for all positive integers $t \in \mathbf{N}$, for directed acyclic graphs whose underlying undirected graph is a caterpillar with hair length one. Some final remarks are made in Sect. 6. In this extended abstract, we describe the intuition behind the proofs; the precise proofs can be found in the version of the paper on arXiv [3].

2 Definitions

All graphs in this paper are considered to be simple and undirected. We assume that the reader is familiar with standard notions from graph theory and fixed parameter complexity (see e.g. [10,11,16]).

P_n denotes the path graph with n vertices. We denote the vertices of P_n by the first n positive integers, $1, 2, \ldots, n$; the edges of P_n are the pairs $\{i, i+1\}$ for $1 \leq i < n$.

A *caterpillar* is a tree such that there is a path that contains all vertices of degree at least three. A caterpillar can be formed by taking a path P_N (the *spine*), and then attaching to vertices of P_N zero or more paths. These latter paths are called the *hairs* of the caterpillar.

A *linear ordering* of a graph $G = (V, E)$ is a bijective function $f : V \rightarrow \{1, 2, \ldots, n\}$. The *bandwidth* of a linear ordering f of G is $\max_{\{v,w\} \in E} |f(v) - f(w)|$. The *bandwidth* of a graph is the minimum bandwidth over its linear orderings.

Let $G = (V, E)$ be an undirected graph, and $w : V \rightarrow \mathbf{Z}^+$ be a function that assigns to each vertex a positive integer weight. An *emulation* of G on a path P_M is a function $f : V \rightarrow \{1, 2, \ldots, M\}$, such that for all edges $\{v, w\} \in E$, $|f(v) - f(w)| \leq 1$. An emulation $f : V \rightarrow \{1, 2, \ldots, M\}$ is said to be *uniform*, if there is an integer c, such that for all $i \in \{1, 2, \ldots, M\}$, $\sum_{v : f(v) = i} w(v) = c$. c is called the *emulation factor*.

For a directed acyclic graph $G = (V, A)$, the *directed bandwidth* of a topological ordering of G is $\max_{(v,w) \in A} f(w) - f(v)$; the *directed bandwidth* of a directed acyclic graph G is the minimum directed bandwidth over all topological orderings of G.

If we have a directed graph $G = (V, A)$, the *underlying undirected graph* of G is the undirected graph $G' = (V, E)$, with $E = \{\{v, w\} \mid (v, w) \in A\}$; i.e., we forget the direction of edges; if we obtain a pair of parallel edges, we take only one.

We consider the following parameterized problems.

BANDWIDTH
Given: An undirected graph $G = (V, E)$, integer k.
Parameter: k.
Question: Is the bandwidth of G at most k?

DIRECTED BANDWIDTH
Given: A directed acyclic graph $G = (V, E)$, integer k.
Parameter: k.
Question: Is the directed bandwidth of G at most k?

WEIGHTED PATH EMULATION
Given: Integers N, M, c, weight function $w : \{1, 2, \ldots, N\} \to \mathbf{Z}^+$, such that $\sum_{i=1}^{N} w(i)/M = c \in \mathbf{Z}^+$.
Parameter: c.
Question: Is there a uniform emulation f of P_N with weight function w on P_M?

Note that in the problem statement above, c is the emulation factor, i.e., we have for a solution f that for each j, $1 \leq j \leq M$, $\sum_{i:f(i)=j} w(i) = c$.

A Boolean formula is said to be *t-normalized*, if it is the conjunction of the disjunction of the conjunction of ... of literals, with t alternations of AND's and OR's. So, a Boolean in Conjunctive Normal Form is 2-normalized. Downey and Fellows [9] consider the following parameterized problem; this is the starting point for our reductions.

WEIGHTED t-NORMALIZED SATISFIABILITY
Given: A t-normalized Boolean formula F and a positive integer $k \in \mathbf{Z}^+$.
Parameter: k
Question: Can F be satisfied by setting exactly k variables to true?

Theorem 1 (Downey and Fellows [9]). WEIGHTED t-NORMALIZED SATISFIABILITY *is* $W[t]$*-complete.*

3 Hardness of WEIGHTED PATH EMULATION

Our first main result is the following; the proof can be found in the full paper [3].

Theorem 2. WEIGHTED PATH EMULATION *is* $W[t]$*-hard for all positive integers* t.

Suppose we are given a t-normalized Boolean expression F over n variables, say $x_1, \ldots, x_n$, and integer k. We let t' be the number of nested levels of disjunction. We consider the problem to satisfy F by making exactly k variables true.

We will define a path P_N with a weight function $w : \{1, \ldots, N\} \to \mathbf{Z}^+$, an emulation factor c, and an integer M, such that P_N has a uniform emulation on a path P_M if and only if F can be satisfied by setting exactly k variables to true. Before giving the proof, we give a high level overview of some main ideas of the proof.

3.1 Intuition and Techniques

In this subsection, we give some ideas behind the construction. The precise construction and formal proofs are given in the next subsection.

We assume we have given a t-normalized Boolean formula F. We transform the formula to a weighted path P_N, such that P_N has a uniform emulation on P_M with c the emulation factor, if and only if F can be satisfied by setting exactly k variables to true.

We can view F as a tree, with internal nodes marked with disjunction or conjunction, and each leaf with a literal, then we alternatingly have a level in the tree with disjunctions, and with conjunctions. We set t' to be the number of levels with disjunctions.

The path P_N is formed by taking, in this order, the following: a part called the 'floor', k 'variable parts', t' 'disjunction parts', and a 'filler path'. t' is the number of levels in the formula tree with disjunctions, and for each 'level' of disjunction we have one disjunction part. E.g., if F is in conjunctive normal form, then $t' = 1$.

The *floor* has M vertices, each with a weight that is larger than $c/2$. Thus, we cannot map two floor vertices to the same element of P_M, and thus, can assume, without loss of generality, that the ith floor vertex is mapped to i. The different weights for floor vertices help to build the further gadgetry of the construction.

The variable and disjunction parts are forced to start at M, then move to 1, and then move (possibly with some 'zigzagging') back to M, where then the next part starts. This is done by giving each part one vertex of large weight that only can fit at vertex 1, and another vertex of even larger weight, that only can fit at vertex M. These large weight vertices are called left and right turning points. See Fig. 1 for an illustration of the construction.

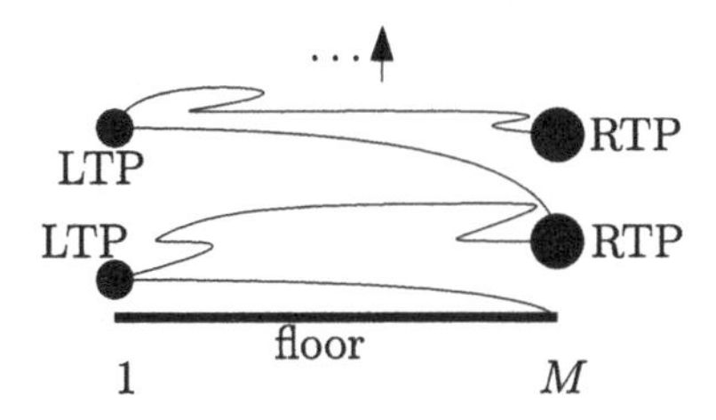

Fig. 1. Impression of a first part of the construction. The ith floor vertex is mapped to i; after this, the path then moves from M to 1 and back, with left turning points (LTR) mapped to 1, and right turning points (RTP) mapped to M. The picture shows the floor and first two variable parts.

We have k *variable parts*. Each models one variable that is set to true. We start with a left turning point, $M - 2$ vertices of weight one, and a right turning point: this is to move back from M to 1. Then, we have $n - 1$ vertices of weight one, $M - 2n - 2$ 'heavy' vertices, and again $n - 1$ vertices of weight one.

The heavy vertices are mapped consecutively (except possibly at the first n and last n positions); the weight one vertices before and after the heavy vertices allows us to shift the sequence of heavy vertices in n ways—each different such shift sets another variable to true. By using two different heavy weights, combined with weight settings for floor vertices and vertices from disjunction parts, we can check that all variable parts select a different variable to be true (which is done at positions $n+2,\ldots,2n+1$), and that F is satisfied (which is done at positions $2n+2,\ldots,M-n-2$).

We have for each level in the formula tree with disjunctions a *disjunction part.* Thus, we have t' disjunction parts. With help of 'anchors' (vertices of large weight that can go only to one specific position), we can ensure that a subpart for a disjunction has to be mapped to the part of the floor that corresponds with this disjunction. Such a subpart consist of a path with weight one vertices, a path with $3m(F')$ *selecting vertices* (which have larger weight), and another path with weight one vertices. Now, each term in the disjunction has an associated interval of size $m(F')$ and between these intervals we have $m(F')$ elements. Then, we can show that the selecting vertices must cover entirely one of the intervals of a term—this corresponds to that term being satisfied. See Fig. 2 below. Say $F' = F_1 \vee F_2 \vee F_3$. In the illustration we see the intervals assigned to F_1, F_2, and F_3, and the space between, before and after these. Each of the seven intervals has size $m(F')$. We can show that the $3m(F')$ selecting vertices must be mapped to consecutive vertices between the left and right anchor of F', and thus these cover the interval of each least one F_i entirely.

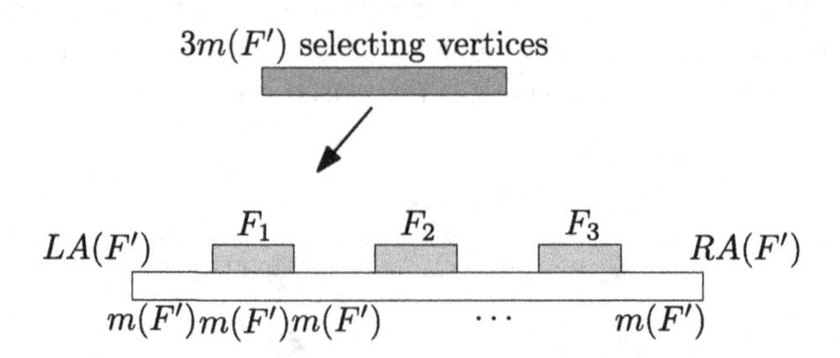

Fig. 2. Illustration: consecutive selecting vertices cover the interval of one term

Heavy vertices of variable parts come in two weights: c^v and $c^c + c^u$. This is used for checking that F is satisfied. As an example, consider a negative literal $\neg x_j$ in F. We have one specific position on P_M, say i, that checks whether this literal is satisfied, in case its satisfaction contributes to the satisfaction of F—that case corresponds to having a selecting vertex mapped to i for each level of disjunction. Now, the weight of the floor vertex mapped to i is such that when this floor vertex and all selecting vertices are mapped to i, then we can only fit k heavy vertices of weight c^v here; if at least one of these heavy vertices has weight $c^v + c^u$, then the total weight mapped to i exceeds c. If this happens, then this heavy vertex belongs to a variable part which corresponds to setting x_i to be true; thus, this enforces that x_i is false. A somewhat similar construction is used for positive literals.

The last part of P_N is the *filler path.* This is a long path with vertices of weight one. This is used to ensure that the mapping becomes uniform: if the total weight of vertices of floor, variable part, and disjunction parts vertices mapped to i is z_i, then we map $c - z_i$ (consecutive) vertices of the filler path to i. See Fig. 3 for an illustration.

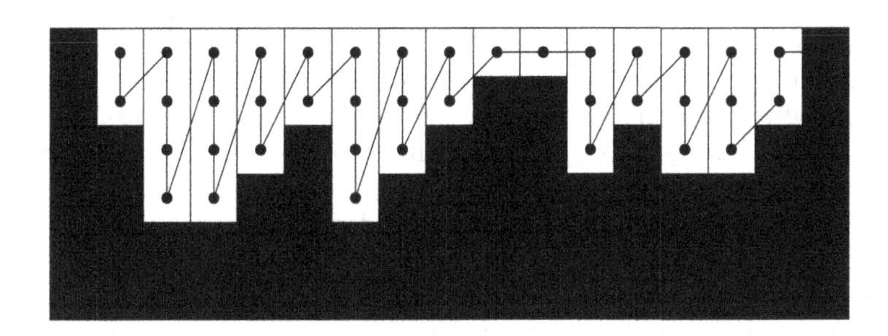

Fig. 3. Illustration of the mapping of the filler path. The black area represents the weights of floor, variable part and disjunction part vertices mapped to the element of P_M

We need in the proof for BANDWIDTH actually a slightly different result (for an easier proof), namely, we require that the first vertex of P_N is mapped to M. From the proof of Theorem 2 we also can conclude the next result.

Corollary 1. WEIGHTED PATH EMULATION WITH $f(1) = 1$ *and* WEIGHTED PATH EMULATION WITH $f(1) = M$ *are* $W[t]$*-hard for all positive integers* t.

We also have the following corollary; see also [3].

Corollary 2. WEIGHTED PATH EMULATION *is strongly NP-complete.*

4 Hardness of BANDWIDTH of Caterpillars

The main result of this section is the following theorem. Many details of the proof can be found in [3].

Theorem 3. BANDWIDTH *is* $W[t]$*-hard for all positive integers* t*, when restricted to caterpillars with hair length at most three.*

To prove Theorem 3, we transform from the WEIGHTED PATH EMULATION WITH $f(1) = M$ problem.

Let P_N and P_M be paths with weight function $w : \{1, \ldots, N\} \to \{1, \ldots, c\}$, and $c = \sum_{i=1}^{N} w(i)/M$ the emulation factor. Thus, $\sum_{i=1}^{N} w(i) = cM$. Recall that we parameterized this problem by the value c.

Assume that $c > 6$; otherwise, obtain an equivalent instance by multiplying all weights by 7.

Let $b = 12c + 6$. Let $k = 9bc + b$. Note that k is even. We give a caterpillar $G = (V, E)$ with hair length at most three, with the property that P_N has a uniform emulation on P_M, if and only if G has bandwidth at most k.

G is constructed in the following way:

- We have a *left barrier*: a vertex p_0 which has $2k - 1$ hairs of length one, and is neighbor to p_1.
- We have a path with $5M - 3$ vertices, $p_1, \ldots, p_{5M-3}$. As written above, p_1 is adjacent to p_0. Each vertex of the form $5i - 2$ or $5i$ $(1 \leq i \leq M - 1)$ receives $2k - 2b$ hairs of length one. See Fig. 4. We call this part the *floor*.
- Adjacent to vertex p_{5M-3}, we add the *turning point* from the proof of Monien [19]. We have vertices $v_a = p_{5M-3}$, v_b, v_c, v_d, v_e, v_f, v_g, which are successive vertices on a path. I.e., we identify one vertex of the turning point (v_a) with the last vertex of the floor p_{5M-3}. To v_c, we add $\frac{3}{2}(k - 2)$ hairs of length one; to v_d, we add k hairs of length three, and to v_f we add $\frac{3}{2}(k - 2)$ hairs of length one. Note that this construction is identical to the one by Monien [19]; vertex names are chosen to facilitate comparison with Moniens proof. See Fig. 5.
- Add a path with $6N - 5$ vertices, say $y_1, \ldots y_{6n-5}$, with y_1 adjacent to v_g. To each vertex of the form y_{6i-5}, add $9b \cdot w(i)$ hairs of length one. We call this part the *weighted path gadget*.
- Note that the number of vertices that we defined so far and that is not part of the turning point equals $2k + 5M - 3 + 2(M-1)(2k-2b) + 6N - 5 + 9b \sum_{i=1}^{N} w_i = 5M + 4Mk - 2k - 4Mb + 4b + 9bcM$. Let this number be α. One easily sees that $\alpha \leq (5M - 2)k - 1$. Add a path with $(5M - 2)k - 1 - \alpha$ vertices and make it adjacent to y_{6n}. We call this the *filler* path.

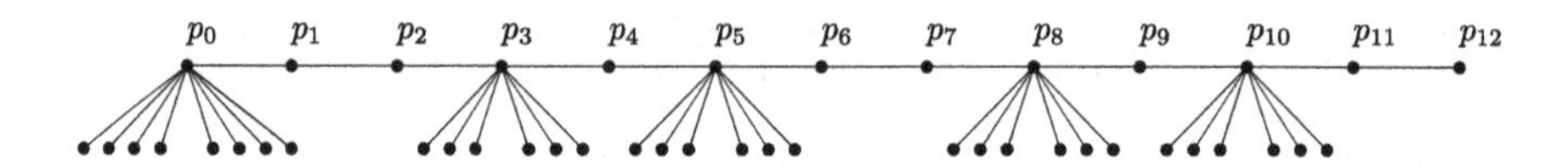

Fig. 4. First part of the caterpillar

Let G be the remaining graph. Clearly, G is a caterpillar with hair length at most three. It is interesting to note that the hair lengths larger than one are only used for the turning point.

The correctness of the construction follows from the following lemma. The proof can be found in the full paper [3].

Lemma 1. *P_N has a uniform emulation g on P_M with emulation factor c with $f(1) = M$, if and only if the bandwidth of G is at most k.*

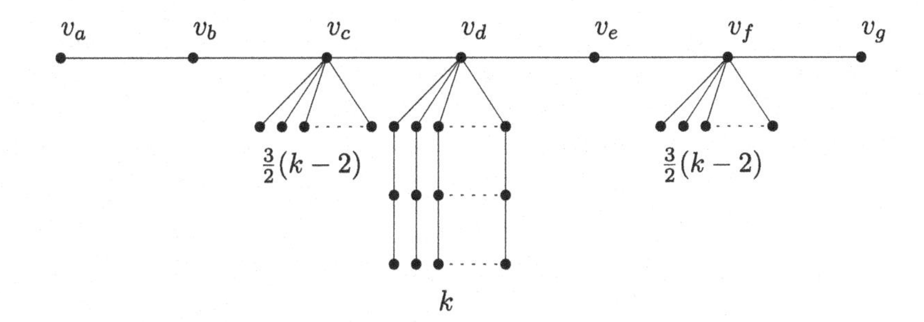

Fig. 5. The Turning Point, after Monien [19]

Fig. 6. Initial part of the floor, with left barrier and two gaps

Some Intuition. We sketch some ideas behind the proof. Suppose we have a linear ordering g of G with bandwidth at most k. We have a number of observations.

- p_0 with hairs form a blockage (called the *left barrier*, in the sense that these must either entirely at the left side or entirely at the right side of the linear ordering.
- In the same way, the turning point forms a blockage; the proof of this is due to Monien [19]. Without loss of generality, we can assume p_0 is at the left side, the turning point is at the right side.
- By considering the total number of vertices, we can show that the successive vertices p_i, p_{i+1} always have distance $k-1$ or k, with $g(p_{i+1}) = g(p_i) + k$ or $g(p_{i+1}) = g(p_i) + k - 1$.
- Vertices p_3, p_5, p_8 have 'many' hairs: these fills most of the nearby positions. E.g., in intervals $[g(p_2), g(p_3)]$, $[g(p_3), g(p_4)]$, $[g(p_4), g(p_5)]$, and $[g(p_5), g(p_6)]$ we have many hairs of the vertices p_i, while the interval $[g(p_6), g(p_7)]$ has not. So, every fifth interval has 'more space', which we call a *gap*. See Fig. 6 for the initial part of the floor with two gaps.
- Vertices of the form y_{6i-5} also have a large number of hairs. We must have that most of these hairs must be mapped to intervals of the form $[g(p_{5j+1}), g(p_{5j+2})]$. In such a case, map the ith vertex of P_N to the jth vertex of P_M. Let f be the resulting mapping
- An interval of the form $[g(p_{5j+1}), g(p_{5j+2})]$ (and the neighboring intervals, after taking hairs of the floor into account) cannot fit $9b(c+1)$ hairs of the weighted path gadget. This implies that the total weight of all vertices mapped by f to j is bounded by c; and, as we have M such intervals, must be exactly c. This shows uniformity of the mapping f.
- As discussed, a vertex of the form y_{6i-5} has hairs mapped to an interval $[g(p_{5j+1}), g(p_{5j+2})]$. Thus, when we map i to j, y_{6i-5} is mapped to an integer between $g(p_{5j})$ and $g(p_{5j+3})$. Then, y_{6i+1} is mapped to an integer between

$g(p_{5j-6})$ and $g(p_{5j+9})$—as there is a path with six edges from y_{6i-5} to y_{6i+1}, they can be mapped at most six intervals apart. This shows that there are hairs of y_{6i+1} that are mapped to the interval $[g(p_{5j-5}), g(p_{5j-4})]$ or to the interval $[(g(p_{5j+1}), g(p_{5j+2})]$ or to the interval $[g(p_{5j+7}), g(p_{5j+8})]$. And thus, vertex $i+1$ from P_N is mapped to $j-1$, j or $j+1$. This shows that the mapping f is an emulation.

An illustration of the construction of a linear ordering, given a uniform emulation is given in Fig. 7.

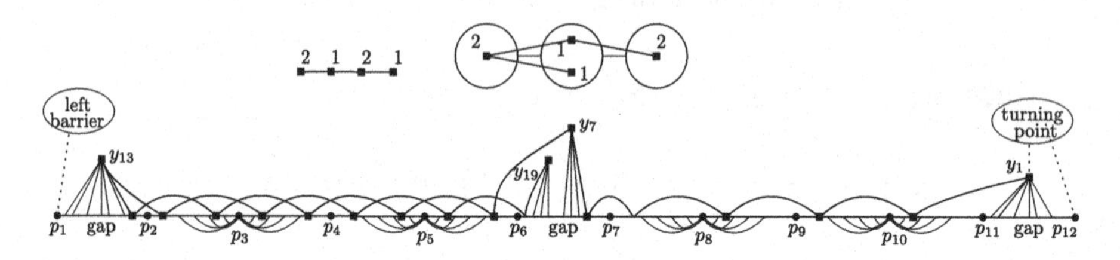

Fig. 7. Illustration of part of the construction. Shown are P_4 with successive vertex weights 2, 1, 2, 1; a uniform emulation on P_3 with emulation factor 2; a layout of a part of G.

As the construction of the caterpillar G can be done in polynomial time, given M, N and w, the main result of this section now follows.

Theorem 4. BANDWIDTH *for caterpillars with hair length at most three is* $W[t]$*-hard for all* $t \in \mathbf{N}$.

5 Directed Bandwidth

A minor variation of the proof of Theorem 3 gives the following result. The details can be found in the full paper [3].

Theorem 5. DIRECTED BANDWIDTH *is hard for* $W[t]$ *for all positive integers* t*, when restricted to directed acyclic graphs whose underlying undirected graph is a caterpillar with hair length at most one.*

6 Conclusions

In this paper, we showed that BANDWIDTH is hard for the complexity class $W[t]$ for all positive integers $t \in N$, even when the input graph is a caterpillar with hairs of length at most three. The proof uses some techniques and gadgets from the NP-completeness proof of BANDWIDTH for this class of graphs by Monien [19]. Monien also shows NP-completeness of BANDWIDTH for caterpillars of maximum degree three (with arbitrary hair length). This raises the question whether BANDWIDTH for caterpillars with maximum degree three is $W[t]$-hard for all t. We conjecture that this is the case; perhaps with a modification of our proof such a result can be achieved?

The intermediate result of the $W[t]$-hardness of WEIGHTED PATH EMULATION is of independent interest. We used this result as a stepping stone for our main result, but expect that the result may also be useful for proving hardness for other problems as well.

It is unlikely that BANDWIDTH belongs to $W[P]$. In [14], Fellows and Rosamond describe an argument, due to Hallett, that gives the intuition behind the conjecture that BANDWIDTH does not belong to $W[P]$. From the works of Bodlaender et al. [4] and Drucker [13], it follows that problems that are AND-compositional do not have a polynomial kernel unless $NP \subseteq coNP/poly$. The intuition behind this methodology is that such a polynomial kernel for an AND-compositional problem would give an unlikely strong compression of information. While BANDWIDTH is not in FPT, assuming $W[t] \not\subseteq FPT$, for some t, and thus has no kernel (of any size), it is AND-compositional. If BANDWIDTH would belong to $W[P]$, it would have a certificate of $O(k \log n)$ bits (namely, the indices of the variables that are set to true), and it is unlikely that an AND-compositional problem has such a small certificate. We thus can formulate the following conjecture, due to Hallett.

Conjecture 1 (Hallett, see also [14]*).* BANDWIDTH does not belong to $W[P]$, unless $NP \subseteq coNP/poly$.

Very recently, the author showed with Groenland, Nederlof and Swennenhuis [6] that WEIGHTED PATH EMULATION is complete for the class of problems that can be solved with a non-deterministic algorithm that uses $f(k)n^c$ time and $f(k) \log n$ space (f a computable function, c a constant). This class is known as $N[fpoly, f \log]$ and denoted as $XNLP$ in [6]. From the observation that the transformation described in Sect. 4 can be carried out in logarithmic space, it follows that BANDWIDTH for caterpillars with hair length at most three is also XNLP-complete. We thus also have that BANDWIDTH does not belong to $W[P]$ unless $W[P] \subseteq XNLP$.

Finally, we conjecture that with modifications of the techniques from this paper, it is possible to show for more problems hardness for the $W[t]$-classes.

Acknowledgements. I thank Michael Fellows and Michael Hallett for discussions and earlier joint work on the parameterized complexity of BANDWIDTH, as reported in [5]. Some techniques in this paper are based upon techniques, underlying the results reported in [5]. I thank Marieke van der Wegen for discussions.

References

1. Assmann, S.F., Peck, G.W., Sysło, M.M., Zak, J.: The bandwidth of caterpillars with hairs of length 1 and 2. SIAM J. Algebraic Discrete Methods **2**, 387–392 (1981)

2. Bodlaender, H.L.: The complexity of finding uniform emulations on paths and ring networks. Inf. Comput. **86**(1), 87–106 (1990)
3. Bodlaender, H.L.: Parameterized complexity of bandwidth of caterpillars and weighted path emulation. arXiv:2012.01226 (2020)
4. Bodlaender, H.L., Downey, R.G., Fellows, M.R., Hermelin, D.: On problems without polynomial kernels. J. Comput. Syst. Sci. **75**(8), 423–434 (2009)
5. Bodlaender, H.L., Fellows, M.R., Hallett, M.: Beyond NP-completeness for problems of bounded width: hardness for the W hierarchy. In: Proceedings of the 26th Annual Symposium on Theory of Computing, STOC 1994, pp. 449–458, New York. ACM Press (1994)
6. Bodlaender, H.L., Groenland, C., Nederlof, J., Swennenhuis, C.M.F.: Parameterized problems complete for nondeterministic FPT time and logarithmic space. arXiv:2105.14882 (2021)
7. Bodlaender, H.L., van der Wegen, M.: Parameterized complexity of scheduling chains of jobs with delays. In: Cao, Y., Pilipczuk, M. (eds.) 15th International Symposium on Parameterized and Exact Computation, IPEC 2020, volume 180 of LIPIcs, pp. 4:1–4:15. Schloss Dagstuhl - Leibniz-Zentrum für Informatik (2020)
8. Bodlaender, H.L., van Leeuwen, J.: Simulation of large networks on smaller networks. Inf. Control **71**(3), 143–180 (1986)
9. Downey, R.G., Fellows, M.R.: Fixed-parameter tractability and completeness I: basic results. SIAM J. Comput. **24**, 873–921 (1995)
10. Downey, R.G., Fellows, M.R.: Parameterized Complexity. Monographs in Computer Science. Springer, New York (1999). https://doi.org/10.1007/978-1-4612-0515-9
11. Downey, R.G., Fellows, M.R.: Fundamentals of Parameterized Complexity. Texts in Computer Science. Springer, London (2013). https://doi.org/10.1007/978-1-4471-5559-1
12. Dregi, M.S., Lokshtanov, D.: Parameterized complexity of bandwidth on trees. In: Esparza, J., Fraigniaud, P., Husfeldt, T., Koutsoupias, E. (eds.) ICALP 2014. LNCS, vol. 8572, pp. 405–416. Springer, Heidelberg (2014). https://doi.org/10.1007/978-3-662-43948-7_34
13. Drucker, A.: New limits to classical and quantum instance compression. SIAM J. Comput. **44**(5), 1443–1479 (2015)
14. Fellows, M.R., Rosamond, F.A.: Collaborating with Hans: some remaining wonderments. In: Fomin, F.V., Kratsch, S., van Leeuwen, E.J. (eds.) Treewidth, Kernels, and Algorithms. LNCS, vol. 12160, pp. 7–17. Springer, Cham (2020). https://doi.org/10.1007/978-3-030-42071-0_2
15. Fishburn, J.P., Finkel, R.A.: Quotient networks. IEEE Trans. Comput. **C-31**, 288–295 (1982)
16. Flum, J., Grohe, M.: Parameterized Complexity Theory. Texts in Theoretical Computer Science. Springer, Heidelberg (2006). https://doi.org/10.1007/3-540-29953-X
17. Gurari, E.M., Sudborough, I.H.: Improved dynamic programming algorithms for bandwidth minimization and the MinCut linear arrangement problem. J. Algorithms **5**, 531–546 (1984)
18. Hung, L.T.Q., Sysło, M.M., Weaver, M.L., West, D.B.: Bandwidth and density for block graphs. Discret. Math. **189**(1–3), 163–176 (1998)
19. Monien, B.: The bandwidth minimization problem for caterpillars with hair length 3 is NP-complete. SIAM J. Algebraic Discrete Methods **7**, 505–512 (1986)
20. Muradian, D.: The bandwidth minimization problem for cyclic caterpillars with hair length 1 is NP-complete. Theor. Comput. Sci. **307**(3), 567–572 (2003)

21. Papadimitriou, C.H.: The NP-completeness of the bandwidth minimization problem. Computing **16**, 263–270 (1976)
22. Saxe, J.B.: Dynamic programming algorithms for recognizing small-bandwidth graphs in polynomial time. SIAM J. Algebraic and Discrete Methods **1**, 363–369 (1980)
23. Sysło, M.M., Zak, J.: The Bandwidth problem: critical subgraphs and the solution for caterpillars. Ann. Discrete Math. **16**, 281–286 (1982)

Block Elimination Distance

Öznur Yaşar Diner[1,2], Archontia C. Giannopoulou[3], Giannos Stamoulis[3,4](✉), and Dimitrios M. Thilikos[5]

[1] Computer Engineering Department, Kadir Has University, Istanbul, Turkey
oznur.yasar@khas.edu.tr

[2] Department of Mathematics, Universitat Politècnica de Catalunya, Barcelona, Spain

[3] Department of Informatics and Telecommunications, National and Kapodistrian University of Athens, Athens, Greece

[4] LIRMM, Univ Montpellier, Montpellier, France
giannos.stamoulis@lirmm.fr

[5] LIRMM, Univ Montpellier, CNRS, Montpellier, France
sedthilk@thilikos.info

Abstract. We introduce the parameter of *block elimination distance* as a measure of how close a graph is to some particular graph class. Formally, given a graph class $\mathcal{G}$, the class $\mathcal{B}(\mathcal{G})$ contains all graphs whose blocks belong to $\mathcal{G}$ and the class $\mathcal{A}(\mathcal{G})$ contains all graphs where the removal of a vertex creates a graph in $\mathcal{G}$. Given a hereditary graph class $\mathcal{G}$, we recursively define $\mathcal{G}^{(k)}$ so that $\mathcal{G}^{(0)} = \mathcal{B}(\mathcal{G})$ and, if $k \geq 1$, $\mathcal{G}^{(k)} = \mathcal{B}(\mathcal{A}(\mathcal{G}^{(k-1)}))$. The *block elimination distance* of a graph G to a graph class $\mathcal{G}$ is the minimum k such that $G \in \mathcal{G}^{(k)}$ and can be seen as an analog of the elimination distance parameter, defined in [*J. Bulian & A. Dawar. Algorithmica, 75(2):363–382, 2016*], with the difference that connectivity is now replaced by biconnectivity. We show that, for every non-trivial hereditary class $\mathcal{G}$, the problem of deciding whether $G \in \mathcal{G}^{(k)}$ is NP-complete. We focus on the case where $\mathcal{G}$ is minor-closed and we study the minor obstruction set of $\mathcal{G}^{(k)}$ i.e., the minor-minimal graphs not in $\mathcal{G}^{(k)}$. We prove that the size of the obstructions of $\mathcal{G}^{(k)}$ is upper bounded by some explicit function of k and the maximum size of a minor obstruction of $\mathcal{G}$. This implies that the problem of deciding whether $G \in \mathcal{G}^{(k)}$ is *constructively* fixed parameter tractable, when parameterized by k. Our results are based on a structural characterization of the obstructions of $\mathcal{B}(\mathcal{G})$, relatively to the obstructions of $\mathcal{G}$. Finally, we give two graph operations that generate members of $\mathcal{G}^{(k)}$ from members of $\mathcal{G}^{(k-1)}$ and we prove that this set of operations is complete for the class $\mathcal{O}$ of outerplanar graphs. This yields the *identification* of all members $\mathcal{O} \cap \mathcal{G}^{(k)}$, for every $k \in \mathbb{N}$ and every non-trivial minor-closed graph class $\mathcal{G}$.

Keywords: Elimination distance · Graph minors · Obstructions · Parameterized algorithms · Biconnected graphs

The first author was supported by the Spanish *Agencia Estatal de Investigacion* project MTM2017-82166-P. The two last authors were supported by the ANR projects DEMOGRAPH (ANR-16-CE40-0028), ESIGMA (ANR-17-CE23-0010), and the French-German Collaboration ANR/DFG Project UTMA (ANR-20-CE92-0027).

L. Kowalik et al. (Eds.): WG 2021, LNCS 12911, pp. 28–38, 2021.
https://doi.org/10.1007/978-3-030-86838-3_3

1 Introduction

Graph distance parameters are typically introduced as measures of "how close" is a graph G to some given graph class $\mathcal{G}$. One of the main motivating factors behind introducing such distance parameters is the following. Let $\mathcal{G}$ be a graph class on which a computational problem Π is tractable and let $\mathcal{G}^{(k)}$ be the class of graphs with distance at most k from $\mathcal{G}$, for some notion of distance. Our aim is to exploit the "small" distance of the graphs in $\mathcal{G}^{(k)}$ from $\mathcal{G}$ in order to extend the tractability of Π in the graph class $\mathcal{G}^{(k)}$. This approach on dealing with computational problems is known as *parameterization by distance from triviality* [11]. Usually, a graph distance measure is defined by minimizing the number of modification operations that can transform a graph G to a graph in $\mathcal{G}$.

The most classic modification operation is the *apex extension* of a graph class $\mathcal{G}$, defined as $\mathcal{A}(\mathcal{G}) = \{G \mid \exists v \in V(G) \quad G \setminus v \in \mathcal{G}\}$ and the associated parameter, the *vertex-deletion distance* of G to $\mathcal{G}$, is defined as $\min\{k \mid G \in \mathcal{A}^k(\mathcal{G})\}$. The vertex-deletion distance has been extensively studied. Other, popular variants of modification operations involve edge removals/additions/contractions or combinations of them [4,8,10].

Elimination Distance. Bulian and Dawar in [5,6], introduced the *elimination distance* of G to a class $\mathcal{G}$ as follows:

$$\mathsf{ed}_{\mathcal{G}}(G) = \begin{cases} 0 & G \in \mathcal{G} \\ \max\{\mathsf{ed}_{\mathcal{G}}(C) \mid C \in \mathsf{cc}(G)\} & \text{if } G \notin \mathcal{G} \text{ and is not connected} \\ 1 + \min\{\mathsf{ed}_{\mathcal{G}}(G \setminus v) \mid v \in V(G)\} & \text{if } G \notin \mathcal{G} \text{ and is connected} \end{cases}$$

where by $\mathsf{cc}(G)$ we denote the connected components of G. Notice that the definition $\mathsf{ed}_{\mathcal{G}}$, apart from vertex deletions, also involves the *connected closure* operation, defined as $\mathcal{C}(\mathcal{G}) = \{G \mid \forall C \in \mathsf{cc}(G),\ C \in \mathcal{G}\}$. Observe that $\mathsf{ed}_{\mathcal{G}}(G) = 0$ iff $G \in \mathcal{G} \cup \mathcal{C}(\mathcal{G})$, while, for $k > 0$, $\mathsf{ed}_{\mathcal{G}}(G) \leq k$ iff $G \in \mathcal{G}' \cup \mathcal{C}(\mathcal{G}')$, where $\mathcal{G}' = \mathcal{A}(\{G \mid \mathsf{ed}_{\mathcal{G}}(G) \leq k-1\})$. Therefore, $\mathsf{ed}_{\mathcal{G}}$ can be seen as as a tree counterpart of the vertex-deletion operation where the "branching effect" is based on operation $\mathcal{C}$, that is, in each level of the recursion, the vertex deletion operation is applied to each of the connected components of the current graph. A motivation of Bulian and Dawar in [5] for introducing $\mathsf{ed}_{\mathcal{G}}$ was the study of the Graph Isomorphism Problem. Indeed, it is easy to see that there are constants c_{α} and c_{κ} such that if Graph Isomorphism can be solved in $O(n^c)$ time in some graph class $\mathcal{G}$, then it can be solved in time $O(n^{c+c_{\alpha}})$ (resp. $O(n^{c+c_{\kappa}})$) in the graph class $\mathcal{A}(\mathcal{G})$ (resp. $\mathcal{C}(\mathcal{G})$) (see [7,12,13]). This implies that Graph Isomorphism can be solved in $n^{O(k)}$ steps in the class of graphs where $\mathsf{ed}_{\mathcal{G}}$ is bounded by k. In [5], Bulian and Dawar improved this implication for the class $\mathcal{G}_d$ of graphs of maximum degree at most d and proved that Graph Isomorphism can be solved in $f(k) \cdot n^{c_d}$ time in the class $\{G \mid \mathsf{ed}_{\mathcal{G}_d}(G) \leq k\}$ (here c_d is a constant depending on d). In other words, for every d, Graph Isomorphism is fixed parameter tractable (in short FPT), when parameterized by $\mathsf{ed}_{\mathcal{G}_d}$.

Computing the Elimination Distance. Typically, the algorithmic results on $\mathsf{ed}_{\mathcal{G}}$ apply for instantiations of $\mathcal{G}$ that are hereditary, i.e., the removal of a vertex of a graph in $\mathcal{G}$ results to a graph that is again in $\mathcal{G}$. Bulian and Dawar in [6] examined the case where $\mathcal{G}$ is minor-closed. One may observe that containment in $\mathcal{G}$ is equivalent to the exclusion of the graphs in the minor-obstruction set of $\mathcal{G}$, that is the set $\mathsf{obs}(G)$ of the minor-minimal graphs not in $\mathcal{G}$. Also the minor-closed property is invariant under the operations $\mathcal{A}$ and $\mathcal{C}$, therefore the class $\{G \mid \mathsf{ed}_{\mathcal{G}}(G) \leq k\}$ is also minor-closed. From the Robertson and Seymour's theorem, $\mathsf{obs}(\{G \mid \mathsf{ed}_{\mathcal{G}}(G) \leq k\})$ is finite, and this implies, using the algorithmic results of [16,18], that for every minor-closed class $\mathcal{G}$, deciding whether $\mathsf{ed}_{\mathcal{G}}(G) \leq k$ is FPT (parameterized by k) by an algorithm that runs in $f(k) \cdot n^2$ time. While this approach is not constructive in general, Bulian and Dawar in [6] proved that there is an algorithm that, with input $\mathsf{obs}(\mathcal{G})$ and k, outputs the set $\mathsf{obs}(\{G \mid \mathsf{ed}_{\mathcal{G}}(G) \leq k\})$. This makes the aforementioned $f(k) \cdot n^2$-time algorithm constructive in the sense that the function f is computable. An explicit estimation of this function f can be derived from the recent results in [20–22]. The computational complexity of $\mathsf{ed}_{\mathcal{G}}$ was also studied for different instantiations of $\mathcal{G}$. In [17] Lindermayr, Siebertz, and Vigny considered the class $\mathcal{G}_d$ of graphs of degree at most d. They proved that, given k, d, and a planar graph G, deciding whether $\mathsf{ed}_{\mathcal{G}_d}(G) \leq k$ is FPT (parameterized by k and d) by designing an $f(k,d) \cdot n^{O(1)}$ time algorithm. Also, in [2] the same result was proved without the planarity restriction. Moreover, in [2], more general hereditary classes where considered: let $\mathcal{F}$ be some finite set of graphs and let $\mathcal{G}_{\mathcal{F}}$ be the class of graphs excluding all graphs in $\mathcal{F}$ as induced subgraphs. It was proved in [2] that for every such $\mathcal{F}$ the problem that, given some graph G and k, deciding whether $\mathsf{ed}_{\mathcal{G}_{\mathcal{F}}}(G) \leq k$ is FPT (parameterized by k) by designing an $f(k) \cdot n^{c_d}$ time algorithm, where c_d is a constant depending on d (see also [3] for earlier results). Approximately optimal decompositions for computing elimination distance $\mathsf{ed}_{\mathcal{G}}(G)$ for some specific graph classes $\mathcal{G}$ are given in a recent paper [15].

Block Elimination Distance. We introduce a more general version of elimination distance where the branching is guided by *biconnectivity* instead of connectivity. The recursive application of the vertex deletion operation is now done on the *blocks* of the current graph instead of its components. That way, the *block elimination distance* of a graph G to a graph class $\mathcal{G}$ is defined as

$$\mathsf{bed}_{\mathcal{G}}(G)=\begin{cases} 0 & G \in \mathcal{G} \\ \max\{\mathsf{bed}_{\mathcal{G}}(C) \mid C \in \mathsf{bc}(G)\} & \text{if } G \notin \mathcal{G} \text{ and is not biconnected} \\ 1+\min\{\mathsf{bed}_{\mathcal{G}}(G \setminus v) \mid v \in V(G)\} & \text{if } G \notin \mathcal{G} \text{ and is biconnected} \end{cases}$$

where by $\mathsf{bc}(G)$ we denote the blocks of the graph G. We stress that the "branching effect" in the above definition is the *biconnected closure* operation, defined as $\mathcal{B}(\mathcal{G}) = \{G \mid \forall B \in \mathsf{bc}(G),\ B \in \mathcal{G}\}$.

The above parameter is more general than $\mathsf{ed}_{\mathcal{G}}$ in the sense that it upper bounds $\mathsf{ed}_{\mathcal{G}}$ but it is not upper bounded by any function of $\mathsf{ed}_{\mathcal{G}}$. For instance, if

G is a connected graph whose blocks belong to $\mathcal{G}$, it follows that $\mathsf{bed}_{\mathcal{G}}(G) = 0$, while $\mathsf{ed}_{\mathcal{G}}(G)$ can be arbitrarily big. Moreover, $\mathsf{bed}_{\mathcal{G}}$, can also serve as a measure for the distance to triviality in the same way as $\mathsf{ed}_{\mathcal{G}}$. For instance, there is a constant $c_{\mathcal{B}}$ such that if GRAPH ISOMORPHISM can be solved in $O(n^c)$ time in some graph class $\mathcal{G}$, then it can be solved in time $O(n^{c+c_{\mathcal{B}}})$ in the graph class $\mathcal{B}(\mathcal{G})$ (using standard techniques, see e.g., [7,12,13]). This implies that GRAPH ISOMORPHISM can be solved in $n^{O(k)}$ steps in the class of graphs where $\mathsf{bed}_{\mathcal{G}}$ is bounded by k. Clearly, all the problems studied so far on the elimination distance have their counterpart for the block elimination distance and this is a relevant line of research, as the new parameter is more general than its connected counterpart.

Our Results. As a first step, we prove that if $\mathcal{G}$ is a non-trivial and hereditary class, then deciding whether $\mathsf{bed}_{\mathcal{G}}(G) \leq k$ is an NP-complete problem (Sect. 3). For our proof we certify yes-instances by using an alternative definition of $\mathsf{bed}_{\mathcal{G}}$ that is based on an (multi)-embedding of G in a rooted forest (Sect. 2). We next focus our study on the case where $\mathcal{G}$ is minor-closed (and non-trivial). As the operation $\mathcal{B}$ maintains minor-closedness, it follows that the class $\mathcal{G}^{(k)} := \{G \mid \mathsf{bed}_{\mathcal{G}}(G) \leq k\}$ is minor-closed for every k, therefore for every minor-closed $\mathcal{G}$, deciding whether $G \in \mathcal{G}^{(k)}$ is FPT (parameterized by k). Following the research line of [6], we make this result *constructive* by proving that it is possible to bound the size of the obstructions of $\mathcal{G}^{(k)}$ by some explicit function of k and the maximum size of the obstructions of $\mathcal{G}$. This bound is based on the results of [1,21] (Sect. 4) and a structural characterization of $\mathsf{obs}(\mathcal{B}(G))$, in terms of $\mathsf{obs}(\mathcal{G})$, implying that no obstruction of $\mathcal{B}(G)$ has size that is more than twice the maximum size of an obstruction of $\mathcal{G}$. In Sect. 5 we take a closer look of the obstructions of $\mathcal{G}^{(k)}$. We give two graph operations, called *parallel join* and *triangular gluing*, that generate members of $\mathcal{G}^{(k)}$ from members of $\mathcal{G}^{(k-1)}$. This yields that the number of obstructions of $\mathcal{G}^{(k)}$ is at least doubly exponential on k. Moreover, we prove that this set of operations is *complete* for the class $\mathcal{O}$ of outerplanar graphs. This implies the *complete identification* of $\mathcal{O} \cap \mathcal{G}^{(k)}$, for every $k \in \mathbb{N}$ and every non-trivial minor-closed graph class $\mathcal{G}$. This yields that the number of obstructions of $\mathcal{G}^{(k)}$ is at least doubly exponential on k. The paper concludes in Sect. 6 with some further observations and open problems.

In this paper we omit most of the proofs of our results. The full version can be accessed at [9].

2 Definitions and Preliminary Results

Basic Concepts on Graphs. All graphs considered in this paper are undirected, finite, and without loops or multiple edges. We use $V(G)$ and $E(G)$ for the sets of vertices and edges of G, respectively. For simplicity, an edge $\{x, y\}$ of G is denoted by xy or yx. We say that H is a *subgraph* of G if $V(H) \subseteq V(G)$ and $E(H) \subseteq E(G)$. Also, if S is a set of vertices we denote by $G \setminus S$ the graph obtained if we remove S from G. We denote by $G[S]$ the subgraph of G induced by the vertices from S, that is the graph $G \setminus (V(G) \setminus S)$, and we say that the graph H is an *induced subgraph* of a graph G if $H = G[S]$ for some $S \subseteq V(G)$. We write $G \setminus v$ instead of $G \setminus \{v\}$ for a single vertex set. Given $e \in E(G)$, we denote $G \setminus e = (V(G), E(G) \setminus \{e\})$. If a vertex is not an endpoint of an edge, then we call it *isolated.*

A graph G is *connected* (resp. *biconnected*) if for every $u, v \in V(G)$, G contains a path (resp. cycle) containing u and v. A *(bi)connected component* of G is a maximally (bi)connected subgraph of G. We denote by $\mathsf{cc}(G)$ the set of all connected components of G. A vertex x is a *cut-vertex* of a graph G if $|\mathsf{cc}(G)| < |\mathsf{cc}(G \setminus v)|$. A *bridge* of a graph G is a connected subgraph on two vertices x, y and the edge $e = xy$ such that $|\mathsf{cc}(G)| < |\mathsf{cc}(G \setminus e)|$. A *block* of a graph is either an isolated vertex of G, or is a bridge of G, or is a biconnected component of G. We denote by $\mathsf{bc}(G)$ the set of all blocks of G. A graph G is a *block-graph* if $\mathsf{bc}(G) = \{G\}$.

Graph Classes. We use the term *graph class* (or simply *class*) for any set of graphs (this set might be finite or infinite). We denote by $\mathcal{E}$ the class of the edgeless graphs. We say that a graph class is *non-trivial* if it contains at least one non-empty graph and does not contain all graphs. We say that a class $\mathcal{G}$ is *hereditary* if every induced subgraph of a graph in $\mathcal{G}$ belongs also to $\mathcal{G}$. Notice that both operations $\mathcal{A}$ and $\mathcal{B}$ maintain the property of being non-trivial and hereditary.

In this paper we consider only classes that are non-trivial and hereditary. This implies that $\mathcal{G} \subseteq \mathcal{A}(\mathcal{G})$. Notice that this assumption is necessary as $\{K_1\} \not\subseteq \mathcal{A}(\{K_1\}) = \{K_2\}$ ($\{K_1\}$ is non-hereditary) and $\{K_0\} \not\subseteq \mathcal{A}(\{K_0\}) = \{K_1\}$ ($\{K_0\}$ is not non-trivial). Also the hereditarity of $\mathcal{G}$ implies that $\mathcal{G} \subseteq \mathcal{B}(\mathcal{G})$ and hereditarity is necessary for this as, for example, $\{P_3\} \not\subseteq \mathcal{B}(\{P_3\}) = \{K_0\}$. However, $\mathcal{G} \subseteq \mathcal{B}(\mathcal{G})$ also holds for the two finite classes that are not non-trivial, i.e., $\mathcal{B}(\{\}) = \{K_0\}$ and $\mathcal{B}(\{K_0\}) = \{K_0\}$. We also exclude the class of all graphs as, in this case, $\mathcal{A}$ and $\mathcal{B}$ do not generate new classes.

Observe also that for every non-trivial and hereditary class $\mathcal{G}$, $\mathcal{B}(\mathcal{G}) = \mathcal{B}(\mathcal{B}(\mathcal{G}))$. This implies that $\mathsf{bed}_{\mathcal{G}}$ and $\mathsf{bed}_{\mathcal{B}(\mathcal{G})}$ are the same parameter.

An Alternative Definition. A *rooted forest* is a pair (F, R) where F is an acyclic graph and $R \subseteq V(F)$ such that each connected component of F contains exactly one vertex of R, its *root.* A vertex $t \in V(F)$ is a *leaf* of F if either it is an isolated vertex in R or it is a non-isolated vertex adjacent to exactly one edge of F. We use $L(F, R)$ in order to denote the leaves of (F, R). Given $t, t' \in V(F)$ we say that $t \leq_{F,R} t'$ if there is a path from t' to some root in R that contains t. If neither $t \leq_{F,R} t'$ nor $t' \leq_{F,R} t$ then we say that t and t' are *incomparable* in (F, R). An (F, R)-*antichain* is a non-empty set C of pairwise incomparable vertices of F. An (F, R)-antichain is *non-trivial* if it contains at least two elements. The *depth* of a rooted forest (F, R) is the maximum number of vertices in a path between a leaf and the root of the connected component of F where this leaf belongs. Let $\mathcal{G}$ be a non-trivial hereditary class and let G be a graph. Let (F, R, τ) be a triple consisting of a rooted forest F whose root set is R and a function τ mapping vertices of G to subsets of $V(F)$. Given a vertex set $S \subseteq V(F)$, we set $\tau^{-1}(S) = \{v \in V(G) \mid \tau(v) \cap S \neq \emptyset\}$. Also, for every $t \in V(F)$, we define $G_t = G[\tau^{-1}(\{t' \in V(F) \mid t \leq_{F,R} t'\})]$. A triple (F, R, τ) is a *$\mathcal{G}$-block tree layout* of G if the following hold:

(1) for every $v \in V(G)$, $\tau(v)$ is an (F, R)-antichain,
(2) for every $t \in V(F)$, G_t is a block-graph,
(3) if $t \notin L(F, R)$, then $|\tau^{-1}(\{t\})| = 1$ and $G_t \notin \mathcal{G}$ or,
(4) if $t \in L(F, R)$, then $G_t \in \mathcal{G}$ and
(5) for every non-trivial (F, R)-antichain C, the union of the graphs in $\{G_t \mid t \in C\}$ is not biconnected.

The *depth* of the $\mathcal{G}$-block tree layout (F, R, τ) is equal to the depth of the rooted forest (F, R). We give the following alternative definition for $\mathsf{bed}_{\mathcal{G}}$.

Lemma 1. *If $\mathcal{G}$ is a non-trivial hereditary class and G is a graph, then the minimum depth of a $\mathcal{G}$-block tree layout of G is equal to* $\mathsf{bed}_{\mathcal{G}}(G) - 1$.

3 NP-completeness

We consider the following family of problems, each defined by some non-trivial and hereditary graph class $\mathcal{G}$. We say that a class $\mathcal{G}$ is *polynomially decidable* if there exists an algorithm that, given an n-vertex graph G, decides whether $G \in \mathcal{G}$ in polynomial, in n, time.

> Block Elimination Distance to $\mathcal{G}$ ($\mathcal{G}$-BED)
> **Instance:** A graph G and a non-negative integer k.
> **Question:** Is the block elimination distance of G to $\mathcal{G}$ at most k?

Lemma 2. *For every polynomially decidable, non-trivial, and hereditary graph class $\mathcal{G}$, the problem $\mathcal{G}$-BED is* NP*-complete.*

The proof of the above result is a (multi) reduction from the problem Balanced Complete Bipartite Subgraph (BCBS). It is based on the alternative definition of block elimination distance (Lemma 1) and has two parts. The first proves the NP-hardness of $\mathcal{E}$-BED (recall that $\mathcal{E}$ is the class of edgeless graphs). The second is a multi-reduction from $\mathcal{E}$-BED to $\mathcal{G}$-BED where the existence of the main gadget is based on the following lemma.

Lemma 3. *Let $\mathcal{G}$ be a non-trivial hereditary class. Then there exists a graph Z with the following properties: (1) Z is a block-graph, (2) $Z \notin \mathcal{B}(\mathcal{G})$ and, (3) $\forall v \in V(Z)$, $Z \setminus v \in \mathcal{B}(\mathcal{G})$.*

We stress that the proof of the above lemma is not constructive in the sense that it does not give any way to construct Z. However, if the non-trivial and hereditary class $\mathcal{G}$ is decidable, then Z is effectively computable and this makes the proof of Lemma 2 constructive.

4 Elimination Distance to Minor-Free Graph Classes

Minors and Obstructions. A graph H is a *minor* of a graph G if H can be obtained by some subgraph of G after contracting edges. Given a set $\mathcal{Q}$ of graphs, we denote by $\mathsf{excl}(\mathcal{Q})$ the class of all graphs excluding every graph in $\mathcal{Q}$ as a minor and by $\mathbf{obs}(\mathcal{Q})$ the class of all minor-minimal graphs that do not belong to $\mathcal{Q}$. Clearly, for every class $\mathcal{G}$, $\mathcal{G} = \mathsf{excl}(\mathsf{obs}(\mathcal{G}))$. Also, according to Robertson and Seymour's theorem [19], for every minor-closed class $\mathcal{G}$, $\mathsf{obs}(\mathcal{G})$ is finite. We call a class *essential* if it is a finite minor-antichain that is non-empty and does not contain the graph K_0 or the graph K_1. Notice that $\mathcal{G}$ is trivial iff $\mathsf{obs}(\mathcal{G})$ is essential. We call an essential class $\mathcal{Z}$ *biconnected* if all graphs in $\mathsf{obs}(\mathcal{Z})$ are block-graphs. Given that $\mathcal{Z}$ is an essential graph class, we define $s(\mathcal{Z}) = \max\{|V(G)| \mid G \in \mathcal{Z}\}$.

The next result, can be seen as the biconnected analog of [6, Lemma 5], where the structure of $\mathsf{obs}(\mathcal{C}(\mathsf{excl}(\mathcal{Z})))$ is studied. The *connected closure* operation in [6] allows for a less complicated proof, since also the structure of graphs in $\mathsf{obs}(\mathcal{C}(\mathsf{excl}(\mathcal{Z})))$ is simpler. However, in Lemma 4 below we deal with the *biconnected closure* operation and thus richer structural properties are revealed, resulting also in a more technical proof.

Lemma 4. *Let $\mathcal{Z}$ be a finite graph class. For every graph $G \in \mathsf{obs}(\mathcal{B}(\mathsf{excl}(\mathcal{Z})))$ there is a graph $H \in \mathcal{Z}$ such that G can be transformed to H after a sequence of at most $|\mathsf{bc}(H)| - 1$ edge deletions and $|\mathsf{bc}(H)| - 1$ edge contractions.*

Proof (of Lemma 4 - sketch). Let $G \in \mathsf{obs}(\mathcal{B}(\mathsf{excl}(\mathcal{Z})))$. We assume that $|V(G)| \geq 4$, since otherwise the lemma holds trivially. Since $G \in \mathsf{obs}(\mathcal{B}(\mathsf{excl}(\mathcal{Z})))$, G is biconnected and also, the fact that $G \notin \mathcal{B}(\mathsf{excl}(\mathcal{Z}))$ implies that there exists a graph $H \in \mathcal{Z}$ that is a minor of G. Moreover, since G is a minor-minimal biconnected graph with the latter property, it holds that no proper minor of G is biconnected and contains H as a minor (this fact is essential for the arguments of all the statements that we expose later). Let M be a (vertex-minimal and, subject to this, edge-minimal) subgraph of G such that H is a contraction of M. We prove that G can be transformed to M after a sequence of at most $|\mathsf{bc}(H)|-1$ edge removals. This fact, combined with the following claim completes the proof of the lemma.

Claim: M can be transformed to H after at most $|\mathsf{bc}(H)| - 1$ edge contractions.

Proof of the Claim: For every $v \in V(H)$, we set X_v be the set of the vertices of M that have been contracted to v. We will prove that $\sum_{v \in V(H)} |E(G[X_v])| \leq |\mathsf{bc}(H)| - 1$, which implies the above Claim. For this we first prove that for every vertex $v \in V(H)$ that is not a cut-vertex of H, it holds that $|X_v| = 1$ and, next, we prove that for every cut-vertex v of H it holds that $|E(G[X_v])| \leq |\mathsf{cc}(G \setminus X_v)| - 1$. To conclude the proof of Claim, for every cut-vertex v of H, we set $\mathsf{blocks}(H, v)$ to be the blocks of H that contain v. Observe that $|\mathsf{cc}(G \setminus X_v)| \leq |\mathsf{blocks}(H, v)|$. We set $\mathsf{cv}(H)$ to be the set of cut-vertices of

H and we notice that $\sum_{v \in \mathsf{cv}(H)} |E(G[X_v])| \leq \sum_{v \in \mathsf{cv}(H)} (|\mathsf{cc}(G \setminus X_v)| - 1) \leq \sum_{v \in \mathsf{cv}(H)} (|\mathsf{blocks}(H, v)| - 1)$. The fact that $\sum_{v \in \mathsf{cv}(H)} (|\mathsf{blocks}(H, v)| - 1) \leq |\mathsf{bc}(H)| - 1$ implies that $\sum_{v \in \mathsf{cv}(H)} |E(G[X_v])| \leq |\mathsf{bc}(H)| - 1$. The latter together with the fact that for every vertex $v \in V(H)$ that is not a cut-vertex of H, it holds that $|X_v| = 1$ completes the proof of Claim. □

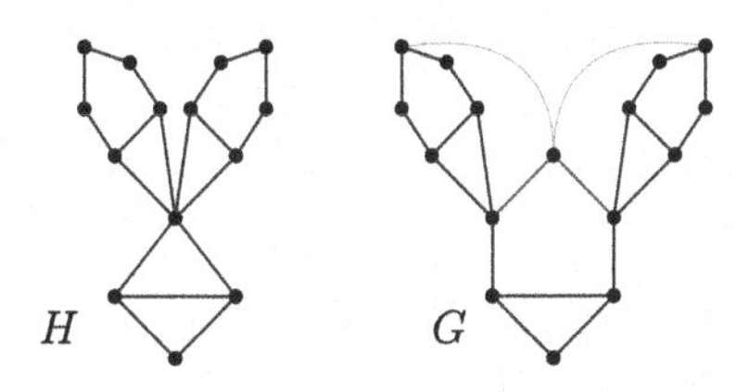

Fig. 1. Example of a graph H (on the left) and a graph $G \in \mathsf{obs}(\mathcal{B}(\mathsf{excl}(H)))$ (on the right) such that G can be transformed to H after exactly $|\mathsf{bc}(H)| - 1$ edge deletions and $|\mathsf{bc}(H)| - 1$ edge contractions.

We stress that the bounds on the number of operations in Lemma 4 are tight in the sense that, given a graph H, there is a graph $G \in \mathsf{obs}(\mathcal{B}(\mathsf{excl}(\{H\})))$ such that G can be transformed to H after *exactly* $|\mathsf{bc}(H)| - 1$ edge deletions and $|\mathsf{bc}(H)| - 1$ edge contractions. For example, in Fig. 1, the graph H has three blocks (i.e., $|\mathsf{bc}(H)| = 3$), the graph G is a graph in $\mathsf{obs}(\mathcal{B}(\mathsf{excl}(\{H\})))$, and to transform G to H one has to remove the two grey edges and contract the two red ones.

An algorithmic consequence of Lemma 4 and the results in [21] is the following.

Lemma 5. *There is an explicit function $f : \mathbb{N} \to \mathbb{N}$ and an algorithm that, given a finite class $\mathcal{Z}$, where $s = s(\mathcal{Z})$, an n-vertex graph G, and an integer k, outputs whether $\mathsf{bed}_{\mathsf{excl}(\mathcal{Z})}(G) \leq k$ in $O(f(s, k) \cdot n^2)$ time. Moreover, if $\mathcal{Z}$ contains some planar graph, then the dependence of the running time on n is linear.*

Also, using Lemma 4 we can prove that, in the definition of $\mathsf{bed}_{\mathcal{G}}$, the class $\mathcal{G}$ can be chosen so that $\mathsf{obs}(\mathcal{G})$ is biconnected and, moreover, such a $\mathcal{G}$ has an explicit obstruction characterization.

5 Outerplanar Obstructions for Block Elimination Distance

In this section we study the set $\mathsf{obs}(\mathcal{G}^{(k)})$ for distinct instantiations of k and $\mathcal{G}$. For a warm up, we mention here that $\mathsf{obs}(\mathcal{E}^{(1)}) = \{K_3\}$ and, as a less trivial example, $\mathsf{obs}(\mathcal{E}^{(2)}) = \{ \}$.

Our objective is to generate obstructions of $\mathcal{G}^{(k+1)}$ using obstructions of $\mathcal{G}^{(k)}$. For this, we define the following two operations. (See also Fig. 2.)

Parallel join: Let G_1 and G_2 be graphs and let $v_1^i, v_2^i \in V(G_i)$, $i \in [2]$. We denote by $\|(G_1, v_1^1, v_2^1, G_2, v_1^2, v_2^2)$ the graph obtained from the disjoint union of G_1 and G_2 after we add the edges $\{v_i^1, v_i^2\}, i \in [2]$ and we call it the *parallel join* of G_1 and G_2 on (v_1^1, v_2^1) and (v_1^2, v_2^2).

Triangular gluing: Let G_1, G_2, and G_3 be graphs and let $v_1^i, v_2^i \in V(G_i)$, $i \in [3]$. We denote by $\triangle(G_1, v_1^1, v_2^1, G_2, v_1^2, v_2^2, G_3, v_1^3, v_2^3)$ the graph obtained from the disjoint union of G_1, G_2, and G_3 after we identify the pairs v_2^1 and v_1^2, v_2^2 and v_1^3, and v_2^3 and v_1^1. We call this graph the *triangular gluing* of G_1, G_2, and G_3 on (v_1^1, v_2^1), (v_1^2, v_2^2), and (v_1^3, v_2^3).

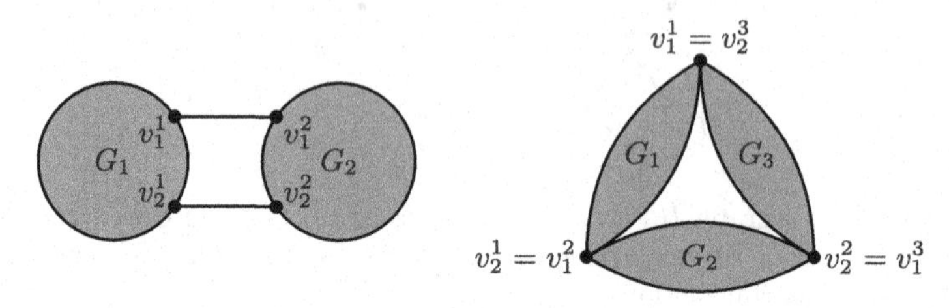

Fig. 2. The parallel join and the triangular gluing operations.

Lemma 6. *Let $\mathcal{G}$ be a non-trivial and minor-closed class and let $k \in \mathbb{N}$. If $G_1, G_2, G_3 \in \mathsf{obs}(\mathcal{G}^{(k)})$ and $v_1^i, v_2^i \in V(G_i), i \in [3]$, then $\|(G_1, v_1^1, v_2^1, G_2, v_1^2, v_2^2) \in \mathsf{obs}(\mathcal{G}^{(k+1)})$ (provided that $\mathcal{G} \neq \mathcal{E}$ or $k \geq 1$) and $\triangle(G_1, v_1^1, v_2^1, G_2, v_1^2, v_2^2, G_3, v_1^3, v_2^3) \in \mathsf{obs}(\mathcal{G}^{(k+1)})$.*

Lemma 6 implies that the set $\bigcup_{i\geq 0} \mathsf{obs}(\mathcal{G}^{(i)})$ is closed under the parallel join and the triangular gluing operations. We denote by $\mathcal{O}$ the class of all outerplanar graphs. We claim that $\mathcal{O} \cap \bigcup_{i\geq 1} \mathsf{obs}(\mathcal{G}^{(i)})$ is complete under these two operations. In particular we prove the following:

Lemma 7. *Let $\mathcal{G}$ be a non-trivial and minor-closed class. For every $k \geq 1$ and for every graph $G \in \mathsf{obs}(\mathcal{G}^{(k+1)}) \cap \mathcal{O}$, there are*

- *either two graphs G_1 and G_2 of $\mathsf{obs}(\mathcal{G}^{(k)}) \cap \mathcal{O}$ and $v_1^i, v_2^i \in V(G_i)$, $i \in [2]$, such that $G = \|(G_1, v_1^1, v_2^1, G_2, v_1^2, v_2^2)$ or*
- *three graphs G_1, G_2 and G_3 of $\mathsf{obs}(\mathcal{G}^{(k)}) \cap \mathcal{O}$ and $v_1^i, v_2^i \in V(G_i)$, $i \in [3]$, such that $G = \triangle(G_1, v_1^1, v_2^1, G_2, v_1^2, v_2^2, G_2, v_1^3, v_2^3)$.*

As $\mathbf{obs}(\mathcal{G}^{(0)}) = \mathsf{obs}(\mathcal{B}(\mathcal{G}))$, Lemma 4, Lemma 6, and Lemma 7 give a complete characterization of $\mathcal{O} \cap \mathcal{G}^{(k)}$, for every $k \in \mathbb{N}$ and every non-trivial minor-closed graph class $\mathcal{G}$. It is easy to verify that for every $\mathcal{G}$, there are at least two obstructions in $\mathsf{obs}(\mathcal{G}^{(3)})$ that are generated by the triangular gluing operation. Moreover, as the operation of trianglular gluing three graphs from a set of q graphs results to $q^2 + \binom{q}{3} \geq q^2$ new graphs, our results imply that, for $k \geq 3$, $|\mathsf{obs}(\mathcal{G}^{(k)})| \geq |\mathsf{obs}(\mathcal{G}^{(k-1)})|^2$. It follows that, for every non-trivial minor-closed class $\mathcal{G}$, $\mathsf{obs}(\mathcal{G}^{(k)})$ contains doubly exponentially many graphs.

6 A Conjecture on the Universal Obstructions

Recently, Huynh et al. in [14] defined the parameter td_2 as follows. A *biconnected centered coloring* of a graph G is a vertex coloring of G such that for every connected subgraph H of G that is a block-graph, some color is assigned to *exactly one* vertex of H. Given a non-empty graph G, $\mathsf{td}_2(G)$ is defined as the minimum number of colors in a biconnected centered coloring of G. Using the alternative definition of Sect. 2, it can easily be verified that, for every non-empty graph G, $\mathsf{td}_2(G) = \mathsf{bed}_{\mathcal{E}}(G) + 1$. We define the *t-ladder* as the $(2 \times t)$-grid (i.e., the Cartesian product of K_2 and a path on t-vertices) and we denote it by L_t. It is easy to check that $\mathsf{td}_2(L_t) = \Omega(\log(t))$. One of the main results of [14] was that there is a function $f : \mathbb{N} \to \mathbb{N}$ such that every graph excluding a t-ladder belongs to $\mathcal{E}^{(f(t))}$. This implies that the t-ladder L_t is a universal minor obstruction for $\mathsf{bed}_{\mathcal{E}}$. This motivates us to make a conjecture on how the results of [14] should be extended for every non-trivial minor-closed class $\mathcal{G}$: Given a positive t, we define $\mathcal{L}_{\mathcal{G},t}$ as the class containing every graph that can be constructed by first taking the disjoint union of two paths $P_i, i \in [2]$, with vertices $v_1^i, \dots, v_t^i$ (ordered the way they appear in P_i) and t graphs $G_1, \dots, G_t$ from $\mathbf{obs}(\mathcal{B}(\mathcal{G}))$ and then, for $i \in [t]$, identify v_i^1 and v_i^2 with two different vertices in G_i. It is easy to check that if $G \in \mathcal{L}_{\mathcal{G},t}$, then $\mathsf{bed}_{\mathcal{G}}(G) = \Omega(\log t)$. We conjecture that $\mathcal{L}_{\mathcal{G},t}$ is a *universal minor obstruction* for $\mathsf{bed}_{\mathcal{G}}$, i.e., there is a function $f : \mathbb{N} \to \mathbb{N}$ such that every graph excluding all graphs in $\mathcal{L}_{\mathcal{G},t}$ as a minor, has block elimination distance to $\mathcal{G}$ bounded by $f(t)$, i.e., $\mathsf{excl}(\mathcal{L}_{\mathcal{G},t}) \subseteq \mathcal{G}^{(f(t))}$. Notice that the two operations in Sect. 5 imply that, when restricted to outerplanar graphs, this conjecture is correct for $f(t) = O(t)$. However we do not believe that the linear upper bound is maintained in the general case.

Acknowledgements. Öznur Yaşar Diner is grateful to the members of the research group GAPCOMB for hosting a research stay at Universitat Politècnica de Catalunya.

References

1. Adler, I., Grohe, M., Kreutzer, S.: Computing excluded minors. In: Proceedings of the 19th annual ACM-SIAM Symposium on Discrete Algorithms (SODA), pp. 641–650 (2008). https://doi.org/10.1145/1347082.1347153
2. Agrawal, A., Kanesh, L., Panolan, F., Ramanujan, M.S., Saurabh, S.: An FPT algorithm for elimination distance to bounded degree graphs. In: Proceedings of the 38th International Symposium on Theoretical Aspects of Computer Science (STACS), volume 187 of LIPIcs, pp. 5:1–5:11. Schloss Dagstuhl - Leibniz-Zentrum für Informatik (2021). https://doi.org/10.4230/LIPIcs.STACS.2021.5
3. Agrawal, A., Ramanujan, M.S.: On the parameterized complexity of clique elimination distance. In: Proceedings of the 15th International Symposium on Parameterized and Exact Computation (IPEC), vol. 180, pp. 1:1–1:13 (2020). https://doi.org/10.4230/LIPIcs.IPEC.2020.1
4. Bodlaender, H.L., Heggernes, P., Lokshtanov, D.: Graph modification problems (Dagstuhl seminar 14071). Dagstuhl Rep. **4**(2), 38–59 (2014). https://doi.org/10.4230/DagRep.4.2.38

5. Bulian, J., Dawar, A.: Graph isomorphism parameterized by elimination distance to bounded degree. Algorithmica **75**(2), 363–382 (2015). https://doi.org/10.1007/s00453-015-0045-3
6. Bulian, J., Dawar, A.: Fixed-parameter tractable distances to sparse graph classes. Algorithmica **79**(1), 139–158 (2016). https://doi.org/10.1007/s00453-016-0235-7
7. Corneil, D.G., Gotlieb, C.C.: An efficient algorithm for graph isomorphism. J. ACM **17**(1), 51–64 (1970). https://doi.org/10.1145/321556.321562
8. Crespelle, C., Drange, P.G., Fomin, F.V., Golovach, P.A.: A survey of parameterized algorithms and the complexity of edge modification. arXiv:2001.06867 (2020)
9. Diner, Ö.Y., Giannopoulou, A.C., Stamoulis, G., Thilikos, D.M.: Block elimination distance. arXiv:2103.01872 (2021)
10. Fomin, F.V., Saurabh, S., Misra, N.: Graph modification problems: a modern perspective. In: Wang, J., Yap, C. (eds.) FAW 2015. LNCS, vol. 9130, pp. 3–6. Springer, Cham (2015). https://doi.org/10.1007/978-3-319-19647-3_1
11. Guo, J., Hüffner, F., Niedermeier, R.: A structural view on parameterizing problems: distance from triviality. In: Downey, R., Fellows, M., Dehne, F. (eds.) IWPEC 2004. LNCS, vol. 3162, pp. 162–173. Springer, Heidelberg (2004). https://doi.org/10.1007/978-3-540-28639-4_15
12. Hopcroft, J.E., Tarjan, R.E.: A V^2 algorithm for determining isomorphism of planar graphs. Inf. Process. Lett. **1**(1), 32–34 (1971). https://doi.org/10.1016/0020-0190(71)90019-6
13. Hopcroft, J.E., Tarjan, R.E.: Isomorphism of planar graphs. In: Complexity of Computer Computations. The IBM Research Symposia Series, pp. 131–152 (1972). https://doi.org/10.1007/978-1-4684-2001-2_13
14. Huynh, T., Joret, G., Micek, P., Seweryn, M.T., Wollan, P.: Excluding a ladder. arXiv:2002.00496 (2020)
15. Jansen, B.M.P., de Kroon, J.J.H., Wlodarczyk, M.: Vertex deletion parameterized by elimination distance and even less. In: Proceedings of the 53rd Annual ACM SIGACT Symposium on Theory of Computing (STOC), pp. 1757–1769. ACM (2021). https://doi.org/10.1145/3406325.3451068
16. Kawarabayashi, K., Kobayashi, Y., Reed, B.: The disjoint paths problem in quadratic time. J. Combinat. Theory Ser. B **102**(2), 424–435 (2011). https://doi.org/10.1016/j.jctb.2011.07.004
17. Lindermayr, A., Siebertz, S., Vigny, A.: Elimination distance to bounded degree on planar graphs. In: Proceedings of the 45th International Symposium on Mathematical Foundations of Computer Science (MFCS), volume 170 of LIPIcs, pp. 65:1–65:12 (2020). https://doi.org/10.4230/LIPIcs.MFCS.2020.65
18. Robertson, N., Seymour, P.D.: Graph minors. XIII. The Disjoint Paths Problem. J. Combinat. Theory Ser. B **63**(1), 65–110 (1995). https://doi.org/10.1006/jctb.1995.1006
19. Robertson, N., Seymour, P.D.: Graph Minors. XX. Wagner's conjecture. J. Combinat. Theory Ser. B **92**(2), 325–357 (2004). https://doi.org/10.1016/j.jctb.2004.08.001
20. Sau, I., Stamoulis, G., Thilikos, D.M.: An FPT-algorithm for recognizing k-apices of minor-closed graph classes. In: Proceedings of the 47th International Colloquium on Automata, Languages, and Programming (ICALP), volume 168 of LIPIcs, pp. 95:1–95:20 (2020). https://doi.org/10.4230/LIPIcs.ICALP.2020.95
21. Sau, I., Stamoulis, G., Thilikos, D.M.: k-apices of minor-closed graph classes. I. Bounding the obstructions. arXiv:2103.00882 (2021)
22. Sau, I., Stamoulis, G., Thilikos, D.M.: k-apices of minor-closed graph classes. II. Parameterized algorithms. arXiv:2004.12692 (2021)

On Fair Covering and Hitting Problems

Sayan Bandyapadhyay[1(✉)], Aritra Banik[2], and Sujoy Bhore[3]

[1] Department of Informatics, University of Bergen, Bergen, Norway
sayan.bandyapadhyay@uib.no
[2] School of Computer Sciences, NISER, Bhubaneswar, India
aritra@niser.ac.in
[3] Indian Institute of Science Education and Research, Bhopal, India
sujoy@iiserb.ac.in

Abstract. In this paper, we study two generalizations of Vertex Cover and Edge Cover, namely Colorful Vertex Cover and Colorful Edge Cover. In the Colorful Vertex Cover problem, given an n-vertex edge-colored graph G with colors from $\{1, \ldots, \omega\}$ and coverage requirements $r_1, r_2, \ldots, r_\omega$, the goal is to find a minimum-sized set of vertices that are incident on at least r_i edges of color i, for each $1 \leq i \leq \omega$, i.e., we need to cover at least r_i edges of color i. Colorful Edge Cover is similar to Colorful Vertex Cover except here we are given a vertex-colored graph and the goal is to cover at least r_i vertices of color i, for each $1 \leq i \leq \omega$, by a minimum-sized set of edges. These problems have several applications in *fair* covering and hitting of geometric set systems involving points and lines that are divided into multiple groups. Here, "fairness" ensures that the coverage (resp. hitting) requirement of every group is fully satisfied.

We obtain a $(2+\epsilon)$-approximation for the Colorful Vertex Cover problem in time $n^{O(\omega/\epsilon)}$, i.e., we obtain an $O(1)$-approximation in polynomial time for constant number of colors. Next, for the Colorful Edge Cover problem, we design an $O(\omega n^3)$ time exact algorithm, via a chain of reductions to a matching problem. For all intermediate problems in this chain of reductions, we design polynomial time algorithms, which might be of independent interest.

1 Introduction

Vertex Cover and Edge Cover are two classical graph problems which have been studied for at least forty years [8]. Vertex Cover is known to be NP-complete and admits a 2-approximation [8]. On the other hand, Edge Cover can be solved in polynomial time using a connection to Maximum Matching [8]. In this paper, we study the following two generalizations of these problems on vertex- or edge-colored graphs.

L. Kowalik et al. (Eds.): WG 2021, LNCS 12911, pp. 39–51, 2021.
https://doi.org/10.1007/978-3-030-86838-3_4

COLORFUL VERTEX COVER
Input: A graph G with n vertices and m edges where every edge is colored by a color from $\{1, 2, \ldots, \omega\}$, and coverage requirements $r_1, r_2, \ldots, r_\omega$.
Question: Find a minimum-sized set of vertices that are incident on at least r_i edges of color i, for each $1 \leq i \leq \omega$.

COLORFUL EDGE COVER
Input: A graph G with n vertices and m edges where every vertex is colored by a color from $\{1, 2, \ldots, \omega\}$, and coverage requirements $r_1, r_2, \ldots, r_\omega$.
Question: Find a minimum-sized set E' of edges such that at least r_i vertices of color i are incident on the edges of E', for each $1 \leq i \leq \omega$.

Bera et al. [3] designed an $O(\log \omega)$-approximation for COLORFUL VERTEX COVER. Indeed, they study a more general "weighted-version" called PARTITION VERTEX COVER. Moreover, they noted that an extension of the greedy algorithm of Slavík [19] gives an $O(\log(\sum_{t=1}^{\omega} r_t))$ approximation for this problem. On the other hand, it is NP-hard to obtain an approximation guarantee asymptotically better than $O(\log \omega)$ [3]. Cohen et al. [5] studied a variant of COLORFUL EDGE COVER where all the requirements are 1 and the solution set of edges E' must form a matching. They gave a polynomial time algorithm for this problem.

Our motivation of studying COLORFUL VERTEX COVER and COLORFUL EDGE COVER partly comes from a series of recent works that study the FAIR k-CENTER problem[1] [1,2,13]. In FAIR k-CENTER, given a set of n points in a metric space where each point is colored by a color from $\{1, 2, \ldots, \omega\}$, coverage requirements $r_1, r_2, \ldots, r_\omega$, and an integer k, the goal is to find k balls of minimum radius whose union contains at least r_t points of color t, for $1 \leq t \leq \omega$. $O(1)$-approximations are known for this problem when the number of colors ω is a constant [1,13]. In particular, one can obtain a 4-approximation in $n^{O(\omega)}$ time [1] and a 3-approximation in $n^{O(\omega^2)}$ time [13].

We note that FAIR k-CENTER can also be seen as a covering problem where the goal is to cover colored points by balls, albeit with some expansion. This leads to the question of studying fair or colorful covering problems with other geometric objects. In particular, suppose we are given a set of axis-parallel lines along with a set of points in the plane. Then, one can similarly study fair variants of the classical point-line covering and hitting problems. In the following, we formally define two such problems.

FAIR COVERING OF POINTS BY LINES (FCPL)
Input: A set of axis-parallel lines $\mathcal{L}$ and a set P of n points in the plane, where each point in P is colored by a color from $\{1, \ldots, \omega\}$, and coverage requirements $r_1, \ldots, r_\omega$.
Question: Find a minimum-sized subset $\mathcal{L}' \subseteq \mathcal{L}$ such that the lines in $\mathcal{L}'$ together contain at least r_t points of color t, for each $1 \leq t \leq \omega$.

[1] The term "fair" stresses on the fact, in an abstract manner, that the resources should be divided evenly among different groups.

Fair Hitting of Lines by Points (FHLP)
Input: A set of axis-parallel lines $\mathcal{L}$ and a set P of n points in the plane, where each line of $\mathcal{L}$ is colored by a color from $\{1, \ldots, \omega\}$ with hitting requirements $r_1, \ldots, r_\omega$.
Question: Find a minimum-sized subset $P' \subseteq P$ such that the points in P' intersect at least r_t lines of color t, for each $1 \leq t \leq \omega$.

Note that, in case of axis-parallel lines, each point can intersect at most two lines. For simplicity, suppose each input point intersects exactly two lines. Then FCPL and FHLP are special cases of Colorful Vertex Cover and Colorful Edge Cover, respectively. Indeed, given such a set of axis-parallel lines and points in the plane, one can construct a graph where the vertices are corresponding to the lines and edges are corresponding to the points. Then, covering points by lines correspond to covering edges by vertices and hitting lines by points correspond to covering vertices by edges in the new graph.

Similar to Colorful Vertex Cover, it is NP-hard to obtain a $o(\log \omega)$-approximation for FCPL, even when the input consists of only horizontal lines. Indeed, the following simple reduction from Set Cover is sufficient. (In Set Cover, given a set system (X, S), where X is a set of n elements, and S is a collection of subsets of X, the goal is to find a minimum size collection $S' \subseteq S$ such that the union of the sets in S' contains the elements of X.) For each set, take a different horizontal line. For each element, place a copy (i.e., a point) of it on each line corresponding to the sets which contain it. Also, for each element, color all of its copies by a unique color. Set all the requirements to 1. Then, there is a solution to Set Cover with k sets if and only if there is a solution to FCPL with k lines. Moreover, FCPL is NP-hard even for a single color, by the following reduction from Partial Vertex Cover in bipartite graphs, which is known to be NP-hard [4]. (In this problem, given an undirected graph $G = (V, E)$, and two integers $k_1, k_2 > 0$, the goal is to decide if there a subset V' of V of size k_1 that covers at least k_2 edges.) Let (V_1, V_2) be the given bipartition of the vertices and p be the number of edges need to be covered. Take a different horizontal line for each vertex in V_1, and a different vertical line for each vertex in V_2. Place a point on the intersection of two lines if the corresponding two vertices share an edge. Set the requirements to p. Then, there is a solution to Partial Vertex Cover with k vertices if and only if there is a solution to FCPL with k lines. To the best of our knowledge, FHLP has not been studied before.

1.1 Our Results

In this work, we achieve $(2+\epsilon)$-approximations for Colorful Vertex Cover and FCPL in time $n^{O(\omega/\epsilon)}$, this means that we obtain $O(1)$-approximations in polynomial time for constant number of colors, matching the result for Fair k-center. Our algorithms are based on LP rounding and construction of a sparse LP. Sparsity of LPs was also used in the works on Fair k-center. However, our approach is very different. Indeed, our rounding scheme is uncomplicated, as in our case each element (e.g., point) can be covered by at most two objects (e.g., lines). In Sect. 2, first, we demonstrate our approach for FCPL, and then describe its trivial extension to COLORFUL VERTEX COVER. For COLORFUL EDGE COVER and FHLP, we design $O(\omega n^3)$ time exact algorithms, via a chain of reductions to a matching problem studied by Cohen et al. [5] (Sect. 3).

1.2 Related Work

POINT COVER is a special case of FCPL with a single color where the goal is to cover all points [11]. Hassin and Megiddo [11] introduced this problem and gave a polynomial time algorithm. Gaur and Bhattacharya [9] showed that one can also obtain such an algorithm using LP rounding. We refer the reader to the following series of combinatorial as well as algorithmic works on the problem involving arbitrary lines; [6,7,14–16,18].

Inamdar and Varadarajan [12] studied a generalization of COLORFUL VERTEX COVER, called PARTITION SET COVER (PSC). They gave an LP-rounding based $O(\beta + \log \omega)$ approximation, where β denotes the approximation guarantee for a related SET COVER instance obtained by rounding the standard LP. See also [3,10,12,19] for a comprehensive understanding of this problem.

2 A $(2+\epsilon)$-Approximation for FCPL and COLORFUL VERTEX COVER

First, we describe an LP-rounding based additive approximation for FCPL, and then show how to convert this to a multiplicative $O(1)$-approximation. Let $\mathcal{L} = \{\ell_1, \ldots, \ell_m\}$ and $P = \{p_1, \ldots, p_n\}$. For $1 \le t \le \omega$, let C_t denote the color class t, i.e., the set of points of color t. A line ℓ is said to *cover* a point p if ℓ contains p. A solution S is said to *cover* a point p, if the lines in S contain p. Next, we describe the natural ILP of FCPL. For each point p_j, we have a 0/1 variable x_j that denotes whether p_j is covered in the solution. For each line ℓ_i, there is a 0/1 variable y_i that denotes whether ℓ_i is chosen in the solution. There are two main constraints in the ILP other than the domain constraints. The first constraint is the coverage constraint which ensures that from each color class t, at least r_t points are covered. The second constraint is the sanity constraint which ensures that if a point p_j is covered in the solution, then at least one line of $\mathcal{L}$ must be in the solution that contains p_j. The LP relaxation of the ILP is as follows.

$$\begin{aligned} \text{minimize} \quad & \sum_{\ell_i \in \mathcal{L}} y_i & & \text{(FCPL-LP)} \\ \text{subject to} \quad & \sum_{j: p_j \in C_t} x_j \ge r_t & \forall 1 \le t \le \omega \quad & (1) \\ & \sum_{i: \ell_i \in \mathcal{L}, p_j \in \ell_i} y_i \ge x_j & \forall p_j \in P \quad & (2) \\ & 0 \le x_j, y_i \le 1 & \forall p_j \in P, \ell_i \in \mathcal{L} \quad & (3) \end{aligned}$$

We denote any solution to FCPL-LP by (x, y). The cost of (x, y) is defined as, $\text{cost}(x, y) = \sum_{\ell_i \in \mathcal{L}} y_i$. Our rounding algorithm consists of two major steps.

First Step. In the first step, we compute a fractional optimal solution $(\bar{x}, \bar{y})$ using any LP solver and modify it to obtain another fractional solution which has a special structure. Let OPT^{LP} denote the cost of $(\bar{x}, \bar{y})$ and OPT the optimal cost. Also, let P_1 and P_2 be the disjoint subsets of points which are covered by exactly one and two input lines, respectively.

Lemma 1. *There is a solution $(\tilde{x}, \tilde{y})$ to FCPL-LP with the following properties: (i) $\text{cost}(\tilde{x}, \tilde{y}) \leq 2 \cdot OPT^{LP}$. (ii) There is a function $\phi : P \to \mathcal{L}$ such that for each point $p_j \in P_2$ and the axis-parallel lines ℓ_{j^1} and ℓ_{j^2} that contain it, either $\phi(p_j) = \ell_{j^1}$ or $\phi(p_j) = \ell_{j^2}$. For each point $p_j \in P_1$ and the line ℓ_{j^1} containing it, $\phi(p_j) = \ell_{j^1}$. Also, $\tilde{x}_j$ is equal to the $\tilde{y}$ value of $\phi(p_j)$. (iii) $(\tilde{x}, \tilde{y})$ can be obtained in polynomial time.*

Proof. We construct $(\tilde{x}, \tilde{y})$ by modifying the solution $(\bar{x}, \bar{y})$. First, we define a function ϕ that assigns each point to a line. For each point $p_j \in P_1$ and the line ℓ_{j^1} containing it, $\phi(p_j) = \ell_{j^1}$. For each point $p_j \in P_2$, let ℓ_{j^1} and ℓ_{j^2} be the respective horizontal and vertical lines that cover p_j. We assign p_j to ℓ_{j^1} or ℓ_{j^2}, whichever has the larger y-value in $(\bar{x}, \bar{y})$. If both y-values are same, we assign p_j to one of the two arbitrarily. This completes the description of the assignment ϕ. Next, we construct a new solution based on ϕ. For each point p_j, we set its new x-value to the minimum of 1 and two times the y-value of $\phi(p_j)$, i.e., $\tilde{x}_j = \min\{1, 2\bar{y}_{i'}\}$ where $\ell_{i'} = \phi(p_j)$. For each line ℓ_i, we set its new y-value to the minimum of 1 and two times of its old y-value, i.e., $\tilde{y}_i = \min\{1, 2\bar{y}_i\}$. Note that $\tilde{x}_j \geq \min\{1, \bar{y}_{j^1} + \bar{y}_{j^2}\} \geq \bar{x}_j$. Thus, it is not hard to see that the new solution satisfies the coverage and sanity constraints. Moreover, $\text{cost}(\tilde{x}, \tilde{y})$ is at most two times the cost of $(\bar{x}, \bar{y})$. Hence, Property (i) is satisfied. Property (ii) is satisfied by construction. Lastly, as the modification takes polynomial time, $(\tilde{x}, \tilde{y})$ can also be obtained in polynomial time. □

By the above lemma, we obtain a separated LP-solution where each point gets its fractional coverage $\tilde{x}_j$ either from a horizontal or from a vertical line. Based on this separation we write a sparse LP (containing only a few constraints) for our instance and use the sparsity of this LP to obtain an integral solution. Next, we describe the details.

Second Step. Consider the solution $(\tilde{x}, \tilde{y})$ and the assignment ϕ in Lemma 1. For each color $1 \leq t \leq \omega$ and for each line $\ell_i \in \mathcal{L}$, let $C_{t,i}$ be the set of points p_j in C_t such that $\phi(p_j) = \ell_i$. Denote the size of $C_{t,i}$ by $n_{t,i}$. Lastly, let $k = \sum_{i:\ell_i \in \mathcal{L}} \tilde{y}_i$. We define the following LP that does not contain too many constraints.

$$\begin{aligned} \text{maximize} \quad & \sum_{i=1}^{m} n_{1,i} z_i && \text{(Sparse-LP)} \\ \text{subject to} \quad & \sum_{i=1}^{m} n_{t,i} z_i \geq r_t \quad \forall 2 \leq t \leq \omega && (4) \\ & \sum_{i=1}^{m} z_i \leq k && (5) \\ & 0 \leq z_i \leq 1 \quad \forall 1 \leq i \leq m && (6) \end{aligned}$$

Lemma 2. *There is a solution to Sparse-LP whose objective function value is at least* r_1.

Proof. For each line $\ell_i \in \mathcal{L}$, set $z_i = \tilde{y}_i$. Constraint 5 is trivially satisfied. Now, fix any $1 \leq t \leq \omega$.

$$\begin{aligned} \sum_{i=1}^{m} n_{t,i} z_i = \sum_{i=1}^{m} n_{t,i} \cdot \tilde{y}_i &= \sum_{i=1}^{m} \sum_{p_j \in \mathcal{C}_{t,i}} \tilde{y}_i && (\text{as } n_{t,i} = |\mathcal{C}_{t,i}|) \\ &= \sum_{i=1}^{m} \sum_{p_j \in \mathcal{C}_{t,i}} \tilde{x}_j && (\text{from the definitions of } \mathcal{C}_{t,i} \text{ and } \phi) \\ &= \sum_{j : p_j \in \mathcal{C}_t} \tilde{x}_j && (\text{as } \mathcal{C}_{t,i} \text{ is a partition of } \mathcal{C}_t) \\ &\geq r_t && (\text{by Constraint 1 of FCPL-LP}) \end{aligned}$$

Hence the lemma follows. □

Next, we compute a fractional optimal solution $\hat{z}$ to Sparse-LP using any LP solver. The above lemma implies the value of this solution is at least r_1. In the following, we argue about some additional properties of this solution. For that we need the following lemma (Lemma 2.1.4 in [17]).

Lemma 3 [17]**.** *In any extreme point feasible solution (or equivalently, a basic feasible solution) to a linear program, the number of linearly independent tight constraints is equal to the number of variables.*

The following lemma is an easy consequence of the above lemma.

Lemma 4. *The number of fractional variables in* $\hat{z}$ *is at most* ω.

Proof. First, note that Sparse-LP has $2m + \omega$ constraints and m variables. Now, by Lemma 3, the number of linearly independent tight constraints in $\hat{z}$ is m. Consider the set S of $2m$ constraints $0 \leq z_i \leq 1$. As there are only ω more constraints in the LP, there must be at least $m - \omega$ many linearly independent constraints in S which are tight in $\hat{z}$. Note that the two constraints $z_i \geq 0$ and

$z_i \leq 1$ cannot be tight simultaneously for any fixed i, as it would imply $z_i = 0$ and $z_i = 1$. Hence, it must be the case that at least $m - \omega$ variables are integral in $\hat{z}$, and the lemma follows. □

Based on the above lemma we compute an integral solution z^* to Sparse-LP by rounding the values of the at most ω fractional variables to 1. Note that this integral solution satisfies all the constraints except Constraint 5. But, as we round at most ω variables, it follows that $\sum_{i=1}^{m} z_i^* \leq k + \omega$. Thus, we obtain a set Γ of at most $k + \omega$ lines in $\mathcal{L}$ that for each $1 \leq t \leq \omega$ contain at least r_t points from $\mathcal{C}_t$. Thus, Γ is a feasible solution for FCPL. By noting that $k = \text{cost}((\tilde{x}, \tilde{y})) \leq 2\text{OPT}^{LP}$, we obtain the following theorem.

Theorem 1. *There is a feasible solution to FCPL with cost at most* $2OPT+\omega$ *that can be computed in polynomial time.*

Next, we show how to convert the above additive approximation to a multiplicative constant approximation, albeit with a time complexity that exponentially depends on ω.

Theorem 2. *For any* $\epsilon > 0$*, there is a* $(2 + \epsilon)$*-approximation for FCPL in* $m^{O(\omega/\epsilon)} n^{O(1)}$ *time.*

Proof. Fix $\epsilon > 0$. First, we enumerate all the solutions of size $\kappa = 1, 2, \ldots, \omega/\epsilon$ in $m^{O(\omega/\epsilon)} n^{O(1)}$ time. We stop the first time we obtain a feasible solution. Thus, if we obtain a feasible solution at some step, it must be an optimal solution, and we are done. Otherwise, the optimal cost OPT is more than ω/ϵ. In this case, we use our additive approximation algorithm based on LP rounding. By Theorem 1, we obtain a feasible solution with cost at most $2\text{OPT}+\omega < 2\text{OPT}+\epsilon \cdot \text{OPT} = (2 + \epsilon) \cdot \text{OPT}$. □

Remarks. The above LP rounding based scheme is much more general in the sense that it also yields a $(2+\epsilon)$-approximation for Colorful Vertex Cover in $n^{O(\omega/\epsilon)} m^{O(1)}$ time. The steps are similar to the ones for FCPL, as here each edge gets covered by exactly two vertices. We can mimic the above mentioned scheme by assigning each edge to the vertex whose variable value is larger leading to the situation where each edge gets its coverage value from exactly one vertex. Thus, again we can write a sparse LP for the instance and achieve the same guarantees. Similarly, this LP rounding based scheme achieves an $(f + \epsilon)$-approximation for Partition Set Cover in $n^{O(\omega/\epsilon)} m^{O(1)}$ time, where each element appears in at most f sets in the input. The idea is again similar: assign each element to a unique set having the largest variable value. The f factor comes from the fact that the variable value of each set is scaled up by f factor to obtain the new LP solution where each element is (fractionally) covered by exactly one set. This implies the general case of FCPL with arbitrary lines admits an $(f + \epsilon)$-approximation in $m^{O(\omega/\epsilon)} n^{O(1)}$ time, where each point is contained in at most f lines in the input.

3 A Polynomial Time Algorithm for FHLP and COLORFUL EDGE COVER

First, we argue that it is sufficient to obtain the algorithm for COLORFUL EDGE COVER, in order to obtain the algorithm for FHLP. Then, we design a polynomial time algorithm for COLORFUL EDGE COVER.

Consider any instance of FHLP. If there is an input point p that intersects or hits exactly one line ℓ, we add another line ℓ' to the instance orthogonal to ℓ such that ℓ' also intersects p. We can also ensure that this new line does not intersect any other input point, potentially by a slight shifting of p. Also, we color all the new lines added in this process by a new color C. Now, it is sufficient to solve FHLP on the new instance by setting the requirement of the color C to 0. Hence, without loss of generality, we can assume that in any instance of FHLP, each point hits exactly two axis-parallel lines. Now, given an instance of FHLP, one can construct a vertex-colored graph where the vertices are corresponding to the lines and edges are corresponding to the points. Hence, we have the following observation.

Observation 1. *If* COLORFUL EDGE COVER *can be solved in $T(n, \omega)$ time, FHLP can be solved in $T(2n, \omega + 1) + O(n + m)$ time.*

In the following, we design an algorithm for COLORFUL EDGE COVER which runs in $O(\omega n^3)$ time. Hence, by the above observation, we have the following theorem.

Theorem 3. *FHLP can be solved in $O(\omega n^3)$ time.*

An edge e is said to cover a vertex v if e is incident on v. A set of edges E' covers the set of vertices $V' = \{v \mid \exists e \in E' \text{ such that } e \text{ covers } v\}$. In the rest of this section, we design the algorithm for COLORFUL EDGE COVER, which is based on an algorithm for the following matching problem.

BUDGETED MATCHING
Input: A graph G with n vertices and m edges where every vertex is colored by a color from $\{1, 2, \ldots, \omega\}$, and coverage requirements $r_1, r_2, \ldots, r_\omega$.
Question: Find a minimum-sized matching which covers at least r_i vertices of color i for each $1 \leq i \leq \omega$.

We design a polynomial time algorithm for BUDGETED MATCHING. But before that, we have the following observation which establishes a connection between COLORFUL EDGE COVER and BUDGETED MATCHING.

Lemma 5. *If* BUDGETED MATCHING *can be solved in time $T(n, \omega)$, then* COLORFUL EDGE COVER *can be solved in time $T(2n, \omega + 1) + O(m + n)$.*

Proof. Suppose we would like to solve COLORFUL EDGE COVER on a given instance $\mathcal{I}$ consisting of a vertex-colored graph $G = (V, E)$ and a set of colors $\{1, 2, \ldots, \omega\}$. WLOG, there is no isolated vertex in G. We construct a new instance $\mathcal{I}'$ of BUDGETED MATCHING consisting of a vertex-colored graph $G' = (V', E')$ and a set of colors $\{1, 2, \ldots, \omega, C\}$ as follows.

For each vertex $v \in V$, we add two vertices, v and an auxiliary vertex $a(v)$, to V'. The color of v in G' is same as the color of v in G and the color of $a(v)$ is C. E' consists of all the edges in E and the edge $(v, a(v))$ for each $v \in V$. For each color $1 \leq i \leq \omega$, the requirement r_i in $\mathcal{I}'$ remains the same as in $\mathcal{I}$. The requirement corresponding to C is set to 0. Note that $|V'| = 2n$ and $|E'| = m + n = O(m)$.

Next, we show that $\mathcal{I}$ has a solution to COLORFUL EDGE COVER with at most k edges if and only if $\mathcal{I}'$ has a solution to BUDGETED MATCHING with at most k edges. First, assume that there is a set of edges $E_1 \subseteq E$ of size k which is a solution to COLORFUL EDGE COVER. We construct a matching M for $\mathcal{I}'$ from E_1. First, note that if there is a path in E_1 consisting of three edges, we can always remove the middle edge from the solution without losing any vertex coverage. Thus, WLOG, we can assume that E_1 is a collection of star graphs. Consider any such star S. We include any arbitrary edge (s, v) of S in M where s is the central vertex of S. For any other edge (s, u) in S, we include $(u, a(u))$ in M. By construction, M is a matching in G' of size at most k. Also, all the requirements are trivially satisfied.

Now, suppose there is a matching M in G' of size k which is a solution to BUDGETED MATCHING. We construct a solution E_1 for $\mathcal{I}$ from M. For each edge $e \in M \cap E$, include e in E_1. For each edge $e \in M \cap (E' \setminus E)$, where $e = (u, a(u))$, include any arbitrary edge (u, v) of E in E_1 (that covers u). Note that such an edge always exist, as there is no isolated vertex in G. Again, by construction, E_1 is a feasible solution to $\mathcal{I}$ of size k.

We solve BUDGETED MATCHING on $\mathcal{I}'$ to obtain a matching M of the minimum size, say k. We return the set of edges E_1 as constructed above as the solution to $\mathcal{I}$. We claim that E_1 is a solution to COLORFUL EDGE COVER of the minimum size. Suppose it is not. Suppose there is a solution E_2 to COLORFUL EDGE COVER of size $k' < k$. Then, by the above discussion, there is a solution to BUDGETED MATCHING of size at most $k' < k$. But, this is a contradiction to the assumption that M is a minimum size solution.

Finally, BUDGETED MATCHING can be solved on $\mathcal{I}'$ in $T(2n, \omega + 1)$ time, and construction of G' can be done in $O(m + n)$ time. Hence, the lemma follows. □

In the following, we design an algorithm for BUDGETED MATCHING which runs in $O(\omega n^3)$ time. Hence, by the above lemma, we have the following theorem.

Theorem 4. COLORFUL EDGE COVER *can be solved in* $O(\omega n^3)$ *time.*

To solve BUDGETED MATCHING, we show that it can be converted to a problem where each color has unit requirement. Essentially we need the following problem definition due to Cohen et al. [5].

> TROPICAL MATCHING
> **Input:** A graph G with n_1 vertices and m_1 edges where every vertex is colored by a color from $\{1, 2, \ldots, \omega\}$.
> **Question:** Find a maximum-sized matching which covers at least one vertex of color i for each $1 \leq i \leq \omega$.

We need the following theorem due to Cohen et al. [5].

Theorem 5. *[5]* TROPICAL MATCHING *can be solved in* $O(n_1 m_1)$ *time.*

The next lemma establishes the connection between BUDGETED MATCHING and TROPICAL MATCHING.

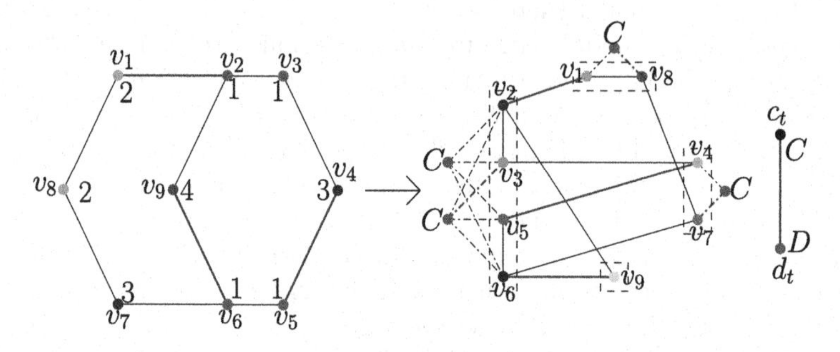

Fig. 1. A sample reduction from Budgeted Matching with budget [2, 1, 1, 1] for colors 1, 2, 3, 4 respectively to Tropical Matching

Lemma 6. *If* TROPICAL MATCHING *can be solved in* $T(n_1, m_1)$ *time,* BUDGETED MATCHING *can be solved in* $T(\alpha n, \beta\omega n^2) + \gamma\omega n^2$ *time for some constants* α, β *and* γ.

Proof. Suppose we would like to solve BUDGETED MATCHING on a given instance $\mathcal{I}$ consisting of a vertex-colored graph $G = (V, E)$ and a set of colors $\{1, 2, \ldots, \omega\}$. Let n be the number of vertices in G. We construct a new instance $\mathcal{I}_t$ of TROPICAL MATCHING consisting of a vertex-colored graph $G_t = (V_t, E_t)$ and a set of colors $\{1, 2, \ldots, n, C, D\}$ as follows.

Let C_x be the set of vertices in G of color x and $n_x = |C_x|$ for $1 \leq x \leq \omega$. V_t contains all the vertices in V and for each color $1 \leq x \leq \omega$, a set of $n_x - r_x$ vertices V^x. Additionally, V_t contains two more auxiliary vertices c_t and d_t. Thus, $V_t = V \cup (\cup_{x=1}^{\omega} V^x) \cup \{c_t, d_t\}$. E_t contains all the edges in E and for each color $1 \leq x \leq \omega$, a set of $(n_x - r_x) \times n_x$ edges $E^x = \{(u, v) \mid u \in V^x \text{ and } v \in C_x\}$. Additionally, the edge (c_t, d_t) is included in E_t. Thus, $E_t = E \cup (\cup_{x=1}^{\omega} E^x) \cup \{(c_t, d_t)\}$. Each vertex $u \in V$ in $\mathcal{I}_t$ gets a unique color j for some $1 \leq j \leq n$. Colors of c_t and d_t are C and D, respectively. Finally, colors of all vertices in $\cup_{x=1}^{\omega} V^x$ are C. See Fig. 1 for an example construction. Note that $|V_t| = O(n)$ and $|E_t| = O(\omega n^2)$.

Next, we show that $\mathcal{I}$ has a solution to BUDGETED MATCHING with k edges if and only if $\mathcal{I}_t$ has a solution to TROPICAL MATCHING with $n-k+1$ edges. First, assume that there is a matching M of size k in G which is a solution to BUDGETED MATCHING. We construct a new matching M_t for G_t. We include all the edges in M and (c_t, d_t) in M_t. For each $1 \leq x \leq \omega$, let U_x be the set of vertices in $\mathcal{C}_x$ which are not matched by M. We also include a matching between U_x and V^x of size $|U_x|$ in M_t. Note that such a matching always exists, as $|U_x| \leq n_x - r_x$ by the definition of M. Now, we argue that M_t is a valid solution to TROPICAL MATCHING of size $(n-k)+1$. First, note that M_t is a matching in G_t which matches all the vertices in V. Thus, there is a matched vertex of color j for each $1 \leq j \leq n$. Now, as (c_t, d_t) is also in M_t, there are matched vertices of colors C and D as well. Thus, M_t is a feasible solution to TROPICAL MATCHING. Note that the size of $\cup_{x=1}^{\omega} U_x$ is exactly $n-2k$, as k edges in M match exactly $2k$ vertices in V. Thus, the size of M_t is $k+1+(n-2k)=(n-k)+1$.

Now, suppose there is a matching M_t of size $(n-k)+1$ in G_t which is a solution to TROPICAL MATCHING. We construct a matching M for G starting from M_t. Indeed, M is the subset of edges in M_t which are contained in E. We argue that M is a feasible solution to BUDGETED MATCHING. Note that M_t must match all the vertices in V, as each such vertex has a unique color which does not appear in any other vertex. Consider any color x for $1 \leq x \leq \omega$. The edges in E^x can match at most $n_x - r_x$ vertices of $\mathcal{C}_x$, as $|V^x| = n_x - r_x$. Thus, there exist at least r_x edges in $M_t \cap E$ which match the vertices in $\mathcal{C}_x$ not matched by the edges in $M_t \cap E^x$. It follows that M matches at least r_x vertices of $\mathcal{C}_x$ for each $1 \leq x \leq \omega$, and hence it is a feasible solution to BUDGETED MATCHING. Next, we show that the size of M is exactly k. Let k_1 and k_2 be the number of edges of M_t which are in $\cup_{x=1}^{\omega} E^x$ and E, respectively. Note that (c_t, d_t) must be included in M_t, as otherwise there will be no vertex of color D in M_t. It follows that $k_1 + k_2 = n - k$, or $n = k_1 + k_2 + k$. Now, the k_1 edges of M_t in $\cup_{x=1}^{\omega} E^x$ match k_1 vertices of V, and the k_2 edges of M_t in E match exactly $2k_2$ vertices of V. As these k_1+k_2 edges match all the vertices of V, $k_1+2k_2 = n = k_1+k_2+k$. It follows that $k_2 = k$ making the size of M exactly k.

We solve TROPICAL MATCHING on $\mathcal{I}_t$ to obtain a matching M_t of the maximum size, say s. We return the matching $M = M_t \cap E$ as the solution to $\mathcal{I}$. We claim that M is a solution to BUDGETED MATCHING of the minimum size. Suppose it is not. From the above discussion, we know that the size of M constructed this way, is $n-s+1$. Suppose there is a solution M' to BUDGETED MATCHING of size $z < n-s+1$. Then, by the above discussion, there is a solution to TROPICAL MATCHING of size $n-z+1 > n+1-n+s-1 = s$. But, this is a contradiction to the assumption that M_t is a maximum size solution.

Finally, TROPICAL MATCHING can be solved on $\mathcal{I}_t$ in $T(\alpha n, \beta\omega n^2)$ time for some constants α, β, and construction of G_t can be done in $\gamma\omega n^2$ time for some constant γ. Hence, BUDGETED MATCHING can be solved on $\mathcal{I}$ in time $T(\alpha n, \beta\omega n^2) + \gamma\omega n^2$. □

From Theorem 5 and Lemma 6, we obtain the following theorem.

Theorem 6. BUDGETED MATCHING *can be solved in* $O(\omega n^3)$ *time.*

Acknowledgements. The authors would like to thank Tanmay Inamdar and Kasturi Varadarajan for helpful discussions. The authors are also thankful to the anonymous reviewers. The research of the first author is partly funded by the European Research Council (ERC) via grant LOPPRE, reference 819416.

References

1. Anegg, G., Angelidakis, H., Kurpisz, A., Zenklusen, R.: A technique for obtaining true approximations for k-center with covering constraints. In: Bienstock, D., Zambelli, G. (eds.) IPCO 2020. LNCS, vol. 12125, pp. 52–65. Springer, Cham (2020). https://doi.org/10.1007/978-3-030-45771-6_5
2. Bandyapadhyay, S., Inamdar, T., Pai, S., Varadarajan, K.R.: A constant approximation for colorful k-center. In: 27th Annual European Symposium on Algorithms, ESA 2019, volume 144 of LIPIcs, pp. 12:1–12:14. Schloss Dagstuhl - Leibniz-Zentrum für Informatik (2019)
3. Bera, S.K., Gupta, S., Kumar, A., Roy, S.: Approximation algorithms for the partition vertex cover problem. Theoret. Comput. Sci. **555**, 2–8 (2014)
4. Caskurlu, B., Mkrtchyan, V., Parekh, O., Subramani, K.: Partial vertex cover and budgeted maximum coverage in bipartite graphs. SIAM J. Discret. Math. **31**(3), 2172–2184 (2017)
5. Cohen, J., Manoussakis, Y., Phong, H., Tuza, Z.: Tropical matchings in vertex-colored graphs. Electron. Notes Discrete Math. **62**, 219–224 (2017)
6. Dell, H., Marx, D.: Kernelization of packing problems. In: Proceedings of the Twenty-Third Annual ACM-SIAM Symposium on Discrete Algorithms, pp. 68–81. SIAM (2012)
7. Dell, H., Van Melkebeek, D.: Satisfiability allows no nontrivial sparsification unless the polynomial-time hierarchy collapses. Journal of the ACM (JACM) **61**(4), 1–27 (2014)
8. Garey, M.R., Johnson, D.S.: Computers and Intractability: A Guide to the Theory of NP-Completeness. W.H Freeman, New York (1979)
9. Gaur, D.R., Bhattacharya, B.: Covering points by axis parallel lines. In: Proceedings 23rd European Workshop on Computational Geometry, pp. 42–45 (2007)
10. Har-Peled, S., Jones, M.: On separating points by lines. In: Proceedings of the Twenty-Ninth Annual ACM-SIAM Symposium on Discrete Algorithms, SODA 2018, pp. 918–932. SIAM (2018)
11. Hassin, R., Megiddo, N.: Approximation algorithms for hitting objects with straight lines. Discret. Appl. Math. **30**(1), 29–42 (1991)
12. Inamdar, T., Varadarajan, K.R.: On the partition set cover problem. CoRR abs/1809.06506 (2018)
13. Jia, X., Sheth, K., Svensson, O.: Fair colorful k-center clustering. In: Bienstock, D., Zambelli, G. (eds.) IPCO 2020. LNCS, vol. 12125, pp. 209–222. Springer, Cham (2020). https://doi.org/10.1007/978-3-030-45771-6_17
14. Kratsch, S., Philip, G., Ray, S.: Point line cover: the easy kernel is essentially tight. ACM Trans. Algorithms (TALG) **12**(3), 1–16 (2016)
15. Kumar, V.S.A., Arya, S., Ramesh, H.: Hardness of set cover with intersection 1. In: Montanari, U., Rolim, J.D.P., Welzl, E. (eds.) ICALP 2000. LNCS, vol. 1853, pp. 624–635. Springer, Heidelberg (2000). https://doi.org/10.1007/3-540-45022-X_53
16. Langerman, S., Morin, P.: Covering things with things. Discrete Comput. Geometry **33**(4), 717–729 (2005)

17. Lau, L.C., Ravi, R., Singh, M.: Iterative Methods in Combinatorial Optimization, vol. 46. Cambridge University Press, Cambridge (2011)
18. Megiddo, N., Tamir, A.: On the complexity of locating linear facilities in the plane. Oper. Res. Lett. **1**(5), 194–197 (1982)
19. Slavik, P.: Improved performance of the greedy algorithm for partial cover. Inf. Process. Lett. **64**(5), 251–254 (1997)

On the Parameterized Complexity of the Connected Flow and Many Visits TSP Problem

Isja Mannens[1], Jesper Nederlof[1], Céline Swennenhuis[2](✉), and Krisztina Szilágyi[1]

[1] Utrecht University, Utrecht, The Netherlands
{i.m.e.mannens,j.nederlof,k.szilagyi}@uu.nl
[2] Eindhoven University of Technology, Eindhoven, The Netherlands
c.m.f.swennenhuis@tue.nl

Abstract. We study a variant of Min Cost Flow in which the flow needs to be connected. Specifically, in the Connected Flow problem one is given a directed graph G, along with a set of demand vertices $D \subseteq V(G)$ with demands $\mathsf{dem} : D \to \mathbb{N}$, and costs and capacities for each edge. The goal is to find a minimum cost flow that satisfies the demands, respects the capacities and induces a (strongly) connected subgraph. This generalizes previously studied problems like the (Many Visits) TSP.

We study the parameterized complexity of Connected Flow parameterized by $|D|$, the treewidth tw and by vertex cover size k of G and provide:

1. NP-completeness already for the case $|D| = 2$ with only unit demands and capacities and no edge costs, and fixed-parameter tractability if there are no capacities,
2. a fixed-parameter tractable $\mathcal{O}^\star(k^{\mathcal{O}(k)})$ time algorithm for the general case, and a kernel of size polynomial in k for the special case of Many Visits TSP,
3. a $|V(G)|^{\mathcal{O}(tw)}$ time algorithm and a matching $|V(G)|^{o(tw)}$ time conditional lower bound conditioned on the Exponential Time Hypothesis.

To achieve some of our results, we significantly extend an approach by Kowalik et al. [ESA'20].

1 Introduction

In the Connected Flow problem we are given a directed graph $G = (V, E)$ with costs and capacities on the edges and a set $D \subseteq V$ such that each $v \in D$ has a fixed demand. We then ask for a minimum cost *connected flow* on the edges that satisfies the demand for each $v \in D$, i.e. we look for a minimum cost *flow*

Supported by the project CRACKNP that has received funding from the European Research Council (ERC) under the European Union's Horizon 2020 research and innovation programme (grant agreement No. 853234) and by the Netherlands Organization for Scientific Research under project no. 613.009.031b.

L. Kowalik et al. (Eds.): WG 2021, LNCS 12911, pp. 52–79, 2021.
https://doi.org/10.1007/978-3-030-86838-3_5

conserving function $f : E \to \mathbb{N}$, such that the set of edges with strictly positive flow f is connected and the total flow coming into $v \in D$ is equal to its demand (see below for a formal definition of the problem).

One arrives (almost) directly at the CONNECTED FLOW problem by adding a natural connectivity constraint to the well known MIN COST FLOW problem (from now on abbreviated with simply 'FLOW', see Appendix A for details). But unfortunately, CONNECTED FLOW has the same fate as many other slight generalizations of FLOW: The additional requirement changes the complexity of the problem from being solvable in polynomial time to being NP-complete (see [9, Section A2.4] for more of such NP-complete generalizations).

The problem generalizes a number of problems, including the MANY VISITS TSP (MVTSP)[1]. This problem has a variety of potential applications in scheduling and computational geometry (see e.g. the discussion by Berger et al. [1]), and its study from the exponential time perspective recently witnessed several exciting results. In particular, Berger et al. [1] improved an old $n^{\mathcal{O}(n)}$ time algorithm by Cosmadakis and Papadimitriou [4] to $\mathcal{O}^\star(5^n)$ time and polynomial space, and recently the analysis of that algorithm was further improved by Kowalik et al. [10] to $\mathcal{O}^\star(4^n)$ time.

The CONNECTED FLOW problem also generalizes other problems studied in parameterized complexity, such as the EULERIAN STEINER SUBGRAPH problem, that was used in an algorithm for HAMILTONIAN INDEX by Philip et al. [11], or the problem of finding 2 short edge disjoint paths in undirected graphs (whose parameterized complexity was for example studied by Cai and Ye [3]).

Based on these connections with existing literature on in particular the MVTSP, its appealing formulation, and it being a direct extension of the well-studied FLOW problem, we initiate the study of the parameterized complexity of CONNECTED FLOW in this paper.

Our Contributions. We first study the (arguably) most natural parameterization: the number of demand vertices for which we require a certain amount of flow. We observe that the problem is NP-hard even for $|D| = 2$ by a reduction from the problem of finding two vertex disjoint paths in a directed graph by Fortune et al. [8]. The reduction heavily relies on the capacities and we show that this is indeed what makes the problem hard: If all capacities are set to ∞, the problem can be solved in $\mathcal{O}^\star(4^{|D|})$ time by combining a simple reduction with the algorithm for MVTSP from Kowalik et al. [10].

Next we study a typically much larger parameterization, the size k of a vertex cover of G. One of our main technical contributions is that CONNECTED FLOW is fixed-parameter tractable, parameterized by k:

Theorem 1. *There is an algorithm solving a given instance* $(G, D, \mathsf{dem}, \mathsf{cost}, \mathsf{cap})$ *of* CONNECTED FLOW *such that* G *has a vertex cover of size* k *in time* $\mathcal{O}^\star(k^{\mathcal{O}(k)})$.

[1] In this problem a minimum length tour is sought that satisfies each vertex a given number of times. The generalization is by setting the demand of a vertex to the number of times the tour is required to visit that vertex and using infinite capacities.

Theorem 1 is interesting even for the special case of MVTSP as it generalizes the $\mathcal{O}^\star(n^n)$ time algorithm from Cosmadakis and Papadimitriou [4], though it is a bit slower than the more recent algorithms from [1,10]. For this special case, we even find a polynomial kernel:

Theorem 2. MVTSP *admits a kernel polynomial in the size k of the vertex cover of G.*

The starting point of the proofs of both Theorem 1 and Theorem 2 is a strengthening of a non-trivial lemma from Kowalik et al. [10] which proves the existence of a solution s' that is 'close' to a solution r of the FLOW problem instance obtained by relaxing the connectivity requirement. Since such an r can be found in polynomial time, it can be used to determine how the optimal solution roughly looks.

This is subsequently used by a dynamic programming algorithm that aims to find such a solution close to r to establish Theorem 1; the restriction to solutions being close to r crucially allows us to evaluate only $\mathcal{O}^\star(k^{\mathcal{O}(k)})$ table entries. Additionally this is used in the kernelization algorithm of Theorem 2 to locate a set of $\mathcal{O}(k^5)$ vertices such that only edges incident to vertices in this set will have a different flow in r and s'.

The last parameter we consider is the *treewidth*, denoted by tw, of G, which is a parameter that is widely used for many graph problems and that is smaller than k. We show that, unfortunately, CONNECTED FLOW most likely cannot be solved by a fixed-parameter tractable algorithm:

Theorem 3. *Assuming the Exponential Time Hypothesis,* MVTSP *cannot be solved in time $f(tw)n^{o(tw)}$ for any computable function $f(\cdot)$.*

Note that since MVTSP is a special case of CONNECTED FLOW this lower bound extends to CONNECTED FLOW.

We also present a Dynamic Programming algorithm for CONNECTED FLOW running in time $n^{\mathcal{O}(tw)}$, matching the lower bound. For the special case of MVTSP our algorithm runs in $M^{\mathcal{O}(tw)}$ time, where M is the maximum demand of a vertex (which can be assumed to be $n^{\mathcal{O}(1)}$ by Kowalik et al. [10]).

Notation and Formal Problem Definitions. We let $\mathcal{O}^\star(\cdot)$ omit factors polynomial in the input size. We assume that all integers are represented in binary, so in this paper the input size will be polynomial in the number of vertices of the input graph and the logarithm of the maximum input integer. For a Boolean b we define $[b]$ to be 1 if b is true and 0 otherwise. For integers a and b we denote $[a, b]$ as the set of all integers i such that $a \leqslant i \leqslant b$. All graphs in this paper are directed unless stated otherwise.

We use the notion of *multisets*, which are sets in which the same element may appear multiple times. Formally, a multiset is an ordered pair (A, m_A) consisting of a set A and a multiplicity function $m_A : A \to \mathbb{Z}^+$. We slightly abuse notation and let $m_A(e) = 0$ if $e \notin A$. We can see flow f as a multiset of directed edges, where each edge appears $f(e)$ number of times. We then say that $f(e)$ is the *multiplicity* of e. Given a function $f : E \to \mathbb{N}$, we define $G_f = (V', E')$ as the multigraph where $e \in E'$ has multiplicity $f(e)$ and V' is the set of vertices incident to at least one $e \in E'$. We let $E(G_f)$ be equal to the multiset E'. We also define $\text{supp}(f) = \{e \in E : f(e) > 0\}$ as the *support* of f. The formal statement of CONNECTED FLOW is as follows:

CONNECTED FLOW
Input: $G = (V, E)$, $D \subseteq V$, $\mathsf{dem} : D \to \mathbb{N}$, $\mathsf{cost} : E \to \mathbb{N}$, $\mathsf{cap} : E \to \mathbb{N} \cup \{\infty\}$
Task: Find a function $f : E \to \mathbb{N}$ such that

- G_f is connected,
- for every $v \in V$ we have $\sum_{(u,v) \in E} f(u,v) = \sum_{(v,u) \in E} f(v,u)$,
- for every $v \in D$ we have $\sum_{(u,v) \in E} f(u,v) = \mathsf{dem}(v)$,
- for every $e \in E : f(e) \leqslant \mathsf{cap}(e)$,

and the value $\mathsf{cost}(f) = \sum_{e \in E} \mathsf{cost}(e) f(e)$ is minimized.

Note that G_f in the above definition is Eulerian (every vertex has the same in and out degree), so it is strongly connected if and only if it is weakly connected. We define FLOW as the CONNECTED FLOW problem without the connectivity requirement, which can be solved in polynomial time[2]. MVTSP is a special case of CONNECTED FLOW, where $D = V$ and capacities are infinite. Formal definitions of these problems can be found in Appendix A.

Organization. The remainder of this paper is organized as follows: In Sect. 2 we first introduce an extension of a lemma from Kowalik et al. [10] that shows that we can transform an optimal solution to the FLOW relaxation to include a specific edge set from an optimal solution of the original CONNECTED FLOW instance, without changing too many edges. This lemma is subsequently used in Sect. 2.2 to prove Theorem 1 and in Sect. 2.3 to prove Theorem 2. In Sect. 3 we discuss the treewidth parameterization and in Sect. 4 we conclude the paper with a discussion on further research opportunities.

In Appendix A we provide problem definitions for FLOW and prove it is equivalent to MIN COST FLOW. In Appendix B we formally support the observations that characterize the complexity of CONNECTED FLOW, parameterized by the number of demand vertices $|D|$ and in Appendix C we prove the lower bound in Theorem 3 and the matching upper bound. Finally, Appendix D contains missing proofs of claims from Sects. 2.1, 2.2 and 2.3.

2 Parameterization by Vertex Cover

In this section, we consider CONNECTED FLOW and MVTSP, parameterized by the cardinality k of a vertex cover of the input graph. We first extend a lemma from Kowalik et al. [10] to instances of CONNECTED FLOW. Then we use this lemma to obtain a fixed-parameter tractable algorithm for CONNECTED FLOW and a polynomial-sized kernel for MVTSP.

[2] In Appendix A we show that FLOW is equivalent to the MIN COST FLOW problem, which is polynomial-time solvable.

2.1 Transforming the Flow Relaxation to Enforce Some Edges

Let s be an optimal solution of CONNECTED FLOW and let $T \subseteq \text{supp}(s)$. We prove that, given any optimal solution r for FLOW, there is always a flow f that is close to r and $T \subseteq \text{supp}(f)$. Furthermore it has cost $\mathsf{cost}(f) \leqslant \mathsf{cost}(s)$. Note that if T connects all demand vertices to each other, this implies that f is connected and thus an optimal solution of CONNECTED FLOW.

The basic idea and arguments are from Kowalik et al. [10], where a similar theorem for MVTSP was proved. We adjusted their proof to the case with capacities and where not all vertices have a demand. Furthermore, we noted that we can restrict the tours $C \in \mathcal{C}$ in the proof to be inclusion-wise minimal, which allows us to conclude a stronger inequality.

Lemma 1. *Fix an input instance* $(G, D, \mathsf{dem}, \mathsf{cost}, \mathsf{cap})$ *with* $G = (V, E)$. *Let* s *be an optimal solution of* CONNECTED FLOW *and let* $T \subseteq \text{supp}(s)$. *For every optimal solution* r *of* FLOW, *there is a flow* f *with* $\mathsf{cost}(f) \leqslant \mathsf{cost}(s)$, *with* $f(e) > 0$ *for all* $e \in T$ *and such that for every* $v \in V$:

$$\sum_{u \in V} |r(u,v) - f(u,v)| \leqslant 2T|, \qquad \text{and} \qquad \sum_{u \in V} |r(v,u) - f(v,u)| \leqslant 2T|.$$

Proof. We follow the structure of the proof of Lemma 3.2 from Kowalik et al. [10]. We build a flow f (not necessarily optimal for FLOW), containing T and with multiplicities close to r. Recall that m_B denotes the multiplicity function of the multiset B. We define the multisets of edges A_s, A_r and A such that for all $e \in E$:

- $m_{A_s}(e) = \max\{s(e) - r(e), 0\}$,
- $m_{A_r}(e) = \max\{r(e) - s(e), 0\}$, and
- $m_A(e) = \max\{m_{A_r}(e), m_{A_s}(e)\} = \max\{s(e) - r(e), r(e) - s(e)\}$.

Note that A is the symmetric difference of s and r, and therefore any $e \in A$, is exactly either in A_r or in A_s, but never in both.

Let H be a tour (i.e. a closed walk) of undirected edges. We then say that $\overrightarrow{H}$ is a *cyclic orientation* of H if it is an orientation of the edges in H such that $\overrightarrow{H}$ forms a directed tour. A directed edge e that overlaps with H is in *positive orientation* with respect to $\overrightarrow{H}$ if it has the same orientation, and negative otherwise. We now define $(s - r)$ directed tours, of which an example is shown in Fig. 1.

Definition 1. *Let* $C = (e_0, \ldots, e_\ell) \subseteq A$ *be a set of edges such that its underlying undirected edge set* H *is a tour. We then say that* C *is an* $(s - r)$ directed tour *if there is an orientation* $\overrightarrow{H}$ *of* H *such that:*

- *if* $e \in C$ *is in positive orientation with respect to* $\overrightarrow{H}$, *then* $e \in A_s$,
- *if* $e \in C$ *is in negative orientation with respect to* $\overrightarrow{H}$, *then* $e \in A_r$,
- *if two subsequent edges* e_i, e_{i+1} *of* C *have the same orientation, then their shared vertex,* v, *is not in* D. *This also holds for the edge pair* (e_ℓ, e_0).

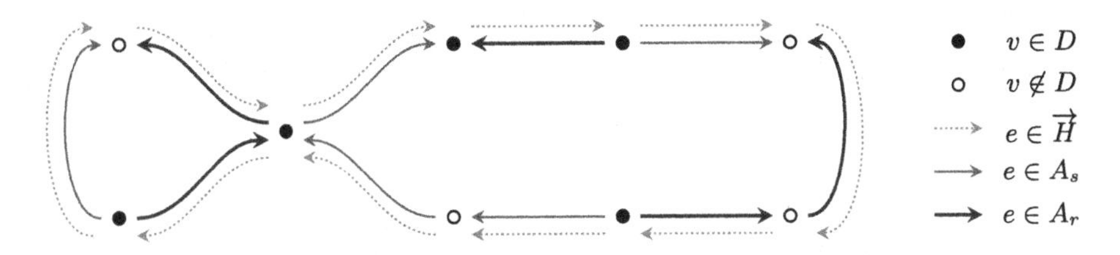

Fig. 1. Example of an $(s-r)$ directed tour. Note that every time the tour visits a vertex $v \in D$, the orientation changes. However if the tour visits a vertex $v \notin D$, the orientation might not change.

We give a construction such that A can be partitioned into a multiset of $(s-r)$ directed tours. We take $(u,v) \in A$ arbitrarily as our first edge of our walk and iteratively add edges until we find an $(s-r)$ directed tour. We assume the current edge (u,v) is in A_s (if the edge is in A_r, the arguments are similar). If $v \in D$, then there exists $(v,w) \in A_r$, because v is visited $\mathsf{dem}(v)$ times by both r and s. If $v \notin D$, there exists either $(v,w) \in A_r$ or $(w,v) \in A_s$ because A is the symmetric different of the flows r and s. We take this edge as the next edge in our $(s-r)$ directed tour. This way we can keep finding the next edges, until we can take our first edge (u,v) as our next edge and we find an $(s-r)$ directed tour. We then remove this tour and inductively find the next until A is empty.

It follows that A can be partitioned into a multiset $\mathcal{C}$ of $(s-r)$ directed tours, i.e.

$$m_A = \sum_{C \in \mathcal{C}} m_C,$$

where m_C is the multiplicity function of $(s-r)$ directed tour C.

We may assume that these $(s-r)$ directed tours are inclusion-wise minimal, i.e. for each $(s-r)$ directed tour $C \in \mathcal{C}$, no subset $C' \subset C$ is an $(s-r)$ directed tour. Otherwise, C can be split into two disjoint $(s-r)$ directed tours C' and $C \backslash C'$.

Claim 1. *For any $v \in V$ and any inclusion-wise minimal $C \in \mathcal{C}$:*

$$\sum_{u \in V} [(u,v) \in C] \leqslant 2 \qquad \text{and} \qquad \sum_{u \in V} [(v,u) \in C] \leqslant 2. \tag{1}$$

The proof of the above claim can be found in Appendix D.

We denote $T^+ = E(T) \backslash \operatorname{supp}(r)$ as the set of edges of T that are not yet covered by r. Hence, if $e \in T^+$, then $e \in A_s$ and there is at least one $C \in \mathcal{C}$ that contains e. We choose for each $e \in T^+$ such an $(s-r)$ directed tour $C_e \in \mathcal{C}$ arbitrarily. Let $\mathcal{C}^+ = \{C_e : e \in T^+\}$ be the set of chosen $(s-r)$ directed tours. We define f as follows: for each $u, v \in V$ we set

$$f(u,v) = r(u,v) + (-1)^{[(u,v) \in A_r]} \sum_{C \in \mathcal{C}^+} [(u,v) \in C]. \tag{2}$$

In other words, f is obtained from r by removing one copy of edges in $C \cap A_r$ and adding one copy of edges in $C \cap A_s$ for all $C \in \mathcal{C}^+$.

Notice that $|\mathcal{C}^+| \leqslant |T^+| \leqslant |T|$. By using (2) and subsequently (1), we get for all $v \in V$:

$$\begin{aligned}\sum_{u \in V} |r(u,v) - f(u,v)| &\leqslant \sum_{u \in V} \sum_{C \in \mathcal{C}^+} [(u,v) \in C] \\ &\leqslant \sum_{C \in \mathcal{C}^+} 2 \leqslant 2T|.\end{aligned}$$

Similarly we can conclude for all $v \in V$ that $\sum_{u \in V} |r(v,u) - f(v,u)| \leqslant 2|T|$.

Claim 2. *For all $e \in T$, $f(e) > 0$ and f is a flow for the given instance.*

The proof can be found in Appendix D.

We are left to prove that $\mathsf{cost}(f) \leqslant \mathsf{cost}(s)$. For any $C \in \mathcal{C}$ define $\delta(C) = \mathsf{cost}(A_s \cap C) - \mathsf{cost}(A_r \cap C)$ as the cost of adding all edges in $A_s \cap C$ and removing all edges in $A_r \cap C$. Notice that $\delta(C) \geqslant 0$ for all tours $C \in \mathcal{C}$, as otherwise r would not have been optimal since we could improve it by augmenting along C. We note that $\sum_{C \in \mathcal{C}^+} \delta(C) \leqslant \sum_{C \in \mathcal{C}} \delta(C)$ as $\mathcal{C}^+ \subseteq \mathcal{C}$. Therefore: $\mathsf{cost}(f) = \mathsf{cost}(r) + \sum_{C \in \mathcal{C}^+} \delta(C) \leqslant \mathsf{cost}(r) + \sum_{C \in \mathcal{C}} \delta(C) = \mathsf{cost}(s)$. □

2.2 Fixed Parameter Tractable Algorithm

Now we use Lemma 1 to show that Connected Flow is fixed-parameter tractable parameterized by the size of a vertex cover of G:

Theorem 1. *There is an algorithm solving a given instance $(G, D, \mathsf{dem}, \mathsf{cost}, \mathsf{cap})$ of* Connected Flow *such that G has a vertex cover of size k in time $\mathcal{O}^\star(k^{\mathcal{O}(k)})$.*

Proof. Let X be a vertex cover of size k of $G = (V, E)$, let s be an arbitrary optimal solution of Connected Flow and let $X' \subseteq X$ be the set of vertices of X that are visited at least once by s. We will guess this set X' as part of our algorithm, i.e. go through all possible sets. Hence we do the following algorithm for all X' such that $(D \cap X) \subseteq X' \subseteq X$, which is at most 2^k times.

For any X', we adjust G such that the vertex cover is an independent set and all $x \in X'$ are visited at least once in any solution as follows. We remove any edge $(x_i, x_j) \in E$ for $x_i, x_j \in X'$ and replace this edge by adding a new vertex y to V. This y has no demand and has edges (x_i, y) and (y, x_j), with capacities equal to the old capacity $\mathsf{cap}(x_i, x_j)$ and $\mathsf{cost}(x_i, y) = \mathsf{cost}(x_i, x_j)$ and $\mathsf{cost}(y, x_j) = 0$. This removes any edges between vertices in the set X', making it an independent set. We note that X' is still a vertex cover of size k.

For all $x \in X' \backslash D$ we add a vertex b_x to V, with $\mathsf{dem}(b_x) = 1$ and we add edges (x, b_x) and (b_x, x) both with 0 cost and a capacity of 1. As b_x has a demand of 1 and has only x as its neighbor, this ensures that x is visited at least once.

We remove all $x \in X \backslash X'$ from V and denote the resulting graph as $G' = (V', E')$. Note that if X' is guessed correctly, the optimal solution s of the original instance, is also optimal for this newly created instance (by adding flow over the newly created edges between x and b_x, and replacing any edge (x_i, x_j) by the edges (x_i, y) and (y, x_j)). We compute a *relaxed* solution r for this newly created instance, which can be done in polynomial time.

Let T be any directed tree of size at most $2k$ such that $T \subseteq \text{supp}(s)$ and all $x \in X'$ are incident to at least one edge $e \in T$. We argue why such tree T exists. Since s is connected and visits all $x \in X'$, we can find a tree $T \subseteq \text{supp}(s)$ that covers all $x \in X'$. If $|T| > 2k$, we remove all the leaves from T not in X'. Since $V' \setminus X'$ is an independent set (as X' is a vertex cover), this means that the size of T is bounded by $2k$. Note that all $x \in X'$ are still incident to an edge $e \in T$.

We apply Lemma 1, to s and T to find that there is a flow f such that $\mathsf{cost}(f) \leqslant \mathsf{cost}(s)$ and for every $v \in V$:

$$\sum_{u \in V'} |r(u,v) - f(u,v)| \leqslant 4k, \qquad \text{and} \qquad \sum_{u \in V'} |r(v,u) - f(v,u)| \leqslant 4k. \quad (3)$$

Since $T \subseteq \text{supp}(f)$, f visits all the vertices in X' at least once. As X' is a vertex cover, this means that f is a connected flow and hence an optimal solution of the instance of Connected Flow. We will use a dynamic programming method to find solution f. Namely, we iteratively add vertices from the independent set $B = V' \setminus X'$ and keep track of the connectedness of our vertex cover X' with a partition π. We later will restrict the number of table entries we actually compute with the help of Eq. (3).

Denote $X' = \{x_1, \dots, x_{k'}\}$ and let $B = \{b_1, \dots, b_n\}$. For $j \in [0, n]$ let B_j be the set of the first j vertices of B, i.e. $B_j = \{b_1, \dots, b_j\}$ and define $V_j = X' \cup B_j$. For any $f : (V_j)^2 \to \mathbb{N}$ and $v \in V_j$ define $f^{\mathsf{out}}(v) = \sum_{u \in V_j} f(v,u)$ and $f^{\mathsf{in}}(v) = \sum_{u \in V_j} f(u,v)$. Let $\mathbf{c}^{\mathsf{in}} = (c_1^{\mathsf{in}}, \dots, c_{k'}^{\mathsf{in}}) \in \mathbb{N}^{k'}$ and $\mathbf{c}^{\mathsf{out}} = (c_1^{\mathsf{out}}, \dots, c_{k'}^{\mathsf{out}}) \in \mathbb{N}^{k'}$ be two vectors of integers and let π be a partition of the vertices of X'.

For $j \in [0, n]$ we define the dynamic programming table entry $T_j(\pi, \mathbf{c}^{\mathsf{in}}, \mathbf{c}^{\mathsf{out}})$ to be equal to the minimal cost of any partial solution $f : (V_j)^2 \to \mathbb{N}$ having the specified in and out degrees ($\mathbf{c}^{\mathsf{in}}$ and $\mathbf{c}^{\mathsf{out}}$) for vertices in X' and connecting all vertices $x \in S$ for each $S \in \pi$. More formally, $T_j(\pi, \mathbf{c}^{\mathsf{in}}, \mathbf{c}^{\mathsf{out}})$ is equal to $\min_f \mathsf{cost}(f)$ over all $f : (V_j)^2 \to \mathbb{N}$ such that the following conditions hold:

1. for all blocks S of the partition π, the block is weakly connected in G'_f,
2. for all $x_i \in X'$: $f^{\mathsf{out}}(x_i) = c_i^{\mathsf{out}}$, $f^{\mathsf{in}}(x_i) = c_i^{\mathsf{in}}$,
3. for all $v \in B_j$: $f^{\mathsf{out}}(v) = f^{\mathsf{in}}(v)$, and if $v \in B_j \cap D$, then $f^{\mathsf{in}}(v) = \mathsf{dem}(v)$,
4. for all $u, v \in V_j : f(u,v) \leqslant \mathsf{cap}(u,v)$.

We set $T_j(\pi, \mathbf{c}^{\mathsf{in}}, \mathbf{c}^{\mathsf{out}}) = \infty$ if no such f exists.

Claim 3. *Each table entry $T_j(\pi, \mathbf{c}^{\mathsf{in}}, \mathbf{c}^{\mathsf{out}})$ can be computed from all table entries T_{j-1}.*

We describe how to compute each table entry in the proof of Claim 3 in Appendix D. We restrict this dynamic program using Eq. (3). As X' is an independent set, there are only edges between $x \in X'$ and $b \in B_j$. Therefore, there exists a solution f such that for every $x \in X'$ and $j \in [0, n]$:

$$\sum_{b\in B_j} |r(b,x) - f(b,x)| \leqslant 4k, \qquad \text{and} \qquad \sum_{b\in B_j} |r(x,b) - f(x,b)| \leqslant 4k \tag{4}$$

We restrict the dynamic program to only compute table entries T_j respecting Eq. (4), by requiring for all $i \in [1, k']$:

$$\begin{aligned} c_i^{\mathsf{out}} &\in \left[\sum_{b\in B_j} r(x_i,b) - 4k, \sum_{b\in B_j} r(x_i,b) + 4k\right], \text{ and} \\ c_i^{\mathsf{in}} &\in \left[\sum_{b\in B_j} r(b,x_i) - 4k, \sum_{b\in B_j} r(b,x_i) + 4k\right]. \end{aligned} \tag{5}$$

Note that the dynamic program is still correct with this added restriction, as $\sum_{b\in B_{j-1}} |r(b,x) - f(b,x)| \leqslant \sum_{b\in B_j} |r(b,x) - f(b,x)| \leqslant 4k$, so any table entry T_j respecting (5) can be computed from all table entries T_{j-1} respecting (5).

The dynamic program returns the minimum value of $T_n(\{X'\}, \mathbf{c}, \mathbf{c})$ for all $\mathbf{c}$ such that $c_i = \mathsf{dem}(x_i)$ for all $x_i \in D \cap X'$. This returns the value of a minimum cost solution f for G', respecting Eq. (4), if one exists. Let $f_{X'}$ be solution the dynamic program found in the iteration using X'. Then $\min\{f_{X'} : (D \cap X) \subseteq X' \subseteq X\}$ is equal to the minimum cost connected flow.

We count the number of different table entries T_j computed by the dynamic program for fixed j. There are at most $(8k)^k$ possible values for both $\mathbf{c}^{\mathsf{in}}$ and $\mathbf{c}^{\mathsf{out}}$ and at most k^k different partitions π of X', so a total of $k^k \cdot (8k)^{2k}$ different entries. To compute one table entry of T_j, we only need (the at most $k^k \cdot (8k)^{2k}$) table entries of T_{j-1}. Note that we compute this dynamic programming table for each X' such that $(D \cap X) \subseteq X' \subseteq X$, that is at most 2^k different X'. Hence the algorithm runs in time $\mathcal{O}^\star(k^{\mathcal{O}(k)})$. □

2.3 Kernel for Many Visits TSP with $\mathcal{O}(k^5)$ vertices

We now present how to find a kernel with $\mathcal{O}(k^5)$ vertices for any instance of MVTSP, where k is the size of a vertex cover of G. We do this by first finding an optimal solution r to the relaxed Flow problem and then fixing the amount of flow on some edges based on this r. We prove that there is an optimal solution s of MVTSP such that for all except $\mathcal{O}(k^5)$ vertices, all edges incident to these vertices have exactly the same flow in r and s, as a consequence of Lemma 1.

Theorem 2. MVTSP *admits a kernel polynomial in the size k of the minimum vertex cover of G.*

Proof. Fix an input instance on MVTSP. Let k be the number of vertices in the vertex cover $X = \{x_1, \ldots, x_k\}$ of G and let n be the size of the independent set $B = V \backslash X$. Let r be an optimal solution of the instance of Flow obtained by relaxing the connectivity constraint from in the given instance of MVTSP.

Define multisets $\overrightarrow{F} = (X \times B) \cap r$ (i.e. all edges in r going from vertices in X to vertices in B) and $\overleftarrow{F} = (B \times X) \cap r$.

Claim 4. *We may assume that for both $\overrightarrow{F}$ and $\overleftarrow{F}$, their underlying undirected edge sets do not contain cycles.*

We describe how adjust r such that Claim 4 holds in Appendix D.

We partition B as follows: $B = Y \cup \left(\bigcup_{i,j\in[1,k]} B_{ij}\right)$, where for each $b \in B_{ij}$: $r(x_i, b) > 0$, $r(b, x_j) > 0$, and

$$r(x_a, b) = 0 \text{ for all } a \neq i \qquad \text{and} \qquad r(b, x_a) = 0 \text{ for all } a \neq j,$$

and $Y = B \backslash \left(\bigcup_{i,j\in[1,k]} B_{ij}\right)$.

We argue that $|Y| \leqslant k$. Recall that m_B denotes the multiplicity function of a multiset B. Let $F = \text{supp}(m_{\overleftarrow{F}}) \cup \text{supp}(m_{\overrightarrow{F}})$ (note that F is a set and not a multiset). Then $|F| \geqslant \sum_{i,j\in[1,k]} 2|B_{ij}| + 3|Y| = 2n + |Y|$, as any vertex $v \in B_{ij}$ must be responsible for exactly 2 edges in F and each vertex in Y must add at least 3 edges to F. Here we use that each vertex has a demand and therefore must have at least one incoming and outgoing edge from r. As F is a union of two forests on $n + k$ vertices, we see that $|F| \leqslant 2(n + k - 1)$. We conclude that $2(n + |Y|) \leqslant 2(n + k - 1)$, i.e. $|Y| \leqslant k$.

Let s be an optimal solution of the MVTSP instance, so s visits every vertex at least once. Hence there exists a directed tree $T \subseteq \text{supp}(s)$, covering all vertices of X, of size at most $2k$. This tree exists by similar arguments as in the proof of Theorem 1. We apply Lemma 1 to s and T, to find that there exists an optimal solution f to the given MVTSP instance such that

$$\sum_{v\in V} (|r(x_i, v) - f(x_i, v)| + |r(v, x_i) - f(v, x_i)|) \leqslant 8k \qquad \forall i \in [1, k]. \tag{6}$$

We note that G_f is connected because $T \subseteq \text{supp}(f)$ and T connects all the vertices of the vertex cover. Equation (6) implies that at most $8k^2$ edges of $\overleftarrow{F}$ and $\overrightarrow{F}$ are different in an optimal solution f of MVTSP that is close compared to r.

For every $i, j, \ell \in [1, k]$, we define $\overrightarrow{A_{ij}}(\ell)$ as the set of $8k^2 + 2$ vertices $v \in B_{ij}$ with the smallest values of $\mathsf{cost}(x_\ell, v) - \mathsf{cost}(x_i, v)$ (arbitrarily breaking ties if needed). Intuitively, the vertices in $\overrightarrow{A_{ij}}(\ell)$ are the vertices for which re-routing the flow sent from x_i to v to go from x_ℓ to v is the least expensive. Similarly we define $\overleftarrow{A_{ij}}(\ell)$ as a set of size $8k^2 + 2$ containing vertices $v \in B_{ij}$ with the smallest values of $\mathsf{cost}(v, x_\ell) - \mathsf{cost}(v, x_j)$.

We also define a set R_{ij} of 'remainder vertices' as follows:

$$R_{ij} = B_{ij} \backslash \left(\left(\bigcup_{\ell\in[1,k]} \overleftarrow{A_{ij}}(\ell) \right) \cup \left(\bigcup_{\ell\in[1,k]} \overrightarrow{A_{ij}}(\ell) \right) \right) \quad \text{for all } i, j \in [1, k].$$

Claim 5. *There exists an optimal solution f' of the* MVTSP *instance such that for all $i, j \in [1, k]$, $b \in R_{ij}$ and $x_\ell \in X$ it holds that $r(x_\ell, b) = f'(x_\ell, b)$ and $r(b, x_\ell) = f'(b, x_\ell)$.*

We build this f' iteratively from f, by removing any edge between $b \in R_{ij}$ and x_ℓ that shouldn't exist. The proof of this claim is deferred to Appendix D.

Therefore, we may assume that, in $G_{f'}$, the vertices in R_{ij} are adjacent only to x_i and x_j for all $i, j \in [1, k]$. This proves that the following reduction rule is correct: contract all vertices in R_{ij} into one vertex r_{ij} with edges only (x_i, r_{ij}) and (r_{ij}, x_j) of cost zero and let the demand $\mathsf{dem}(r_{ij}) = \sum_{v \in R_{ij}} \mathsf{dem}(v)$. Hence, we require any solution to use the vertices in r_{ij} exactly the number of times that we would traverse all the vertices of R_{ij}. By applying this rule, we get a kernel with the vertices from the sets X, Y, $\overleftarrow{A_{ij}}(\ell)$, $\overrightarrow{A_{ij}}(\ell)$, and r_{ij}, which is of size

$$|X|+|Y|+\sum_{i,j,\ell \in [1,k]} \left(\left|\overleftarrow{A_{ij}}(\ell)\right| + \left|\overrightarrow{A_{ij}}(\ell)\right|\right)+k^2 \leqslant k+k+k^3 \cdot (8k^2+2)+k^2 = \mathcal{O}(k^5).$$

To subsequently reduce all costs to be at most $2^{k^{\mathcal{O}(1)}}$ we can use a method from Etscheid et al. [7] in a standard manner.

We check that one can construct this kernel in polynomial time. First, compute a relaxed solution r and remove any cycles in $\overrightarrow{F}$ and $\overleftarrow{F}$ in polynomial time. Next for each $i, j, \ell \in [1, k]$, compute in polynomial time what the sets $\overrightarrow{A_{ij}}(\ell)$ and $\overleftarrow{A_{ij}}(\ell)$ should be, by computing the values of $\mathsf{cost}(x_\ell, v) - \mathsf{cost}(x_i, v)$ and sorting. Finally we can contract all vertices in R_{ij} into a vertex r_{ij} polynomial time for all $i, j \in [1, k]$. □

3 Treewidth

When parameterized by treewidth, we get a tight classification of the problem by heavily using (fairly standard) methodology from previous papers. In particular, we have a dynamic programming algorithm which establishes the following result:

Theorem 4. *Let M be an upper bound on the demands in the input graph G, and suppose a tree decomposition of width tw of G is given. Then a* Connected Flow *instance with G can be solved in time $|V(G)|^{\mathcal{O}(tw)}$ and an* MVTSP *instance with G can be solved in time $\min\{|V(G)|, M\}^{\mathcal{O}(tw)}|V(G)|^{\mathcal{O}(1)}$.*

The algorithm builds a table with entries describing partial solutions. These entries are indexed by the in and out degrees of vertices of bags in a tree decomposition, as well as a partition of the bag which describes the connectivity. It turns out that this running time is essentially optimal:

Theorem 5. *Let M be an upper bound on the demands in a graph G. Then* MVTSP *cannot be solved in time $f(pw)\min\{|V(G)|, M\}^{o(pw)}|V(G)|^{\mathcal{O}(1)}$, unless ETH fails.*

Note that this theorem implies Theorem 3. We prove this upper bound using a reduction from 3-CNF-SAT. The main idea is to encode assignments of sets of variables as flow through a path, where the amount of flow encodes a partial assignment of a group of variables. We then build so-called scanner gadgets which detect whether a clause is satisfied. This reduction is based on ideas from Cygan et al. [6]. Details are deferred to Appendix C.

4 Conclusion and Further Research

We initiated the study of the parameterized complexity of the CONNECTED FLOW problem and showed that the problem behaves very differently when parameterized by the number of demand vertices, the size of the vertex cover of the graph, or treewidth of the input graph.

While we essentially settled the complexity of the variants of the problem parameterized by the number of demands or by the treewidth, we still leave the following questions open for the vertex cover parameterization:

Can CONNECTED FLOW be solved in $\mathcal{O}^\star(c^{\mathcal{O}(k)})$ time, with c a constant and k the size of the vertex cover of the input graph? Such an algorithm would be a strong generalization of the algorithms from [1,10]. While we believe our approach from Theorem 1 makes significant progress towards solving this question affirmatively, it seems that non-trivial ideas are required.

Does CONNECTED FLOW admit a kernel polynomial in k where k is the size of the vertex cover if the input graph? It seems that especially the capacities can make the problem a lot harder. It would be interesting to see if our arguments for Theorem 2 can be extended to kernelize this more general problem as well.

A Problem Definitions

In this section we formally introduce and discuss a number of computational problems that are relevant for this paper.

Formally, we define the FLOW problem as follows.

FLOW
Input: Given digraph $G = (V, E)$, $D \subseteq V$, $\mathsf{dem} : D \to \mathbb{N}$, $\mathsf{cost} : E \to \mathbb{N}$, $\mathsf{cap} : E \to \mathbb{N} \cup \{\infty\}$
Task: Find a function $f : E \to \mathbb{N}$ such that

- for every $v \in V$ we have $\sum_{u \in V} f(u, v) = \sum_{u \in V} f(v, u)$,
- for every $v \in D$ we have $\sum_{u \in V} f(u, v) = \mathsf{dem}(v)$,
- for every $e \in E : f(e) \leqslant \mathsf{cap}(e)$,

and the value $\mathsf{cost}(f) = \sum_{e \in E} \mathsf{cost}(e) f(e)$ is minimized.

From the definition it is clear that apart from the connectivity requirement, it is indeed equivalent to CONNECTED FLOW.

We will use the following standard definition of MIN COST FLOW.

Min Cost Flow
Input: Digraph $G = (V, E)$ with source node set $S \subseteq V$ and sink nodes $T \subseteq V$, $\mathsf{cost} : E \to \mathbb{N}$, $\mathsf{cap} : E \to \mathbb{N} \cup \infty$
Task: Find a function $f : E \to \mathbb{N}$ such that

- for every $v \in V \setminus (T \cup S)$ we have $\sum_{u \in V} f(u, v) = \sum_{u \in V} f(v, u)$,
- for every $e \in E : f(e) \leqslant \mathsf{cap}(e)$,
- the value of $\sum_{v \in S} \sum_{u \in V} f(v, u)$ is maximal,

and the value $\mathsf{cost}(f) = \sum_{e \in E} \mathsf{cost}(e) f(e)$ is minimized.

Equivalence of Flow *and* Min Cost Flow. We argue that Flow is equivalent to Min Cost Flow by simple reductions. First we reduce in the forward way. For each $d \in D$, create vertices $d_{\mathsf{out}}, d_{\mathsf{in}}$ where d_{out} is a source node with outgoing flow $\mathsf{dem}(d)$ and d_{in} is a sink node with ingoing flow $\mathsf{dem}(d)$. For all other vertices in $V \backslash D$, create a node and connect to all its neighbors, where all outgoing edges to a vertex in D go to d_{in} and all ingoing edges from a vertex in D connect to d_{out}.

For the other direction, let S be the set of source nodes and T be the set of sink nodes of the Min Cost Max Flow problem. Then add one 'big' node x to the graph, with demand equal to the outgoing flow from all the source nodes. Then add (t, x) for all $t \in T$ with $\mathsf{cost}(t, x) = 0$, $\mathsf{cap}(t, x) = \mathsf{out}(t)$. Furthermore add (x, s) for all $s \in S$ with $\mathsf{cost}(x, s) = 0$, $\mathsf{cap}(x, s) = \mathsf{in}(s)$.

Since Min Cost Flow is well-known to be solvable in polynomial time, we can therefore conclude that Flow is solvable in polynomial time as well.

In Kowalik et al. [10], the Many Visit TSP (MVTSP) is defined as follows.

Many Visits TSP (MVTSP)
Input: Digraph $G = (V, E)$, $\mathsf{dem} : V \to \mathbb{N}$, $\mathsf{cost} : V^2 \to \mathbb{N}$
Task: Find a minimal cost tour c, such that each $v \in V$ is visited exactly $\mathsf{dem}(v)$ times.

Note that MVTSP is a special case of Connected Flow, where $D = V$ and the capacities of all edges are infinite.

B Parameterization by Number of Demand Vertices

In this section we study the parameterized complexity of Connected Flow with parameter $|D|$, the number of vertices with a demand. We first prove that the problem is NP-hard, even for $|D| = 2$, by a reduction from the problem of finding two vertex disjoint paths in a directed graph. Next we show that, if $\mathsf{cap}(e) = \infty$ for all $e \in E$, the problem can be reduced to an instance of MVTSP, and hence solved in time $\mathcal{O}^\star(4^{|D|})$.

Theorem 6. Connected Flow *with* 2 *demand vertices is NP-complete.*

Proof. We give a reduction from the problem of finding two vertex-disjoint paths in a directed graph to CONNECTED FLOW with demand set D of size 2. The directed vertex-disjoint paths problem has been shown to be NP-hard for fixed $k = 2$ by Fortune et al. [8], so this reduction will prove our theorem for $|D| = 2$. Note that the case of $|D| > 2$ is harder, since we can view $|D| = 2$ as a special case, by adding isolated vertices with demand 0.

Given a graph G and pairs (s_1, t_1) and (s_2, t_2), we construct an instance $(G', D, \mathsf{dem}, \mathsf{cost}, \mathsf{cap})$ of CONNECTED FLOW. Let $V_0 = V \backslash \{s_1, s_2, t_1, t_2\}$, we define

$$V(G') = \{s_1, s_2, t_1, t_2\} \cup \{v_{\mathsf{in}} : v \in V_0\} \cup \{v_{\mathsf{out}} : v \in V_0\}$$

We let $D = \{s_1, s_2\}$ and set $\mathsf{dem}(s_1) = \mathsf{dem}(s_2) = 1$. We also define

$$\begin{aligned} E(G') = &\{(v_{\mathsf{in}}, v_{\mathsf{out}}) : v \in V_0\} \\ &\cup \{(s_i, v_{\mathsf{in}}) : (s_i, v) \in E(G), i = 1, 2\} \\ &\cup \{(v_{\mathsf{out}}, t_i) : (v, t_i) \in E(G), i = 1, 2\} \\ &\cup \{(u_{\mathsf{out}}, v_{\mathsf{in}}) : u, v \in V_0, (u, v) \in E(G)\} \\ &\cup \{(t_1, s_2), (t_2, s_1)\}. \end{aligned}$$

We now set $\mathsf{cost}(u, v) = 0$ and $\mathsf{cap}(u, v) = 1$ for every $(u, v) \in E(G')$. We prove that G has two vertex-disjoint paths (from s_1 to t_1 and from s_2 to t_2) if and only if $(G', D, \mathsf{dem}, \mathsf{cost}, \mathsf{cap})$ has a connected flow of cost 0.

Let P_1 and P_2 be two vertex disjoint paths in G, from s_1 to t_1 and from s_2 to t_2 respectively. Intuitively we will simply walk through the same two paths in G' and then connect the end of one to the start of the other. More formally, we construct a flow f in G' as follows. Let $P_1 = s_1, v^1, \ldots, v^l, t_1$, we set $f(s_1, v^1_{\mathsf{in}}) = f(v^l_{\mathsf{out}}, t_1) = 1$ as well as $f(v^i_{\mathsf{in}}, v^i_{\mathsf{out}}) = f(v^i_{\mathsf{out}}, v^{i+1}_{\mathsf{in}}) = 1$ for all $i \in [1, l]$. We do the same for P_2. Finally we set $f(t_1, s_2) = f(t_2, s_1) = 1$ and set f to 0 for all other edges. We note that all capacities have been respected and all demands have been met. The resulting flow is connected, since the paths were connected and $f(t_1, s_2) = 1$.

For the other direction, let f be a connected flow for $(G', D, \mathsf{dem}, \mathsf{cost}, \mathsf{cap})$. Since $\mathsf{dem}(s_1) = \mathsf{dem}(s_2) = 1$ and s_1 and s_2 only have one incoming edge, we have that $f(t_1, s_2) = f(t_2, s_1) = 1$. We argue that $G_f - \{(t_1, s_2), (t_2, s_1)\}$ consists of two vertex disjoint paths in G', one from s_1 to t_1 and the other from s_2 to t_2. First we note that for every vertex in G', it has either in-degree 1 or out-degree 1 (or possibly both). This means that since we have $\mathsf{cap}(u, v) = 1$ for every $(u, v) \in E(G')$, every vertex in $V(G_f)$ has in- and out-degree 1 in G_f. Since G_f is connected we find that G_f is a single cycle and thus $G_f - \{(t_1, s_2), (t_2, s_1)\}$ is the union of two vertex-disjoint paths. We now find two vertex-disjoint paths in G by contracting the edges $(v_{\mathsf{in}}, v_{\mathsf{out}})$ in $G_f - \{(t_1, s_2), (t_2, s_1)\}$. □

Lemma 2. *Given an instance* $(G, D, \mathsf{dem}, \mathsf{cost}, \mathsf{cap})$ *of* CONNECTED FLOW *where* $\mathsf{cap}(e) = \infty$ *for all* $e \in E$*, we can construct an equivalent instance of* MVTSP *on* $|D|$ *vertices.*

Proof. We construct an equivalent instance $(G', \mathsf{dem}, \mathsf{cost}')$ of MVTSP as follows. First we let $V(G') = D$ and for $u, v \in D$ we let $(u, v) \in E(G')$ if and only if there is a $u - v$ path in G, disjoint from other vertices in D. We then set $\mathsf{cost}(u, v)$ to be the total cost of the shortest such path. We keep $\mathsf{dem}(v)$ the same.

We now show equivalence of the two instances. Let $s' : E(G') \to \mathbb{N}$ be a valid tour on $(G', \mathsf{dem}, \mathsf{cost}')$. We construct a connected flow f on $(G, D, \mathsf{dem}, \mathsf{cost}, \mathsf{cap})$ by, for each $(u, v) \in E(G')$ adding $s'(u, v)$ copies of the shortest D-disjoint u-v-path in G to the flow. Note that the demands are met, since the demands in both instances are the same. Also note that by definition the total cost of $s'(u, v)$ copies of the shortest D-disjoint $u - v$ path is equal to $s'(u, v) \cdot \mathsf{cost}'(u, v)$ and thus the total cost of f is equal to that of s'. Finally we note that the capacities are trivially met.

In the other direction, let $f : E(G) \to \mathbb{N}$ be an optimal connected flow on $(G, D, \mathsf{dem}, \mathsf{cost}, \mathsf{cap})$. Note that G_f is connected and that every vertex in this multigraph has equal in- and out-degrees. This means we can find some Eulerian tour on G_f. We now construct an MVTSP tour s' on G' by adding the edge (u, v) every time v is the first vertex with demand to appear after an appearance of u in the Eulerian tour. Again it is easy to see that s' is connected and that the demands are met. The total cost of s' is the same as f, namely if it is larger, then there is some pair $u, v \in D$ such that the cost of some path in the Eulerian tour from u to v is less than $\mathsf{cost}'(u, v)$, which contradicts the definition of cost'. If it were smaller, then there is some D-disjoint path in the Eulerian tour from some u to some v which is longer than $\mathsf{cost}'(u, v)$. We can then find a cheaper flow by replacing this path with the shortest path, contradicting the optimality of f. □

Since MVTSP can be solved in $\mathcal{O}^\star(4^n)$ time by Kowalik et al. [10], we get as a direct consequence:

Theorem 7. *Any instance instance* $(G, D, \mathsf{dem}, \mathsf{cost}, \mathsf{cap})$ *of* Connected Flow *where* $\mathsf{cap}(e) = \infty$ *for all* $e \in E(G)$ *can be solved in time* $\mathcal{O}^\star(4^{|D|})$.

C Parameterisation by Treewidth

In this section we consider the complexity of Connected Flow, when parameterized by the treewidth tw of G. We first give a $|V(G)|^{\mathcal{O}(tw)}$ time dynamic programming algorithm for Connected Flow. Subsequently, we give a matching conditional lower bound on the complexity of MVTSP parameterized by the pathwidth of G. Since MVTSP is a special case of Connected Flow this shows that our dynamic programming algorithm is in some sense optimal.

C.1 An XP Algorithm for Connected Flow

In this subsection we show the following:

Theorem 4. *Let M be an upper bound on the demands in the input graph G, and suppose a tree decomposition of width tw of G is given. Then a* Connected Flow *instance with G can be solved in time $|V(G)|^{\mathcal{O}(tw)}$ and an* MVTSP *instance with G can be solved in time $\min\{|V(G)|, M\}^{\mathcal{O}(tw)}|V(G)|^{\mathcal{O}(1)}$.*

Proof. The algorithm is based on a standard dynamic programming approach; we only describe the table entries and omit the recurrence to compute table entries since it is standard. We assume we have a tree decomposition $\mathcal{T} = (\{X_i\}, R)$ on the given graph. For a given bag X_i, let π be a partition on X_i. Furthermore let $\mathbf{d}^{\mathsf{in}} = (d_v^{\mathsf{in}})_{v \in X_i} \in \mathbb{N}^{X_i}$ and $\mathbf{d}^{\mathsf{out}} = (d_v^{\mathsf{out}})_{v \in X_i} \in \mathbb{N}^{X_i}$ be two vectors of integers, indexed by X_i. We define the dynamic programming table entry $T(X_i, \pi, \mathbf{d}^{\mathsf{in}}, \mathbf{d}^{\mathsf{out}})$ to be the cost of the cheapest partial solution on the graph 'below' the bag X_i, among solutions whose connected components agree with the partition π and whose in and out degrees agree with the vectors $\mathbf{d}^{\mathsf{in}}$ and $\mathbf{d}^{\mathsf{out}}$. More formally, for $r \in V(R)$ the root of the tree decomposition, we consider a bag X_j to be below another bag X_i if one can reach j from i by a directed path in the directed tree obtained from R by orienting every edge away from r. We will denote this as $X_j \preccurlyeq X_i$ and define $Y_i = \cup_{X_j \preccurlyeq X_i} X_j$. For each bag X_i, a partition π of X_i and sequences $\mathbf{d}^{\mathsf{in}}$ and $\mathbf{d}^{\mathsf{out}}$ satisfying (i) $0 \leqslant d_v^{\mathsf{in}}, d_v^{\mathsf{out}} \leqslant \mathsf{dem}(v)$ for each $v \in D$ and (ii) $0 \leqslant d^{\mathsf{in}}(v), d^{\mathsf{out}}(v) \leqslant M|V(G)|$ for each $v \notin D$, define $T(X_i, \pi, \mathbf{d}^{\mathsf{in}}, \mathbf{d}^{\mathsf{out}}) = \min_s \mathsf{cost}(s)$ over all $s : Y_i^2 \to \mathbb{N}$ such that the following conditions hold:

1. $\sum_{u \in Y_i} s(u,v) = \sum_{u \in Y_i} s(v,u) = \mathsf{dem}(v)$ for all $v \in D \cap (Y_i \backslash X_i)$
2. $\sum_{u \in Y_i} s(u,v) = d_v^{\mathsf{in}}$ for all $v \in X_i$
3. $\sum_{u \in Y_i} s(v,u) = d_v^{\mathsf{out}}$ for all $v \in X_i$
4. All blocks of the partition π are weakly connected in G_s.
5. $s(u,v) \leqslant \mathsf{cap}(u,v)$ for all $(u,v) \in E(G[Y_i])$

We can compute the table starting at the leaves of R and work our way towards the root.

Let us examine the necessary size of this dynamic programming table. First we note that there are at most $|V(G)|^{\mathcal{O}(1)}$ bags in the tree decomposition. Next we consider the values d_v^{in} and d_v^{out}. Note that we can assume that an optimal solution only visits any vertex without demand at most $M|V(G)|$ times: Any solution can be decomposed into a collection of paths between vertices with demand. Each such path can be assumed to not visit any vertex more than once (except possibly in the end points of the path) since the solution is of minimum weight and all costs are non-negative. We find that each vertex gets visited at most $M|V(G)|$ times and thus we only need to consider $M|V(G)|$ many values of d_v^{in} and d_v^{out}. Thus the degree values of the partial solutions contribute a factor of $(M|V(G)|)^{\mathcal{O}(tw)}$ to the overall running time of the algorithm if the given instance is a Connected Flow instance, and only $M^{\mathcal{O}(tw)}$ if the given instance is an MVTSP instance (in which all vertices are demand vertices).

We argue that we may assume that $M = |V(G)|^{\mathcal{O}(1)}$. Together with the fact that the number of possibilities for π is $tw^{\mathcal{O}(tw)} \leqslant |V(G)|^{\mathcal{O}(tw)}$, the claimed result for CONNECTED FLOW follows. We support this assumption using a variation on the proof of Theorem 3.4 from Kowalik et al. [10]. Let r be some optimal solution to FLOW, then by applying Lemma 1 with T being some subtree of G_s spanning all demand vertices, we find that there is some optimal solution s of CONNECTED FLOW such that $|r(u,v) - s(u,v)| \leqslant 2n$ at every edge (u,v).

We now construct a flow f from r by subtracting simple directed cycles from r. Note that each time that we subtract such a cycle, the result is again a flow. We start with $f = r$ everywhere. Now if there is an edge $(u,v) \in E$ for which $f(u,v) > \max\{r(u,v) - 2n - 1, 0\}$, we can find a simple directed cycle $C \in G_f$, containing (u,v), as f is a flow and thus G_f is Eulerian. Then define $f'(u,v) = f(u,v) - [(u,v) \in C]$. Note that f' is again a flow. Set $f = f'$. We then repeat this process of subtracting simple directed cycles from f until $f(u,v) \leqslant \max\{r(u,v) - 2n - 1, 0\}$ for every edge (u,v).

Note that $0 \leqslant s(u,v) - f(u,v)$ for all $(u,v) \in E$. Then define the instance with $\mathsf{dem}'(v) = \mathsf{dem}(v) - \sum_{u \in V} f(u,v)$ and $\mathsf{cap}'(u,v) = \mathsf{cap}(u,v) - f(u,v)$ for which $s(u,v) - f(u,v)$ is an optimal connected flow. If $\mathsf{dem}'(v) \leqslant 2n^2 + n$ we are done. Otherwise let r' be a relaxed solution for the new instance. Note that there is some edge (u',v') for which $r'(u',v') > 2n + 1$ and thus we can repeat the previous argument to find a non-zero flow f' such that $f'(u,v) \leqslant \max\{r'(u,v) - 2n - 1, 0\}$ on every edge and define a corresponding new instance. Since each time we subtract a non-zero flow, after some number of repetitions we find $\mathsf{dem}'(v) \leqslant 2n^2 + n$.

For the result for MVTSP, the above approach would give a running time of $\min\{|V(G)|, M\}^{\mathcal{O}(tw)} tw^{\mathcal{O}(tw)} |V(G)|^{O(1)}$. However, the factor $tw^{\mathcal{O}(tw)}$ in the running time needed to keep track of all partitions π can be reduced to $2^{\mathcal{O}(tw)}$ via a standard application of the rank based approach (see e.g. [5, Section 11.2.2] or [2,6]). □

C.2 Lower Bound

We now present a modified version of a reduction from 3-CNF-SAT to HAMILTONIAN CYCLE parameterized by pathwidth from Cygan et al. [6]. We modify it to be a reduction to MVTSP instead.

We will produce an instance of MVTSP that is symmetric in the sense that the graph G is undirected, hence we denote edges as unordered pairs of vertices (i.e. $\{u,v\} = \{v,u\}$). As a consequence, when c is a tour on G, then we say $c(u,v) = c(v,u)$. The general proof strategy is as follows. For a given 3-CNF-SAT formula ϕ on n variables[3] we will construct an equivalent MVTSP instance (G,d). This graph will consist of n/s paths, for some value s, with each path propagating some information encoding the value of s variables of ϕ. For each clause of ϕ we will add a gadget which checks if the assignment satisfies the clause. We then bound the size and the pathwidth of the constructed graph G. This allows us to conclude a lower bound based on this reduction.

[3] In this section, we will only use n to refer to the number of variables of a 3-CNF-SAT instance.

Gadgets. We start by borrowing the following gadget from Cygan et al. [6], called a 2-label gadget (Fig. 2).

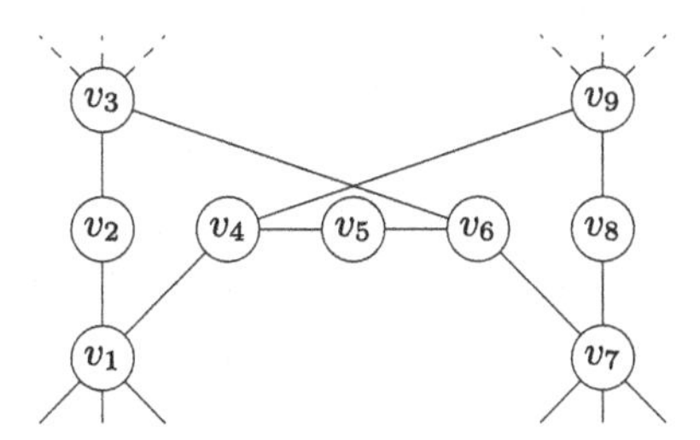

Fig. 2. A 2-label gadget.

The key feature of this gadget is that if all vertices in the gadget have demand equal to 1, then if a solution tour enters the gadget at v_3, it has to leave the gadget at v_9 and vice versa. A similar relation holds for v_1 and v_7. We will refer to any edge connected to either v_1 or v_7 as having label 1 and any edge connected to v_3 or v_9 as having label 2. We will use this gadget to construct a gadget that can detect certain multisets of edges in a part of a graph. In this construction we will chain 2-label gadgets together using label 1 edges. Whenever we do this, we always connect the vertex v_7 of one gadget to the vertex v_1 in the next. To keep things concise, in the rest of this section we will refer to any 2-label gadget as if it were a single vertex.

This next gadget is also inspired by a construction from Cygan et al. [6].

Definition 2. *A scanner gadget in an unweighted* MVTSP *instance* (G, d) *is described by a tuple* $(X, a, b, \mathcal{F})$, *where* $X \subseteq V$, $a, b \in V \backslash X$ *with* $\mathsf{dem}(a) = \mathsf{dem}(b) = 1$, $\mathcal{F}$ *is a family of multisets of edges in*[4] $E(X, X)$ *and* $\emptyset \notin \mathcal{F}$. *A tour* c *of* G *is* consistent with $(X, a, b, \mathcal{F})$ *if its restriction* $c_{E(X,X)}$ *is in* $\mathcal{F}$ *and if* $c(a, b) > 0$.

When refering to the gadget as a subgraph, we will use $G_{\mathcal{F}}$. We implement the scanner gadget using the following construction, obtaining a different instance (G', dem') of MVTSP.

- Remove the edges in $E(X, X)$.
- Add an independent set $I = \{s_1, \ldots, s_\ell\}$ and edges $\{a, s_1\}$ and $\{s_\ell, b\}$, for $\ell = |\mathcal{F}|$.
- Let $\mathcal{F} = \{F_1, \ldots, F_\ell\}$. For $i = 1, \ldots, \ell$ we do the following.
  - Let $F_i = \{e_1^{q_1}, \ldots, e_z^{q_z}\}$, that is F_i contains q_i copies of e_i.
  - Add a path $P_i = \{p_i^1, \ldots, p_i^{t_i}\}$ of 2-label gadgets, where $t_i = |F_i| = \sum_{i=j}^{z} q_j$. We connect the gadgets in a chain using label 1 edges.

[4] Here $E(X, X)$ are all edges with both endpoints in X and the restriction c_Y are all edges in c in Y (keeping multiplicities).

  - Connect p_i^1 to s_{i-1} and s_i using label 1 edges (green edges in Fig. 3) and connect $p_i^{|F_i|}$ to s_i and s_{i+1} using label 1 edges (blue edges in Fig. 3).
  - For all $j = 1, \ldots, z$ add label 2 edges from x $p_i^{j'}$ and from y to $p_i^{j'}$ for $e_j = \{x, y\}$ and for q_j different, previously unused values of j' (red edges in Fig. 3).
- We set the demand of all added vertices to 1.

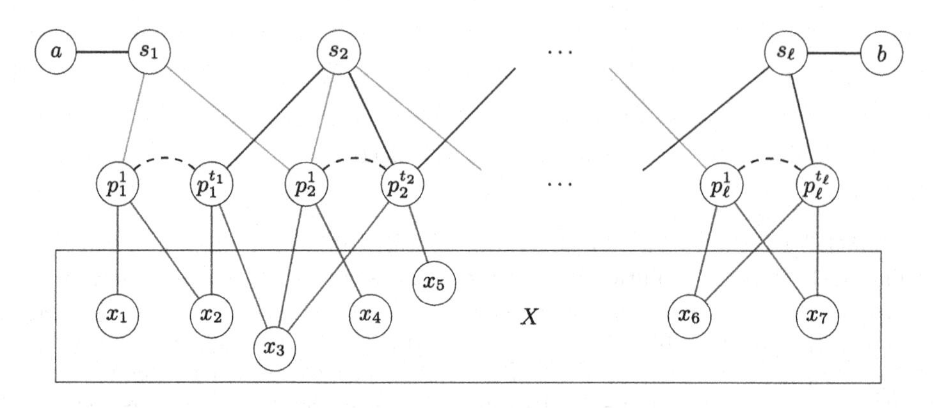

Fig. 3. Example of the scanner gadget.

The function of the gadget is captured by the following lemma.

Lemma 3. *There exists an tour on* (G, dem) *that is consistent with* $(X, a, b, \mathcal{F})$ *if and only if there exists a tour on* (G', dem').

The proof will closely follow that in Cygan et al. [6].

Proof. Suppose we have a tour on (G, dem) which is consistent with a gadget $(X, a, b, \mathcal{F})$. Let $F_i \in \mathcal{F}$ be the restriction of the tour on $E(X, X)$. Then the tour on (G, dem) can be extended to a tour on (G', dem') by replacing the q_j instances of an edge $\{u, v\} \in F_i$ with two edges $\{u, p_i^{j'}\}$ and $\{v, p_i^{j'}\}$ for q_j different values of j'. We also replace the edge $\{a, b\}$ by the path

$$a, s_1, P_1, \ldots, P_{i-1}, s_i, P_{i+1}, s_{i+1}, \ldots, P_\ell, s_\ell, b.$$

Since the obtained tour visits all vertices in the gadget exactly once and since the restriction of the adjusted tour connects the same pairs of vertices in X as the restriction of the original tour, the obtained tour will be a solution for the instance (G', dem').

For the other direction, suppose we have a tour c' on (G', dem'). Note that by the nature of the 2-label gadgets any tour cannot cross from some s_i into X through one of the 2-label gadgets in one of the paths P_i. Thus the tour can only travel from outside the gadget to s_i, by going through a or b. Therefore the tour must include the edges $\{a, s_1\}$ and $\{s_\ell, b\}$. Furthermore s_1 and s_ℓ must be connected by some path P' in the tour. Because I is an independent set, P' has to jump back and forth between the P_i's and the s_i's and has to include every s_i, since this is the only way to reach a vertex s_i with a tour.

This means that there will be exactly one path P_{i_0} which is not covered by P'. We can now obtain a tour c of (G, dem) by first setting $c(u, v) = c'(u, v)$ for $\{u, v\} \neq \{a, b\}$ for u or v not in X. We then include any edge in X a number of times according to its multiplicity in F_{i_0} i.e. we set $c(u, v) = F_{i_0}(u, v)$. Finally we set $c(a, b) = 1$. Note that since $c(a, b) > 0$ and $F_{i_0} \in \mathcal{F}$, we find that c is consistent with $(X, a, b, \mathcal{F})$. □

The following lemma will allow us to implement the gadget without increasing the pathwidth of the graph too much.

Lemma 4. *The scanner gadget has pathwidth at most* $|X| + 21$.

Proof. We define the bags of the decomposition as follows

$$\begin{aligned} B_a &:= X \cup \{a, s_1\} \\ B_{i,j} &:= X \cup \{s_{i-1}, s_i, s_{i+1}, p_i^j, p_i^{j+1}\} \\ B_b &:= X \cup \{b, s_\ell\}. \end{aligned}$$

We now have the following path decomposition of $G_{\mathcal{F}}$

$$B_a, B_{1,1}, B_{1,2}, \dots B_{1,t_1}, B_{2,1}, \dots B_{\ell,t_\ell}, B_b.$$

It is easy to check that every vertex/edge is covered by some bag and that for every vertex v the set of bags containing v form an interval in the decomposition. □

Construction. Suppose we are given a 3-CNF-SAT formula $\phi = C_1 \wedge \dots \wedge C_m$. We will construct an equivalent unweighted MVTSP instance Γ_ϕ using scanner gadgets. We will interpret a tuple $(q, j) \in \{1, \dots, 2^s\} \times \{1, \dots, n/s\}$ as an assignment of $x_{(j-1)s+1}, \dots, x_{js}$ by first decomposing

$$q - 1 = \sum_{i=1}^{s} c_i 2^{i-1}$$

and setting $x_{(j-1)s+i}$ as true if $c_i = 1$ and false if $c_i = 0$. We say a clause C is satisfied by a set Q of such tuples, if $j \neq j'$ for all $(q, j), (q, j') \in Q$, and if the partial assignment collectively given by the tuples satisfies C (Fig. 4).

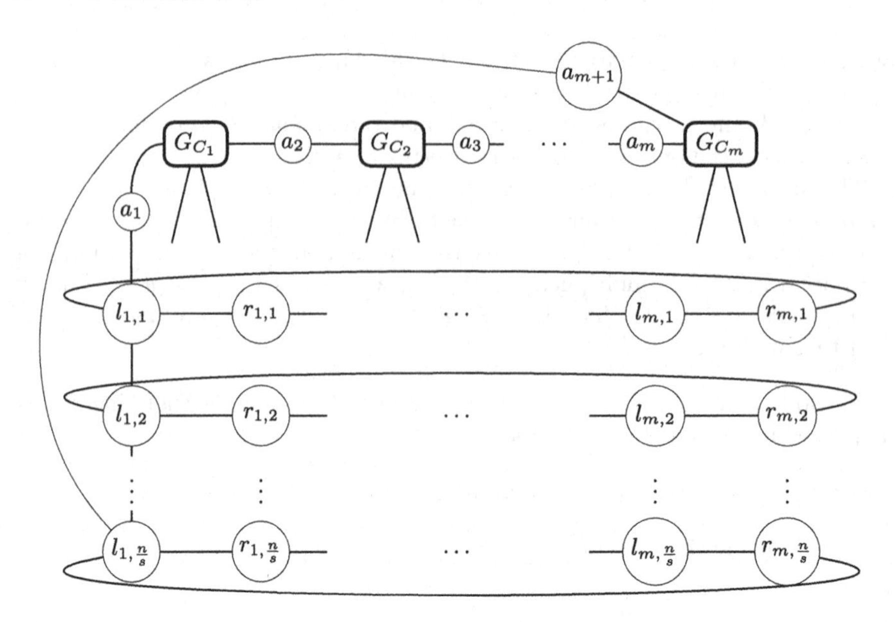

Fig. 4. Construction of the graph Γ_ϕ.

- We start by creating vertices $l_{i,1}, \ldots, l_{i,n/s}$ and $r_{i,1}, \ldots, r_{i,n/s}$ for $i = 1, \ldots, m$ and some constant s to be determined later[5].
- We set the demand of $l_{1,j}$ to $2^s + 1$ for $j = 1, \ldots, n/s$ and add edges $\{l_{1,j}, l_{1,j+1}\}$ for $j = 1, \ldots, n/s - 1$.
- We set the demand of every other $l_{i,j}$ and every $r_{i,j}$ to 2^s and add edges $\{l_{i,j}, r_{i,j}\}$, $\{r_{i,j}, l_{i+1,j}\}$ and $\{r_{m,j}, l_{1,j}\}$ for $i = 1, \ldots, m-1$ and $j = 1, \ldots, n/s$.
- We connect $l_{1,1}$ to $l_{1,n/s}$ using a path $a_1, \ldots, a_{m+1}$.
- For $i = 1, \ldots, m$ let x_a, x_b, x_c be the variables appearing in C_i. We set $j_1 = \lceil a/s \rceil, j_2 = \lceil b/s \rceil, j_3 = \lceil c/s \rceil$. Let

  $$X = \{l_{i,j_1}, l_{i,j_2}, l_{i,j_3}, r_{i,j_1}, r_{i,j_2}, r_{i,j_3}\}$$

  and let $\mathcal{F}_{C_i}$ be the set of all

  $$F = \{\{l_{i,j_1}, r_{i,j_1}\}^{q_1}, \{l_{i,j_2}, r_{i,j_2}\}^{q_2}, \{l_{i,j_3}, r_{i,j_3}\}^{q_3}, \}$$

  such that $Q = \{(q_1, j_1), (q_2, j_2), (q_3, j_3)\}$ satisfies C_i.
- For $i = 1, \ldots, m$ we implement a scanner gadget G_{C_i} using the tuple $(X_i, a_i, a_{i+1}, \mathcal{F}_{C_i})$

We prove the following useful facts about this graph.

Lemma 5. *Γ_ϕ is a yes instance of* MVTSP *if and only if ϕ has a satisfying assignment.*

[5] If n is not divisible by s, we may either add dummy variables until it is, or lower the demand of $l_{i,n/s}$ and $r_{i,n/s}$.

Proof. Let $x_1, \ldots, x_n$ be the variables used in the formula ϕ. Let $\chi_1, \ldots, \chi_n$ be some satisfying assignment. We first define the tour on the construction before implementing the scanner gadgets, which we will refer to as Γ'_ϕ, and then use Lemma 3 to find the desired tour on Γ_ϕ. Set $c(l_{i,j}, l_{i+1,j}) = c(l_{1,1}, a_1) = c(a_{m+1}, l_{1,n/s}) = c(a_i, a_{i+1}) = 1$. We choose

$$c'(l_{i,j}, r_{i,j}) = 1 + \sum_{k=1}^{s} 2^{k-1} \chi_{(j-1)s+k}$$

for $i = 1, \ldots, m$ and $j = 1, \ldots, n/s$. Due to the chosen demands we need to choose

$$c'(r_{i,j}, l_{i+1,j}) = 2^{s+1} - c'(l_{i,j}, r_{i,j})$$

for $i = 1, \ldots, m$ and $j = 1, \ldots, n/s$, where we interpret i modulo m, i.e. $m+1 \equiv 1$. Note that c' is connected and satisfies the demands on Γ'_ϕ. Also note that since χ is a satisfying assignment, c' is consistent with all the scanner gadgets G_{C_i} and thus by Lemma 3 we there is some valid tour c on Γ_ϕ.

Now suppose we find a valid tour c on Γ_ϕ. Then by Lemma 3 there exists a tour c' on Γ'_ϕ consistent with each gadget G_{C_i}. By definition of G_{C_i} the values of $c'(l_{i,j}, r_{i,j})$ encode an assignment satisfying C_i for $i = 1, \ldots, m$. Since for $i \geqslant 2$ the demands of $l_{i,j}$ and $r_{i,j}$ equal 2^s we have that $c'(l_{i,j}, r_{i,j}) = 2^{s+1} - c'(r_{i,j}, l_{i+1,j}) = c'(l_{i+1,j}, r_{i+1,j})$ and therefore the values of $c'(l_{1,j}, r_{1,j})$ encode an assignment satisfying all clauses $C_1, \ldots, C_m$, which means we find an assignment which satisfies ϕ. □

Lemma 6. *Γ_ϕ has pathwidth at most $3n/s + 21$.*

Proof. We define the bags of the decomposition as follows. First we add

$$A = \{l_{1,1}, \ldots, l_{1,n/s}\}$$

to every bag. Let $W_1, \ldots, W_{l_i}$ be a path decomposition of G_{C_i}. We define bag $X_{i,j}$ as follows

$$X_{i,j} = A \cup \{l_{i,k}\}_{k=1}^{n/s} \cup \{r_{i,k}\}_{k=1}^{n/s} \cup W_j.$$

We then define Y_i as $\{l_{i+1,k}\}_{k=1}^{n/s} \cup \{r_{i,k}\}_{k=1}^{n/s}$ The final path decomposition then becomes

$$X_{1,1}, \ldots, X_{1,l_1}, Y_1, X_{2,1}, \ldots, X_{i,l_i}, Y_i, X_{i+1,1}, \ldots, X_{m,l_m}.$$

Note that all vertices and edges are covered by the decomposition. The set of bags containing any of the vertices of A gives the whole decomposition. The set of bags containing any $l_{i,j}$ or $r_{i,j}$ for $i \geqslant 2$ gives the path $X_{i,1} \ldots X_{i,l_1}$ with Y_i at the end for $r_{i,j}$ and Y_{i-1} at the beginning for $l_{i,j}$. Any vertex in the gadgets gives a single set $X_{i,j}$. By Lemma 4 the width of this path decomposition is at most[6]

$$3\frac{n}{s} + 21.$$

□

[6] We don't include the term $|X|$, since $X \subseteq \{l_{i,k}\}_{k=1}^{n/s} \cup \{r_{i,k}\}_{k=1}^{n/s}$.

Now we use our reduction to prove the following lower bound:

Theorem 5. *Let M be an upper bound on the demands in a graph G. Then* MVTSP *cannot be solved in time $f(pw)\min\{|V(G)|, M\}^{o(pw)}|V(G)|^{\mathcal{O}(1)}$, unless ETH fails.*

Proof. We start by proving the following claim.

Claim 6 $|V(G_{C_i})| = \mathcal{O}(2^{3s})$ *for* $i = 1, \ldots, m$.

Proof of Claim. Note that $\mathcal{F}_{C_i}$ is defined on at most three unique edges with each edge being chosen at most 2^s times[7]. Therefore we can represent $\mathcal{F}_{C_i}$ by tuples $(z_1, z_2, z_3) \in [2^s]^3$. Since each tuple contributes a path of $z_1 + z_2 + z_3$ vertices, we find that

$$\begin{aligned}|V(G_{C_i})| &= 8 + |\mathcal{F}_{C_i}| + \sum_{(z_1,z_2,z_3)\in\mathcal{F}_{C_i}} z_1 + z_2 + z_3 \\ &\leqslant 2^{3s+1} + \sum_{z_1,z_2=1}^{2^s}\left(2^s(z_1+z_2) + \sum_{z_3=1}^{2^s} z_3\right) \\ &\leqslant 2^{3s+1} + \sum_{z_1,z_2=1}^{2^s}\left(2^s(z_1+z_2) + 2^{s+1}\right) \\ &\leqslant 2^{3s+1} + \sum_{z_1=1}^{2^s}\left(2^{2s}z_1 + 2^{2s+1} + 2^{2s+1}\right) \\ &\leqslant 2^{3s+3}.\end{aligned}$$

■

Note that by Lemma 5, solving a 3-CNF-SAT instance ϕ reduces to solving MVTSP on Γ_ϕ for some choice of s. We remark that

$$\mathcal{O}(f(pw)\min\{|V(G)|, M\}^{o(pw)}|V(G)|^{\mathcal{O}(1)}) \leqslant \mathcal{O}(f(pw)M^{o(pw)}|V(G)|^{\mathcal{O}(1)}).$$

It is therefore sufficient to show that there is no $\mathcal{O}(f(pw)M^{o(pw)}|V(G)|^{\mathcal{O}(1)})$ time algorithm for MVTSP, unless ETH fails.

Suppose we have a $\mathcal{O}\left(f(pw)M^{o(pw)}|V(G)|^{\mathcal{O}(1)}\right)$ time algorithm for MVTSP. Let $s = 4n/g(n)$ for some strictly increasing function $g(n) = 2^{o(n)}$ such that $f(g(n)) = 2^{o(n)}$. Note that $s = o(n)$ and $pw \leqslant g(n)$ for large enough n. We construct the instance Γ_ϕ as previously described. We first note that by Claim 6

$$|V(\Gamma_\phi)| = 2m\frac{n}{s} + \sum_{i=1}^{m}|V(G_{C_i})| = \mathcal{O}\left(m\left(\frac{n}{s} + 2^{3s}\right)\right)$$

[7] Due to the way we interpret the multiplicities as truth assignments (in particular the '−1') we know each edge gets chosen at least once.

and by Lemma 6 we have that for any choice of s and large enough n, Γ_ϕ has pathwidth at most $4n/s$. By applying our hypothetical algorithm for MVTSP to Γ_ϕ we now find an algorithm for 3-CNF-SAT running in time

$$\begin{aligned}\mathcal{O}\left(f(pw)M^{o(pw)}|V(\Gamma_\phi)|^{\mathcal{O}(1)}\right) &= \mathcal{O}\left(f(4n/s)(2^s)^{o(n/s)}\left(m\left(\frac{n}{s}+2^{3s}\right)\right)^{\mathcal{O}(1)}\right)\\ &= \mathcal{O}\left(f(g(n))\cdot 2^{o(n)}\cdot\left(m\left(g(n)/4+2^{o(n)}\right)\right)^{\mathcal{O}(1)}\right).\end{aligned}$$

We may assume that $m = 2^{o(n)}$ by the sparsification lemma. Using this and the fact that $g(n) = 2^{o(n)}$ we find

$$\begin{aligned}&= \mathcal{O}\left(2^{o(n)}\cdot\left(2^{o(n)}\right)^{\mathcal{O}(1)}\right)\\ &= \mathcal{O}\left(2^{o(n)}\right).\end{aligned}$$

This contradicts ETH, completing our proof. □

D Proofs Deferred to Appendix

D.1 Proofs of Claims from Subsection 2.1

Claim 1. *For any $v \in V$ and any inclusion-wise minimal $C \in \mathcal{C}$:*

$$\sum_{u\in V}[(u,v)\in C] \leqslant 2 \qquad \text{and} \qquad \sum_{u\in V}[(v,u)\in C] \leqslant 2. \tag{1}$$

Proof. We only prove the first inequality. The second inequality can be proved with an analogous argumentation. Assume not, i.e. assume there exists $C \in \mathcal{C}$ and $v \in V$ such that there exist $x_1, x_2, x_3 \in V$ with $(x_i, v) \in C$ for $i = 1, 2, 3$. Each of these edges must be either in A_s or A_r. Assume without loss of generality that $(x_1, v), (x_2, v) \in A_s$. (We will only need the fact that at least two of these edges are either both in A_r or both in A_s. The case of at least two edges in A_r has equivalent reasoning.) Both (x_1, v) and (x_2, v) can be paired with the edge it traverses v with, i.e. its subsequent edge in the tour, as $(x_1, v), (x_2, v) \in A_s$ are positively oriented. Let e_1, e_2 be these subsequent edges. Then note that C can be split into two smaller $(s-r)$ directed tours C_1 and C_2, with C_1 starting with edge e_1 and ending with (x_2, v), and C_2 starting with edge e_2 and ending with (x_1, v). This contradicts the assumption that C was inclusion-wise minimal. □

Claim 2. *For all $e \in T$, $f(e) > 0$ and f is a flow for the given instance.*

Proof. We first show that for all $e \in E$:

$$\min\{r(e), s(e)\} \leqslant f(e) \leqslant \max\{r(e), s(e)\}.$$

If $e \notin A$, Eq. (2) implies that $r(e) = f(e) = s(e)$. If $e \in A_s$, by definition $r(e) < s(e)$ and we can see from (2) that if the multiplicity of e changes, it is because copies of e are added to r to form f (and none are removed). Because $m_A(e) \leqslant s(e) - r(e)$, at most this many copies of e can be added to r to form f. Hence $r(e) \leqslant f(e) \leqslant r(e) + (s(e) - r(e)) = s(e)$. Similarly, if $e \in A_r$, $r(e) > s(e)$ and at most $r(e) - s(e)$ copies of e are removed from r to form f.

Next we prove that f is an allowed solution to the given FLOW instance. Let $C \in \mathcal{C}^+$ and let e, e' be two subsequent edges from C with common vertex v. If $e \in A_s$ and $e' \in A_r$, then the in- and out-degrees of v do not change while adding a copy of e and removing a copy of e' (in Eq. (2)), because the orientation of e and e' is different. This is also true if $e \in A_r$ and $e' \in A_s$. If $e, e' \in A_r$, then both e and e' have a copy removed in Eq. (2). Since the orientation of e and e' is the same, both the in- and out-degree of v go down by one. We remark that this situation only happens if $v \notin D$ by definition of $(s-r)$ directed tours. Similarly, if $e, e' \in A_s$, the in- and out-degree of v increases by one. Since r was an allowed solution, this implies that the number of incoming- and outgoing edges of v in f are equal, in other words, the flow is preserved. Since for $v \in D$, the total incoming (and total outgoing) edges do not change, the demands are satisfied by f. Furthermore, the capacity constraints are satisfied since $f(e) \leqslant \max\{r(e), s(e)\} \leqslant \mathsf{cap}(e)$.

We show that $T \subseteq \text{supp}(f)$. Let $e \in T^+$, then $e \in A_s$ so copies of e are added to r to form f in Eq. (2). Since $e \in T^+$, at least one tour $C \in \mathcal{C}^+$ contains e. Hence, $f(e) > 0$. If $e \in T \backslash T^+$, then $r(e) > 0$ because $T^+ = T \backslash \text{supp}(r)$ by definition. We also see that $s(e) > 0$ because $T \subseteq \text{supp}(s)$ by assumption. Using our earlier result that $\min\{r(e), s(e)\} \leqslant f(e)$, we conclude that $f(e) > 0$. □

D.2 Proof of Claim 3 from Subsection 2.2

Claim 3. *Each table entry $T_j(\pi, \mathbf{c}^{\mathsf{in}}, \mathbf{c}^{\mathsf{out}})$ can be computed from all table entries T_{j-1}.*

Proof. Compute table entries for $j = 0$ as follows. Set $T_0(\{\{x_1,\}, \ldots, \{x_{k'}\}\}, \mathbf{0}, \mathbf{0})$ to 0, and all other entries of T_0 to ∞, as $V_0 = X'$ is an independent set and so a flow of zero on every edge is the only possible flow.

Now assume $j > 0$. We compute the values of $T_j(\pi, \mathbf{c}^{\mathsf{in}}, \mathbf{c}^{\mathsf{out}})$ as the minimum of the following value over all suitable $\mathbf{h}_i^{\mathsf{in}} = (h_1^{\mathsf{in}}, \ldots, h_{k'}^{\mathsf{in}}) \in \mathbb{N}^{k'}$, $\mathbf{h}_i^{\mathsf{out}} = (h_1^{\mathsf{out}}, \ldots, h_{k'}^{\mathsf{out}}) \in \mathbb{N}^{k'}$, and all suitable partitions π' of X':

$$T_{j-1}(\pi', \mathbf{c}^{\mathsf{in}} - \mathbf{h}^{\mathsf{in}}, \mathbf{c}^{\mathsf{out}} - \mathbf{h}^{\mathsf{out}}) + \sum_{i=1}^{k'} \left(h_i^{\mathsf{in}} \cdot \mathsf{cost}(b_j, x_i) + h_i^{\mathsf{out}} \cdot \mathsf{cost}(x_i, b_j)\right).$$

Here we interpret h_i^{in} as the multiplicity of the edge (b_j, x_i) and h_i^{out} as the multiplicity of the edge (x_i, b_j). Therefore, we require $h_i^{\mathsf{in}} \leqslant \mathsf{cap}(b_j, x_i)$ and $h_i^{\mathsf{out}} \leqslant \mathsf{cap}(x_i, b_j)$ so that the capacity constraints hold. Furthermore, we require that the solution is flow preserving in b_j, i.e. $\sum_{i=1}^{k'} h_i^{\mathsf{in}} = \sum_{i=1}^{k'} h_i^{\mathsf{out}}$ and $\sum_{i=1}^{k'} h_i^{\mathsf{in}} =$

$\mathsf{dem}(b_j)$ if $b_j \in D$. For π' we require for all $S \in \pi$ that either $S \in \pi'$ or there exist $S'_1, \dots S'_\ell \in \pi'$ such that $S_1 \cup \dots \cup S_\ell = S$ and $S'_1, \dots, S'_\ell$ are all connected to b_j. This latter can be formalized by requiring that for each $t \in [1, \ell]$, there is an $x_i \in S'_t$ such that $h_i^{\mathsf{in}} + h_i^{\mathsf{out}} > 0$.

Notice that with this recurrence, the table entries are computed correctly as only the vertex b_j was added, compared to the table entries T_{j-1}. Therefore we may assume that only the edges incident to b_j were added to another solution for some table entry in T_{j-1}. □

D.3 Proofs of Claims from Subsection 2.3

Claim 4. *We may assume that for both $\overrightarrow{F}$ and $\overleftarrow{F}$, their underlying undirected edge sets do not contain cycles.*

Proof. We change r such that for both $\overrightarrow{F}$ and $\overleftarrow{F}$, their underlying undirected edge sets do not contain cycles. Assume that there is an *alternating* cycle $C \subseteq \overleftarrow{F}$, meaning that its underlying edge set is a cycle and (hence) the edges alternate between being in positive and negative orientation. We can then create solutions r' and r'' of Flow by alternatingly adding and removing edges from C. Note that we can start by either adding or removing, giving us these two different solutions r' and r''. Since the edges added to r to form r' are exactly the edges that were removed from r to form r'', and vice versa, it holds that $\mathsf{cost}(r) - \mathsf{cost}(r') = -(\mathsf{cost}(r) - \mathsf{cost}(r''))$. Since r is an optimal solution, we conclude $\mathsf{cost}(r) = \mathsf{cost}(r') = \mathsf{cost}(r'')$. We can therefore choose either r' or r'' to replace r, such that $\overleftarrow{F}$ now has one alternating cycle less without changing any of the edges of r outside C. Hence we can iteratively remove the cycles from $\overleftarrow{F}$ and $\overrightarrow{F}$ and obtain an optimal solution r to the Flow instance in which both $\overleftarrow{F}$ and $\overrightarrow{F}$ are forests in polynomial time. □

Claim 5. *There exists an optimal solution f' of the* MVTSP *instance such that for all $i, j \in [1, k]$, $b \in R_{ij}$ and $x_\ell \in X$ it holds that $r(x_\ell, b) = f'(x_\ell, b)$ and $r(b, x_\ell) = f'(b, x_\ell)$.*

Proof. We build this f' iteratively from f, by removing any edges $(x_{i'}, b)$ and $(b, x_{j'})$ for $i' \neq i$ and $j' \neq j$ for each $b \in R_{ij}$. In particular, this implies that $r(x_i, b) = f'(x_i, b)$ and $r(b, x_j) = f'(b, x_j)$, as b then only has edges coming from x_i and to x_j and since b has a fixed demand.

Throughout the process we retain optimality and connectivity for f'. Furthermore, after each step, the solutions r and f' differ at at most $8k^2$ edges. We start by setting $f' = f$.

Let us consider $b \in R_{ij}$ and suppose that $f'(x_\ell, b) > 0$ for some $\ell \neq i$. Note that we can tackle the case where $f'(b, x_\ell) > 0$ for some $\ell \neq j$ with similar steps. We remark that $|\overrightarrow{A_{ij}}(\ell)| = 8k^2 + 2$ as $R_{ij} \neq \emptyset$. As at most $8k^2$ edges are different between r and f', there are vertices $v, w \in \overrightarrow{A_{ij}}(\ell)$ such that all of the edges adjacent to v and w have the same multiplicities in r and f', i.e. $f'(x,v) = r(x,v)$ and $f'(x,w) = r(x,w)$ for all $x \in X$.

Define flow f'' with at most the same costs as f' by removing one copy of the edges (x_ℓ, b) and (x_i, v) and adding one copy of the edges (x_i, b) and (x_ℓ, v), see Fig. 5. As $b \notin \overrightarrow{A_{ij}}(\ell)$ and $v \in \overrightarrow{A_{ij}}(\ell)$, the cost of f'' is indeed at most the cost of f' by definition of the set $\overrightarrow{A_{ij}}(\ell)$.

We now argue that f'' is connected. As f' is a solution to MVTSP, it must be connected. Since we removed (x_ℓ, b) and (x_i, v) from f' to form f'', proving that the pairs x_ℓ, b and x_i, v are connected in f'' proves f'' to be connected. The edges (x_i, w), (w, x_j) and (v, x_j) in f'' connect x_i and v. As a consequence, x_ℓ and b are also connected, because of the edges (x_ℓ, v) and (x_i, b).

We remark that the number of edges that differ between f'' and r has not changed. Hence, we continue with setting $f' = f''$ and repeat until f' has the required properties. □

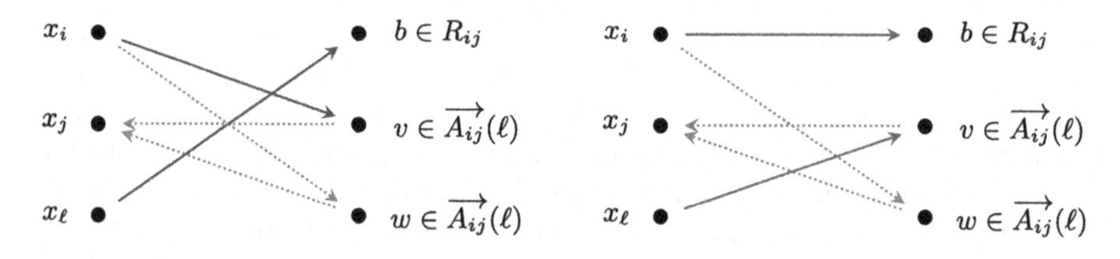

Fig. 5. Adjusting flow f', depicted on the left, to get flow f'', depicted on the right. The blue edges are replaced by the red edges, the rest of the solutions are equal. The vertex w assures the new solution remains connected

References

1. Berger, A., Kozma, L., Mnich, M., Vincze, R.: Time- and space-optimal algorithm for the many-visits TSP. ACM Trans. Algorithms **16**(3), 35:1–35:22 (2020)
2. Bodlaender, H.L., Cygan, M., Kratsch, S., Nederlof, J.: Deterministic single exponential time algorithms for connectivity problems parameterized by treewidth. Inf. Comput. **243**, 86–111 (2015)
3. Cai, L., Ye, J.: Finding two edge-disjoint paths with length constraints. In: Heggernes, P. (ed.) WG 2016. LNCS, vol. 9941, pp. 62–73. Springer, Heidelberg (2016). https://doi.org/10.1007/978-3-662-53536-3_6
4. Cosmadakis, S.S., Papadimitriou, C.H.: The traveling salesman problem with many visits to few cities. SIAM J. Comput. **13**(1), 99–108 (1984)
5. Cygan, M., et al.: Parameterized Algorithms. Springer, Cham (2015). https://doi.org/10.1007/978-3-319-21275-3
6. Cygan, M., Kratsch, S., Nederlof, J.: Fast Hamiltonicity checking via bases of perfect matchings. J. ACM **65**(3), 12:1–12:46 (2018)
7. Etscheid, M., Kratsch, S., Mnich, M., Röglin, H.: Polynomial kernels for weighted problems. J. Comput. Syst. Sci. **84**, 1–10 (2017)

8. Fortune, S., Hopcroft, J.E., Wyllie, J.: The directed subgraph homeomorphism problem. Theor. Comput. Sci. **10**, 111–121 (1980)
9. Garey, M.R., Johnson, D.S.: Computers and Intractability: A Guide to the Theory of NP-Completeness. W. H. Freeman (1979)
10. Kowalik, Ł., Li, S., Nadara, W., Smulewicz, M., Wahlström, M.: Many visits TSP revisited. In: Grandoni, F., Herman, G., Sanders, P. (eds.) (ESA 2020), Volume 173 of Leibniz International Proceedings in Informatics (LIPIcs), Dagstuhl, Germany, pp. 66:1–66:22. Schloss Dagstuhl-Leibniz-Zentrum für Informatik (2020)
11. Philip, G., Rani, M.R., Subashini, R.: On computing the Hamiltonian index of graphs. In: Fernau, H. (ed.) CSR 2020. LNCS, vol. 12159, pp. 341–353. Springer, Cham (2020). https://doi.org/10.1007/978-3-030-50026-9_25

FPT Algorithms to Compute the Elimination Distance to Bipartite Graphs and More

Bart M. P. Jansen and Jari J. H. de Kroon(✉)

Eindhoven University of Technology, Eindhoven, The Netherlands
{b.m.p.jansen,j.j.h.d.kroon}@tue.nl

Abstract. For a hereditary graph class $\mathcal{H}$, the $\mathcal{H}$-elimination distance of a graph G is the minimum number of rounds needed to reduce G to a member of $\mathcal{H}$ by removing one vertex from each connected component in each round. The $\mathcal{H}$-treewidth of a graph G is the minimum, taken over all vertex sets X for which each connected component of $G - X$ belongs to $\mathcal{H}$, of the treewidth of the graph obtained from G by replacing the neighborhood of each component of $G - X$ by a clique and then removing $V(G) \setminus X$. These parameterizations recently attracted interest because they are simultaneously smaller than the graph-complexity measures treedepth and treewidth, respectively, and the vertex-deletion distance to $\mathcal{H}$. For the class $\mathcal{H}$ of bipartite graphs, we present non-uniform fixed-parameter tractable algorithms for testing whether the $\mathcal{H}$-elimination distance or $\mathcal{H}$-treewidth of a graph is at most k. Along the way, we also provide such algorithms for all graph classes $\mathcal{H}$ defined by a finite set of forbidden induced subgraphs.

Keywords: Elimination distance · FPT · Odd cycle transversal

1 Introduction

Background. Assuming some structure on the input of a computational problem can greatly decrease its difficulty. For instance, it is well known that many NP-hard graph problems can be solved efficiently on graphs of bounded *treewidth* using dynamic programming over so-called tree decompositions [4]. The analysis of computational problems in terms of the input size and an additional parameter such as treewidth is the main objective in the field of parameterized complexity [10,11]. A parameter similar to treewidth is *treedepth* [26, §6.4]. It can be

This project has received funding from the European Research Council (ERC) under the European Union's Horizon 2020 research and innovation programme (grant agreement No 803421, ReduceSearch).

The original version of this chapter was previously published non-open access. A Correction to this chapter is available at https://doi.org/10.1007/978-3-030-86838-3_32

L. Kowalik et al. (Eds.): WG 2021, LNCS 12911, pp. 80–93, 2021.
https://doi.org/10.1007/978-3-030-86838-3_6

defined as the minimum number of rounds needed to get to the empty graph, where in each round we can delete one vertex from each connected component (formal definitions in the preliminaries). Some NP-hard graph problems become solvable in polynomial time if the input graph is restricted to be in a certain class. For instance the NP-hard VERTEX COVER can be solved in polynomial time in chordal graphs; those graphs without induced cycles of length at least four. A parameter that naturally follows from this observation is the minimum cardinality of a set of vertices whose deletion results in a graph contained in graph class $\mathcal{H}$. Such a set is called an $\mathcal{H}$-deletion set. This parameter essentially indicates how far the problem is from being a trivial case (cf. [18]). The size of a feedback vertex set [20,23] or vertex cover number [13,14] of the graph are often used examples of such parameters, where $\mathcal{H}$ is the class of forests and edgeless graphs respectively.

Recently there has been a push [12,16,17] in obtaining parameterized algorithm where the parameter is a hybrid of some overall structure of the graph, like treewidth and treedepth, and some distance to triviality. One such example introduced by Bulian and Dawar is $\mathcal{H}$-elimination distance ($\mathbf{ed}_{\mathcal{H}}$) [6,7], which can be defined as the minimum number of deletion rounds needed to obtain a graph in $\mathcal{H}$ by removing one vertex from each connected component in each round; recall that in the elimination-based definition of treedepth, the goal is to eliminate the entire graph. Hence $\mathbf{ed}_{\mathcal{H}}$ is never larger than the treedepth or the (vertex-)deletion distance to $\mathcal{H}$. Bulian and Dawar showed that $\mathbf{ed}_{\mathcal{H}}$ can be computed in FPT time when $\mathcal{H}$ is minor-closed [7].

A related hybrid variant of treewidth was introduced by Eiben et al. [12], namely $\mathcal{H}$-treewidth ($\mathbf{tw}_{\mathcal{H}}$). The $\mathcal{H}$-treewidth of a graph can be defined as the minimum treewidth of the *torso graph* of a vertex set whose removal ensures each component belongs to $\mathcal{H}$. This gives rise to tree decompositions in which each bag has size at most $k+1$, apart for an arbitrarily large set of vertices that occurs in no other bags and induces a subgraph from $\mathcal{H}$. Similarly as before, $\mathbf{tw}_{\mathcal{H}}(G)$ is not larger than $\mathbf{tw}(G)$ or the deletion distance from G to $\mathcal{H}$. For minor-closed graph classes $\mathcal{H}$ it can be shown that graphs of $\mathcal{H}$-treewidth at most k are minor-closed and therefore characterized by a finite set of forbidden minors. This leads to non-uniform algorithms to recognize graphs of $\mathcal{H}$-treewidth at most k for minor-closed $\mathcal{H}$ using the Graph Minor algorithm [29].

Apart from minor-closed families $\mathcal{H}$, some isolated results are known about FPT algorithms to compute $\mathbf{ed}_{\mathcal{H}}$ and $\mathbf{tw}_{\mathcal{H}}$ exactly, parameterized by the parameter value. In recent work, Agrawal and Ramanujan [2] give an FPT algorithm to compute the elimination distance to a cluster graph, as part of a kernelization result using the corresponding structural parameterization. Eiben et al. [12] show that when $\mathcal{H}$ is the class of graphs of rankwidth at most c for some constant c, then $\mathbf{tw}_{\mathcal{H}}$ is FPT. Bulian and Dawar [6] considered the elimination distance to graphs of bounded degree d and gave an FPT approximation algorithm. Lindermayr et al. [24] showed that the elimination distance of a *planar* graph to a bounded-degree graph can be computed in FPT time. Very recently, Agrawal et al. [1] obtained non-uniform FPT algorithms for computing the elimination distance to any family $\mathcal{H}$ defined by a finite number of forbidden induced subgraphs, thereby settling the case of bounded-degree graphs as well.

Results and Techniques. We show that $\mathbf{tw}_{\mathcal{H}}$ and $\mathbf{ed}_{\mathcal{H}}$ are non-uniformly fixed parameter tractable parameterized by the solution value when $\mathcal{H}$ is the class of bipartite graphs. As a side-product of our proof, we show that $\mathbf{tw}_{\mathcal{H}}$ is non-uniformly FPT when $\mathcal{H}$ is defined by a finite number of forbidden induced subgraphs, generalizing the results of Agrawal et al. [1] for $\mathbf{ed}_{\mathcal{H}}$. The non-uniformity of our algorithms stems from the use of a meta-theorem by Lokshtanov et al. [25, Theorem 23] which encapsulates the technique of *recursive understanding*. This theorem essentially states that for any problem expressible in *Counting Monadic Second Order* (CMSO) logic, the effort of classifying whether the problem is in FPT is reduced to inputs that are (s, c)-unbreakable (formally defined later). The theorem allows us to use the technique of recursive understanding in a black box manner, leading to a streamlined proof at the expense of obtaining non-uniform algorithms. We believe that uniform algorithms can be obtained using the same approach by implementing the recursive understanding step from scratch and deriving an explicit bound on the sizes of representatives for the canonical congruence for $\mathbf{ed}_{\mathcal{H}}$ and $\mathbf{tw}_{\mathcal{H}}$ on t-boundaried graphs. As the running times would not be practical in any case, we did not pursue this route.

Our proof is independent of that of Agrawal et al. [1], but is based on an older approach inspired by the earlier work of Ganian et al. [17] that contains similar ideas. The key ingredient for our work is the insight that the approach based on recursive understanding used by Ganian et al. [17] to compute a hybrid parameterization for instances of constraint satisfaction problems, can be applied more generally to aid in the computation of $\mathbf{ed}_{\mathcal{H}}$ and $\mathbf{tw}_{\mathcal{H}}$. We can lift one of their main lemmas to a more general setting, where it roughly shows that given an $(s(k), 2k)$-unbreakable graph G and an $\mathcal{H}$-deletion set X in G that is a *subset* of some (unknown) structure that witnesses the value of $\mathbf{tw}_{\mathcal{H}}$ or $\mathbf{ed}_{\mathcal{H}}$, we can determine in FPT time whether such a witness exists. This allows $\mathbf{ed}_{\mathcal{H}}$ and $\mathbf{tw}_{\mathcal{H}}$ to be computed in FPT time if we can efficiently find a deletion set with the stated property. For families $\mathcal{H}$ defined by finitely many forbidden induced subgraphs, a simple bounded-depth branching algorithm suffices. Our main contribution is for bipartite graphs, where we show that the relation between odd cycle transversals and graph separators that lies at the heart of the iterative compression algorithm for OCT [27], can be combined with the fact that there are only few minimal (u, v)-separators of size at most $2k$ in $(s(k), 2k)$-unbreakable graphs, to obtain an $\mathcal{H}$-deletion set with the crucial property described above.

Related Work. Hols et al. [19] used parameterizations based on elimination distance to obtain kernelization algorithms for Vertex Cover.

In recent work [22], a superset of the authors gave FPT algorithms to *approximate* $\mathbf{ed}_{\mathcal{H}}$ and $\mathbf{tw}_{\mathcal{H}}$ for several classes $\mathcal{H}$, including bipartite graphs and all classes defined by a finite set of forbidden induced subgraphs. That work employed completely different techniques than used here, and left open the question whether the parameters can be computed exactly in FPT time.

2 Preliminaries

We consider simple undirected graphs without self-loops. The vertex and edge set of a graph G are denoted by $V(G)$ and $E(G)$ respectively. When the graph is clear from context, we denote $|V(G)|$ by n and $|E(G)|$ by m. For each $X \subseteq V(G)$, the graph induced by X is denoted by $G[X]$. We denote $G[V(G) \setminus X]$ by $G - X$, and write $G - v$ instead of $G - \{v\}$. The open and closed neighborhoods of $v \in V(G)$ are denoted $N_G(v)$ and $N_G[v]$ respectively. For $X \subseteq V(G)$, $N_G[X] = \bigcup_{v \in X} N_G[v]$ and $N_G(X) = N_G[X] \setminus X$. The subscript G is omitted if it is clear from context. The graph obtained from G by contracting an edge $e = \{u, v\} \in E(G)$ is the graph obtained by deleting u and v and inserting a new vertex that is adjacent to all of $(N_G(u) \cup N_G(v)) \setminus \{u, v\}$. A graph H is a *minor* of G, if it can be obtained from a subgraph of G by a number of edge contractions. A parameter is a function that assigns an integer to each graph. A parameter f is minor-closed if $f(H) \leq f(G)$ for each minor H of G. The connected components of G are denoted by $\mathrm{cc}(G)$. A set $Y \subseteq V(G)$ is an $\mathcal{H}$-deletion set if $G - Y \in \mathcal{H}$. A graph class $\mathcal{H}$ is hereditary if it is closed under vertex deletion, that is, if $G \in \mathcal{H}$, then for every induced subgraph F of G it holds that $F \in \mathcal{H}$. In this work we restrict ourselves to hereditary graph classes. A proper c-coloring of a graph is a function $f\colon V(G) \to [c]$ such that for every $\{u, v\} \in E(G)$ it holds that $f(u) \neq f(v)$. A graph is bipartite if and only if it has a proper 2-coloring. For sets $X, Y \subseteq V(G)$, we say that $S \subseteq V(G)$ is an (X, Y)-separator if the graph $G - S$ does not contain a vertex $u \in X \setminus S$ and $v \in Y \setminus S$ in the same connected component. Whenever we refer to the *size* of a graph, we mean the cardinality of its vertex set.

A parameterized problem Π is a subset $\Sigma^* \times \mathbb{N}$ for some finite alphabet Σ. A parameterized problem is *non-uniformly fixed-parameter tractable* (FPT) if there exists a fixed d such that for every fixed $k \in \mathbb{N}$, there exists an algorithm that determines whether $(x, k) \in \Pi$ in $\mathcal{O}(|x|^d)$ time. (Hence there is a different algorithm for each value of k.)

Due to space limitations, proofs of statements marked by (★) are deferred to the full version [21].

2.1 $\mathcal{H}$-treewidth and $\mathcal{H}$-elimination Distance

Definition 1 *[17, Definition 4]. Let G be a graph and $X \subseteq V(G)$. The* torso *of X, denoted by $\boldsymbol{T}_G(X)$, is the graph obtained by turning the neighborhood of every connected component of $G - X$ into a clique, followed by deleting all of $V(G) \setminus X$.*

Eiben et al. [12] use the term of *collapsing* $V(G) \setminus X$ instead of the torso of X. Since our algorithms try to identify X, the torso terminology is more natural. The *treewidth* of a graph G is denoted by $\mathbf{tw}(G)$ (cf. [3], [10, §7.2]).

Definition 2 *[12, Definition 3]. The $\mathcal{H}$-treewidth of a graph G is the smallest integer k such that there exists a set $X \subseteq V(G)$ with $\mathbf{tw}(\boldsymbol{T}_G(X)) \leq k$ and for each connected component $C \in cc(G - X)$ we have $C \in \mathcal{H}$. We call X an $\mathbf{tw}_{\mathcal{H}}$ witness of width k.*

Definition 3 *[6, 7]. The $\mathcal{H}$-elimination distance of G for a hereditary graph class $\mathcal{H}$, denoted by $\mathbf{ed}_{\mathcal{H}}(G)$, is defined recursively. If G is disconnected, then $\mathbf{ed}_{\mathcal{H}}(G) = \max_{C \in cc(G)} \mathbf{ed}_{\mathcal{H}}(C)$. If G is connected and belongs to $\mathcal{H}$, then $\mathbf{ed}_{\mathcal{H}}(G) = 0$. Otherwise, $\mathbf{ed}_{\mathcal{H}}(G) = 1 + \min_{v \in V(G)} \mathbf{ed}_{\mathcal{H}}(G - v)$. The treedepth of a graph, denoted $\mathbf{td}(G)$, is equivalent to $\mathbf{ed}_{\mathcal{H}}(G)$ where $\mathcal{H}$ only contains the empty graph.*

Note that the definition above is well defined when $\mathcal{H}$ is hereditary, since each hereditary graph class contains the empty graph. We argue that $\mathcal{H}$-elimination distance has an equivalent definition similar to that of $\mathcal{H}$-treewidth.

Proposition 1 (★). *A graph has $\mathbf{ed}_{\mathcal{H}}(G) \leq k$ if and only if there exists $X \subseteq V(G)$ such that $\mathbf{td}(\boldsymbol{T}_G(X)) \leq k$ and $C \in \mathcal{H}$ for each $C \in cc(G - X)$.*

Similar to $\mathbf{tw}_{\mathcal{H}}$ witnesses, we call X an $\mathbf{ed}_{\mathcal{H}}$ witness of depth k. Since the torso operation on X turns the neighborhood of each connected component of $G - X$ into a clique, the following note follows.

Note 1. If X is a $\mathbf{tw}_{\mathcal{H}}$ witness of width $k-1$ (respectively $\mathbf{ed}_{\mathcal{H}}$ witness of depth k), then $|N(C)| \leq k$ for every $C \in \text{cc}(G - X)$.

We are ready to introduce the main problem we try to solve.

$\mathcal{H}$-TREEWIDTH ($\mathbf{tw}_{\mathcal{H}}$) / $\mathcal{H}$-ELIMINATION DISTANCE ($\mathbf{ed}_{\mathcal{H}}$) **Parameter:** k
Input: A graph G, an integer k.
Question: Decide whether $\mathbf{tw}_{\mathcal{H}}(G) \leq k - 1$ / $\mathbf{ed}_{\mathcal{H}}(G) \leq k$.

Definition 4 *[25]. Let G be a graph and $s, c \in \mathbb{N}$. A partition (X, C, Y) of $V(G)$ is an (s, c)-separation in G if:*

- *C is a separator, that is, no edge has one endpoint in X and one in Y,*
- *$|C| \leq c$, $|X| \geq s$, and $|Y| \geq s$.*

A graph G is (s, c)-unbreakable if there is no (s, c)-separation in G.

The following proposition is similar to Lemma 21 of Ganian et al. [17].

Proposition 2 (★). *Let G be an (s, c)-unbreakable graph for $s, c \in \mathbb{N}$ and $\mathcal{H}$ be a graph class such that $\mathbf{tw}_{\mathcal{H}}(G) \leq k - 1$ (resp. $\mathbf{ed}_{\mathcal{H}}(G) \leq k$) and $c \geq k$. Then at least one of the following holds:*

1. *$\mathbf{tw}(G) \leq s + k - 1$ (resp. $\mathbf{td}(G) \leq s + k - 1$),*
2. *each $\mathbf{tw}_{\mathcal{H}}$ (resp. $\mathbf{ed}_{\mathcal{H}}$) witness X of G satisfies the following:*
   - *$G - X$ has exactly one connected component C of size at least s, and*
   - *$|V(G) \setminus N[C]| < s$ and $|X| \leq s + k - 1$.*

The following lemma bounds the number of small connected vertex sets with a small neighborhood. It was originally stated for connected sets of exactly b vertices with an open neighborhood of exactly f vertices.

Lemma 1 *[15, cf. Lemma 3.1]. Let G be a graph. For every $v \in V(G)$ and $b, f \geq 0$, the number of connected vertex sets $B \subseteq V(G)$ such that (a) $v \in B$, (b) $|B| \leq b + 1$, and (c) $|N(B)| \leq f$ is at most $b \cdot f \cdot \binom{b+f}{b}$. Furthermore they can be enumerated in $\mathcal{O}(n \cdot b^2 \cdot f \cdot (b + f) \cdot \binom{b+f}{b})$ time using polynomial space.*

2.2 CMSO

We use the formalism of Counting Monadic Second Order Logic (CMSO) as treated by Lokshtanov et al. [25]. For a more complete introduction we refer to the book of Courcelle and Engelfriet [9].

Let $\mathcal{H}$ be a graph class. We say that containment in $\mathcal{H}$ is expressible in CMSO if there exists a CMSO formula $\varphi_{\mathcal{H}}$ such that for any graph G it holds that $G \models \varphi_{\mathcal{H}}$ if and only if $G \in \mathcal{H}$.

Lemma 2 (★). *There exist CMSO-formulas with the following properties:*

1. *For any graph H, there exists a formula $\varphi_{\text{H-MINOR}}(X)$ such that for any graph G and any $X \subseteq V(G)$ it holds that $(G, X) \models \varphi_{\text{H-MINOR}}(X)$ if and only if H is a minor of $G[X]$.*
2. *For any graph class $\mathcal{H}$ characterized by a finite set of forbidden induced subgraphs, there exists a formula $\varphi_{\mathcal{H}}$ such that for any graph G it holds that $G \models \varphi_{\mathcal{H}}$ if and only if graph $G \in \mathcal{H}$.*
3. *There exists a formula φ_{BIP} such that for any graph G it holds that $G \models \varphi_{BIP}$ if and only if graph G is bipartite.*
4. *For each $k \in \mathbb{N}$, for each graph class $\mathcal{H}$ such that containment in $\mathcal{H}$ is CMSO expressible, and for each minor-closed parameter f, there exists a formula $\varphi_{(k,\mathcal{H},f)}(X)$ such that for any graph G and any $X \subseteq V(G)$ we have $(G, X) \models \varphi_{(k,\mathcal{H},f)}(X)$ if and only if $f(\boldsymbol{T}_G(X)) \le k$ and $C \in \mathcal{H}$ for each $C \in cc(G - X)$.*

Since both treewidth and treedepth are minor-closed parameters, we note the following from the lemma above.

Note 2. For each $k \in \mathbb{N}$ and graph class $\mathcal{H}$ such that containment in $\mathcal{H}$ is CMSO-expressible, there exists a formula $\varphi_{(k,\mathcal{H},\mathbf{tw})}$ (respectively $\varphi_{(k,\mathcal{H},\mathbf{td})}$) such that (G, k) is a YES-instance of $\mathcal{H}$-TREEWIDTH (respectively $\mathcal{H}$-ELIMINATION DISTANCE) if and only if $G \models \varphi_{(k,\mathcal{H},\mathbf{tw})}$ (respectively $G \models \varphi_{(k,\mathcal{H},\mathbf{td})}$).

CMSO formulas can have free variables. A graph together with an evaluation of free variables is called a *structure*. We denote the problem of evaluating a CMSO formula φ on a structure by CMSO[φ]. The following theorem is the main tool used to achieve our algorithms, we apply it only to formulas without free variables. As the formulation differs slightly from its original form, we provide a proof in the full version.

Theorem 1 (★) *[25, Theorem 23]. Let $\hat{\varphi}$ be a CMSO formula. For all $\hat{c}\colon \mathbb{N}_0 \to \mathbb{N}_0$, there exists $\hat{s}\colon \mathbb{N}_0 \to \mathbb{N}_0$ such that if* CMSO[$\hat{\varphi}$] *parameterized by k is FPT on $(\hat{s}(k), \hat{c}(k))$-unbreakable structures, then* CMSO[$\hat{\varphi}$] *parameterized by k is FPT on general structures.*

3 Algorithms for Computing $\mathbf{ed}_{\mathcal{H}}$ and $\mathbf{tw}_{\mathcal{H}}$

In this section we present our algorithms. In Sect. 3.1 we present a key lemma. In Sect. 3.2 we use it to deal with $\mathcal{H}$ characterized by a finite number of forbidden induced subgraphs, and in Sect. 3.3 we deal with bipartite graphs.

3.1 Extracting Witnesses from Deletion Sets Contained in Them

Our strategy for solving $\mathcal{H}$-TREEWIDTH and $\mathcal{H}$-ELIMINATION DISTANCE is similar to that of lemmas 9 and 10 of Ganian et al. [17] and is based on Proposition 2. Given an $(s(k), c(k))$-unbreakable graph, either the treewidth of the graph is bounded (1) and we can solve the problem directly using Courcelle's Theorem, or each witness is of bounded size and introduces some structure (2).

In the following lemma we assume we are in the latter case (hence the $\mathbf{tw}(G) > s(k) + k$ condition) and are given some $\mathcal{H}$-deletion set Y. We show that given an $(s(k), c(k))$-unbreakable graph, in FPT time we can find a witness X such that $Y \subseteq X$ if such a witness exists.

Lemma 3. *Consider some $k \in \mathbb{N}$ and $c\colon \mathbb{N} \to \mathbb{N}$ such that $c(k) \geq k$. Let $\mathcal{H}$ be a graph class such that containment in $\mathcal{H}$ is solvable in polynomial time. There is an algorithm that runs in FPT time that, given an $(s(k), c(k))$-unbreakable graph for any $s\colon \mathbb{N} \to \mathbb{N}$ with $\mathbf{tw}(G) > s(k) + k$ and an $\mathcal{H}$-deletion set Y of size at most $s(k)+k$, decides whether there is an $\mathbf{tw}_{\mathcal{H}}(G)$ witness X of width at most $k-1$ (respectively $\mathbf{ed}_{\mathcal{H}}(G)$ witness X of depth at most k) such that $Y \subseteq X$.*

Proof. We refer to a witness as either being an $\mathbf{tw}_{\mathcal{H}}$ witness of width at most $k-1$ or an $\mathbf{ed}_{\mathcal{H}}$ witness of depth at most k. Given a set $X \subseteq V(G)$, we can verify that it is a witness by testing whether $\mathbf{tw}(\mathbf{T}_G(X)) \leq k-1$ (respectively $\mathbf{td}(\mathbf{T}_G(X)) \leq k$) in FPT time [3,28] and verifying that each connected component $C \in \mathrm{cc}(G - X)$ is contained in $\mathcal{H}$, which can be done in polynomial time by assumption.

We show that we can find a witness if it exists, by doing the above verification for FPT many vertex subsets $D \subseteq V(G)$, as follows.

1. For each $y \in Y$, let $\mathcal{C}_y$ be the set of connected vertex sets S with $y \in S$, $|S| \leq s(k)$ and $|N(S)| \leq k$. For each $B \subseteq Y$ with $|B| \leq k$, a choice tuple t_B contains an entry for each $y \in Y \setminus B$, where entry $t_B[y]$ is some set $C_y \in \mathcal{C}_y$.
2. For each $B \subseteq Y$ with $|B| \leq k$ and each choice tuple t_B, if $G - (Y \cup \bigcup_{y \in Y \setminus B} N(t_B[y]))$ has exactly one connected component C of size at least $s(k)$ and $|V(G) \setminus N[C]| < s(k)$, apply the witness verification test to $D = Y \cup \bigcup_{y \in Y \setminus B} N(t_B[y]) \cup Q$ for each $Q \subseteq V(G) \setminus N[C]$.
3. Return the logical or of all witness verification tests.

We argue that the algorithm runs in FPT time. Note that as $|Y| \leq s(k) + k$, there are at most $\binom{s(k)+k}{k}$ choices for B. Furthermore $\mathcal{C}_y$ can be computed in FPT time using Lemma 1, hence the number of choice tuples is also FPT many. For each choice for B and each choice tuple t_B, there are at most $2^{s(k)}$ choices for Q. Since each vertex set can be verified to be a witness in FPT time, the running time claim follows.

Finally we argue correctness of the algorithm. Since $\mathbf{tw}(G) > s(k) + k$ (and also $\mathbf{td}(G) > s(k) + k$ as $\mathbf{tw}(G) \leq \mathbf{td}(G) - 1$), by Proposition 2 any witness X is of size at most $s(k) + k - 1$, the graph $G - X$ has exactly one large connected component C of size at least $s(k)$, and $|V(G) \setminus N[C]| < s(k)$.

Suppose G has a witness that is a superset of Y. Fix some witness X of minimal cardinality with $Y \subseteq X$ and let C be the unique component of size at least $s(k)$ of $G - X$. Note that since $C \cap X = \emptyset$, we have $C \cap Y = \emptyset$.

Let $B = N(C) \cap Y$. By Note 1 we have $|N(C)| \leq k$, hence the branching algorithm makes this choice for B at some point. For each $y \in Y \setminus B$, let C_y be the connected component of $G - N[C]$ containing y. Since $|V(G) \setminus N[C]| < s(k)$ and $|N(C)| \leq k$, we have that $|V(C_y)| < s(k)$ and $|N(C_y)| \leq k$. Note that $N(C_y) \subseteq N(C) \subseteq X$. The branching algorithm at some point tries the choice tuple t_B where $t_B[y] = C_y$ for each $y \in Y \setminus B$. Consider the set $A = Y \cup \bigcup_{y \in Y \setminus B} N(t_B[y])$. Note that $A \subseteq X$ by construction.

If $N(C) \subseteq A$, then the single large component of $G - A$ of size at least $s(k)$ is exactly C. Since $|V(G) \setminus N[C]| < s(k)$, it follows that $X = A \cup Q$ for some $Q \subseteq V(G) \setminus N[C]$. It follows that the algorithm correctly identifies X in this case.

The only remaining case is $N(C) \not\subseteq A$. We argue that this cannot happen when witness X is of minimal cardinality. Suppose $N(C) \not\subseteq A$ and let $v \in N(C) \setminus A$. Let $Z = Y \cup \bigcup_{y \in Y \setminus B} N[C_y]$ and note that we take the *closed* neighborhoods of the components, instead of the *open* neighborhoods as in the definition of A. Let C^*_v be the connected component of $G - (C \cup Z)$ that contains v. We argue that $X \setminus C^*_v$ is a witness. Note that $C^*_v \cap Y = \emptyset$ by construction as $Y \subseteq Z$. Because Y is an $\mathcal{H}$-deletion set, it follows that for each connected component C' in $G - (X \setminus C^*_v)$ we have $C' \in \mathcal{H}$. We argue that $N(C^*_v) \subseteq N[C]$. Since C^*_v is a connected component of $G - (C \cup Z)$ we have $N(C^*_v) \subseteq C \cup Z$, so it suffices to show that $N(C^*_v) \cap Z \subseteq N(C)$. Assume for a contradiction that C^*_v contains a vertex v' adjacent to some $z \in Z \setminus N(C)$; note that $v' \notin Z$. If $z \in Y$, then $z \in Y \setminus N(C) = Y \setminus B$ and the connected component C_z of $G - N[C]$ is adjacent to v', implying $v' \in N[C_z]$ and therefore $v' \in Z$; a contradiction. If $z \notin Y$, then by definition of Z we have $z \in N[C_y]$ for some $y \in Y \setminus B$. Since $N(C_y) \subseteq N(C)$ this implies $z \in C_y$. But then $v' \notin C \cup Z$ is adjacent to a vertex of the component C_y of $G - N[C]$, so $v' \in Z$ by definition of Z; a contradiction. Since $N(C^*_v) \subseteq N[C]$ and v is adjacent to at least one vertex in C as $v \in N(C)$, it follows that $C \cup C^*_v$ is a connected component of $G - (X \setminus C^*_v)$ with $N(C \cup C^*_v) \subseteq N(C)$. Therefore $\mathbf{T}_G(X \setminus C^*_v)$ is an induced subgraph of $\mathbf{T}_G(X)$. We conclude that $X \setminus C^*_v \supseteq Y$ is a witness. Since X was assumed to be of minimal cardinality, we arrive at a contradiction and hence $A \supseteq N(C)$. □

3.2 Classes $\mathcal{H}$ with Finitely Many Forbidden Induced Subgraphs

Theorem 2. *Let $\mathcal{H}$ be a graph class characterized by a finite set of forbidden induced subgraphs. Then $\mathcal{H}$-TREEWIDTH and $\mathcal{H}$-ELIMINATION DISTANCE are non-uniformly fixed-parameter tractable.*

Proof. By Lemma 2 containment in $\mathcal{H}$ is CMSO expressible, therefore by Note 2 there exists a formula $\varphi_{(k,\mathcal{H},f)}$ for each $f \in \{\mathbf{tw}, \mathbf{td}\}$ such that an instance (G, k) of $\mathcal{H}$-TREEWIDTH (respectively $\mathcal{H}$-ELIMINATION DISTANCE) is a YES-instance if and only if $G \models \varphi_{(k,\mathcal{H},f)}$. Furthermore, containment in $\mathcal{H}$ is polynomial time solvable, as we can verify that a graph does not contain any of the finitely many forbidden induced subgraphs.

We argue that both problems are in FPT when the input graph G is $(s(k), k)$-unbreakable for any $s\colon \mathbb{N} \to \mathbb{N}$. If $\mathbf{tw}(G) \leq s(k) + k$, we solve the problems directly using Courcelle's Theorem [8] using $\varphi_{(k,\mathcal{H},f)}$. Otherwise by Proposition 2 each witness X is of size at most $s(k) + k - 1$. We can enumerate all minimal $\mathcal{H}$-deletion sets $\mathcal{Y}$ of size at most $s(k)+k-1$ in FPT time by finding a forbidden induced subgraph and branching in all finitely many ways of destroying it. Since any witness X is an $\mathcal{H}$-deletion set, for some $Y \in \mathcal{Y}$ we have $Y \subseteq X$. Hence we solve the problem by calling Lemma 3 for each $Y \in \mathcal{Y}$. Applying Theorem 1 concludes the proof. □

Using known characterizations by a finite number of forbidden induced subgraphs (cf. [5]) we obtain the following corollary to Theorem 2.

Corollary 1. *Let $\mathcal{H}$ be set of graphs that are either (1) cliques, (2) claw-free, (3) of degree at most d for fixed d, (4) cographs, or (5) split graphs.* $\mathcal{H}$-TREEWIDTH *and* $\mathcal{H}$-ELIMINATION DISTANCE *are non-uniformly fixed-parameter tractable.*

3.3 Bipartite Graphs

We use shorthand bip to denote the class of bipartite graphs. The problem of deleting k vertices to obtain a bipartite graph is better known as the ODD CYCLE TRANSVERSAL (OCT) problem. The problem was shown to be FPT for the first time by Reed et al. [27]. We use some of their ingredients to show the following.

Lemma 4. *The* bip-TREEWIDTH *and* bip-ELIMINATION DISTANCE *problems are non-uniformly fixed-parameter tractable.*

Proof. By Lemma 2 containment in the class of bipartite graphs is CMSO expressible, therefore by Note 2 there exists a formula $\varphi_{(k,\text{bip},f)}$ for each $f \in \{\mathbf{tw}, \mathbf{td}\}$ such that an instance (G, k) of bip-TREEWIDTH (respectively bip-ELIMINATION DISTANCE) is a YES-instance if and only if $G \models \varphi_{(k,\text{bip},f)}$. We argue that both problems are FPT in $(s(k), 2k)$-unbreakable graphs for any $s\colon \mathbb{N} \to \mathbb{N}$. Note that the theorem then follows by Theorem 1.

Let G be an $(s(k), 2k)$-unbreakable graph. As before, we use the term witness to either refer to an $\mathbf{tw}_{\mathcal{H}}$ witness of width at most $k-1$ or an $\mathbf{ed}_{\mathcal{H}}$ witness of depth at most k, depending on the problem being solved. We first test whether $\mathbf{tw}(G) \leq s(k)+k$, in FPT time [3]. If so, then we can solve the problems directly using Courcelle's Theorem [8] using $\varphi_{(k,\text{bip},f)}$. Otherwise by Proposition 2 the size of each witness in G is at most $s(k)+k-1$, and for each witness X there is a unique connected component of $G-X$ of at least $s(k)$ vertices, henceforth called the *large component*. We use a two-step process to find an odd cycle transversal that is a subset of some witness (if a witness exists), so that we may invoke Lemma 3 to find a witness.

For a witness X^* in G and an odd cycle transversal W of G, we say that a partition (W_L, W_I) of W is *weakly consistent* with X^* if for the unique large component C of $G - X^*$ we have that $W \cap C = W_L$, $|W_L| \leq k$, and $W \subseteq C \cup X^*$. An odd cycle transversal W is *strongly consistent* with X^* if $W \subseteq X^*$.

The following claim encapsulates the connection between odd cycle transversals and separators that forms the key of the iterative-compression algorithm for OCT due to Reed, Smith, and Vetta [27].

Claim 1 (★). For each partitioned OCT $W = (W_L, W_I)$ of G, for each partition of $W_L = W_{L,1} \cup W_{L,2}$ into two independent sets, for each proper 2-coloring c of $G - W$, we have the following equivalence for each $X \subseteq V(G) \setminus W$: the graph $(G - W_I) - X$ has a proper 2-coloring with $W_{L,1}$ color 1 and $W_{L,2}$ color 2 *if and only if* the set X separates A from R in the graph $G - W$, with:

$$A = (N_{G-W_I}(W_{L,1}) \cap c^{-1}(1)) \cup (N_{G-W_I}(W_{L,2}) \cap c^{-1}(2))$$
$$R = (N_{G-W_I}(W_{L,1}) \cap c^{-1}(2)) \cup (N_{G-W_I}(W_{L,2}) \cap c^{-1}(1)).$$

Observe that $c^{-1}(i) \subseteq V(G - W)$ for each $i \in [2]$, so that $A \cup R \subseteq V(G - W)$, and that the separator X is allowed to intersect $A \cup R$.

The next two claims show that certain types of OCTs can be computed efficiently in the $(s(k), 2k)$-unbreakable input graph G.

Claim 2 (★). There is an FPT algorithm that outputs a list of partitioned OCTs in G with the guarantee that for each witness X, there is a partitioned OCT on the list that is weakly consistent with X.

Claim 3. There is an FPT algorithm that, given a partitioned OCT that is weakly consistent with some (unknown) witness X in G, outputs a list of OCTs in G such that at least one is strongly consistent with X.

Proof. Let (W_L, W_I) be the given partitioned OCT, where $W_L \cup W_I = W$. If $|W| > s(k) + k - 1$, then no witness is strongly consistent with W by Proposition 2, hence we may assume $|W| \leq s(k) + k - 1$.

1. Initialize an empty list $\mathcal{W}$. For each $y \in V(G)$, let $\mathcal{C}_y$ be the set of connected vertex sets S with $y \in S$, $|S| \leq s(k)$ and $|N(S)| \leq 2k$. Let c^* be an arbitrary proper 2-coloring of $G - W$ and let $B_i^* = (c^*)^{-1}(i)$ for each $i \in [2]$.
2. For each partition (W_1, W_2) of W_L, let $B_1 = N(W_2) \setminus W$ and $B_2 = N(W_1) \setminus W$. Let $A = (B_1 \cap B_2^*) \cup (B_2 \cap B_1^*)$ and $R = (B_1 \cap B_1^*) \cup (B_2 \cap B_2^*)$.
   (a) For each choice $Q \in \{A, R\}$ with $|Q| \leq s(k) + k$, for each $D \subseteq Q$ with $|D| \leq k$, choice tuple $t_{Q,D}$ has an entry for each $y \in Q \setminus D$, where entry $t_{Q,D}[y]$ is some vertex set $C_y \in \mathcal{C}_y$.
   (b) For each choice $Q \in \{A, R\}$ with $|Q| \leq s(k) + k$, for each $D \subseteq Q$ with $|D| \leq k$, and for each choice tuple $t_{Q,D}$, add $(W \cup D \cup \bigcup_{y \in Q \setminus D} N(t_{Q,D}[y])) \setminus W_L$ to $\mathcal{W}$ in case it is an OCT.

The resulting list $\mathcal{W}$ is given as the output of the algorithm. The running time follows from Lemma 1 and the fact that there are FPT many choices for (W_1, W_2), D, and tuple $t_{Q,D}$.

We argue the correctness of the algorithm. Note that each set in the output list is an OCT by construction. Consider some witness X with (W_L, W_I) weakly consistent with X and let C be the unique large component of $G - X$, which is bipartite by definition of witness. Let $Y = (W \setminus W_L) \cup N(C) \subseteq X$, note that Y is an OCT of G. Let $c\colon V(G) \setminus Y \to [2]$ be a proper 2-coloring of $G - Y$. For some partition (W_1, W_2) of W_L we have $W_i \subseteq c^{-1}(i)$ for each $i \in [2]$. Note that since $Y \setminus W_I \subseteq N(C)$, we have that $|Y \setminus W_I| \leq k$.

By Claim 1, it follows that $Y \setminus W_I \subseteq N(C)$ separates A and R in $G - W$. Note that $B_i \subseteq N[C]$ for each $i \in [2]$ since $W_L \subseteq C$, therefore $A \subseteq N[C]$ and $R \subseteq N[C]$. Observe that $W_L \cup N(C)$ is an (A, R)-separator of size at most $2k$ in G. Therefore, since G is $(s(k), 2k)$-unbreakable, for at least one $Q \in \{A, R\}$ the vertex set reachable from $Q \setminus (W_L \cup N(C))$ in $G - (W_L \cup N(C))$ has size at most $s(k)$. Since A and R are disjoint from $W \supseteq W_L$ by definition, this implies $|Q| \leq s(k) + k$. Hence the algorithm tries this choice as $|Q| \leq s(k) + k$ is satisfied. Let $D = N(C) \cap Q$. For each $y \in Q \setminus D$, let C_y be the connected component of $G - (N(C) \cup W_L)$ containing y. Note that $|C_y| \leq s(k)$ and $|N(C_y)| \leq 2k$. Let the choice tuple $t_{Q,D}$ be such that $t_{Q,D}[y] = C_y$ for each $y \in Q \setminus D$. Observe that $(D \cup \bigcup_{y \in Q \setminus D} N(t_{Q,D}[y])) \setminus W_L \subseteq N(C)$ is an (A, R)-separator in $G - W$. Therefore $(W_I \cup D \cup \bigcup_{y \in Q \setminus D} N(t_{Q,D}[y])) \setminus W_L$ is an OCT by Claim 1 contained in X, concluding the proof.

With the two claims above, we can solve the problem as follows. Compute a list of partitions $\mathcal{W}$ using Claim 2 and use each $W \in \mathcal{W}$ as input to Claim 3. Using the output $\mathcal{U}$ of Claim 3, call Lemma 3 for each $U \in \mathcal{U}$. By the output guarantee of the claims, for each witness X we call the lemma with $U \subseteq X$ at some point, thus solving the problem. □

4 Conclusion

We have shown that $\mathcal{H}$-elimination distance and $\mathcal{H}$-treewidth are non-uniformly fixed-parameter tractable for $\mathcal{H}$ being the class of bipartite graphs, and whenever $\mathcal{H}$ is defined by a finite set of forbidden induced subgraphs. While the algorithms presented here solve the decision variant of the problem, by self-reduction they can be used to identify a witness if one exists. The main observation driving such a self-reduction is the following: if $\mathbf{tw}_{\mathcal{H}}(G) \leq k$, then for an arbitrary $v \in V(G)$ there exists a $\mathbf{tw}_{\mathcal{H}}(G)$-witness that contains v if and only the graph G' obtained from G by inserting a minimal forbidden induced subgraph into $\mathcal{H}$ and identifying one of its vertices with v, still satisfies $\mathbf{tw}_{\mathcal{H}}(G') \leq k$. Hence an iterative process can identify all vertices of a witness in this way.

While we have focused on the established notions of $\mathbf{tw}_{\mathcal{H}}$ and $\mathbf{ed}_{\mathcal{H}}$, the ideas presented here can be generalized using minor-closed graph parameters f other than treewidth and treedepth. As long as f can attain arbitrarily large values, implying its value on a clique grows with the size of the clique, and $\mathcal{H}$ is characterized by a finite set of forbidden induced subgraphs, we believe our approach can be generalized to answer questions of the form: does G have an $\mathcal{H}$-deletion set X for which $f(\mathbf{T}_G(X)) \leq k$?

References

1. Agrawal, A., Kanesh, L., Panolan, F., Ramanujan, M., Saurabh, S.: An FPT algorithm for elimination distance to bounded degree graphs. In: Proceedings of the 38th International Symposium on Theoretical Aspects of Computer Science, STACS 2021 (2021). https://doi.org/10.4230/LIPIcs.STACS.2021.50
2. Agrawal, A., Ramanujan, M.S.: On the parameterized complexity of clique elimination distance. In: Cao, Y., Pilipczuk, M. (eds.) 15th International Symposium on Parameterized and Exact Computation, IPEC 2020, Hong Kong, China, 14–18 December 2020 (Virtual Conference). LIPIcs, vol. 180, pp. 1:1–1:13. Schloss Dagstuhl - Leibniz-Zentrum für Informatik (2020). https://doi.org/10.4230/LIPIcs.IPEC.2020.1
3. Bodlaender, H.L.: A linear-time algorithm for finding tree-decompositions of small treewidth. SIAM J. Comput. **25**(6), 1305–1317 (1996). https://doi.org/10.1137/S0097539793251219
4. Bodlaender, H.L., Koster, A.M.C.A.: Combinatorial optimization on graphs of bounded treewidth. Comput. J. **51**(3), 255–269 (2008). https://doi.org/10.1093/comjnl/bxm037
5. Brandstädt, A., Le, V.B., Spinrad, J.P.: Graph Classes: A Survey. Society for Industrial and Applied Mathematics, Philadelphia (1999)
6. Bulian, J., Dawar, A.: Graph isomorphism parameterized by elimination distance to bounded degree. Algorithmica **75**(2), 363–382 (2016). https://doi.org/10.1007/s00453-015-0045-3
7. Bulian, J., Dawar, A.: Fixed-parameter tractable distances to sparse graph classes. Algorithmica **79**(1), 139–158 (2017). https://doi.org/10.1007/s00453-016-0235-7
8. Courcelle, B.: The monadic second-order logic of graphs. I. recognizable sets of finite graphs. Inf. Comput. **85**(1), 12–75 (1990). https://doi.org/10.1016/0890-5401(90)90043-H
9. Courcelle, B., Engelfriet, J.: Graph Structure and Monadic Second-Order Logic: A Language-Theoretic Approach. Encyclopedia of Mathematics and Its Applications, Cambridge University Press (2012). https://doi.org/10.1017/CBO9780511977619
10. Cygan, M., et al.: Parameterized Algorithms (2015). https://doi.org/10.1007/978-3-319-21275-3
11. Downey, R.G., Fellows, M.R.: Fundamentals of Parameterized Complexity. Texts in Computer Science. Springer, London (2013). https://doi.org/10.1007/978-1-4471-5559-1
12. Eiben, E., Ganian, R., Hamm, T., Kwon, O.: Measuring what matters: a hybrid approach to dynamic programming with treewidth. In: Rossmanith, P., Heggernes, P., Katoen, J. (eds.) 44th International Symposium on Mathematical Foundations of Computer Science, MFCS 2019, Aachen, Germany, 26–30 August 2019. LIPIcs, vol. 138, pp. 42:1–42:15. Schloss Dagstuhl - Leibniz-Zentrum für Informatik (2019). https://doi.org/10.4230/LIPIcs.MFCS.2019.42

13. Fellows, M.R., Lokshtanov, D., Misra, N., Rosamond, F.A., Saurabh, S.: Graph layout problems parameterized by vertex cover. In: Hong, S.-H., Nagamochi, H., Fukunaga, T. (eds.) ISAAC 2008. LNCS, vol. 5369, pp. 294–305. Springer, Heidelberg (2008). https://doi.org/10.1007/978-3-540-92182-0_28
14. Fluschnik, T., Niedermeier, R., Schubert, C., Zschoche, P.: Multistage S-T path: confronting similarity with dissimilarity in temporal graphs. In: Cao, Y., Cheng, S., Li, M. (eds.) 31st International Symposium on Algorithms and Computation, ISAAC 2020, Hong Kong, China, 14–18 December 2020 (Virtual Conference). LIPIcs, vol. 181, pp. 43:1–43:16. Schloss Dagstuhl - Leibniz-Zentrum für Informatik (2020). https://doi.org/10.4230/LIPIcs.ISAAC.2020.43
15. Fomin, F.V., Villanger, Y.: Treewidth computation and extremal combinatorics. Combinatorics **32**(3), 289–308 (2012). https://doi.org/10.1007/s00493-012-2536-z
16. Ganian, R., Ordyniak, S., Szeider, S.: A join-based hybrid parameter for constraint satisfaction. In: Schiex, T., de Givry, S. (eds.) CP 2019. LNCS, vol. 11802, pp. 195–212. Springer, Cham (2019). https://doi.org/10.1007/978-3-030-30048-7_12
17. Ganian, R., Ramanujan, M.S., Szeider, S.: Combining treewidth and backdoors for CSP. In: Vollmer, H., Vallée, B. (eds.) 34th Symposium on Theoretical Aspects of Computer Science, STACS 2017, Hannover, Germany, 8–11 March 2017. LIPIcs, vol. 66, pp. 36:1–36:17. Schloss Dagstuhl - Leibniz-Zentrum für Informatik (2017). https://doi.org/10.4230/LIPIcs.STACS.2017.36
18. Guo, J., Hüffner, F., Niedermeier, R.: A structural view on parameterizing problems: distance from triviality. In: Downey, R., Fellows, M., Dehne, F. (eds.) IWPEC 2004. LNCS, vol. 3162, pp. 162–173. Springer, Heidelberg (2004). https://doi.org/10.1007/978-3-540-28639-4_15
19. Hols, E.C., Kratsch, S., Pieterse, A.: Elimination distances, blocking sets, and kernels for vertex cover. In: Paul, C., Bläser, M. (eds.) 37th International Symposium on Theoretical Aspects of Computer Science, STACS 2020, Montpellier, France, 10–13 March 2020. LIPIcs, vol. 154, pp. 36:1–36:14. Schloss Dagstuhl - Leibniz-Zentrum für Informatik (2020). https://doi.org/10.4230/LIPIcs.STACS.2020.36
20. Jansen, B.M.P., Bodlaender, H.L.: Vertex cover kernelization revisited - upper and lower bounds for a refined parameter. Theory Comput. Syst. **53**(2), 263–299 (2013). https://doi.org/10.1007/s00224-012-9393-4
21. Jansen, B.M.P., de Kroon, J.J.H.: FPT algorithms to compute the elimination distance to bipartite graphs and more. CoRR abs/2106.04191 (2021). http://www.arxiv.org/abs/2106.04191
22. Jansen, B.M.P., de Kroon, J.J.H., Wlodarczyk, M.: Vertex deletion parameterized by elimination distance and even less. In: Khuller, S., Williams, V.V. (eds.) STOC 2021: 53rd Annual ACM SIGACT Symposium on Theory of Computing, Virtual Event, Italy, 21–25 June 2021, pp. 1757–1769. ACM (2021). https://doi.org/10.1145/3406325.3451068
23. Kratsch, S., Schweitzer, P.: Isomorphism for graphs of bounded feedback vertex set number. In: Kaplan, H. (ed.) SWAT 2010. LNCS, vol. 6139, pp. 81–92. Springer, Heidelberg (2010). https://doi.org/10.1007/978-3-642-13731-0_9
24. Lindermayr, A., Siebertz, S., Vigny, A.: Elimination distance to bounded degree on planar graphs. In: Esparza, J., Král', D. (eds.) 45th International Symposium on Mathematical Foundations of Computer Science, MFCS 2020, Prague, Czech Republic, 24–28 August 2020. LIPIcs, vol. 170, pp. 65:1–65:12. Schloss Dagstuhl - Leibniz-Zentrum für Informatik (2020). https://doi.org/10.4230/LIPIcs.MFCS.2020.65

25. Lokshtanov, D., Ramanujan, M.S., Saurabh, S., Zehavi, M.: Reducing CMSO model checking to highly connected graphs. In: Chatzigiannakis, I., Kaklamanis, C., Marx, D., Sannella, D. (eds.) 45th International Colloquium on Automata, Languages, and Programming, ICALP 2018, Prague, Czech Republic, 9–13 July 2018. LIPIcs, vol. 107, pp. 135:1–135:14. Schloss Dagstuhl - Leibniz-Zentrum für Informatik (2018). https://doi.org/10.4230/LIPIcs.ICALP.2018.135
26. Nesetril, J., de Mendez, P.O.: Sparsity - Graphs, Structures, and Algorithms, Algorithms and Combinatorics, vol. 28. Springer, Heidelberg (2012). https://doi.org/10.1007/978-3-642-27875-4
27. Reed, B.A., Smith, K., Vetta, A.: Finding odd cycle transversals. Oper. Res. Lett. **32**(4), 299–301 (2004). https://doi.org/10.1016/j.orl.2003.10.009
28. Reidl, F., Rossmanith, P., Villaamil, F.S., Sikdar, S.: A faster parameterized algorithm for treedepth. In: Esparza, J., Fraigniaud, P., Husfeldt, T., Koutsoupias, E. (eds.) ICALP 2014. LNCS, vol. 8572, pp. 931–942. Springer, Heidelberg (2014). https://doi.org/10.1007/978-3-662-43948-7_77
29. Robertson, N., Seymour, P.D.: Graph minors. XIII. The disjoint paths problem. J. Comb. Theory Ser. B **63**(1), 65–110 (1995). https://doi.org/10.1006/jctb.1995.1006

Disjoint Stable Matchings in Linear Time

Aadityan Ganesh[1], H. V. Vishwa Prakash[1(✉)], Prajakta Nimbhorkar[1,2], and Geevarghese Philip[1,2]

[1] Chennai Mathematical Institute, Chennai, India
vishwa@cmi.ac.in
[2] UMI ReLaX, Chennai, India

Abstract. We show that given a Stable Matching instance G as input, we can find a *largest collection* of pairwise edge-disjoint *stable matchings* of G in time linear in the input size. This extends two classical results:

1. The Gale-Shapley algorithm, which can find at most two ("extreme") pairwise edge-disjoint stable matchings of G in linear time, and
2. The polynomial-time algorithm for finding a largest collection of pairwise edge-disjoint *perfect matchings* (without the stability requirement) in a bipartite graph, obtained by combining König's characterization with Tutte's f-factor algorithm.

Moreover, we also give an algorithm to enumerate all maximum-length chains of disjoint stable matchings in the lattice of stable matchings of a given instance. This algorithm takes time polynomial in the input size for enumerating each chain. We also derive the expected number of such chains in a random instance of Stable Matching.

Keywords: Stable matching · Disjoint matchings

1 Introduction

All our graphs are finite, undirected, and simple. We use $V(G), E(G)$ to denote the vertex and edge sets of a graph G, respectively. A *matching* in a graph G is any subset $M \subseteq E(G)$ of edges of G such that no two edges in M have a common end-vertex. An input instance of the Stable Matching problem contains a bipartite graph G with the vertex partition $V(G) = \mathcal{M} \uplus \mathcal{W}$ where the two sides $\mathcal{M}, \mathcal{W}$ are customarily called "the set of men" and "the set of women", respectively. Each woman has a strictly ordered preference list containing her neighbors—a woman prefers to be matched with a man who comes earlier in her list, than with one who comes later—and each man similarly has a strictly ordered preference list containing all his neighbors.

Definition 1 (Blocking Pair). *A man-woman pair* $(m, w) \in E$ *is said to be a* blocking pair *with respect to a matching* M *of* G *if both* m *and* w *prefer each other over their matched partner in* M.

Definition 2 (Stable Matching). *A matching* M *of* G *is said to be* stable *if there is no blocking pair in* G *with respect to* M.

Ł. Kowalik et al. (Eds.): WG 2021, LNCS 12911, pp. 94–105, 2021.
https://doi.org/10.1007/978-3-030-86838-3_7

A matching M that is not stable is said to be *unstable*. The STABLE MATCHING instance consists of a bipartite graph G with vertex partition $\mathcal{M} \uplus \mathcal{W}$ and the associated preference lists. The STABLE MATCHING problem involves deciding if G has a stable matching, and outputting one if it exists.

The STABLE MATCHING problem models a number of real-world applications where two disjoint sets of entities—fresh graduates and intern positions; students and hostel rooms; internet users and CDN servers; and so on—need to be matched based on strict preferences. Gale and Shapley famously proved that *every* instance of STABLE MATCHING indeed has a stable matching, and that one such matching can be found in linear time [3]. The Gale-Shapley algorithm for STABLE MATCHING follows a simple—almost simplistic—greedy strategy: in turn, each unmatched man proposes to the most preferred woman who has not rejected him so far, and each woman holds on to the best proposal (as per her preference) that she has got so far. Gale and Shapley proved that this algorithm invariably finds a stable matching, which is said to be a *man-optimal* stable matching. Of course, the algorithm also works if the women do the proposing; a stable matching found this way is said to be *woman-optimal.*

It is not difficult to come up with instances of STABLE MATCHING where the man-optimal and women-optimal stable matchings are identical, as also instances where they differ. A rich theory about the combinatorial structure of stable matchings has been developed over the years. In particular, it is known that the set of all stable matchings of a STABLE MATCHING instance forms a *distributive lattice* under a certain natural partial order, and that the woman-optimal and man-optimal stable matchings form the maximum and minimum elements of this lattice. It follows that each instance has exactly one man-optimal stable matching and one woman-optimal stable matching, and that if these two matchings are identical, then the instance has exactly one stable matching in total.

The Gale-Shapley algorithm can thus do a restricted form of *counting* stable matchings: it can correctly report that an instance has exactly one stable matching, or that it has *at least* two, in which case it can output two different stable matchings. The *maximum* number of stable matchings that an instance can have has also received quite a bit of attention. Irving and Leather [6] discovered a method for constructing instances with exponentially-many stable matchings; these instances with n men and n women have $\Omega(2.28^n)$ stable matchings. This is the current best lower bound on the maximum number of stable matchings. After a series of improvements, the current best *upper* bound on this number is $\mathcal{O}(c^n)$ for some constant c [8,15].

Our focus in this work is on finding a large collection of *pairwise edge-disjoint* stable matchings:

> DISJOINT STABLE MATCHINGS
> **Input:** A STABLE MATCHING instance G and an integer k.
> **Task:** Decide if G has at least k pairwise disjoint stable matchings, and output such a collection of stable matchings if it exists.

Finding such a collection of disjoint stable matchings is clearly useful in situations which involve *repeated* assignments. For instance, when assigning people to tasks—drivers to bus routes, medical professionals to wards, cleaning staff to locations—this helps in avoiding monotony without losing stability. As another example, consider a business school program which has a series of projects on which the students are supposed to work in teams of two. Using a different stable matching from a disjoint collection to pair up students for each project will help with their collaborative skills while still avoiding problems of instability.

Even in those cases where only one stable matching suffices—such as when assigning medical students to hospitals once a year—a disjoint collection can still be very useful. Given such a collection, an administrator in charge of deciding the residencies can evaluate each stable matching based on other relevant considerations—such as gender or racial diversity, or costs of relocation—to choose an assignment which optimizes these other factors while still being stable.

Our main result is that DISJOINT STABLE MATCHINGS can be solved in *linear* time:

Theorem 1. *There is an algorithm which takes an instance G of* STABLE MATCHING, *runs in time linear in the size of the input, and outputs a pairwise disjoint collection of stable matchings of G of the largest size.*

This immediately yields:

Corollary 1. DISJOINT STABLE MATCHINGS *can be solved in linear time.*

To the best of our knowledge there is no published work about finding disjoint *stable* matchings. Finding disjoint *matchings* (without the stability requirement) has received a lot of attention over the years, and a number of structural and algorithmic results are known [1,12,14]; we mention just one, for perfect matchings in bipartite graphs.

Observe that a bipartite graph G has a perfect matching only if both sides have the same size, say n. Also, any collection of pairwise disjoint perfect matchings of such a graph G can have size at most n. This is because deleting the edges of one perfect matching from G decrements the degree of each vertex by exactly one, and the maximum degree of G is not more than n. A graph is said to be k-regular if each of its vertices has degree exactly k. König proved that a bipartite graph G contains k pairwise edge-disjoint perfect matchings if and only if G has a k-regular subgraph [10]. Tutte's polynomial-time algorithm for finding the so-called f-factors [18] can be used to find a k-regular subgraph of G. Putting these together we get a polynomial-time algorithm for finding a largest collection of edge-disjoint perfect matchings in bipartite graphs.

In stark contrast, checking if a non-bipartite graph has *two* disjoint perfect matchings is already NP-hard even in 3-regular graphs [2,5].

Relation to Lattice Structure. It is known that the set of stable matchings in a given instance forms a distributive lattice [9]. We show that there is always a solution to DISJOINT STABLE MATCHINGS that is a chain in this lattice.

We give an algorithm to enumerate all the chains of disjoint stable matchings. The algorithm takes time polynomial in the size of the input for outputting each such chain. We also show that the number of such chains in a random instance is at most quasi-polynomial with high probability.

2 Preliminaries

We recall the Gale-Shapley algorithm and the lattice structure of stable matchings here for the sake of completeness. The classical Gale-Shapley algorithm [4, Fig. 1.3] solves the STABLE MATCHING problem by a deferred acceptance mechanism. Each man proposes the women on his list in decreasing order of preference until some woman accepts his proposal. A woman w accepts a proposal from a man m if either w is unmatched or she prefers m over her current partner. The extended version of the Gale-Shapley algorithm [4, Fig. 1.7] *reduces* the preference lists by eliminating certain pairs that do not belong to any stable matching. By *deleting* a (*man-woman*) pair (m, w), we mean deleting m from w's preference list and w from that of m.

Lattice Structure of Stable Matchings. We need the following results about the lattice structure of stable matchings [4]. For a given stable marriage instance, a *dominance relation* on stable matchings is defined as follows:

Definition 3 (Dominance). *A stable matching M is said to* dominate *a stable matching M', written $M \preceq M'$, if every man has at least as good a partner in M as he has in M'; i.e., every man either prefers M to M' or is indifferent between them.*

Lemma 1 [4, Lemma 1.3.1]. *For a given stable marriage instance, let M and M' be two (distinct) stable matchings. If each man is given the better (or poorer) of his partners in M and M' denoted as $M \wedge M'$ (denoted as $M \vee M'$), then the result is a stable matching that dominates (dominated by) both M and M'.*

With the help of the above lemmas, it is easy to see that the set of all stable matchings forms a distributive lattice and the man-optimal matching and the woman-optimal matching represent the minimum and maximum elements of the lattice [4, Theorem 1.3.2]. Moreover, $M \wedge M'$ represents the *greatest lower bound* and $M \vee M'$ represents *least upper bound* of M and M' in the lattice of all the stable matchings.

3 Finding Disjoint Stable Matchings

In this section we describe and analyze our algorithm for finding a largest collection of disjoint stable matchings in a given instance of STABLE MATCHING.

Given a stable marriage instance, two matchings M_1 and M_2 are said to be *disjoint stable matchings* if both M_1 and M_2 are stable and they do not share a common edge. Throughout this section, we denote the man-optimal and woman-optimal stable matchings by M_o and M_z respectively.

The following lemma gives a necessary condition for the existence of two or more disjoint stable matchings for a given marriage instance.

Lemma 2 [4, Section 1.2.2]. *Let (m, w) be a pair in $M_o \cap M_z$. Then (m, w) is contained in every stable matching.*

The algorithm first finds the man-optimal and woman-optimal stable matchings (M_o and M_z respectively) by executing GS-Extended. If these matchings share an edge, the algorithm stops. Otherwise it modifies the instance by deleting all the edges that appear in M_o. It then computes a man-optimal matching M' of the new instance using GS-Extended. If M' is disjoint from the woman-optimal matching M_z then it deletes the edges of M' from the instance. The algorithm repeats this procedure as long as GS-Extended keeps returning a stable matching which is disjoint from M_z. It stores all the M_z-disjoint matchings obtained during this process in a set S. We note that this is a stronger version of the *BreakMarriage* algorithm of McVitie and Wilson [13].

Algorithm 1. Disjoint Stable Matchings

```
    Input  : A stable matching instance G
    Output: A maximum size set S of disjoint stable matchings.
 1: procedure DISJOINT STABLE MATCHINGS(G)
 2:     S ← ∅
 3:     M_z ←STABLEMATCHING(G, woman-optimal)   ▷ Woman-proposing GS Algorithm
 4:     X ←GS-EXTENDED(G)                       ▷ This modifies preference lists
 5:     while X ∩ M_z = ∅ do
 6:         S ← S ∪ {X}
 7:         for every man m do
 8:             Delete the first woman w on m's list   ▷ m's partner in X
 9:             Delete the last man on w's list        ▷ w's partner in X
10:         end for
11:         X ←GS-EXTENDED(G)                   ▷ Get a new disjoint matching as X
12:     end while
13:     S ← S ∪ {X}
        return S
14: end procedure
```

We first show that the matchings in the set S constructed by Algorithm 1 are stable. They are clearly disjoint by construction, since each step starts off by deleting *every* matched pair in the matching computed in the previous step.

Lemma 3. *All the matchings in the set S are stable matchings.*

Proof. For the sake of contradiction, let (m, w) be a blocking pair for a matching $M_i \in S$. Then, m prefers w to $p_{M_i}(m)$, where $p_{M_i}(m)$ is the partner of m in M_i. That is, w appears before $p_{M_i}(m)$ in m's preference list. As m is matched to $p_{M_i}(m)$ in the matching M_i, w would have been deleted from m's preference list *before* the call to GS-Extended that returned the matching M_i. This deletion can happen in two ways. Either in one of the calls to the Extended GS algorithm, or in one of the iterations of the for loop in line 7 of the algorithm. We know that in both the cases, after the deletion of w from m's preference list, w gets a strictly better partner than m in the subsequent matching. Therefore, w does not prefer m to $p_{M_i}(w)$. This contradicts our assumption. □

Building on the notion of dominance from Definition 3, we say that M *strictly dominates* M', denoted by $M \prec M'$, if $M \preceq M'$ and $M \cap M' = \emptyset$. The strict dominance relation imposes a partial order on the set of stable matchings in G. We call a set of stable matchings a *chain* if it forms a chain under the (non-strict) dominance relation of Definition 3. Let M_i be the matching included in S at the end of iteration i of the algorithm, and let $|S| = k$.

Lemma 4. *The stable matchings in the set S form a chain $M_o = M_1, \ldots, M_k$.*

Proof. Each iteration of the algorithm modifies the given instance by deleting the edges of the matching constructed. Let the instance considered at the beginning of iteration i be G_i. Thus $G_1 = G$. Since M_i is constructed by executing the extended Gale-Shapley algorithm on the instance G_i, it follows that M_i is the man-optimal matching in G_i. Further, all the men get strictly better partners in M_i compared to M_j, $j > i$ and all the women get strictly worse partners in M_i compared to M_j, for $j > i$. □

We now show that among all the chains of disjoint stable matchings, the one output by Algorithm 1 is a longest chain.

Lemma 5. *Algorithm 1 outputs a longest chain of disjoint stable matchings.*

Proof. Let $C : M_o = M_1 \prec M_2 \prec \cdots \prec M_k$ be the chain of disjoint matchings obtained by running Algorithm 1. For the sake of contradiction, let $C'' : M'_1 \prec M'_1 \prec \cdots \prec M'_\ell$ be a longest chain of disjoint matchings such that $\ell > k$.

We know that the matching $M_1 = M_o$ dominates *every* stable matching [4, Theorem 1.2.2]. Matching M'_1 cannot be disjoint with M_1, as otherwise, $M_1 \prec M'_1 \prec M'_2 \prec \cdots \prec M'_\ell$ would be a longer chain of disjoint stable matchings. Therefore, M'_1 shares some edges with M_1. As $M_1 \preceq M'_1 \prec M'_2$, we have $M_1 \prec M'_2$. Therefore we can replace M'_1 in $M'_1 \prec M'_2 \prec \cdots \prec M'_\ell$ with M_1 to get another chain of disjoint stable matchings $M_1 \prec M'_2 \prec \cdots \prec M'_\ell$ of length ℓ.

We know that M_2 dominates all the stable matchings which are disjoint with M_1. Matching M'_2 cannot be disjoint with M_2, as otherwise, we can get a longer chain $M_1 \prec M_2 \prec M'_2 \prec M'_3 \prec \cdots \prec M'_\ell$. Therefore, M'_2 shares edges with M_2. As $M_2 \preceq M'_2 \prec M'_3$, we have $M_2 \prec M'_3$. Therefore we can replace M'_2 with M_2 to get another chain of disjoint stable matchings $M_1 \prec M_2 \prec M'_3 \prec \cdots \prec M'_\ell$ of length ℓ.

In this way, we successively replace each M'_i of the chain C'' with M_i from the chain C to get the ℓ-length chain $M_1 \prec M_2 \prec \ldots M_k \prec M'_{k+1} \prec \cdots \prec M'_\ell$ of disjoint stable matchings. But this implies that there exists a stable matching M'_{k+1} which satisfies the strict relation $M_k \prec M'_{k+1}$, which is a contradiction since M_k has non zero interection with the woman-optimal matching M_z . □

We have shown that among all the chains of disjoint stable matchings, the one output by Algorithm 1 is of maximum length. We still need to prove that there is no larger set of disjoint stable matchings which is possibly not a chain. We use the following result due to Teo and Sethuraman to show that any such set of disjoint stable matchings has a corresponding chain of disjoint stable matchings. Moreover, the length of this chain is same as the size of the set.

Theorem 2 [17]. *Let $S = \{M_1, M_2, \cdots, M_k\}$ be a set of stable matchings for a particular stable matchings instance. For each man m, let S_m be the* sorted *multiset $\{p_{M_1}(m), p_{M_2}(m), \cdots, p_{M_k}(m)\}$, sorted according to the preference order of m. For every $i \in \{1, 2, \cdots, k\}$ let $M'_i = \{(m, w) \mid m \in \mathcal{M}$ and w is the ith woman in $S_m\}$. Then for each $i \in \{1, 2, \cdots, k\}$, M'_i is a stable matching.*

The following is an immediate corollary of Theorem 2:

Corollary 2. *Let $M_1, \ldots, M_k$ and $M'_1, \ldots, M'_k$ be as defined in Theorem 2. If $M_1, \ldots, M_k$ are pairwise disjoint, then $M'_1, \ldots, M'_k$ form a k-length chain of disjoint stable matchings.*

The following theorem now completes the correctness of Algorithm 1.

Theorem 3. *For a given stable marriage instance, Algorithm 1 gives a maximum size set of disjoint stable matchings.*

Proof. Let $S = \{M_1 = M_o, M_2, \cdots, M_k\}$ be the set of disjoint stable matchings output by Algorithm 1. For the sake of contradiction, Let $S' = \{M'_1, M'_2, \cdots, M'_\ell\}$ be a maximum size set of disjoint stable matchings such that $\ell > k$. Then, from Corollary 2 of Theorem 2, we know that there exists an ℓ-length chain of disjoint stable matchings. This contradicts Lemma 5, that the $k < \ell$ matchings from S form a *longest* chain of disjoint stable matchings. □

Time complexity: Each edge of G is visited exactly once during the course of the algorithm. Hence the time complexity is $O(m + n)$ where $2n$ is the number of vertices in G and m is the number of edges in G. This completes the proof of Theorem 1.

4 Enumerating All Max-Length Chains

Algorithm 1 gives one maximum-length chain of disjoint stable matchings. It is an interesting question whether such a chain is unique.

We now give an algorithm to enumerate all such chains. For the enumeration, we exploit the lattice structure of stable matchings described in Sect. 2.

The $\#P$-hardness of counting all the maximum-length chains can be easily deduced from the $\#P$-hardness of counting all the stable matchings in a given instance [7]. For a given instance G, if we construct a new instance G' by adding a new man-woman pair (m, w) such that both prefer each other over all the others, then every stable matching in G' contains the pair (m, w). Hence the length of a maximum-length chain of disjoint stable matchings is 1, and each stable matching in the given instance is such a chain.

Algorithm 2 on page 9 describes the enumeration procedure. We need some notation and definitions. Let A_0 be the man-optimal matching. Define the set $A = \{A_0, A_1, \ldots A_k\}$ such that for $1 \leq i \leq k$, $A_i = \bigvee\{M | A_{i-1} \prec M\}$, that is, A_i is the least upper bound of the set of all the stable matchings which are *strictly dominated* by A_{i-1} Similarly, let B_0 be the woman-optimal stable matching. Define the set $B = \{B_0, B_1, \ldots, B_t\}$ such that for $1 \leq i \leq t$, $B_i = \bigwedge\{M | B_{i-1} \succ M\}$, that is, B_i is the greatest lower bound of the set of all the stable matchings which *strictly dominate* B_{i-1}. We note that A and B are the chains returned by Algorithm 1 with man-proposing and woman-proposing versions respectively. Since both are maximum-length chains of disjoint stable matchings, $t = k$.

Let $X = \{X_0, \cdots X_k\}$ be a maximum-length chain of disjoint stable matchings i.e. $X_0 \prec X_1 \prec \cdots \prec X_k$. We note the following property of the matchings in X.

Lemma 6. *For* $0 \leq i \leq k$, $A_i \preceq X_i \preceq B_{k-i}$

Proof. By induction on i, we prove $A_i \preceq X_i$ for $0 \leq i \leq k$. Proving $X_i \preceq B_{k-i}$ is analogous.
As A_0 is the man-optimal matching, $A_0 \preceq X_0$. Assume for some i, $A_i \preceq X_i$. Hence $A_i \preceq X_i \prec X_{i+1}$. Therefore X_{i+1} is strictly dominated by A_i. Since A_{i+1} is the greatest lower bound of all such stable matchings which are strictly dominated by A_i, $A_{i+1} \preceq X_{i+1}$. □

Corollary 3. *For each* i, $A_i \preceq B_{k-i}$. *Moreover,* $\{X_0, \ldots, X_{i-1}, X_i, B_{k-i-1}, \ldots, B_0\}$ *is also a maximum chain of disjoint stable matchings given that* $A_j \preceq X_j \preceq B_{k-j}$ *for* $0 \leq j \leq i$.

Outline of the Algorithm. An algorithm to enumerate all the stable matchings in a given instance is known in the literature [4, Sect.3.5]. We use this result to construct the sub-lattice L of all the stable matchings N which are *in between* two matchings M and M' (i.e. $M \preceq N \preceq M'$), where M, M' are any two stable matchings such that $M \preceq M'$. To construct the sub-lattice L, we construct a new instance as follows:

1. Delete every woman in m's list better than his partner in M and worse than his partner in M'. Delete every man in w's list better than her partner in M' and worse than her partner in M.
2. Update the preference list so that m is in w's list iff w is in m's list.

In the new instance, M and M' are man-optimal and woman-optimal matchings respectively. The set of stable matchings in this instance is precisely L, which can be enumerated by the algorithm for enumeration of stable matchings.

In Algorithm 2, we first compute the sublattice L_0 between A_0 and B_k. Then we recursively call Algorithm 2 for every $X_0 \in L$. From Corollary 3 we know that given a partial list $X_0, X_1 \ldots, X_i$ of disjoint stable matchings, we can find the next matching in the chain. The algorithm first finds the man-optimal matching Y_{i+1} after deleting X_i from the given instance. In Algorithm 2, this method is referred to as NextBestDisjointMatching. Then it constructs the sub-lattice $\alpha_{Y_{i+1}}$ between Y_{i+1} and $B_{k-(i+1)}$. Now, for every stable matching M in $\alpha_{Y_{i+1}}$, it appends the input list as $X_0, X_1 \ldots, X_i, M$ and recursively calls itself to extend each list further. The correctness of the algorithm can be seen from the fact that it picks exactly one stable matching from each of the k sublattices, and they are disjoint by construction.

Algorithm 2. Enumeration($X_0, X_1, \cdots, X_i$)

Input: A stable matching instance G,
the output of a man-oriented version of Algorithm 1 $A = \{A_0, A_1, \ldots, A_k\}$,
the output of a woman-oriented version of Algorithm 1 $B = \{B_0, B_1, \ldots, B_k\}$ and
a list $(X_0, \cdots, X_i)$ such that $A_j \preceq X_j \preceq B_{k-j}$ for $0 \leq j \leq i$

Output: Print all maximum size chains of disjoint stable matchings in G.

```
1: if (X_i ∩ B_0 ≠ ∅) then
2:     print (X_0, X_1, ··· , X_i)
3:     return
4: end if
5: if Next[X_i] = ∅ then                                  ▷ Global Memoization
6:     Next[X_i] ← NextBestDisjointMatching(X_i)
7: end if
8: Y_{i+1} ← Next[X_i]
9: if S[Y_{i+1}] = ∅ then                                 ▷ Global Memoization
10:    S[Y_{i+1}] ← GetSubLatticeBetween(Y_{i+1}, B_{k-(i+1)})
11: end if
12: for X_{i+1} in S[Y_{i+1}] do
13:    Enumeration(X_0, X_1, ··· , X_i, X_{i+1})
14: end for
15: return
16:
17: procedure NextBestDisjointMatching(M)
18:    for every man m do
19:        Delete the first woman w on m's list            ▷ m's partner in M
20:        Delete the last man on w's list                 ▷ w's partner in M
21:    end for
22:    return GaleShapley(M)                  ▷ with modified preference list
23: end procedure
```

Lemma 7. *Algorithm 2 terminates in* $O(n^3 + n^2(|L| + |P|))$ *time, where* P *is the set of maximum-length chains of disjoint stable matchings and* L *is the set of all stable matchings featuring in the enumeration.*

Proof. If we do not consider the time taken to perform line 6 and line 10, the algorithm takes $O(n)$ time for every longest chain of pairwise disjoint stable matchings. Let L be the set of all stable matchings featuring in the enumeration. Let P be the set of all solutions (longest chains of pairwise disjoint stable matchings). Every execution of line 6 takes $O(n^2)$ time. Since we remember $\textsc{NextBestDisjointMatching}(X_i)$, we need to compute line 6 at most $|L|$ times. So, line 6 takes $O(n^2|L|)$ time.

Performing line 10 once takes $O(n^2|S[Y_{i+1}]|)$ time. Hence, the total time spent on line 10 is

$$O(n^2 \sum_{\substack{Y=Next[X],\\ X\in L}} |S[Y]|)$$

Let the summation be equal to S. Every stable matching M featuring in $S[Y]$ ($Y = \textsc{NextBestDisjointMatching}(X_i)$) features in the solution

$$(A_0, A_1, \cdots, X, M, B_{k-i}, \cdots, B_0)$$

Therefore, as the set mentioned above is unique given M,$S \leq |P| + 2n$. Thus, the total time complexity for line 6 to line 10 is $O(n^2|L| + n^2|P| + n^3)$. Printing the output would take $Max(|L|, |P|)$ time. □

We analyze the number of maximum-length chains of disjoint stable matchings in a random stable matchings instance with complete lists.

Lemma 8. *The probability of the number of maximum size chains of disjoint stable matchings exceeding* $(\frac{n}{\ln n})^{\ln n}$ *is at most* $O(\frac{(\ln n)^2}{n^2})$.

Proof. Let S be the random variable denoting the number of stable matchings in a random stable matching instance. Pittel [16] showed that $\mathbb{E}[S] = \Theta(n \ln n)$. Thus, there exist non-negative reals m_1, m_2 such that $m_1 n \ln n \leq \mathbb{E}[S] \leq m_2 n \ln n$ for sufficiently large n. Further, Lennon and Pittel [11] established that $Var(S) = \sigma^2 = O((n \ln n)^2)$. Thus, for sufficiently large n, there exists a non-negative real number c such that $Var(S) \leq c^2(n \ln n)^2$.

Thus, for a parameter k, we have

$$\begin{aligned} Pr(S \geq m_1 n \ln n + kcn \ln n) &\leq Pr(S \geq m_1 n \ln n \\ &\quad + kcn \ln n \cup S \leq m_2 n \ln n - kcn \ln n) \\ &\leq Pr(|S - \mathbb{E}[S]| \geq kcn \ln n) \\ &\leq Pr(|S - \mathbb{E}[S]| \geq k\sigma) \\ &\leq \frac{1}{k^2} \end{aligned}$$

where the last inequality follows from Chebyshev's inequality. Thus, if $f(k) = m_1 n \ln n + kcn \ln n$, then $Pr(S \geq f(k)) \leq \frac{1}{k^2}$.

Let $L_0, L_1, \ldots, L_{t-1}$ be the sub-lattices constructed in Algorithm 2 where $t-1=k$. Let $S_i = |L_i|$ for $0 \le i \le k$. Let $p = |P|$, the number of maximum-length chains of disjoint stable matchings in the given instance. We have $p \le \Pi_{i=0}^{k} S_i \le (\frac{\sum_{i=0}^{t} S_i}{t})^t$, where the first inequality follows from Lemma 6 and the last inequality follows from the AM-GM inequality. Since $\sum_{i=0}^{k} S_i \le S$, $p \le (\frac{S}{t})^t$.

From the above discussions, $Pr(p \ge (\frac{n}{\ln n})^{\ln n}) \le Pr((\frac{S}{t})^t \ge (\frac{n}{\ln n})^{\ln n}) \le Pr(S \ge n^2) + Pr(t \ge \ln n)$.

Observe that there exists a positive real m such that $f(\frac{n}{m \ln n}) \le n^2$. Thus, $Pr(S \ge n^2) \le Pr(S \ge f(\frac{n}{m \ln n})) \le \frac{m^2 (\ln n)^2}{n^2}$. [Knuth et al. 90] establishes that the probability of some person having more than $\ln n$ stable partners is super-polynomially small. Clearly, no one can have less than t stable partners since each person features alongide a distinct partner in each matching in a maximum size chain of disjoint stable matchings. Hence, $Pr(t \ge \ln n)$ is also super-polynomially small.

Thus, $Pr(p \ge (\frac{n}{\ln n})^{\ln n}) \le \frac{m_1^2 (\ln n)^2}{n^2}$ for some positive constant m_1. Thus, $Pr(p \ge (\frac{n}{\ln n})^{\ln n}) \le O(\frac{(\ln n)^2}{n^2})$. □

Corollary 4. *Algorithm 2 terminates in* $O(n^4 + n^{2 \ln n + 2})$ *time with probability* 1 *as* $n \to \infty$.

Proof. As established in the previous lemma (notation carrying over from the proof of the previou lemma), $Pr(S \ge n^2) \le O(\frac{(\ln n)^2}{n^2})$ and $Pr(p \ge (\frac{n}{\ln n})^{\ln n}) \le O(\frac{(\ln n)^2}{n^2})$ and hence, a simple union bound returns $Pr(S \ge n^2 \cup p \ge (\frac{n}{\ln n})^{\ln n}) \le O(\frac{(\ln n)^2}{n^2})$.

Plugging in $S = O(n^2)$ and $p = O(\frac{n}{\ln n})^{\ln n})$ in the run-time of algorithm 1, algorithm 1 terminates in $O(n^4 + n^{2 \ln n + 2})$ time with probability $1 - \Omega(\frac{(\ln n)^2}{n^2})$ which tends to 1 as $n \to \infty$. □

5 Conclusion

We consider the classical Stable Matching problem and address the question of finding a largest pairwise disjoint collection of solutions to this problem. We show that such a collection can in fact be found in time *linear* in the input size. The collection of stable matchings that our algorithm finds has the additional property that they form a *chain* in the distributive lattice of stable matchings. To the best of our knowledge this is the first work on finding pairwise disjoint *stable* matchings, though this question has received much attention for bipartite matchings without preferences.

A natural next question is what happens when we allow *small* intersections between the stable matchings. In particular: is the problem of finding a collection of k stable matchings such that no two of them share more than *one* edge, solvable in polynomial time? Or is this already NP-hard? Another interesting problem is whether we can find a largest edge-disjoint collection of stable matchings for the related STABLE ROOMMATES problem, in polynomial time.

References

1. Faudree, R.J., Gould, R.J., Lesniak, L.M .: Neighborhood conditions and edge-disjoint perfect matchings. Dis. Math. **91**(1):33–43 (1991)
2. Fomin, F.V., Golovach, P.A., Jaffke, L., Philip, G., Sagunov, D.: Diverse pairs of matchings. CoRR, abs/2009.04567 (2020). https://arxiv.org/abs/2009.04567
3. Gale, D., Shapley. L.S.: College admissions and the stability of marriage. Am. Math. Monthly **69**(1), 9–15 (1962). https://doi.org/10.1080/00029890.1962.11989827
4. Gusfield, D., Irving, R.W.: The Stable marriage problem - structure and algorithms. MIT Press, Foundations of computing series (1989)
5. Holyer, I.: The NP-completeness of edge-coloring. SIAM J. Comput. **10**(4), 718–720 (1981)
6. Irving, R.W., Leather, P.: The complexity of counting stable marriages. SIAM J. Comput. **15**(3), 655–667 (1986). https://doi.org/10.1137/0215048
7. Irving, R.W., Leather, P.: The complexity of counting stable marriages. SIAM J. Comput. **15**(3), 655–667 (1986)
8. Karlin, A.R., Gharan, S.O., Weber, R.: A simply exponential upper bound on the maximum number of stable matchings. In: Proceedings of the 50th Annual ACM SIGACT Symposium on Theory of Computing (STOC 2018). pp. 920–925, New York, NY, USA (2018). Association for Computing Machinery. https://doi.org/10.1145/3188745.3188848
9. Knuth, D.: Mariages stables et leurs relations avec d'autres problèmes combinatoires?: introduction à l'analyse mathèmatique des algorithmes. Presses de l'Universitè de Montrèal, Montrèal (1976)
10. König, D.: Über graphen und ihre anwendung auf determinantentheorie und mengenlehre. Mathematische Annalen **77**(4), 453–465 (1916)
11. Lennon, G., Pittel, B.: On the likely number of solutions for the stable marriage problem. Comb. Probab. Comput. **18**(3), 371–421 (2009)
12. Lu, H., Wang, D.G.L.: The number of disjoint perfect matchings in semi-regular graphs. Appl. Anal. Discrete Math. **11**(1), 11–38 (2017)
13. McVitie, D.G., Wilson, L.B.: The stable marriage problem. Commun. ACM **14**(7), 486–490 (1971)
14. Mkrtchyan, V.V., Musoyan, M.I., Tserunyan, A.V.: On edge-disjoint pairs of matchings. Discrete Math. **308**(23), 5823–5828 (2008)
15. Palmer, C., Pálvölgyi, D.: At most 4.47^n stable matchings (2020). http://arxiv.org/abs/2011.00915
16. Pittel, B.G.: The average number of stable matchings. SIAM J. Discret. Math. **2**(4), 530–549 (1989)
17. Teo, C.-P., Sethuraman, J.: The geometry of fractional stable matchings and its applications. Math. Oper. Res. **23**(4), 874–891 (1998)
18. William Thomas Tutte: A short proof of the factor theorem for finite graphs. Can. J. Math. **6**, 347–352 (1954)

Complementation in T-perfect Graphs

Yixin Cao and Shenghua Wang(✉)

Department of Computing, Hong Kong Polytechnic University, Hong Kong, China
shenghua.wang@connect.polyu.hk

Abstract. Inspired by applications of perfect graphs in combinatorial optimization, Chvátal defined t-perfect graphs in 1970s. The long efforts of characterizing t-perfect graphs started immediately, but embarrassingly, even a working conjecture on it is still missing after nearly 50 years. Unlike perfection, t-perfection is not closed under substitution or complementation. A full characterization of t-perfection with respect to substitution has been obtained in the Ph.D. thesis of Benchetrit. Through this work we attempt to understand t-perfection with respect to complementation. In particular, we show there are only five pairs of graphs such that both the graphs and their complements are minimally t-imperfect.

1 Introduction

Partly motivated by Shannon's work on communication theory, Berge proposed the concept of perfect graphs, and the two perfect graph conjectures [3], both settled now. Chvátal [9] and Padberg [19] independently showed that the independent set polytope of a perfect graph (the convex hull of incidence vectors of independent sets of the graph) is determined by non-negativity and clique inequalities, and this is part of efforts trying to characterize the independent set polytope of a graph [9,17–19]. Chvátal [9] went further to propose a class of graphs directly defined by the properties of their independent set polytopes. A graph is *t-perfect* if its independent set polytope can be fully described by non-negativity, edge, and odd-cycle inequalities. These two classes of perfection are incomparable: C_5 is t-perfect but not perfect, while K_4 is perfect but not t-perfect. Similar as perfect graphs, the maximum independent set problem can be solved in polynomial time in t-perfect graphs [15]; see also [11].

The progress toward understanding t-perfection has been embarrassingly slow. While the original paper of Berge [2] on perfect graphs already contains several important subclasses of perfect graphs, thus far, only few graph classes are known to be t-perfect. Because of the absence of odd cycles, bipartite graphs are trivial examples, and this can be generalized to almost bipartite graphs [12]. Another class is the series-parallel graphs [4]. Extending these two classes, Gerards [13] showed that any graph contains no *odd-*K_4 (a subdivision of K_4 in which every triangle of K_4 becomes an odd cycle) is t-perfect.

Supported by RGC grants 15201317 and 15226116, and NSFC grant 61972330.

Ł. Kowalik et al. (Eds.): WG 2021, LNCS 12911, pp. 106–117, 2021.
https://doi.org/10.1007/978-3-030-86838-3_8

The strong perfect graph theorem states that a graph G is perfect if and only if G does not contain any odd hole (an induced cycle of length at least four) or its complement [8]. In other words, the minimally imperfect graphs (imperfect graphs whose proper induced subgraphs are all perfect) are odd holes and their complements. Naturally, we would like to know minimally t-imperfect graphs, whose definition is more technical and is left to the next section. However, even a working conjecture on them is still missing. Full success has only been achieved on special graphs, e.g., claw-free graphs [7] and P_5-free graphs [5]. In summary, known minimally t-imperfect graphs include (3, 3)-partitionable graphs [6,10] (Fig. 1), odd wheels [20], even Möbius ladders [21], and the complements of some cycle powers ($\overline{C_7}$, $\overline{C_{13}^3}$, $\overline{C_{13}^4}$, $\overline{C_{19}^7}$).

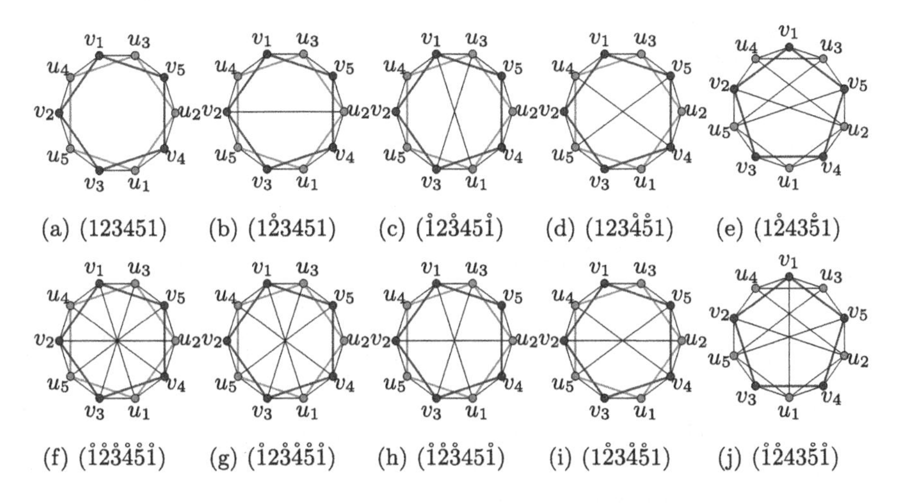

(a) (123451) (b) ($1\mathring{2}3451$) (c) ($\mathring{1}2\mathring{3}45\mathring{1}$) (d) ($123\mathring{4}\mathring{5}1$) (e) ($1\mathring{2}43\mathring{5}1$)

(f) ($\mathring{1}\mathring{2}\mathring{3}\mathring{4}\mathring{5}\mathring{1}$) (g) ($\mathring{1}2\mathring{3}\mathring{4}\mathring{5}\mathring{1}$) (h) ($\mathring{1}\mathring{2}\mathring{3}45\mathring{1}$) (i) ($1\mathring{2}3\mathring{4}\mathring{5}1$) (j) ($\mathring{1}\mathring{2}43\mathring{5}\mathring{1}$)

Fig. 1. The (3, 3)-partitionable graphs (the notation will be introduced in Sect. 2).

A particularly nice property of perfect graphs is that they are closed under complementation [16]. The key step of proving it is the Replication Lemma: The class of perfect graphs is closed under (clique) substitution. As evidenced by K_4, t-perfection is closed under neither substitution nor complementation, and this may partially explain the difficulty in characterizing t-perfect graphs. Benchetrit [1] has fully characterized t-perfection with respect to substitution. The purpose of this paper is to understand t-perfection with respect to complementation. In particular, we want to know whether there exist minimally t-imperfect graphs whose complements are also minimally t-imperfect. Our main result is as follows.

Theorem 1. *Let G be a minimally t-imperfect graph. The complement of G is minimally t-imperfect if and only if G is a* (3, 3)*-partitionable graph.*

We start from an easy observation that such a graph G contains a 5-hole. Fixing a 5-hole C, we study the connection between C and other vertices. We show that the order of G is at most ten, and G is of one of few simple patterns.

A careful inspection of these patterns leads to the main result. As a byproduct, we characterize all self-complementary t-perfect graphs that are not perfect.

Theorem 2. *Let G be a self-complementary graph that is not perfect. Then G is t-perfect if and only if G is one of the five graphs in Fig. 2.*

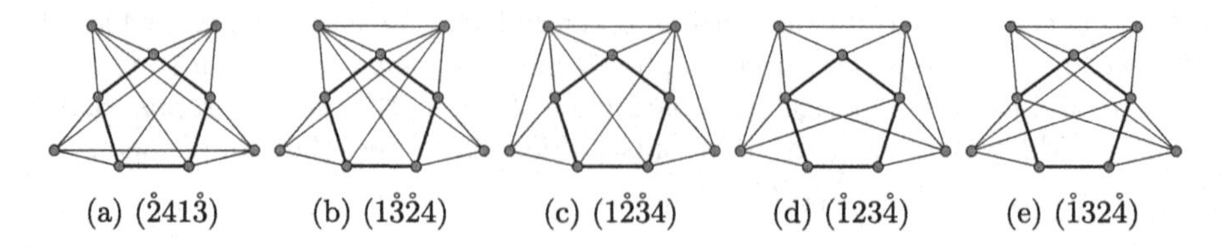

Fig. 2. Self-complementary t-perfect graphs that contain a C_5 (thick lines).

2 Core Graphs

We only consider undirected and simple graphs. Throughout the paper, we use n to denote $|V(G)|$, the *order* of G. Let $G[U]$ denote the subgraph of G induced by U, and let $G - U = G[V(G)\backslash U]$, which is simplified as $G - v$ if U comprises of a single vertex v. A *clique* is a set of pairwise adjacent vertices, and an *independent set* is a set of vertices that are pairwise nonadjacent. The complement $\overline{G}$ of a graph G is defined on the same vertex set as G and two distinct vertices of $\overline{G}$ are adjacent if and only if they are not adjacent in G. A graph is *almost bipartite* if there is a vertex whose deletion leaves the graph bipartite.

For $\ell \geq 1$, we use P_ℓ and K_ℓ to denote the path graph and complete graph, respectively, on ℓ vertices. For $\ell \geq 3$, we use C_ℓ and W_ℓ to denote, respectively, the ℓ-cycle and the *ℓ-wheel*, which is obtained from a C_ℓ by adding a new vertex and making it adjacent to all vertices on the cycle; note that W_3 is precisely K_4. For $\ell \geq 4$, an induced ℓ-cycle is also called an *ℓ-hole*. An ℓ-cycle, ℓ-hole, or ℓ-wheel is *odd* if ℓ is odd. For integers $p, q \geq 2$, a graph G is *(p, q)-partitionable* if $n = pq + 1$ and for every vertex v, the set $V(G)\backslash\{v\}$ can be partitioned into q independent sets of order p and can be partitioned into p cliques of order q.

The *independent set polytope* of a graph G is the convex hull of the characteristic vectors of all independent sets in G. For a graph G, let $P(G)$ denote the polytope defined by

$$
\begin{aligned}
&0 \leq x_v \leq 1 && \text{for every vertex } v \in V(G),\\
&x_u + x_v \leq 1 && \text{for every edge } uv \in E(G),\\
&x(V(C)) \leq (|V(C)| - 1)/2 && \text{for every induced odd cycle } C \text{ in } G.
\end{aligned}
$$

Since the characteristic vector of every independent set of G is in $P(G)$, the independent set polytope of a graph G is contained in $P(G)$, while the other direction is not true in general. A graph G is *t-perfect* if $P(G)$ is precisely the independent set polytope of G. It is not difficult to see that every vector in the independent set polytope of G also satisfies the clique constraints

$$\sum_{v \in K} x_v \leq 1 \qquad \text{for every clique } K \text{ of } G.$$

Since the vector $(\frac{1}{3}, \frac{1}{3}, \frac{1}{3}, \frac{1}{3})^\mathsf{T}$ is in $P(K_4)$ but does not satisfy the clique constraint, K_4 is not t-perfect.

A graph G is *perfect* if neither G nor $\overline{G}$ contains an odd hole [2,8]. The independent set polytope of a perfect graph is determined by non-negativity and clique inequalities [9,19]. If a perfect graph G contains no K_4, then every clique of G has order at most three, and hence any clique constraint in G is one of the three in the definition of $P(G)$.

Proposition 1. *Every K_4-free perfect graph is t-perfect.*

It is easy to verify that t-perfection is preserved under vertex deletions: For every $v \in V(G)$, the polytope $P(G - v)$ is the intersection of $P(G)$ with the face $x_v = 0$. Moreover, t-perfection is also preserved under *t-contractions* at a vertex v with $N(v)$ being an independent set—contracting $N(v) \cup \{v\}$ into a single vertex [14]. Any graph H that can be obtained from G by a sequence of vertex deletions and t-contractions is a *t-minor* of G, and H is a *proper t-minor* of G if H has fewer vertices than G. Therefore, t-perfection is closed under taking t-minors. A graph is *minimally t-imperfect* if it is t-imperfect but all its proper t-minors are t-perfect, e.g., K_4.

We say that a graph G is a *core graph* if neither G nor its complement contains a t-imperfect graph as a proper t-minor. By definition, any t-minor of a core graph is also a core graph. Moreover, if G is a core graph, then G is either t-perfect or minimally t-imperfect, and so is $\overline{G}$; it is possible that G is t-perfect while $\overline{G}$ is minimally t-imperfect, e.g., C_7 and $\overline{C_7}$. However, there are t-perfect graphs that are not core graphs, e.g., C_9 and $\overline{K_5}$.

Proposition 2. *A core graph of order at least five cannot contain a K_4 or $\overline{K_4}$.*

By Proposition 1, any $\{K_4, \overline{K_4}\}$-free perfect graph is a core graph. Therefore, we focus on core graphs that are not perfect. Such a graph cannot contain an odd hole longer than seven or its complement as a proper induced subgraph.

Proposition 3. *Let G be a core graph different from C_7 and $\overline{C_7}$. Every odd hole in G is a C_5. Moreover, if G is t-imperfect, then G contains a C_5.*

Proof. For the first assertion, note that $\overline{C_7}$ is t-imperfect, so the only core graph that contains C_7 as an induced subgraph is C_7 itself; and for $k \geq 4$, the hole C_{2k+1} contains a $\overline{K_4}$. For the second assertion, note that if G does not contain a C_5, then G is perfect, hence t-perfect by Propositions 1 and 2. □

As shown in the following two propositions, 5-holes are pivotal in core graphs.

Proposition 4. *Let G be a core graph different from W_5 and its complement. If G contains a 5-hole C, then for every $u \in V(G)\setminus V(C)$, either*

i) u has exactly two neighbors on C, and they are consecutive on C; or
ii) u has exactly three neighbors on C, and they are not consecutive on C.

Proof. We consider the subgraph G' of G induced by u and the five vertices on C. If u is adjacent to all vertices on C, then G' is a W_5. Since W_5 is t-imperfect, $G = G'$, a contradiction. If u is adjacent to four vertices or three consecutive vertices on C, then K_4 is a proper t-minor of G', with t-contraction at a non-neighbor of u on C. Noting that the complement of C is a C_5, we end with the same contradictions on $\overline{G}$ if u has zero or one neighbor on C, or its two neighbors on C are not consecutive. □

Proposition 5. *In a core graph, every pair of consecutive vertices on a 5-hole has at most one common neighbor.*

Proof. Let G be a core graph, and let $v_1v_2v_3v_4v_5$ be a 5-hole in G. Suppose for contradiction that there are two vertices $x, y \in N(v_2) \cap N(v_3)$. By Proposition 4, neither of x and y is adjacent to v_1 or v_4. But then dependent on whether they are adjacent, x and y either form a K_4 with $\{v_2, v_3\}$, or a $\overline{K_4}$ with $\{v_1, v_4\}$, both contradicting Proposition 2. The same argument applies to other edges on the 5-cycle. □

Propositions 4 and 5 together imply an upper bound on the order of core graphs.

Corollary 1. *If a core graph contains a C_5, then it has at most ten vertices.*

Let G be a core graph that contains a 5-hole, and we use the following notations for its vertices and edges, where the indices are always understood as modulo 5. We fix a 5-hole C and number its vertices as $v_1, \ldots, v_5$ in order, and let $U = V(G)\setminus V(C)$. According to Proposition 4, each vertex in U is adjacent to two consecutive vertices on C. If a vertex in U is adjacent to v_i and v_{i+1}, $i = 1, \ldots, 5$, then we denote it as u_{i+3}; by Proposition 5, this is well defined. The five edges on C are all the edges among $v_1, \ldots, v_5$. For each u_i, the two edges u_iv_{i+2} and u_iv_{i+3} must exist in G. Apart from these $2|U|+5$ edges, by Proposition 4, the other possible edges are among U or u_iv_i, $i = 1, \ldots, 5$; they are called *potential edges*. Shown in Figs. 3(a, b) are two *pattern graphs*, from which we can obtain different particular graphs, with different materializations of potential edges. We use (1324) to denote the graph of pattern Fig. 3(b) in which U induces a path, with edges u_1u_3, u_2u_3, and u_2u_4. In case that $G[U]$ is not connected, we use $\|$ to separate its components, e.g., $(14\|23)$ in Fig. 3(d). Moreover, we cap an index i with $\circ$ to denote the presence of the edge u_iv_i, e.g., $(1\mathring{3}\mathring{2}4)$ in Fig. 3(c).

Proposition 6 ($\star$[1])**.** *The following graphs are t-perfect:* (12), $(1\|\mathring{2})$, $(1\mathring{2})$, $(\mathring{1}\|\mathring{2})$, $(\mathring{1}\mathring{2})$, $(1\|23)$, $(\mathring{3}1\mathring{2})$, $(\mathring{1}\mathring{3}\|\mathring{2})$, $(1\|\mathring{2}4)$, $(14\mathring{2})$, $(1\|\mathring{2}\mathring{4})$, $(1\mathring{4}\mathring{2})$, $(\mathring{1}\|\mathring{2}\|\mathring{4})$, $(\mathring{1}\mathring{2}4)$, $(\mathring{1}\mathring{2}\mathring{4})$, $(1\mathring{3}\mathring{4}\mathring{2})$, $(1\mathring{3}\mathring{4}2)$, $(\mathring{1}\mathring{3}\mathring{4}2)$, $(\mathring{1}\mathring{2}\mathring{4}3)$, $(2\mathring{3}14)$, $(2\mathring{3}\mathring{1}4)$, $(23\mathring{1}4)$, $(1\mathring{4}32)$, $(\mathring{1}\mathring{2}\mathring{4}\mathring{3}1)$, $(1\mathring{2}\mathring{4}3)$, $(1\mathring{3}\mathring{2}41)$, $(1\mathring{3}\mathring{2}4)$, $(14\|23)$, $(1\mathring{4}\mathring{3}2)$, $(1\mathring{3}\|\mathring{2}4)$, *and* $(\mathring{2}41\mathring{3})$.

[1] Proofs of propositions marked with a $\star$ is deferred to the full version.

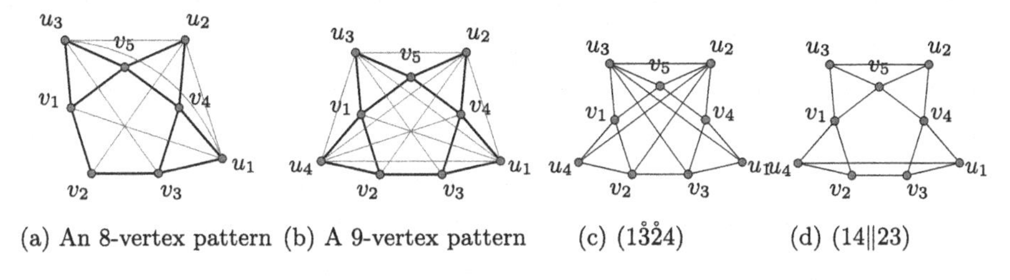

(a) An 8-vertex pattern (b) A 9-vertex pattern (c) $(1\mathring{3}\mathring{2}4)$ (d) $(14\|23)$

Fig. 3. Two patterns (a, b) and two particular graphs (c, d) of the second pattern. In the patterns, potential edges are depicted as thin green lines, while normal ones as thick black lines; no other edges can exist. (Color figure online)

The following observations are easy consequences of Proposition 4. Here $i = 1, \ldots, 5$.

Obs.1) If both $u_i v_i$ and $u_{i+1}u_{i+2}$ are in $E(G)$, then at least one of $u_i u_{i+1}$ and $u_i u_{i+2}$ is in $E(G)$; otherwise, v_{i+4} has four neighbors on the 5-cycle $u_i v_i u_{i+2} u_{i+1} v_{i+3}$. By symmetry, if both $u_i v_i$ and $u_{i-1}u_{i-2}$ are in $E(G)$, then at least one of $u_i u_{i-2}$ and $u_i u_{i-1}$ is in $E(G)$.

Obs.2) If both $u_i u_{i+1}$ and $u_i u_{i+3}$ are in $E(G)$, then at least one of $u_i v_i$ and $u_{i+1}u_{i+3}$ is in $E(G)$; otherwise, v_{i+3} has three consecutive neighbors on the 5-cycle $u_i u_{i+1} v_{i+4} v_i u_{i+3}$. By symmetry, if both $u_i u_{i-1}$ and $u_i u_{i-3}$ are in $E(G)$, then at least one of $u_i v_i$ and $u_{i-1}u_{i-3}$ is in $E(G)$.

Obs.3) Suppose, all of $u_{i-2}u_{i-1}$, $u_{i-1}u_{i+1}$, and $u_{i+1}u_{i+2}$ are in $E(G)$. If $u_{i-1}v_{i-1}$ or $u_{i+1}v_{i+1}$ is in $E(G)$, then at least one of $u_{i-1}u_{i+2}$, $u_{i-2}u_{i+1}$, and $u_{i-2}u_{i+2}$ is in $E(G)$; otherwise, v_{i-1} or v_{i+1} has four neighbors on the 5-cycle $u_{i-2}u_{i-1}u_{i+1}u_{i+2}v_i$.

Obs.4) If $u_{i-1}u_{i+1} \in E(G)$ and $u_{i-1}v_{i-1}, u_{i+1}v_{i+1} \notin E(G)$, then $u_{i+1}u_{i+2}$, $u_{i-1}u_{i-2} \notin E(G)$, and $u_i u_{i-1}, u_i u_{i+1} \in E(G)$; otherwise, the neighbors of u_{i-2}, u_{i+2}, or, respectively, u_i on the cycle $u_{i-1}u_{i+1}v_{i-1}v_i v_{i+1}$ does not satisfy Proposition 4.

Obs.5) If $u_i u_{i+1} \notin E(G)$ and at least one of u_i and u_{i+1} is adjacent to u_{i+3}, then at most one of $u_i v_i$ and $u_{i+1}v_{i+1}$ can be in $E(G)$; otherwise, u_{i+3} has three consecutive neighbors on the 5-cycle $u_i v_i v_{i+1} u_{i+1} v_{i+3}$.

Obs.6) If $u_{i+1}v_{i+1} \in E(G)$ and none of $u_{i+1}u_{i+2}$, $u_{i+2}u_{i-2}$, and $u_{i-1}u_{i-2}$ is, then $u_{i+1}u_{i-2}$, $u_{i+2}u_{i-1}$, and $u_{i+1}u_{i-1}$ cannot be all present in G; otherwise, v_{i+1} has four neighbors on the 5-cycle $u_{i-1}u_{i+2}v_i u_{i-2}u_{i+1}$. By symmetry, if $u_{i-1}v_{i-1}$ is in $E(G)$ and none of $u_{i+1}u_{i+2}$, $u_{i+2}u_{i-2}$, and $u_{i-1}u_{i-2}$ is in $E(G)$, then $u_{i+1}u_{i-2}$, $u_{i+2}u_{i-1}$, and $u_{i+1}u_{i-1}$ cannot be all in $E(G)$.

All graphs of pattern Fig. 3(a) are summarized in Table 1 and Lemma 1.

Lemma 1. *Let G be a core graph of order eight. At least one of G and $\overline{G}$ i) is t-perfect; or ii) has a degree-2 vertex in U.*

Proof. Note that if the degree of a vertex is five in G, then its degree in $\overline{G}$ is two. According to Table 1, it suffices to show that graphs $(\mathring{1}\mathring{3}\|\mathring{2})$, $(\mathring{1}3\|\mathring{2})$, $(\mathring{1}2\mathring{3})$, $(\mathring{3}1\mathring{2})$,

Table 1. Graphs of pattern Fig. 3(a). The columns are for combinations of edges among U; the cases with only u_2u_3 and only $\{u_1u_3, u_2u_3\}$ are omitted because they are symmetric to respectively, u_1u_2 and $\{u_1u_2, u_1u_3\}$. The rows are possible combinations of edges between U and C.

	All	$\{u_1u_2, u_1u_3\}$	$\{u_1u_2, u_2u_3\}$	$\{u_1u_2\}$	$\{u_1u_3\}$
All	$d(u_1) = 5$	$d(u_1) = 5$	$d(u_2) = 5$	Obs.1 ($i = 3$)	$(\mathring{1}\mathring{3}\Vert\mathring{2})$
$\{u_1v_1, u_2v_2\}$	$d(u_1) = 5$	$d(u_1) = 5$	$d(u_2) = 5$	$d(u_3) = 2$	$(\mathring{1}3\Vert\mathring{2})$
$\{u_1v_1, u_3v_3\}$	$d(u_1) = 5$	$d(u_1) = 5$	$(\mathring{1}2\mathring{3})$	Obs.1 ($i = 3$)	$d(u_2) = 2$
$\{u_2v_2, u_3v_3\}$	$d(u_2) = 5$	$(\mathring{3}1\mathring{2})$	$d(u_2) = 5$	Obs.1 ($i = 3$)	$\cong (\mathring{1}3\Vert\mathring{2})$
$\{u_1v_1\}$	$d(u_1) = 5$	$d(u_1) = 5$	$(\mathring{1}23)$	$d(u_3) = 2$	$d(u_2) = 2$
$\{u_2v_2\}$	$d(u_2) = 5$	Obs.4 ($i = 2$)	$d(u_2) = 5$	$d(u_3) = 2$	Obs.4 ($i = 2$)
$\{u_3v_3\}$	$d(u_3) = 5$	$(\mathring{3}12)$	$\cong (\mathring{1}23)$	Obs.1 ($i = 3$)	$d(u_2) = 2$
None	$\overline{G} \cong (\mathring{1}\Vert\mathring{2}\Vert\mathring{4})$	Obs.4 ($i = 2$)	(123)	$d(u_3) = 2$	$d(u_2) = 2$

$(\mathring{1}23)$, $(\mathring{3}12)$, $(\mathring{1}\|\mathring{2}\|\mathring{4})$, and (123) are t-perfect. We have seen in Proposition 6 that $(\mathring{1}\|\mathring{2}\|\mathring{4})$, $(\mathring{3}1\mathring{2})$, and $(\mathring{1}\mathring{3}\|\mathring{2})$ are t-perfect. The graph $(\mathring{1}3\|\mathring{2})$ is t-perfect because $(\mathring{1}3\|\mathring{2})$ is isomorphic to $(\mathring{2}41\mathring{3}) - u_1$, and $(\mathring{2}41\mathring{3})$ is t-perfect. On the other hand, $(\mathring{3}12)$, $(\mathring{1}2\mathring{3})$, $(\mathring{1}23)$, and (123) are isomorphic to, respectively, $(1\mathring{2}43\mathring{5}1) - \{u_1, u_2\}$, $(1\mathring{2}43\mathring{5}1) - \{u_3, u_4\}$, $(1\mathring{2}3451) - \{u_1, u_5\}$, and $(123451) - \{u_1, u_5\}$, all t-perfect. □

3 Degree-Bounded Core Graphs of Order Nine

By Propositions 4 and 5, every core graph of order nine is of the pattern in Fig. 3(b). In this section, let G denote a core graph of order nine where the degree of every vertex is between 3 and 5. (The reason of imposing degree constraints will become clear shortly.) We consider whether edges u_iu_{i+1}, $i = 1, 2, 3$ are present in G.

Proposition 7. *Let G be a degree-bounded core graph on nine vertices. If for all $i = 1, 2, 3$, the edge u_iu_{i+1} is in $E(G)$, then G is an induced subgraph of a $(3, 3)$-partitionable graph.*

Proof. We argue first that none of u_1u_4, u_1u_3, and u_2u_4 can be present in G; i.e., $u_1u_2u_3u_4$ is an induced path in G. Suppose that $u_1u_4 \in E(G)$, then by Obs.4 (with $i = 5$), at least one of u_4v_4 and u_1v_1 is in $E(G)$. We may assume that $u_4v_4 \in E(G)$, and the other case is symmetric. Since $\{u_1, u_4, u_2, v_4\}$ is not a clique, $u_2u_4 \notin E(G)$. By Obs.2 (with $i = 1$), $u_1v_1 \in E(G)$, and then since $\{u_1, u_3, u_4, v_1\}$ is not a clique, u_1u_3 cannot be present. But then $G - \{v_2, v_3\}$ is isomorphic to $\overline{C_7}$, a contradiction. Thus, $u_1u_4 \notin E(G)$. By Obs.2 (with $i = 2$), (noting $u_1u_2 \in E(G)$,) the presence of u_2u_4 would imply the presence of u_2v_2, but then $d(u_2) = 6$. Thus, $u_2u_4 \notin E(G)$, and by a symmetric argument, $u_1u_3 \notin E(G)$.

Now that none of u_1u_4, u_1u_3, and u_2u_4 is present, we consider all possible combinations of edges $\{u_iv_i \mid i = 1, \dots, 4\} \cap E(G)$. If none of them is in $E(G)$, then G is isomorphic to $(123451) - u_1$. If all of them are in $E(G)$, then G is isomorphic to $(\mathring{1}\mathring{2}\mathring{3}\mathring{4}\mathring{5}\mathring{1}) - u_1$. If only one $u_i v_i$ is in $E(G)$, then G is isomorphic to $(\mathring{1}2\mathring{3}45\mathring{1}) - u_3$ or $(123\mathring{4}\mathring{5}1) - u_4$. If only one $u_i v_i$ is absent, then G is isomorphic to $(\mathring{1}\mathring{2}\mathring{3}45\mathring{1}) - u_4$ or $(1\mathring{2}3\mathring{4}\mathring{5}1) - u_1$. Otherwise, exact two of edges $u_i v_i$ are in $E(G)$, then G is isomorphic to one of $(\mathring{1}2\mathring{3}45\mathring{1}) - u_4$, $(\mathring{1}\mathring{2}\mathring{3}45\mathring{1}) - u_1$, $(\mathring{1}2\mathring{3}45\mathring{1}) - u_2$, and $(123\mathring{4}\mathring{5}1) - u_2$. □

In the rest, at least one of u_1u_2, u_2u_3, u_3u_4 is absent from G. In the second case, we assume that both u_1u_2 and u_2u_3 are absent from G; see Fig. 4(b).

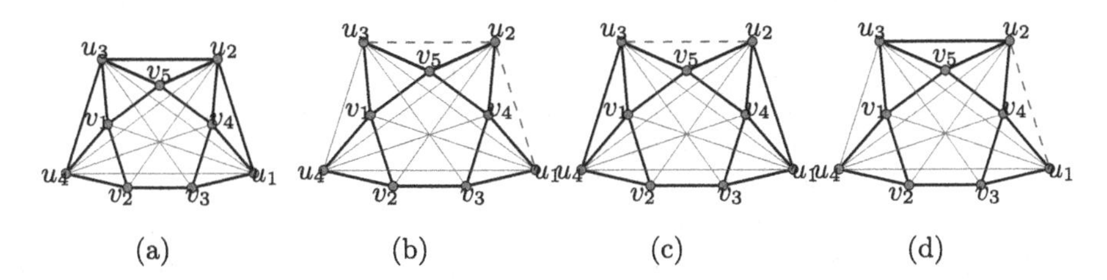

Fig. 4. Refined patterns on nine vertices, the potential edges in Fig. 3(b) but absent here are emphasized by red dashed lines. (a) all the three edges u_1u_2, u_2u_3, and u_3u_4 are present; (b) both u_1u_2 and u_2u_3 are absent; (c) u_2u_3 is absent but both u_1u_2 and u_3u_4 are present; (d) u_1u_2 is absent but u_2u_3 is present. (Color figure online)

Proposition 8 (⋆). *Let G be a degree-bounded core graph on nine vertices. If both u_1u_2 and u_2u_3 are absent from G, then G is isomorphic to one of* $(1\mathring{3}\mathring{4}\mathring{2})$, $(1\mathring{3}\mathring{4}2)$, $(\mathring{1}\mathring{3}\mathring{4}2)$, $(1\mathring{3}\|\mathring{2}4)$, *and* $(\mathring{2}41\mathring{3})$.

It is symmetric to Proposition 8 if both u_2u_3 and u_3u_4 are absent. Next we consider the situation that u_2u_3 is absent but both u_1u_2 and u_3u_4 are present; see Fig. 4(c).

Proposition 9 (⋆). *Let G be a degree-bounded core graph on nine vertices. If both u_1u_2 and u_3u_4 are in $E(G)$ but u_2u_3 is not, then G is isomorphic to one of* $(1\mathring{2}\mathring{4}3)$, $(\mathring{1}\mathring{2}\mathring{4}\mathring{3}\mathring{1})$, $(\mathring{1}\mathring{2}\mathring{4}3)$, $(1\mathring{2}43\mathring{5}1) - u_2$, *and* $(\mathring{1}\mathring{2}43\mathring{5}\mathring{1}) - u_5$.

In the last case, u_2u_3 is in $E(G)$, but at least one of u_1u_2 and u_3u_4 is not. We may assume without loss of generality that u_1u_2 is absent; see Fig. 4(d).

Proposition 10 (⋆). *Let G be a degree-bounded core graph on nine vertices. If u_2u_3 is in $E(G)$ but u_1u_2 is not, then G is isomorphic to one of* $(2\mathring{3}14)$, $(2\mathring{3}\mathring{1}4)$, $(23\mathring{1}4)$, $(1\mathring{4}32)$, $(1\mathring{3}\mathring{2}41)$, $(1\mathring{3}\mathring{2}4)$, $(14\|23)$, $(1\mathring{4}\mathring{3}2)$, $(\mathring{1}\mathring{2}43\mathring{5}\mathring{1}) - u_3$, $(1\mathring{2}43\mathring{5}1) - u_3$, *and* $(1\mathring{2}43\mathring{5}1) - u_1$.

We are now ready to summarize Propositions 7–10 and prove Theorem 2.

Corollary 2. *All degree-bounded core graphs of order nine are t-perfect. Only* $(\mathring{1}2\mathring{3}45\mathring{1}) - u_2$, $(123\mathring{4}\mathring{5}1) - u_2$, $(1\mathring{2}43\mathring{5}1) - u_1$, $(1\mathring{3}\mathring{2}4)$, *and* $(\mathring{2}41\mathring{3})$ *of them are self-complementary graphs.*

Proof (of Theorem 2). The sufficiency is quite obvious. It is easy to verify that C_5, $(\mathring{2}41\mathring{3})$, $(1\mathring{3}\mathring{2}4)$, $(1\mathring{2}\mathring{3}4)$, $(\mathring{1}23\mathring{4})$, and $(\mathring{1}32\mathring{4})$ are self-complementary. Since all of them contain a C_5, they are not perfect. We have seen that $(\mathring{2}41\mathring{3})$ and $(1\mathring{3}\mathring{2}4)$ are t-perfect, while $(1\mathring{2}\mathring{3}4)$, $(\mathring{1}23\mathring{4})$, and $(\mathring{1}32\mathring{4})$ are isomorphic to $(123\mathring{4}\mathring{5}1) - u_2$, $(\mathring{1}2\mathring{3}45\mathring{1}) - u_2$, and $(1\mathring{2}43\mathring{5}1) - u_1$ respectively, hence t-perfect as well.

For the necessity, suppose that G is a self-complementary t-perfect graph and not perfect. Since both G and $\overline{G}$ are t-perfect, G is a core graph. By Corollary 1, $5 \leq n \leq 10$. Since the order of a self-complementary graph is either $4k$ or $4k+1$ for some $k \geq 0$, we can have $n \in \{5, 8, 9\}$. Since G is not perfect, it contains an odd hole, and by Proposition 3, every odd hole in G is a 5-cycle. If $n = 5$, then G is C_5.

If $n = 9$, then G is of pattern Fig. 3(b). We argue that G is degree bounded. Every vertex in C has degree at least three and at most five. Suppose that one vertex $u \in U$ has degree two, then it is not adjacent to any other vertex in U. But then the degree of u in $\overline{G}$ is six; thus there is a degree-6 vertex, which has to be in U. But then we have a vertex in U that is nonadjacent to others in U, and another vertex in U that is adjacent to all of the others in U, a contradiction. By Corollary 2, G is one of $(\mathring{1}23\mathring{4})$, $(\mathring{2}41\mathring{3})$, $(\mathring{1}32\mathring{4})$, $(1\mathring{2}\mathring{3}4)$, and $(1\mathring{3}\mathring{2}4)$.

It remains to show that there is no graph of order 8 satisfying the conditions. Let G be a core graph of order 8. We may assume that the indices for the three vertices in U are not consecutive: If G is of pattern Fig. 3(a), then we can consider its complement. (With different choices of 5-cycles, a core graph may be of more than one patterns.) If there is a vertex x of degree 2, then $x \in U$, and the two neighbors of x are adjacent. Then in $\overline{G}$, every vertex in U has degree at least three, which means x is mapped to a vertex y in C. However, if y has degree two, then its two neighbors are not adjacent in $\overline{G}$, a contradiction. Therefore, the minimum degree is at least three, and since G is self-complementary, the maximum degree is at most four. By Lemma 1, G can only be one of $(\mathring{1}\mathring{3}\|\mathring{2})$, $(\mathring{1}3\|\mathring{2})$, $(\mathring{1}2\mathring{3})$, $(\mathring{3}1\mathring{2})$, $(\mathring{1}23)$, $(\mathring{3}12)$, $\overline{(\mathring{1}\|\mathring{2}\|\mathring{4})}$, and (123), but none of them is self-complementary. □

4 Proof of Theorem 1

It is known that the $(3, 3)$-partitionable graphs are minimally t-imperfect [6]. Thus, we only need to show the sufficiency in Theorem 1. We say that a clique K of a connected graph G is a *clique separator* of G if $G - K$ is not connected.

Lemma 2 ([9,14]). *No minimally t-imperfect graph contains a clique separator.*

Throughout this section, we assume that both G and its complement $\overline{G}$ are minimally t-imperfect graphs. By Lemma 2, neither G nor $\overline{G}$ can have a clique separator. Thus, for each vertex $u \in U$, we have

$$2 < d(u) < n - 3. \tag{1}$$

If $d(u) = n - 3$, then u has two neighbors in $\overline{G}$, which is a clique separator.

Note that G is a core graph. By Proposition 3 and Corollary 1, the order of G is between five and ten. Recall that every almost bipartite graph is t-perfect [12]. The only core graph of order five is C_5. Both core graphs of order six, (1) and ($\mathring{1}$), are almost bipartite, e.g., removing v_3. There are 16 core graphs of order seven, (12), ($\mathring{1}$2), (1$\mathring{2}$), ($\mathring{1}\mathring{2}$), (1‖2), ($\mathring{1}$‖2), (1‖$\mathring{2}$), ($\mathring{1}$‖$\mathring{2}$), and their complements. All the listed eight graphs become bipartite after removing v_4, hence almost bipartite. By Lemma 1 and the degree requirements (1), G cannot have order eight either. Likewise, by Corollary 2, all core graphs of order nine satisfying (1) are t-perfect. Therefore, we are only left with $n = 10$.

In the rest of this section, the order of G is ten. Our analysis is based on whether (123451) is a (not necessarily induced) subgraph of G. Let us start with the case that all the five edges u_iu_{i+1} for $i = 1, \ldots, 5$ are in $E(G)$; see Fig. 5(a).

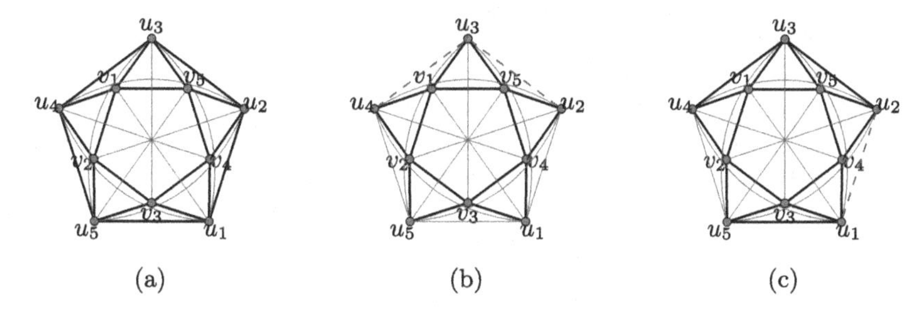

Fig. 5. (a) All the edges u_iu_{i+1} for $i = 1, \ldots, 5$ are present; (b) both u_2u_3 and u_3u_4 are absent; (c) u_1u_2 is absent, while only u_2u_3, u_3u_4, and u_1u_5 are present.

Proposition 11. *If for all $i = 1, \ldots, 5$, the edge u_iu_{i+1} is in $E(G)$, then G is one of the $(3,3)$-partitionable graphs.*

Proof. We first argue that U induces a cycle. Suppose for contradiction that u_1u_3 is present. By Obs.4 (with $i = 2$), at least one of u_1v_1 and u_3v_3 is in $E(G)$. Since they are symmetric, we consider $u_1v_1 \in E(G)$. Since $\{u_1, u_3, u_4, v_1\}$ is not a clique, $u_1u_4 \notin E(G)$. Then by Obs.2 (with $i = 3$), $u_3v_3 \in E(G)$, and since $\{u_1, u_3, u_5, v_3\}$ is not a clique, $u_3u_5 \notin E(G)$. But then $G - \{v_4, v_5, u_2\}$ is isomorphic to $\overline{C_7}$, and G is not minimally t-imperfect. Now that $G[U]$ is a C_5, dependent on the combination of edges $u_iv_i, i = 1, \ldots, 5$, we are in one of the $(3,3)$-partitionable graphs that contain (123451). □

The next proposition states that there cannot be two consecutive missing edges in U; see Fig. 5(b).

Proposition 12 (⋆). *For $i = 1, \ldots, 5$, at least one of u_iu_{i-1}, u_iu_{i+1} is in $E(G)$.*

In the remaining case, u_iu_{i+1} for some $i = 1, \ldots, 5$ is absent, but both $u_{i+1}u_{i+2}$ and u_iu_{i-1} are present. Moreover, by Proposition 12, at least one of $u_{i+2}u_{i+3}$ and $u_{i-1}u_{i-2}$ is in $E(G)$. See Fig. 5(d).

Proposition 13. *If there is an* $i = 1, \ldots, 5$ *such that* $u_i u_{i+1}$ *is not in* $E(G)$*, then* G *is one of the* $(3, 3)$*-graphs.*

Proof. Without loss of generality, let $i = 1$. Then $u_1u_2 \notin E(G)$, u_2u_3 and u_1u_5 are in $E(G)$, and at least one of u_3u_4 and u_4u_5 is in $E(G)$. We show by contradiction that u_3u_4 and u_4u_5 cannot be both in $E(G)$. In particular, we show that none of u_1v_1, u_2v_2, u_1u_3, u_2u_5, and u_3u_5 is in $E(G)$, and then $u_1v_4u_2u_3v_1v_2u_5$ is a 7-cycle.

- If u_3u_5 is in $E(G)$, then by Obs.4 (with $i = 4$), at least one of u_3v_3 and u_5v_5 is in $E(G)$. If u_3v_3 is in $E(G)$ but u_5v_5 is not, then by Obs.2 (with $i = 5$), $u_1u_3 \in E(G)$, which means $d(u_3) = 7$, a contradiction. A symmetric argument applies if u_5v_5 is in $E(G)$ but u_3v_3 is not. Hence, both u_3v_3 and u_5v_5 are in $E(G)$. As a result, neither u_1u_3 nor u_2u_5 can be in $E(G)$, as otherwise $\{u_3, u_1, u_5, v_3\}$ or, respectively, $\{u_5, u_2, u_3, v_5\}$ forms a clique. However, the vertex v_3 has four neighbors on a 5-cycle $u_3u_5u_1v_4u_2$. Therefore, $u_3u_5 \notin E(G)$.
- If u_1v_1 is in $E(G)$, then by Obs.1 (with $i = 1$), $u_1u_3 \in E(G)$. Note that $u_1u_4 \notin E(G)$, as otherwise $\{u_1, u_3, u_4, v_1\}$ forms a cliqued. By Obs.2 (with $i = 3$), $u_3v_3 \in E(G)$. But then $G - \{v_4, v_5, u_2\}$ is isomorphic to $\overline{C_7}$. Therefore, $u_1v_1 \notin E(G)$. By a symmetric argument, $u_2v_2 \notin E(G)$.
- Now that none of u_1v_1, u_2v_2, and u_3u_5 is in $E(G)$, from Obs.2 (with $i = 1$) it can be inferred $u_1u_3 \notin E(G)$, and then by Obs.2 (with $i = 2$), $u_2u_5 \notin E(G)$.

Thus, at most one of u_3u_4 and u_4u_5 is in $E(G)$. We may assume without loss of generality that u_3u_4 is in $E(G)$ and u_4u_5 is not; the other case is symmetric.

We argue that none of u_1u_3, u_3u_5, u_1v_1, and u_5v_5 can be in $E(G)$. Suppose that u_1u_3 is in $E(G)$. By Obs.2 (with $i = 1$), at least one of u_1v_1 and u_3u_5 is in $E(G)$. If $u_3u_5 \in E(G)$, then $u_3v_3 \notin E(G)$, as otherwise $\{u_3, u_1, u_5, v_3\}$ forms a clique. On the other hand, by Obs.4 (with $i = 2$), at least one of u_1v_1 and u_3v_3 is in $E(G)$. Therefore, we always have $u_1v_1 \in E(G)$. Then $u_1u_4 \notin E(G)$, as otherwise $\{u_1, u_3, u_4, v_1\}$ forms a clique. By Obs.2 (with $i = 3$), $u_3v_3 \in E(G)$, which further implies $u_3u_5 \notin E(G)$ because $d(u_3) < 6$. But then all of u_1u_5, u_1u_3, u_3u_4, and u_3v_3 are in $E(G)$ and none of u_1u_4, u_3u_5, and u_4u_5 is in $E(G)$, contradicting Obs.3 (with $i = 2$). Therefore, $u_1u_3 \notin E(G)$. By a symmetric argument, we can conclude that u_3u_5 cannot be in $E(G)$ either. Now that none of u_1u_3, u_3u_5, u_1u_2, and u_4u_5 is in $E(G)$, together with the fact that both u_2u_3 and u_3u_4 are in $E(G)$, from Obs.1 (with $i = 1$ and $i = 5$), it can be inferred that both u_1v_1 and u_5v_5 cannot be in $E(G)$.

At least one of u_4v_4 and u_1u_4 is in $E(G)$, as otherwise $u_1v_4v_5u_3u_4v_2u_5$ is a 7-cycle. If u_4v_4 is in $E(G)$, then Obs.1 (with $i = 4$) will force u_1u_4 in $E(G)$ as well. On the other hand, u_1u_4 is in $E(G)$ and Obs.4 (with $i = 5$) will force u_4v_4 in $E(G)$ as well. Therefore, both u_4v_4 and u_1u_4 are in $E(G)$. Moreover, at least one of u_2v_2 and u_2u_5 is in $E(G)$, as otherwise $u_5v_2v_1u_3u_2v_4u_1$ is a 7-cycle. By a symmetric argument, both u_2v_2 and u_2u_5 are in $E(G)$. Note that u_2u_4 cannot be in $E(G)$, as otherwise $G - \{v_1, v_5, u_3\}$ is isomorphic to $\overline{C_7}$. Dependent on whether u_3v_3 is in $E(G)$, the graph is isomorphic to either $(1\mathring{2}43\mathring{5}1)$ or its complement. □

The discussion on the order of G and Propositions 11–13 imply Theorem 1.

References

1. Benchetrit, Y.: Geometric properties of the chromatic number: polyhedra, structure and algorithms. Ph.D. thesis, Université de Grenoble (2015)
2. Berge, C.: Les problmès de coloration en théorie des graphes. Publ. Inst. Stat. Univ. Paris **9**, 123–160 (1960)
3. Berge, C.: Perfect graphs. In: Six Papers on Graph Theory, pp. 1–21. Indian Statistical Institute, Calcutta (1963)
4. Boulala, M., Uhry, J.: Polytope des indépendants d'un graphe série-parallèle. Discrete Math. **27**(3), 225–243 (1979). https://doi.org/10.1016/0012-365X(79)90160-2
5. Bruhn, H., Fuchs, E.: t-perfection in P_5-free graphs. SIAM J. Discrete Math. **31**(3), 1616–1633 (2017). https://doi.org/10.1137/16M1059874
6. Bruhn, H., Stein, M.: t-perfection is always strong for claw-free graphs. SIAM J. Discrete Math. **24**(3), 770–781 (2010). https://doi.org/10.1137/090769508
7. Bruhn, H., Stein, M.: On claw-free t-perfect graphs. Math. Program. **133**(1–2), 461–480 (2012). https://doi.org/10.1007/s10107-010-0436-9
8. Chudnovsky, M., Robertson, N., Seymour, P.D., Thomas, R.: The strong perfect graph theorem. Ann. Math. **164**, 51–229 (2006). https://doi.org/10.4007/annals.2006.164.51
9. Chvátal, V.: On certain polytopes associated with graphs. J. Comb. Theory Ser. B **18**, 138–154 (1975). https://doi.org/10.1016/0095-8956(75)90041-6
10. Chvátal, V., Graham, R.L., Perold, A.F., Whitesides, S.: Combinatorial designs related to the strong perfect graph conjecture. Discrete Math. **26**(2), 83–92 (1979). https://doi.org/10.1016/0012-365X(79)90114-6
11. Eisenbrand, F., Funke, S., Garg, N., Könemann, J.: A combinatorial algorithm for computing a maximum independent set in a t-perfect graph. In: SODA, pp. 517–522 (2003)
12. Fonlupt, J., Uhry, J.: Transformations which preserve perfectness and h-perfectness of graphs. North-Holland Math. Stud. **16**, 83–95 (1982). https://doi.org/10.1016/S0304-0208(08)72445-9
13. Gerards, A.M.H.: A min-max relation for stable sets in graphs with no odd-K_4. J. Comb. Theory Ser. B **47**, 330–348 (1989). https://doi.org/10.1016/0095-8956(89)90032-4
14. Gerards, A.M.H., Shepherd, F.B.: The graphs with all subgraphs t-perfect. SIAM J. Discrete Math. **11**(4), 524–545 (1998). https://doi.org/10.1137/S0895480196306361
15. Grötschel, M., Lovász, L., Schrijver, A.: Relaxations of vertex packing. J. Comb. Theory Ser. B **40**(3), 330–343 (1986). https://doi.org/10.1016/0095-8956(86)90087-0
16. Lovász, L.: Normal hypergraphs and the perfect graph conjecture. Discrete Math. **2**(3), 253–267 (1972). https://doi.org/10.1016/0012-365X(72)90006-4
17. Nemhauser, G.L., Trotter, L.E., Jr.: Properties of vertex packing and independence system polyhedra. Math. Program. **6**(1), 48–61 (1974). https://doi.org/10.1007/BF01580222
18. Nemhauser, G.L., Trotter, L.E., Jr.: Vertex packings: structural properties and algorithms. Math. Program. **8**(1), 232–248 (1975). https://doi.org/10.1007/BF01580444
19. Padberg, M.W.: Perfect zero-one matrices. Math. Program. **6**(1), 180–196 (1974). https://doi.org/10.1007/BF01580235
20. Schrijver, A.: Combinatorial Optimization: Polyhedra and Efficiency. Algorithms and Combinatorics, vol. 24. Springer, Berlin (2003)
21. Shepherd, F.B.: Applying Lehman's theorems to packing problems. Math. Program. **71**, 353–367 (1995). https://doi.org/10.1007/BF01590960

On Subgraph Complementation to H-free Graphs

Dhanyamol Antony[1(✉)], Jay Garchar[2], Sagartanu Pal[2(✉)], R. B. Sandeep[2], Sagnik Sen[2], and R. Subashini[1]

[1] National Institute of Technology Calicut, Calicut, India
{dhanyamol_p170019cs,suba}@nitc.ac.in
[2] Indian Institute of Technology Dharwad, Dharwad, India
{170010001,183061001,sandeeprb,sen}@iitdh.ac.in

Abstract. For a class $\mathcal{G}$ of graphs, the problem SUBGRAPH COMPLEMENT TO $\mathcal{G}$ asks whether one can find a subset S of vertices of the input graph G such that complementing the subgraph induced by S in G results in a graph in $\mathcal{G}$. We investigate the complexity of the problem when $\mathcal{G}$ is H-free for H being a complete graph, a star, a path, or a cycle. We obtain the following results:

- When H is a K_t (a complete graph on t vertices) for any fixed $t \geq 1$, the problem is solvable in polynomial-time. This applies even when $\mathcal{G}$ is a subclass of K_t-free graphs recognizable in polynomial-time, for example, the class of $(t-2)$-degenerate graphs.
- When H is a $K_{1,t}$ (a star graph on $t+1$ vertices), we obtain that the problem is NP-complete for every $t \geq 5$. This, along with known results, leaves only two unresolved cases - $K_{1,3}$ and $K_{1,4}$.
- When H is a P_t (a path on t vertices), we obtain that the problem is NP-complete for every $t \geq 7$, leaving behind only two unresolved cases - P_5 and P_6.
- When H is a C_t (a cycle on t vertices), we obtain that the problem is NP-complete for every $t \geq 8$, leaving behind four unresolved cases - C_4, C_5, C_6, and C_7.

Further, we prove that these hard problems do not admit subexponential-time algorithms (algorithms running in time $2^{o(|V(G)|)}$), assuming the Exponential Time Hypothesis. A simple complementation argument implies that results for $\mathcal{G}$ are applicable for $\overline{\mathcal{G}}$, thereby obtaining similar results for H being the complement of a complete graph, a star, a path, or a cycle. Our results generalize two main results and resolve one open question by Fomin et al. (Algorithmica, 2020).

Keywords: Subgraph complementation · Graph modification · Graph classes

Partially supported by SERB Grant SRG/2019/002276: "Complexity Dichotomies for Graph Modification Problems".

Ł. Kowalik et al. (Eds.): WG 2021, LNCS 12911, pp. 118–129, 2021.
https://doi.org/10.1007/978-3-030-86838-3_9

1 Introduction

For a class $\mathcal{G}$ of graphs, a general graph modification problem can be defined as follows: Given a graph G, is there a set of modifications applying which on G results in a graph in $\mathcal{G}$? Based on the type and the number of modifications allowed, there are various kinds of graph modification problems. Among them, the most studied problems are vertex deletion problems and edge modification problems. As the names suggest, the allowed modifications in them are vertex deletions and edge modifications (deletion, completion, or editing) respectively. In these types of graph modification problems, the size of the set of modifications is bounded by an additional integer input. For example, in the CLUSTER EDITING problem, given a graph G and an integer k, the task is to find whether there exists a set of at most k pairs of vertices of G such that changing the adjacencies of the pairs in G results in a cluster graph–that is, a vertex-disjoint union of cliques.

In this paper, we deal with a graph modification known as subgraph complementation. In subgraph complementation problems, the objective is to check whether the given graph G has a subset S of vertices such that complementing the subgraph induced by S in G, results in a graph in $\mathcal{G}$. Here, the adjacency of a pair u, v of vertices is flipped only if both u and v are in S. The graph thus obtained, denoted by $G \oplus S$, is known as a subgraph complement of G. Unlike the vertex/edge modification problems, the operation is allowed only once but there is no restriction on the size of S. This operation is introduced by Kamiński et al. [13] as an attempt to generalize different kinds of complementations such as graph complementation and local complementation (replacing the closed neighborhood of a vertex by its complement) in their study of the clique-width of a graph. Recently, a systematic algorithmic study of this problem has been started by Fomin et al. [8]. They proved that the problem is polynomial-time solvable when the graph class $\mathcal{G}$ is triangle-free graphs, or $\mathcal{G}$ is d-degenerate, or $\mathcal{G}$ is of bounded clique-width and expressible in MSO_1 (for example, P_4-free graphs), or when $\mathcal{G}$ is the class of split graphs. They also obtained a hardness result in which they proved that the problem is NP-complete if $\mathcal{G}$ is the class of regular graphs.

We focus on subgraph complementation problems for $\mathcal{G}$ being H-free graphs. For a graph H, in the problem SUBGRAPH COMPLEMENT TO H-free graphs (SC TO H-free graphs), the task is to find whether there exists a subset S of vertices of the input graph G such that $G \oplus S$ is H-free, i.e., $G \oplus S$ does not contain any induced subgraph isomorphic to H. There are numerous algorithmic studies on graph modification problems where the target graph class $\mathcal{G}$ is H-free [1–5,7,9,10,12,17]. We add on to this list by studying H-free graphs with respect to subgraph complementation. A class $\mathcal{G}$ of graphs is hereditary (on induced subgraphs) if for every $G \in \mathcal{G}$, every induced subgraph of G is in $\mathcal{G}$. It is well known that every hereditary class of graphs can be characterized by a set $\mathcal{H}$ of forbidden induced subgraphs. Therefore, studying a graph modification problem with target graph class H-free (i.e., $|\mathcal{H}| = 1$) can be seen as a first step toward understanding the complexity of the problem for hereditary properties.

We consider four classes of graphs H - complete graphs, stars, paths, and cycles - and their complement classes. In all these cases, we obtain complete polynomial-time/NP-complete dichotomies, except for a few cases. The results are summarized below:

- When H is a K_t (a complete graph on t vertices), for any fixed $t \geq 1$, we obtain that SC TO H-free graphs can be solved in polynomial-time. We obtain this result by generalizing the technique used in [8] for SC TO triangle-free graphs and using results on generalized split graphs from [14]. Our result applies to any subclass of K_t-free graphs recognizable in polynomial-time. This result, as far as the existence of polynomial-time algorithms is concerned, subsumes two main results in [8] - the results when $\mathcal{G}$ is triangle-free and when $\mathcal{G}$ is d-degenerate (d-degenerate graphs are K_{d+2}-free).
- When H is a $K_{1,t}$ (a star graph on $t+1$ vertices), we obtain that SC TO H-free graphs is NP-complete for every fixed $t \geq 5$. When $t = 1$ (i.e., $H = K_2$), the problem can be solved trivially - a graph G is a yes-instance of SC TO K_2-free graphs if and only if G is a $K_s \cup tK_1$ (disjoint union of a clique and isolated vertices), which can be recognized in polynomial-time. When $t = 2$ (i.e., $H = P_3$), the problem admits a polynomial-time algorithm as the class of P_3-free graphs has bounded clique-width and can be expressed in MSO_1 – see Sect. 6 in [8]. Therefore, the only remaining cases to be solved among the star graphs are $K_{1,3}$ and $K_{1,4}$.
- When H is a P_t (a path on t vertices), we obtain that SC TO H-free graphs is NP-complete for every fixed $t \geq 7$. It is known from [8] that the problem can be solved in polynomial-time for all $t \leq 4$. Therefore, the only remaining cases to be solved here are P_5 and P_6.
- When H is a C_t (a cycle on t vertices), we obtain that SC TO H-free graphs is NP-complete for every $t \geq 8$. Therefore, the only remaining unknown cases among cycles are C_4, C_5, C_6, and C_7.
- We prove that a graph G is a yes-instance of SC TO $\mathcal{G}$ if and only if $\overline{G}$ is a yes-instane of SC TO $\overline{\mathcal{G}}$, where $\overline{\mathcal{G}}$ is the set of complements of graphs in $\mathcal{G}$. This implies that SC TO $\mathcal{G}$ can be solved in polynomial-time if and only if SC TO $\overline{\mathcal{G}}$ can be solved in polynomial-time. This resolves an open question in [8]. Further, it implies that SC TO H-free graphs is polynomially equivalent to SC TO $\overline{H}$-free graphs. Therefore, all our results for H-free graphs are applicable for $\overline{H}$-free graphs as well.
- We observe that SC TO $\mathcal{G}$, for any polynomial-time recognizable class $\mathcal{G}$ of graphs, can be solved in time $2^{O(|V(G)|)}$ by checking whether every subset S of vertices in the input graph G is a solution or not. We obtain that one cannot hope for much better results for all the cases in which we prove the NP-completeness. To be precise, we prove that, assuming the Exponential Time Hypothesis, there exists no subexponential-time algorithm (algorithm running in time $2^{o(|V(G)|)}$) for every problem for which we prove the NP-completeness.

For the hardness results, we employ two types of reductions. One type, from variants of SAT problems, is used to solve base cases - for example $K_{1,5}$. The other type of reductions acts as an inductive step - for example from SC TO $K_{1,t}$-free graphs to SC TO $K_{1,t+1}$-free graphs. This scheme of obtaining hardness results was used to obtain a complete polynomial-time/NP-complete dichotomy in [1], and a conditional kernelization complexity dichotomy in [16] for H-free edge modification problems. We believe that our results and techniques will come in handy for an eventual complexity dichotomy for SC TO H-free graphs.

The paper is organized as follows: Preliminaries are given in Sect. 2, structural results are obtained in Sect. 3, polynomial-time algorithms are discussed in Sect. 4, and the hardness results are proved in Sect. 5. All proofs have been moved to a full version of this paper due to space constraints.

2 Preliminaries

A simple graph is a pair $G = (V, E)$, where V is a set of vertices and $E \subseteq \binom{V}{2}$ is a set of edges. For a graph G, we refer to its vertex set as $V(G)$ and its edge set as $E(G)$. For a graph G and a set $S \subseteq V(G)$, the *induced subgraph* $G[S]$ is a graph whose vertex set is S and whose edge set contains all the edges in E that have both endpoints in S. For a vertex $v \in V(G)$, the open neighborhood of v, denoted by $N(v)$, is the set of all the vertices adjacent to v, i.e., $N(v) := \{w \mid vw \in E(G)\}$, and the closed neighborhood of v, denoted by $N[v]$, is defined as $N(v) \cup \{v\}$. The degree of a vertex v is the size of its open neighborhood. A vertex v is a degree-k if v has degree k. By $G - X$, we denote the graph obtained from G by removing the vertices in X, i.e., $G - X = G[V(G) \setminus X]$. A set X of vertices in a graph G is said to be a *module* if every vertex in X has the same set of neighbors outside of X. An *empty graph* is a graph without any edges and a *null graph* is a graph without any vertices. Let H be any graph. Then a graph G is called *H-free*, if G does not contain H as an induced subgraph. In a graph G, two sets of vertices are said to be *all-adjacent*, if each vertex in one set is adjacent to every vertex in the other set. Similarly, two sets of vertices are said to be *nonadjacent*, if there are no edges between them. A complete graph, an empty graph, a star, a cycle, and a path with t vertices are denoted by $K_t, tK_1, K_{1,t-1}, C_t$, and P_t respectively. By I_t, we denote an independent set of size t. The center vertex of a star graph $K_{1,t}$ (for any $t \geq 2$), is the vertex having degree t. For a class $\mathcal{G}$ of graphs, by $\overline{\mathcal{G}}$, we denote the class of complements of graphs in $\mathcal{G}$. A graph property Π is *nontrivial* if it is true for infinitely many graphs and false for infinitely many graphs. The property is said to be *trivial* otherwise. The *disjoint union* of two graphs G_1 and G_2, denoted by $G_1 \cup G_2$, is the graph G such that $V(G) = V(G_1) \cup V(G_2)$ and $E(G) = E(G_1) \cup E(G_2)$. The disjoint union of t copies of a graph G is denoted by tG. The *cross product* $H \times H'$ of two graphs H and H' is a graph G such that the vertex set $V(G) = V(H) \times V(H')$ and two vertices (u, u') and (v, v') are adjacent in G if and only if either $u = v$ and u' is adjacent to v' in H', or $u' = v'$ and u is adjacent to v in H. A *k-degenerate* graph is an undirected graph in which every subgraph has a vertex of degree at most k. For a graph G, the *degeneracy* of G is the smallest value of k for which it is k-degenerate.

To generalize split graphs, Gyárfás [11] introduced the notion of (p, q)-split graphs. For positive integers p, q, a graph is a (p, q)-split graph if its vertices can be partitioned into two sets P and Q such that the clique number of $G[P]$ is at most p and the independence number of $G[Q]$ is at most q, i.e., $G[P]$ is K_{p+1}-free and $G[Q]$ is $(q+1)K_1$-free. Clearly, a split graph is a $(1,1)$-split graph. Every such partition of the form (P,Q) will be called a (p,q)-split partition. For integers p,q, by $R(p,q)$ we denote the Ramsey number, i.e., $R(p,q)$ is the minimum integer n such that every graph G with at least n vertices has either a clique of size p or an independent set of size q.

We say that boolean formula is a k-SAT formula if it is in conjunctive normal form (CNF) and every clause contains exactly k literals of distinct variables. We denote the n variables and m clauses of a k-SAT formula Φ by $\{X_1,\ldots,X_n\}$, and $\{C_1,C_2,\ldots,C_m\}$ respectively. For each variable X_i, we denote the positve literal by x_i and the negative literal by $\overline{x_i}$. Each clause C_i is a disjunction of exactly k literals $\ell_{i,1},\ell_{i,2},\ldots\ell_{i,k}$, i.e., $C_i=\ell_{i,1}\vee\ell_{i,2}\vee\cdots\vee\ell_{i,k}$. The problem k-SAT$_{\geq k-2}$ is defined below.

> k-SAT$_{\geq k-2}$: Given a boolean formula Φ with n variables and m clauses in conjunctive normal form (CNF), where each clause contains exactly k literals of distinct variables, find whether there exists a satisfying assignment for Φ with at least $k-2$ true literals per clause.

The Exponential-Time Hypothesis (ETH) along with the Sparsification Lemma imply that 3-SAT cannot be solved in time $2^{o(n+m)}$, where n is the number of variables and m is the number of clauses in the input formula. To show that a graph problem does not admit an algorithm running in time $2^{o(|V(G)|)}$ (where G is the input graph), it is sufficient to give a polynomial-time reduction from 3-SAT such that the resultant graph has only $O(n+m)$ vertices. We can show the same by a reduction from another graph problem (which does not admit a $2^{o(n)}$ algorithm, where n is the number of vertices in the input graph), such that the resultant instance has only at most $O(n)$ vertices. Such reductions, where the blow-up in the input size (with respect to an appropriate measure – the number of vertices in our case) is only linear, are known as linear reductions. We refer to Chap. 14 of [6] for an exposition to these topics. Since all the problems discussed in this paper are trivially in NP, we will not state the same explicitly while proving the NP-completeness of the problems.

Proposition 1 (folklore). *For every $k\geq 3$, k-SAT$_{\geq k-2}$ is NP-Complete. Further, the problem cannot be solved in time $2^{o(n+m)}$, assuming the ETH.*

A *subgraph complement* of a graph G is a graph G' obtained from G, for any $S\subseteq V(G)$, by complementing the subgraph induced by S in G. More formally, $V(G')=V(G)$ and two vertices u,v are adjacent in G' if and only if at least one of the following conditions hold true - (i) u and v are adjacent in G and either u or v is not in S; (ii) u and v are nonadjacent in G and both u and v are in S.

> Subgraph Complement to $\mathcal{G}$ (SC to $\mathcal{G}$): Given a graph G, is there a subgraph complement G' of G such that $G'\in\mathcal{G}$?

By $\mathcal{G}^{(1)}$, we denote the class of graphs where each graph in it can be subgraph complemented to a graph in $\mathcal{G}$, i.e., $\mathcal{G}^{(1)} = \{G : \exists S \subseteq V(G) \text{ such that } G \oplus S \in \mathcal{G}\}$.

3 Structural Results

For a graph H, let Π_H be the property defined as follows: A graph G has property Π_H if there exists a set $S \subseteq V(G)$ such that $G \oplus S$ is H-free. That is, the class of graphs satisfying Π_H is the set of all yes-instances of SC TO H-free graphs. We prove that Π_H is nontrivial if and only if H has at least two vertices. This result guarantees that the pursuit of obtaining optimal complexities of SC TO H-free graphs is meaningful for all nontrivial graphs H.

Lemma 1. *For a graph H, the property Π_H is nontrivial if and only if H has at least two vertices.*

How is a subgraph complement of a graph G related to a subgraph complement of $\overline{G}$ with respect to the same subset S of vertices of the graphs? Lemma 2 answers this question.

Lemma 2. *Let G be a graph and $S \subseteq V(G)$. Then $G \oplus S = \overline{\overline{G} \oplus S}$.*

Lemma 3. *For a class $\mathcal{G}$ of graphs, $\overline{\mathcal{G}^{(1)}} = \overline{\mathcal{G}}^{(1)}$.*

Lemma 3 implies Corollary 1, which tells us that SC TO $\mathcal{G}$ is polynomially equivalent to SC TO $\overline{\mathcal{G}}$. The first statement in Corollary 1 was an open problem raised in [8], whereas the second statement implies that all results obtained in this paper for H-free graphs are applicable for $\overline{H}$-free graphs as well.

Corollary 1. *For a class $\mathcal{G}$ of graphs, $\mathcal{G}^{(1)}$ can be recognized in polynomial time if and only if $\overline{\mathcal{G}}^{(1)}$ can be recognized in polynomial time. In particular,* SC TO *H-free graphs is polynomial-time solvable if and only if* SC TO *$\overline{H}$-free graphs is polynomial-time solvable.*

4 Polynomial-Time Algorithms

In this section, we obtain a polynomial-time algorithm for SC TO $\mathcal{G}$, when $\mathcal{G}$ is a subclass of K_t-free graphs, for any fixed integer $t \geq 1$, such that $\mathcal{G}$ is recognizable in polynomial-time. To obtain this result, we generalize the technique used for SC TO triangle-free graphs in [8]. Assume that a graph G has a solution S, i.e., $G \oplus S \in \mathcal{G}$. Further, assume that S has at least two vertices u and v. Then, we prove that each of the sets $N(u) \cap N(v)$, $\overline{N[u]} \cap \overline{N[v]}$, $N(u) \setminus \overline{N[v]}$, and $N(v) \setminus \overline{N[u]}$ induces a (p,q)-split graph (for $p = q = t-1$), and for each of them, S contains exactly the set Q of some (p,q)-split partition (P,Q) of the corresponding induced subgraph. Then the algorithm boils down to recognizing (p,q)-split graphs and enumerating all (p,q)-split partitions of them in polynomial-time. All the tools required for this task has already been obtained in [14], see the full version [15] for the proofs.

Proposition 2 ([14,15]). *Let G be a (p,q)-split graph. Let (P,Q) and (P',Q') be two (p,q)-split partitions of G. Then $|P \cap Q'| \leq R(p+1,q+1)-1$ and $|P' \cap Q| \leq R(p+1,q+1)-1$.*

Proposition 2 implies that two distinct (p,q)-partitions of a (p,q)-split graph cannot differ too much. This helps us to enumerate all (p,q)-split partitions in polynomial-time, given one of them. Proposition 3 says that there is a polynomial-time algorithm which recognizes a (p,q)-split graph and gives a (p,q)-split partition of the same.

Proposition 3 ([14,15]). *For any fixed constants p and q, there is an algorithm which takes a graph G (with n vertices) as an input, runs in time $O(n^{2R(p+1,q+1)+p+q+4})$, and decides whether G is a (p,q)-split graph. Furthermore, if G is a (p,q)-split graph, then the algorithm outputs a (p,q)-split partition of G.*

The running time in Proposition 3 is not explicitly given in [14,15], but can be easily derived from the proof given in [15]. The proof of the following lemma is very similar to that of Proposition 3.

Lemma 4. *Let G be a (p,q)-split graph with n vertices. Then there are at most $n^{2R(p+1,q+1)}$ (p,q)-split partitions of G. Given a (p,q)-split partition (P,Q) of G, all (p,q)-split partitions of G can be computed in polynomial-time, specifically in time $O(n^{2R(p+1,q+1)+p+q+3})$.*

Proposition 3 and Lemma 4 directly imply Corollary 2.

Corollary 2. *For any fixed constants $p \geq 1$ and $q \geq 1$, there is an algorithm which takes a graph G (with n vertices) as an input, runs in time $O(n^{2R(p+1,q+1)+p+q+4})$, and decides whether G is a (p,q)-split graph. Furthermore, if G is a (p,q)-split graph, then the algorithm outputs all (p,q)-split partitions of G.*

Let G be a yes-instance of SC TO $\mathcal{G}$, where $\mathcal{G}$ is a subclass of K_t-free graphs recognizable in polynomial-time. Assume that G is not a trivial yes-instance. Let $S \subseteq V(G)$ be such that $G \oplus S \in \mathcal{G}$. Clearly, $|S| \geq 2$. Let u, v be two vertices in S. With respect to S, u, v, we partition the vertices in $V(G) \setminus \{u,v\}$ into eight sets as given below. This is depicted in Fig. 1.

(i) $S_{uv} = S \cap N(u) \cap N(v)$
(ii) $S_{\overline{uv}} = S \cap \overline{N[u]} \cap \overline{N[v]}$
(iii) $S_{u\overline{v}} = S \cap (N(u) \setminus N[v])$
(iv) $S_{\overline{u}v} = S \cap (N(v) \setminus N[u])$
(v) $T_{uv} = (N(u) \cap N(v)) \setminus S$
(vi) $T_{\overline{uv}} = (\overline{N[u]} \cap \overline{N[v]}) \setminus S$
(vii) $T_{u\overline{v}} = (N(u) \setminus N[v]) \setminus S$
(viii) $T_{\overline{u}v} = (N(v) \setminus N[u]) \setminus S$

Clearly, $S = S_{uv} \cup S_{\overline{uv}} \cup S_{u\overline{v}} \cup S_{\overline{u}v} \cup \{u,v\}$, and $V(G) \setminus S = T_{uv} \cup T_{\overline{uv}} \cup T_{u\overline{v}} \cup T_{\overline{u}v}$.

Lemma 5. *Let G be a yes-instance of SC TO $\mathcal{G}$, where $\mathcal{G}$ is a subclass of K_t-free graphs, for any fixed integer $t \geq 2$. Let $S \subseteq V(G)$ be such that $|S| \geq 2$ and $G \oplus S \in \mathcal{G}$. Let u and v be any two vertices in S. Then the following statements hold true:*

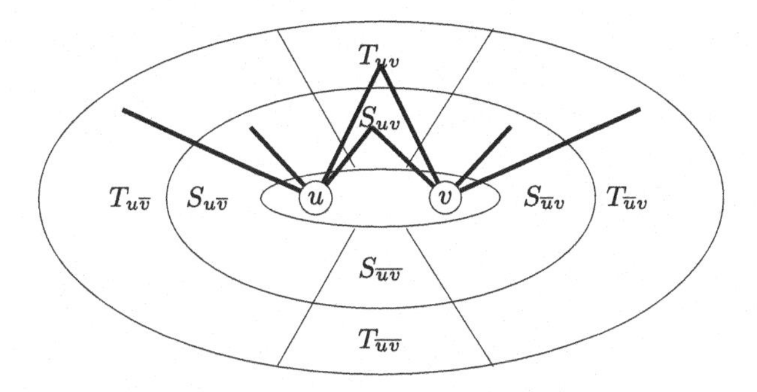

Fig. 1. Partitioning of vertices of a yes-instance G of SC TO $\mathcal{G}$, based on a solution S and two vertices u, v in S. The bold lines represent the adjacency of vertices u and v.

(i) $N(u) \cap N(v)$ *induces a* $(t-1, t-1)$*-split graph with a* $(t-1, t-1)$*-split partition* (T_{uv}, S_{uv})*;*
(ii) $\overline{N[u]} \cap \overline{N[v]}$ *induces a* $(t-1, t-1)$*-split graph with a* $(t-1, t-1)$*-split partition* $(T_{\overline{uv}}, S_{\overline{uv}})$*;*
(iii) $N(u) \setminus N[v]$ *induces a* $(t-1, t-1)$*-split graph with a* $(t-1, t-1)$*-split partition* $(T_{u\overline{v}}, S_{u\overline{v}})$*;*
(iv) $N(v) \setminus N[u]$ *induces a* $(t-1, t-1)$*-split graph with a* $(t-1, t-1)$*-split partition* $(T_{\overline{u}v}, S_{\overline{u}v})$*;*

Algorithm for SC TO $\mathcal{G}$**, where** $\mathcal{G}$ **is a subclass of** K_t**-free graphs**
Input: A graph G
Output: If G is a yes-instance of SC TO $\mathcal{G}$, then returns a set $S \subseteq V(G)$ such that $G \oplus S \in \mathcal{G}$; returns 'None' otherwise.
Step 0 : If $G \in \mathcal{G}$, then return $\emptyset$.
Step 1 : For every unordered pair of vertices $\{u, v\}$ in G:
(i) If any of $N(u) \cap N(v)$, $\overline{N[u]} \cap \overline{N[v]}$, $N(u) \setminus N[v]$, $N(v) \setminus N[u]$ does not induce a $(t-1, t-1)$-split graph, then continue (with Step 1).
(ii) Compute L_{uv}, the list of all (p, q)-split partitions of $N(u) \cap N(v)$.
(iii) Compute $L_{\overline{uv}}$, the list of all (p, q)-split partitions of $\overline{N[u]} \cap \overline{N(v)}$.
(iv) Compute $L_{u\overline{v}}$, the list of all (p, q)-split partitions of $N(u) \setminus N[v]$.
(v) Compute $L_{\overline{u}v}$, the list of all (p, q)-split partitions of $N(v) \setminus N[v]$.
(vi) For every (P_a, Q_a) in L_{uv}, for every (P_b, Q_b) in $L_{\overline{uv}}$, for every (P_c, Q_c) in $L_{u\overline{v}}$, and for every (P_d, Q_d) in $L_{\overline{u}v}$:
(a) Let $S = Q_a \cup Q_b \cup Q_c \cup Q_d \cup \{u, v\}$
(b) If $G \oplus S \in \mathcal{G}$, then return S.
Step 2 : Return 'None'

Theorem 1. *For any fixed* $t \geq 1$*, let* $\mathcal{G}$ *be a subclass of* K_t*-free graphs such that there is a recognition algorithm for* $\mathcal{G}$ *running in time* $O(f(n))$*, for some polynomial function* f*. Then* SC TO $\mathcal{G}$ *can be solved in polynomial-time, specifically in time* $O(f(n) \cdot (n^{8R(t,t)+4}))$*. In particular,* SC TO K_t*-free graphs can be solved in polynomial-time, specifically in time* $O(n^{8R(t,t)+t+4})$*.*

Since graphs of bounded degeneracy have bounded clique number and can be recognized in polynomial-time, Theorem 1 implies that SC TO d-degenerate graphs can be solved in polynomial-time, a result obtained in [8].

5 Hardness Results

In this section, we prove hardness results for SC TO H-free graphs when H is a star, a path, or a cycle. Specifically, we prove that, if H is a star with at least six vertices, or a path with at least seven vertices, or a cycle with at least eight vertices, the problem is NP-complete. For all these hard cases, we prove something stronger: these hard problems cannot be solved in time $2^{o(|V(G)|)}$, assuming the ETH.

The proofs are by induction on the number of vertices in H. For base cases, we give reductions from variants of k-SAT problem (for example, from 4-SAT$_{\geq 2}$ to SC TO $K_{1,5}$-free graphs). For the inductive step, we give reductions from problems of the same kind (for example, from SC TO $K_{1,t}$-free graphs to SC TO $K_{1,t+1}$-free graphs).

Construction 1 will be used for an *inductive* reduction for SC TO $K_{1,t}$-free graphs.

Construction 1. *Let (G', t) be the input to the construction, where G' is a graph and $t \geq 1$ is an integer. For every vertex u of G', introduce $(t+2)$ vertices denoted by the set W_u, which includes a special vertex u'. Each of these sets induces a clique (K_{t+2}). Further, every vertex $u \in V(G')$ is adjacent to every vertex in W_u except u'. Let the resultant graph be G and let W, which induces a cluster graph, be the union of all newly introduced vertices.*

An example of the construction is shown in Fig. 2.

Lemma 6. *Let $\mathcal{G}'$ and $\mathcal{G}$ be the classes of $K_{1,t}$-free graphs and $K_{1,t+1}$-free graphs respectively for any $t \geq 2$. If* SC TO $\mathcal{G}'$ *is NP-complete, then so is* SC TO $\mathcal{G}$*. Further, if* SC TO $\mathcal{G}'$ *cannot be solved in time $2^{o(|V(G)|)}$, then so is* SC TO $\mathcal{G}$.

Now, to utilize Lemma 6, we need a hardness result when $H = K_{1,t}$ for some $t \geq 2$, the smaller the t, the better the implications will be. We obtain such a result for $t = 5$ using Construction 2.

Construction 2. *Let Φ be the input to the construction, where Φ, with n variables and m clauses, is a* 4-SAT *formula. We construct $G = (V, E)$ in the following way:*

- *For each variable X_i in Φ, introduce two vertices - one vertex, denoted by u_i, for the positive literal x_i and one vertex, denoted by u'_i, for the negative literal $\overline{x_i}$. The vertex u_i is adjacent to u'_i. Further, for each variable X_i introduce four sets $U_{i,1}$, $U_{i,2}$, $U_{i,3}$ and $U_{i,4}$ of five vertices each. Each of these sets induces a K_5. The vertices u_i and u'_i are all-adjacent to $U_{i,1}$. Further, $U_{i,1}$ is all-adjacent to $U_{i,2}$, $U_{i,3}$ and $U_{i,4}$. Thus, the total number of vertices corresponding to a variable of Φ is 22.*

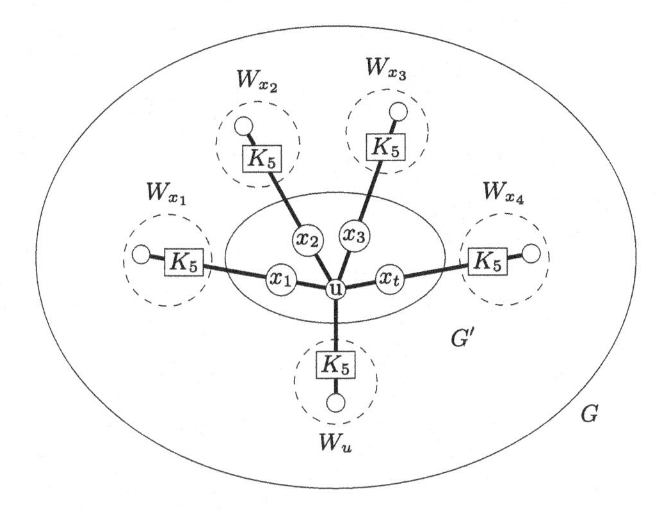

Fig. 2. An example of Construction 1 for $t = 4$. The lines connecting a circle and a rectangle indicate that the vertex corresponding to the circle is adjacent to all vertices in the rectangle.

- *For each clause C_i of the form $\ell_{i,1} \vee \ell_{i,2} \vee \ell_{i,3} \vee \ell_{i,4}$ introduce a set V_i of five vertices each which induces a K_5. All the vertices of each V_i together form a big clique, denoted by V', of size $5{\cdot}m$. Let the four vertices introduced (in the previous step) for the literals $\ell_{i,1}, \ell_{i,2}, \ell_{i,3}$, and $\ell_{i,4}$ be denoted by $y_{i,1}, y_{i,2}, y_{i,3}$ and $y_{i,4}$ respectively. We note that, if $\ell_{i,1} = x_j$, then $y_{i,1} = u_j$, and if $\ell_{i,1} = \overline{x_j}$, then $y_{i,1} = u'_j$. Similarly, let the four vertices introduced for the negation of these literals be denoted by $z_{i,1}, z_{i,2}, z_{i,3}$, and $z_{i,4}$ respectively. We note that, if $\ell_{i,1} = x_j$, then $z_{i,1} = u'_j$, and if $\ell_{i,1} = \overline{x_j}$, then $z_{i,1} = u_j$. Further, every vertex in V_i is adjacent to the vertices $y_{i,1}$, $y_{i,2}$, $y_{i,3}$ and $y_{i,4}$.*

This completes the construction (refer Fig. 3).

For convenience, we call each set V_i introduced in the construction a *clause set* with each of them containing five vertices. For each variable X_i, the vertices u_i and u'_i are called *literal vertices*. The union of all literal vertices is denoted by U. Further, for each variable X_i, the sets $U_{i,s}$ (for $1 \leq r \leq 4$) are called *hanging sets* and the union of which is denoted by U_i^h. By U^h, we denote the union of all U_i^hs. The vertices in the hanging sets are called *hanging vertices*.

Construction 2 will be used for a reduction from 4-SAT$_{\geq 2}$ to SC TO $K_{1,5}$-free graphs. Whenever we introduce a K_5 in the construction, the objective is to forbid a solution S to have all the vertices of that K_5. If all the five vertices in a K_5 is in S, then those vertices along with a vertex adjacent to them (but not in S) or a vertex nonadjacent to them (but in S) form a $K_{1,5}$ in $G \oplus S$, where G is the output of the construction. This makes sure that both u_i and u'_i are not in S together (otherwise, they form a $K_{1,5}$ along with their hanging vertices not in S – note that each hanging set forms a K_5). Further, for every C_i, at least two literal

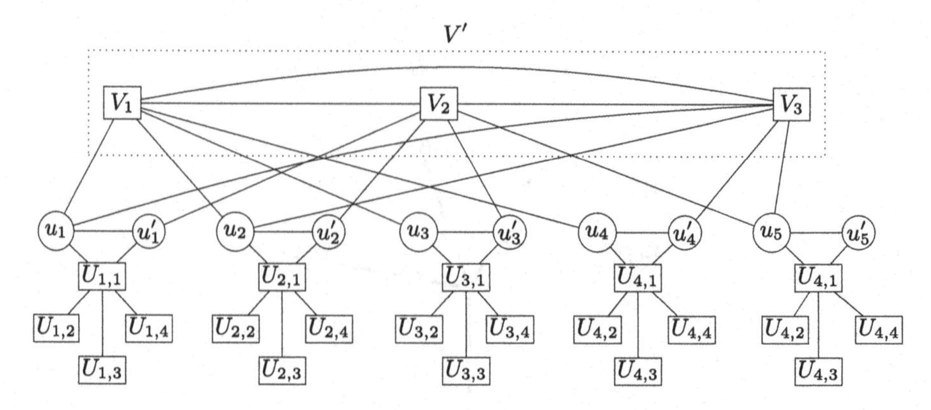

Fig. 3. An example of Construction 2 for the formula $\phi = C_1 \wedge C_2 \wedge C_3$ where $C_1 = x_1 \vee x_2 \vee x_3 \vee x_4$, $C_2 = \overline{x_1} \vee \overline{x_2} \vee \overline{x_3} \vee x_5$, and $C_3 = x_1 \vee x_2 \vee \overline{x_4} \vee x_5$. Each rectangle represents a K_5. The lines connecting two rectangles indicate that each vertex in one rectangle is adjacent to all vertices in the other rectangle. Lines connecting a rectangle and a vertex have a similar meaning.

vertices corresponding to the literals in C_i must be in S, otherwise there will be $K_{1,5}$ where the center vertex is a vertex from V_i. These observations help us to get a valid truth assignment which satisfies the formula Φ, an instance of 4-SAT$_{\geq 2}$.

Theorem 2. SC TO *$K_{1,5}$-free graphs is NP-complete. Further, the problem cannot be solved in time $2^{o(|V(G)|)}$, assuming the ETH.*

Theorem 3 is a direct implication of Lemma 6 and Theorem 2.

Theorem 3. *Let $t \geq 5$ be any integer. Then* SC TO *$K_{1,t}$-free graphs is NP-complete. Further, the problem cannot be solved in time $2^{o(|V(G)|)}$, unless the ETH fails.*

The following two theorems can be proved in a similar fashion.

Theorem 4. *Let $t \geq 7$ be any integer. Then* SC TO *P_t-free graphs is NP-complete. Further, the problem cannot be solved in time $2^{o(|V(G)|)}$, unless the ETH fails.*

Theorem 5. *Let $t \geq 8$ be any integer. Then* SC TO *C_t-free graphs is NP-complete. Further, the problem cannot be solved in time $2^{o(|V(G)|)}$, unless the ETH fails.*

References

1. Aravind, N.R., Sandeep, R.B., Sivadasan, N.: Dichotomy results on the hardness of H-free edge modification problems. SIAM J. Discre. Math. **31**(1), 542–561 (2017)
2. Cai, L., Cai, Y.: Incompressibility of H-free edge modification problems. Algorithmica **71**(3), 731–757 (2014). https://doi.org/10.1007/s00453-014-9937-x

3. Cai, Y.: Polynomial kernelisation of H-free edge modification problems. M .Phil. thesis, Department of Computer Science and Engineering, The Chinese University of Hong Kong, Hong Kong SAR, China (2012)
4. Cao, Y., Chen, J.: Cluster editing: Kernelization based on edge cuts. Algorithmica **64**(1), 152–169 (2012)
5. Cao, Y., Ke, Y., Yuan, H.: Polynomial Kernels for paw-free edge modification problems. In: Chen, J., Feng, Q., Xu, J. (eds.) TAMC 2020. LNCS, vol. 12337, pp. 37–49. Springer, Cham (2020). https://doi.org/10.1007/978-3-030-59267-7_4
6. Cygan, M., et al.: Parameterized Algorithms. Springer, Cham (2015). https://doi.org/10.1007/978-3-319-21275-3
7. Eiben, E., Lochet, W., Saurabh, S.: A polynomial kernel for paw-free editing. In: Cao, Y., Pilipczuk, M. (eds.) 15th International Symposium on Parameterized and Exact Computation (IPEC 2020), December 14–18, 2020, Hong Kong, China (Virtual Conference). LIPIcs, vol. 180, pp. 10:1–10:15. Schloss Dagstuhl - Leibniz-Zentrum für Informatik (2020). https://doi.org/10.4230/LIPIcs.IPEC.2020.10
8. Fomin, F.V., Golovach, P.A., Strømme, T.J.F., Thilikos, D.M.: Subgraph complementation. Algorithmica **82**(7), 1859–1880 (2020)
9. Guillemot, S., Havet, F., Paul, C., Perez, A.: On the (non-)existence of polynomial kernels for P_ℓ-free edge modification problems. Algorithmica **65**(4), 900–926 (2013)
10. Guo, J.: A more effective linear kernelization for cluster editing. Theor. Comput. Sci. **410**(8–10), 718–726 (2009)
11. Gyárfás, A.: Generalized split graphs and ramsey numbers. J. Comb. Theory, Ser. A **81**(2), 255–261 (1998). https://doi.org/10.1006/jcta.1997.2833
12. Jelínková, E., Kratochvíl, J.: On switching to h-free graphs. J. Graph. Theory **75**(4), 387–405 (2014)
13. Kaminski, M., Lozin, V.V., Milanic, M.: Recent developments on graphs of bounded clique-width. Discret. Appl. Math. **157**(12), 2747–2761 (2009)
14. Kolay, S., Panolan, F.: Parameterized algorithms for deletion to (r, ℓ)-graphs. In: Harsha, P., Ramalingam, G. (eds.) 35th IARCS Annual Conference on Foundation of Software Technology and Theoretical Computer Science (FSTTCS 2015). LIPIcs, vol. 45, pp. 420–433. Schloss Dagstuhl - Leibniz-Zentrum für Informatik (2015). https://doi.org/10.4230/LIPIcs.FSTTCS.2015.420
15. Kolay, S., Panolan, F.: Parameterized algorithms for deletion to (r, l)-graphs. arXiv preprint arXiv:1504.08120 (2015)
16. Marx, D., Sandeep, R.B.: Incompressibility of H-free edge modification problems: towards a dichotomy. In: 28th Annual European Symposium on Algorithms (ESA 2020), LIPIcs, vol. 173, pp. 72:1–72:25. Schloss Dagstuhl - Leibniz-Zentrum für Informatik (2020). https://doi.org/10.4230/LIPIcs.ESA.2020.72
17. Yannakakis, M.: Edge-deletion problems. SIAM J. Comput. **10**(2), 297–309 (1981)

Odd Cycle Transversal in Mixed Graphs

Avinandan Das[1], Lawqueen Kanesh[2(✉)], Jayakrishnan Madathil[5], and Saket Saurabh[3,4]

[1] Institut de Recherche en Informatique Fondamontale, Paris, France
avinandan.das@irif.fr
[2] School of Computing, National University of Singapore, Singapore, Singapore
lawqueen@comp.nus.edu.sg
[3] The Institute of Mathematical Sciences, HBNI, Chennai, India
saket@imsc.res.in
[4] Department of Informatics, University of Bergen, Bergen, Norway
[5] Chennai Mathematical Institute, Chennai, India
jayakrishnan@cmi.ac.in

Abstract. An odd cycle transversal (oct, for short) in a graph is a set of vertices whose deletion will leave a graph without any odd cycles. The Odd Cycle Transversal (OCT) problem takes an undirected graph G and a non-negative integer k as input, and the objective is to test if G has an oct of size at most k. The directed counterpart of the problem, Directed Odd Cycle Transversal (DOCT), where the input is a digraph and k, is defined analogously. When parameterized by k, OCT is known to be FPT [Reed et al., Oper. Res. Lett., 2004] whereas DOCT was recently shown to be W[1]-hard [Lokshtanov et al., SODA, 2020].

A mixed graph is a graph that contains both directed and undirected edges. In this paper, we study the Mixed Odd Cycle Transversal (MOCT) problem, i.e., OCT on mixed graphs. And we show that MOCT admits a fixed-parameter tractable algorithm when parameterized by $k+\ell$, where ℓ is the number of directed edges in the input mixed graph. In the course of designing our algorithm for MOCT, we also design a fixed-parameter tractable algorithm for a variant of the well-known Multiway Cut problem, which might be of independent interest.

1 Introduction

A mixed graph is a graph that contains both directed and undirected edges. For convenience, we call directed edges arcs and undirected edges simply edges.

Lawqueen Kanesh is supported in part by NRF Fellowship for AI grant [R-252-000-B14-281] and by Defense Service Organization, Singapore. Saket Saurabh is supported by the European Research Council (ERC) under the European Union's Horizon 2020 research and innovation programme (grant agreement no. 819416), and Swarnajayanti Fellowship (no. DST/SJF/MSA01/2017-18). Jayakrishnan Madathil is supported by the Chennai Mathematical Institute and the Infosys Foundation. Part of this work was done while Jayakrishnan Madathil was at the Indian Institute of Technology Gandhinagar, supported by an IITGN-Early Career Fellowship. The authors would like to thank Komal Muluk for several helpful discussions on the MOCT problem, as well as for several suggestions that improved the presentation of this article.

Ł. Kowalik et al. (Eds.): WG 2021, LNCS 12911, pp. 130–142, 2021.
https://doi.org/10.1007/978-3-030-86838-3_10

An odd cycle transversal (oct, for short) in a mixed/undirected/directed graph is a set of vertices whose deletion will leave a graph without any odd cycles. In this paper, we study the Mixed Odd Cycle Transversal (MOCT) problem, where we are given a mixed graph G and an integer k, and the goal is to test if G has an oct of size at most k. The problem is formally defined below.

Mixed Odd Cycle Transversal (MOCT) **Parameter:** $k+\ell$
Input: An n-vertex mixed graph G with ℓ arcs and a non-negative integer k.
Question: Does G have an oct of size at most k?

We show that MOCT admits a fixed-parameter tractable algorithm when parameterized by $k+\ell$, where ℓ is the number of arcs in the input graph. Specifically, we prove the following theorem. (Due to space constraints, the proofs of statements marked with a $\star$ have been omitted.)

Theorem 1 ($\star$). MOCT *admits an algorithm that runs in time* $(2\ell+k)^{2\ell+k}2^{\mathcal{O}(k^5)}n^{\mathcal{O}(1)}$.

When parameterized by the solution size k, the Odd Cycle Transversal problem is known to be FPT on undirected graphs [10–12,15,20] and W[1]-hard on digraphs [17]. Notice that for the algorithm in Theorem 1, if the number of arcs $\ell \leq k$, then the algorithm runs in time $2^{\text{poly}(k)}n^{\mathcal{O}(1)}$. Otherwise $k \leq \ell$, and thus the dependence of the algorithm's runtime on ℓ is $\ell^{\mathcal{O}(\ell)}$, and for values of ℓ up to $\mathcal{O}(\log n/\log\log n)$, the algorithm runs in time $2^{\mathcal{O}(k^5)}n^{\mathcal{O}(1)}$. That is, MOCT is fixed-parameter tractable, when parameterized by k alone, as long as the number of arcs in the input mixed graph is $\mathcal{O}(\log n/\log\log n)$.

In the course of proving Theorem 1, we also design an FPT algorithm for a variant of the well-known Multiway Cut problem, which we call Bipartition Multiway Cut For Sets (BMC For Sets). Consider an undirected graph G. For a pair of disjoint vertex subsets $X, S \subseteq V(G)$, $R_G(X,S)$ denotes the set of vertices in G that are reachable from X in the subgraph $G-S$, i.e., $R_G(X,S) = \{v \in V(G) \mid \text{there exists a path in } G-S \text{ from } x \text{ to } v \text{ for some } x \in X\}$. If $S=\emptyset$, we simply write $R_G(X)$ instead of $R_G(X,\emptyset)$. We formally define the BMC For Sets problem as follows.

Bipartition Multiway Cut For Sets (BMC For Sets) **Parameter:** k
Input: An undirected n-vertex graph G, pairwise disjoint vertex sets $W_1,\ldots,W_p$ such that $G-W$ is bipartite, where $W=\bigcup_{i=1}^{p} W_i$, and a non-negative integer k.
Question: Does there exist a set of vertices $S \subseteq V(G)\setminus W$ such that $\lvert S\rvert \leq k$, for every $i \in [p]$, $G[R_G(W_i,S)]$ is bipartite and for every distinct $i,j \in [p]$, there is no path from a vertex in W_i to a vertex in W_j in $G-S$?

We show that BMC For Sets is fixed-parameter tractable, when parameterized by the solution size k. Specifically, we prove the following theorem.

Theorem 2 (⋆). BMC FOR SETS *admits an algorithm running in time* $2^{\mathcal{O}(k^5)}n^{\mathcal{O}(1)}$.

We use the algorithm in Theorem 2 as a sub-routine to prove Theorem 1. In particular, we show that solving MOCT amounts to solving $(2\ell+k)^{(2\ell+k)}$ many instances of BMC FOR SETS, where ℓ is the number of arcs in the input mixed graph. We first briefly discuss the results that motivated our study of MOCT and BMC FOR SETS before laying out an overview of our strategy for proving Theorems 1 and 2.

The MOCT problem has three widely-studied antecedents in the literature—the ODD CYCLE TRANSVERSAL (OCT) problem as well as its directed counterpart, the DIRECTED ODD CYCLE TRANSVERSAL (DOCT) problem, and the DIRECTED FEEDBACK VERTEX SET (DFVS) problem. In OCT (resp. DOCT) the input consists of an undirected graph (resp. digraph) G and an integer k, and the objective is to decide if G has an odd cycle transversal of size at most k. And in DFVS, the input consists of a digraph D and an integer k, and the objective is to decide if D has a feedback vertex set of size at most k. A feedback vertex set of a (mixed/undirected/directed) graph is a set of vertices whose deletion will render the graph acyclic. Observe that MOCT generalises both OCT and DOCT, as both undirected and directed graphs are mixed graphs. As is the case with several problems in the realm of parameterized complexity, when parameterized by k, the undirected variant of the problem, i.e., OCT, is fixed-parameter tractable [10–12,15,20], whereas the directed variant, i.e., DOCT, is W[1]-hard [17]. But unlike DOCT, DFVS is FPT [5]. And it was observed in [17] that DFVS reduces to DOCT, and thus MOCT generalises DFVS as well. In fact, the parameterized complexity (with k as the parameter) of OCT, DFVS and DOCT were deemed to be major open problems, until they were settled respectively in 2003, 2008 and 2020 [5,17,20].

The W[1]-hardness of DOCT forecloses the possibility of a fixed-parameter tractable algorithm for MOCT, parameterized by k alone, unless FPT = W[1]. But this jump in complexity from the undirected variant to the directed variant of the problem does raise the following question: To what extent does the presence of arcs impede the tractability of the problem? For example, does the problem become fixed-parameter tractable if there are only at most $o(n)$ arcs in the input graph? We show that this is indeed the case when the input graph has only $\mathcal{O}(\log n/\log\log n)$ arcs. In what follows, (G,k) is an instance of MOCT, where G is a mixed graph with n vertices and ℓ arcs. Observe that if $\ell = 1$, i.e., if G has only one arc, then every cycle in the underlying undirected graph of G is a cycle in G and vice versa. Thus, we can forget the orientation of the unique arc in G, and simply treat (G,k) as an instance of the (undirected) OCT problem. But note that this argument does not work even when $\ell = 2$. For example, consider the mixed graph G on 3 vertices x, y and z with an edge xy and arcs (x,z) and (y,z). Note that G is odd cycle-free, whereas the underlying undirected graph of G is not.

When parameterized by k alone, the FVS problem is FPT on all three types of graphs—directed, undirected and mixed [1–3,5–7,9,13,14], whereas OCT is FPT on undirected graphs [17,20] and W[1]-hard on digraphs. As mentioned earlier, the parameterized complexity of DFVS was deemed a major open problem until its resolution by Chen et al. [5], who showed that the problem is FPT. This result was soon followed by the work of Bonsma and Lokshtanov [1], who studied the MFVS problem, i.e., FVS on mixed graphs. The inquiry of Bonsma and Lokshtanov [1] into the complexity of MFVS was prompted by the following reason: while a number of problems reduces to their directed variants from the undirected variants simply by replacing the edges by a pair of directed arcs, this obvious strategy does not work for FVS. In fact, replacing every edge of an undirected graph G by arcs in opposing directions would create several spurious directed 2-cycles. And thus, as noted by Bonsma and Lokshtanov, "every feedback vertex set of the resulting [directed] graph is a vertex cover of G and vice versa." Thus, they observed that "FVS problems on undirected and directed graphs are different problems; one is not a generalization of the other." And they sought to "*bridge the gap between the parameterized algorithms for* FEEDBACK VERTEX SET *by giving one algorithm that works for both,*" and gave a fixed-parameterized algorithm for MFVS.

Reed et al. [20] introduced the technique of *iterative compression* and used it to demonstrate that, OCT, at its core, is a cut problem—hitting all odd cycles in a graph, in fact, amounts to hitting all paths between two appropriately defined vertex subsets. Chen et al. [5] applied iterative compression, and showed that solving DFVS amounts to solving a cut problem as well, for which they used the idea of *important separators* [18]. The concept of important separators was first introduced by Marx [18] to solve the MULTIWAY CUT problem, a generalisation of the classic $s-t$ cut problem. In MULTIWAY CUT, the input consists of a graph G, a set of vertices $T \subseteq V(G)$ and a non-negative integer k, and the task is to determine if there exists a set of vertices $S \subseteq V(G)$ such that $|S| \leq k$ and for every distinct $t, t' \in T$, there is no $t-t'$ path in $G-S$. The vertices in the set T are called terminals. When $|T| = 2$, MULTIWAY CUT is equivalent to the classic $s-t$ cut problem. Marx showed that MULTIWAY CUT admits an algorithm running in time $2^{\mathcal{O}(k^3)}n^{\mathcal{O}(1)}$, which has been improved to $4^k n^{\mathcal{O}(1)}$ [4], and later to $2^k n^{\mathcal{O}(1)}$ [8]. Lokshtanov and Ramanujan [16] studied a parity constrained version of MULTIWAY CUT called PARITY MULTIWAY CUT (PMWC), in which there are two sets of terminals, T_e and T_o, and for every $v \in T_e$ and for every $w \in T_o$, only even paths between v and $(T_e \cup T_o) \setminus \{v\}$, and only odd paths between w and $(T_e \cup T_o) \setminus \{w\}$ need to be disconnected. Note that PMWC with $T_e = T_o$ is precisely the MULTIWAY CUT problem. Lokshtanov and Ramanujan showed that PMWC admits an algorithm that runs in time $2^{2^{\mathcal{O}(k)}}n^{\mathcal{O}(1)}$ time, which was subsequently improved to $2^{\mathcal{O}(k^3)}n^{\mathcal{O}(1)}$ [19]. These works [16,19] stand out, not just on account of their resolving a nontrivial generalisation of MULTIWAY CUT, but because of their adaptation of important separators to reach the results. In particular, Ramanujan [19] augmented the definition of important separators to accommodate certain structural restrictions on the reachability sets of terminals, refined the notion of a separator's being dominated by another separator, and thus provided a template based on important separators for designing algorithms for "cut problems." In this paper, we adapt the techniques in [16,19] to design our algorithms for MOCT and BMC FOR SETS. Finally, Lokshtanov et al. [17] studied the parameterized complexity of DOCT, parameterized by the solution size, and showed that the problem is W[1]-hard. But on the positive side, they showed that DOCT admits a factor-2 FPT approximation algorithm that runs in time $2^{\mathcal{O}(k^2)}n^{\mathcal{O}(1)}$. They also showed that under the Parameterized Inapproximability Hypothesis, and assuming that $\mathsf{W}[1] \neq \mathsf{FPT}$, there exists an $\epsilon > 0$ such that DOCT does not admit a $(1+\epsilon)$-factor approximation algorithm.

Overview of Our Results: As in the case of OCT and DFVS, we show that solving MOCT amounts to solving (multiple instances of) a cut problem. Unsurprisingly, to design our algorithm for MOCT, we borrow and adapt ideas and techniques from the three results mentioned above—the algorithms for OCT, Multiway Cut and Parity Multiway Cut. Specifically, we make use of the following three ingredients from these results.

1. "Compression" from the algorithm of Reed et al. [20] for OCT. That is, start with a larger solution, and then use the structure imposed by that larger solution to obtain a smaller solution of the required size.
2. Important separators and the accompanying "Pushing Lemma"—the idea that while looking for an $X - Y$ separator, push the separator as far away from X as possible—from the works of Marx et al. [18] and Cygan et al. [8]
3. The ideas of augmenting the definition of important separators and using a set of $2^{k^{\mathcal{O}(1)}}$ vertices that intersects the separator that we are looking for from the work of Ramanujan [19].

We must note that while we have all these previous results mentioned above at our disposal, there does not seem to be a straightforward reduction from MOCT or any appropriate annotated variant of MOCT to any of the problems discussed above. A non-trivial application of a combination of techniques has thus become necessary to derive our result. Observe that if we were to imitate the strategy for OCT to design an algorithm for MOCT, we would hit a snag as soon as we start, for the simple reason that odd cycle-free mixed graphs (and odd cycle-free digraphs, for that matter) lack the extremely useful structural characterisation that odd cycle-free undirected graphs have: An undirected graph is odd cycle-free if and only if it is bipartite. Note that this property does not hold for mixed graphs. For example, a mixed graph G on three vertices x, y and z with an edge xy and arcs (y, z) and (x, z) has no odd cycles, but is not bipartite. We say that a mixed graph is bipartite if its underlying undirected graph is bipartite. Nonetheless, we can make do with the following weaker claim that applies to mixed graphs (and digraphs).

Lemma 1 ($\star$). *A strongly connected mixed graph is odd cycle-free if and only if it is bipartite.*

In light of Lemma 1, it might be tempting to try the following approach: Decompose the given mixed graph into strongly connected components, and then apply the algorithm for OCT on each component. Note that this approach would fail too for it is not necessary to make each strongly connected component bipartite to make the graph odd cycle-free; it might just be enough to destroy the strong connectivity of each component to make the graph odd cycle-free. For example,

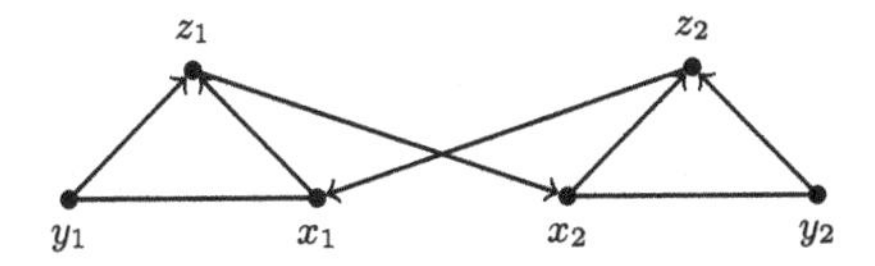

Fig. 1. An example of a mixed graph that is strongly connected, has an oct of size 1, but any minimum-sized oct of the underlying undirected graph has size 2.

consider the mixed graph G, defined by $V(G) = \{x_i, y_i, z_i \mid i \in [2]\}$, $E(G) = \{x_i y_i \mid i \in [2]\}$ and $A(G) = \{(x_i, z_i), (y_i, z_i) \mid i \in [2]\} \cup \{(z_1, x_2), (z_2, x_1)\}$. (See Fig. 1.) Note that G is strongly connected. To make G odd cycle-free, we only need to delete 1 vertex (any one of the vertices x_1, x_2, z_1, z_2 will do), whereas to make G bipartite, we need to delete 2 vertices because $\{x_1, y_1, z_1\}$ and $\{x_2, y_2, z_2\}$ induce two vertex-disjoint triangles in the underlying undirected graph of G.

Outline for Our Algorithm: Our algorithm for MOCT has two main steps. Let (G, k) be the given instance of MOCT where $|A(G)| = \ell$. In the first step, we reduce (G, k) to at most $(2\ell + k)^{2\ell+k}$ many instances of the BMC For Sets problem such that (G, k) is a yes-instance if and only if at least one of the $(2\ell + k)^{2\ell+k}$ instances of BMC For Sets is a yes-instance. And in the second step we solve BMC For Sets on each of the $(2\ell + k)^{2\ell+k}$ instances.

Consider an instance $(G, \{W_1, W_2, \ldots, W_p\}, k)$ of the BMC For Sets problem. Let $T_1 = W_1$ and $T_2 = \bigcup_{i=2}^{p} W_i$. Observe first that the solution Z that we are looking for must contain a minimal $T_1 - T_2$ separator, say, $Z_1 \subseteq Z$, as the W_is need to be disconnected from each other. This idea is the same as in the Multiway Cut problem. But in addition, we want $R_G(W_i, Z)$ to induce a bipartite graph. Therefore, it is not enough to enumerate all important $T_1 - T_2$ separators and branch on them. And to deal with this, we use the augmented definition due to Ramanujan [19] of the domination of a separator by another separator: For two $T_1 - T_2$ separators S_1 and Z_1, we say that S_1 well-dominates Z_1 if (i) $|S_1| \leq |Z_1|$, (ii) $R(T_1, S_1) \supset R(T_1, Z_1)$ and (iii) the minimum size of an oct for $G[R(T_1, S_1)]$ does not exceed the minimum size of an oct for $G[R(T_1, Z_1)]$. Thus, if S_1 and Z_1 are two $T_1 - T_2$ separators of size at most k, and S_1 well-dominates Z_1, then we might as well pick S_1. And if S_1 is minimal and there exists no other separator that well-dominates S_1, then we say that S_1 is well-domination maximal. But unlike in the case of important separators, we cannot enumerate all well-domination maximal separators of size k. Instead we use a result from [19], which says that given G, T_1, T_2 and k, we can either construct a set, $J(T_1, T_2, k) \subseteq V(G) \setminus (T_1 \cup T_2)$, of size at most $2^{\mathcal{O}(k^3)}$ that intersects all well-domination maximal $T_1 - T_2$ separators of size at most k, or conclude that no well-domination maximal T_1-T_2 separator of size at most k exists. Thus, the set $J(T_1, T_2, k)$ must intersect the solution that we are looking for, and therefore, we branch on all subsets of $J(T_1, T_2, k)$ of size at most k, which leads to an algorithm with the claimed runtime.

2 Preliminaries

For a natural number n, we use $[n]$ to denote the set $\{1, 2, \ldots, n\}$. For a mixed graph G, $V(G)$ denotes the vertex set of G. And $A(G)$ and $E(G)$ respectively denote the arc set and edge set of G. By an arc we mean a directed edge, and by an edge we mean an undirected edge. Cycles, odd cycles and strongly connected components in mixed graphs are defined analogously to their respective counterparts in digraphs. For a mixed graph G, by the underlying undirected graph of G, we mean the undirected graph obtained from G by removing the orientations of the arcs in G.

Reed et al. [20] showed that OCT admits an algorithm that runs in time $2^{\mathcal{O}(k)}n^{\mathcal{O}(1)}$. In fact, their algorithm can be adapted to output an oct of size at most k, if it exists. Now, consider the following variant of OCT: We are given G, a set of vertices $B \subseteq V(G)$ and k as input, and the goal is to check if G has an oct $S \subseteq V(G)$ such that $|S| \leq k$ and $S \cap B = \emptyset$. (An instance with $B = \emptyset$ is precisely the OCT problem.) To solve this variant of the problem, notice that we can make B "undeletable" by adding k extra copies of B to G, and then use the algorithm of Reed et al. [20] on the modified graph. We record this fact below.

Lemma 2 (⋆). *There is an algorithm that, given an n-vertex undirected graph G, a set $B \subseteq V(G)$ and a non-negative integer k as input, runs in time $2^{\mathcal{O}(k)}n^{\mathcal{O}(1)}$, and correctly decides if G has an oct $S \subseteq V(G)$ such that $|S| \leq k$ and $S \cap B = \emptyset$. Moreover, the algorithm outputs such an oct S, if it exists.*

3 Solving Mixed Odd Cycle Transversal

In this section, assuming that Theorem 2 holds, we show that MOCT admits an FPT algorithm, and thus prove Theorem 1. Recall that in the Mixed Odd Cycle Transversal (MOCT) problem, given a mixed graph G with ℓ arcs and an integer k, the question is to decide whether G has an oct of size at most k. To design our FPT algorithm, we use the "compression" idea from the classic iterative compression technique (see, e.g., [6, Chapter 4]). The ingredients for iterative compression are a *large* solution (if it exists) and a *disjoint-variant* of the problem that compresses the large solution to a solution of a desired size. To begin with, we find an oct of graph $G - V(A(G))$ of size at most k using the algorithm for OCT by Reed et al. on the instance $(G - V(A(G)), k)$ [20]. If an oct of size k does not exist for the instance, then it directly implies that the instance (G, k) of MOCT is also a no-instance as $G - V(A(G))$ is a subgraph of G.

Next, as is common in algorithms based on iterative compression, instead of designing an algorithm for MOCT, we design an algorithm for a disjoint-variant of MOCT. We call this disjoint-variant Disjoint-Mixed OCT (D-MOCT, for short). In D-MOCT, together with the mixed graph G and an integer k, we are also given a set $W \subseteq V(G)$ of size at most $k + 2\ell$ such that $G[W]$ is bipartite and contains all the arcs in G (i.e., for every arc $(u,v) \in A(G)$, we have $u,v \in W$), and $G - W$ is (undirected and) bipartite, and the question is whether G has an oct $S \subseteq V(G) \setminus W$ of size at most k. We prove the following lemma.

Lemma 3 (⋆). *Let (G,W,k) be an instance of* D-MOCT. *There exists an algorithm for* D-MOCT *running in time $(2\ell + k)^{(2\ell+k)} \cdot \psi(k,n)$, where $\psi(k,n)$ is the running time of an algorithm for* Bipartition Multiway Cut For Sets.

Equipped with the above lemma, we can prove Theorem 1. We now show that by a simple procedure we can further reduce D-MOCT to $k + 2\ell$ many instances of the Bipartition Multiway Cut For Sets problem. Given an instance (G,W,k) of D-MOCT, if (G,W,k) is a yes-instance of D-MOCT, then we show that there exists a partition $\{W_1,\dots,W_p\}$ of W such that it is sufficient to solve BMC For Sets on instance $(G^*,(W_1,\dots,W_p),k)$ where G^* is a subgraph of the underlying undirected graph of G. If we manage to show the existence of such a "good" partition, then as $|W| \leq k+2\ell$, we can guess this partition by going over all possible $(k+2\ell)^{k+2\ell}$ partitions of W. We formalise this idea in the following lemma, which can then be used to prove Lemma 3.

Lemma 4 (⋆). *Let (G,W,k) be an instance of* D-MOCT. *If there exists an odd cycle transversal $S \subseteq V(G) \setminus W$ of size at most k of G, then there exists an ordered partition $\{W_1,\dots,W_p\}$ of W such that the following holds:*

1. *For every $i \in [p]$, W_i is contained in a unique strongly connected component C of $G - S$.*
2. *For every $i \in [p]$, the underlying undirected graph of $G[R(W_i,S)]$ is bipartite.*
3. *For every $i,j \in [p], i \neq j$, W_i and W_j are contained in different strongly connected components of $G - S$.*
4. *For each arc (u,v) in $G - S$, there exist $i,j \in [p]$ such that $i \leq j$, $u \in W_i$ and $v \in W_j$.*

4 Solving BMC for Sets

In this section, we design an algorithm for BMC for Sets that runs in time $2^{\mathcal{O}(k^5)} n^{\mathcal{O}(1)}$. All graphs considered in this section are undirected. We start with a few definitions and preliminary results that we will be using to design our algorithm.

Notation and Terminology: Consider a graph G. We use $\text{oct}(G)$ to denote the size of a minimum-sized oct of G. Let $X^*, S \subseteq V(G)$ be disjoint sets. Recall that by $R_G(X^*,S)$, we denote the set of vertices of G that are reachable from X^* in the graph $G - S$. We drop the subscript G if the graph is clear from the context. Let $X,Y \subseteq V(G)$ be two disjoint sets. We say that $S \subseteq V(G) \setminus (X \cup Y)$ is an $X - Y$ separator in G if there is no path from X to Y in $G - S$, i.e., if $R_G(X,S) \cap Y = \phi$. If no proper subset of S is an $X - Y$ separator, then we say that S is *inclusion wise minimal* (or minimal, for short). Also, we use $\lambda_G(X,Y)$ to denote the cardinality of a minimum-sized X-Y separator. With a slight abuse of terminology, for pairwise disjoint vertex subsets $W_1, W_2, \ldots, W_p \subseteq V(G)$, we say that $S \subseteq V(G)$ is a $\{W_1, W_2, \ldots, W_p\}$ separator if S is a $W_i - W_j$ separator for every distinct $i, j \in [p]$. Let $X, Y \subseteq V(G)$ be two disjoint sets, and let S_1 and S_2 be two $X - Y$ separators. We say that S_1 *dominates* S_2 with respect to X if S_1 is minimal, $|S_1| \leq |S_2|$ and $R(X, S_1) \supset R(X, S_2)$. And we say that S_1 *well-dominates* S_2 if S_1 dominates S_2 and $\text{oct}(G[R(X, S_1)]) \leq \text{oct}(G[R(X, S_2)])$. If the set X is clear from the context, then we just say that S_1 dominates (or well-dominates) S_2.

Definition 1. *Let G be a graph, and let $X, Y \subseteq V(G)$ be disjoint sets. An X-Y separator $S \subseteq V(G) \setminus (X \cup Y)$ is said to be* well-domination maximal *if no other $X - Y$ separator well-dominates S.*

Lemma 5 (Pushing lemma for Multiway Cut **[16]).** *Consider a graph G and pairwise-disjoint vertex subsets $W_1, W_2, \ldots, W_p \subseteq V(G)$. Let $S \subseteq V(G)$ be a $\{W_1, W_2, \ldots, W_p\}$ separator and $S_1 \subseteq S$ be any minimal $W_1 - W'$ separator, where $W' = \bigcup_{i=2}^{p} W_i$. Then, for any $W_1 - W'$ separator $S_1' \subseteq V(G)$ that dominates S_1, $S' = (S \setminus S_1) \cup S_1'$ is also a $\{W_1, W_2, \ldots, W_p\}$ separator.*

Lemma 6 ($\star$). *Consider an instance $(G, \{W_1, \ldots, W_p\}, k)$ of* BMC For Sets. *Suppose that $S \subseteq V(G) \setminus \bigcup_{i=1}^{p} W_i$ is a solution for this instance. Let $S_1 \subseteq S$ be a minimal W_1-W' separator, where $W' = \bigcup_{i=2}^{p} W_i$, and let $S_2 = R(W_1, S_1) \cap S$. Then S_2 is an oct for $G[R(W_1, S_1)]$.*

Lemma 7 (Pushing lemma for BMC For Sets**).** *Consider an instance $(G, \{W_1, \ldots, W_p\}, k)$ of* BMC for Sets. *Suppose that $S \subseteq V(G) \setminus \bigcup_{i=1}^{p} W_i$ is a solution for this instance. Let $S_1 \subseteq S$ be a minimal $W_1 - W'$ separator, where $W' = \bigcup_{i=2}^{p} W_i$, and let $S_2 = S \cap R(W_1, S_1)$. Then, for any minimal $W_1 - W'$ separator S_1' such that S_1' well-dominates S_1, and for any minimum-sized oct S_2' for $G[R(W_1, S_1')]$, the set $S' = (S \setminus (S_1 \cup S_2)) \cup (S_1' \cup S_2')$ is also a solution for the* BMC for Sets *instance $(G, \{W_1, \ldots, W_p\}, k)$.*

Proof. To prove that S' is a solution for the instance $(G, \{W_1, \ldots, W_p\}, k)$ of BMC for Sets, we have to prove that (i) $|S'| \leq k$, (ii) S' is a $\{W_1, W_2, \ldots, W_p\}$ separator and (iii) $R(W_i, S')$ is bipartite for every $i \in [p]$. The fact that S' is a separator for $\{W_1, \ldots, W_p\}$ follows directly from Lemma 5. By the definition of well-domination, we have $|S_1'| \leq |S_1|$. By Lemma 6, S_2 is an oct for $G[R(W_1, S_1)]$. And by the definition of well-domination, we have $|S_2'| = \text{oct}(G[R(W_1, S_1')]) \leq \text{oct}(G[R(W_1, S_1)]) \leq |S_2|$. Since $S_1 \cap S_2 = \emptyset$, and $S_1' \cap S_2' = \emptyset$, we get $|S_1' \cup S_2'| = |S_1'| + |S_2'| \leq |S_1| + |S_2| = |S_1 \cup S_2|$. We thus get that $|S'| = |(S \setminus (S_1 \cup S_2)) \cup (S_1' \cup S_2')| \leq |S|$. And since S is a solution we have $|S| \leq k$, which implies that $|S'| \leq k$. Also, since S' contains S_2', which is an oct for $G[R(W_1, S_1')]$ and $R(W_1, S_1') \supseteq R(W_1, S')$, we also get that $G[R(W_1, S')]$ is bipartite. All that remains to be shown is that $R(W_i, S')$ is bipartite for each i with $2 \leq i \leq p$. To prove that, we first prove the following claims.

Claim 1. *For every $i \in [p] \setminus \{1\}$, S_1' is a $(S_1 \setminus S_1') - W_i$ separator.*

Proof. Fix $i \in [p] \setminus \{1\}$. Assume that the claim is not true for i. Then there exists a path, say P, in $G - S_1'$, from s to w for some $s \in S_1 \setminus S_1'$ and some $w \in W_i$. As S_1 is a minimal $W_1 - W'$ separator, there exists a path from W_1 to W', say P', that intersects S_1 only in s. Let P'' denote the $W_1 - s$ subpath of P'. Consider a vertex $x \in V(P'') \setminus \{s\}$. Note that we have $x \in R(W_1, S_1)$. Now, since S_1' dominates S_1, we have $R(W_1, S_1) \subset R(W_1, S_1')$, and therefore, $x \in R(W_1, S_1')$. Now, since $s \in S_1 \setminus S_1'$, we also have $s \notin S_1'$. That is, no vertex of the path P'' belongs to S_1', or in other words, P'' is a path in $G - S_1'$. Thus, both P and P'' are paths in $G - S_1'$, and the vertex s is common to both of them. Then the path P'' followed by the path P is a path in $G - S_1'$ from W_1 to $W_i \subseteq W'$, which contradicts the assumption that S_1' is a $W_1 - W'$ separator. □

Claim 2. *For every $i \in [p] \setminus \{1\}$, $R(W_i, S_1') \subseteq R(W_i, S_1)$.*

Proof. Fix $i \in [p] \setminus \{1\}$. Consider $z \in R(W_i, S_1')$. Then there exists a path in $G - S_1'$, say Q, from w' to z for some $w' \in W_i$. Now, assume that $z \notin R(W_i, S_1)$. Then, S_1 must intersect the path Q. Let $u \in V(Q) \cap S_1$. Let $Q_{u,w'}$ denote the $u - w$ subpath of Q. Then, $Q_{u,w'}$ is an $(S_1 \setminus S_1') - W_i$ path in $G - S_1'$. But this implies that S_1' is not an $(S_1 \setminus S_1') - W_i$ separator, a contradiction to Claim 1. □

Claim 3 *For every $i \in [p] \setminus \{1\}$, $R(W_i, S') \subseteq R(W_i, S)$.*

Proof. Fix $i \in [p] \setminus \{1\}$. Assume that the claim is not true for i. Then there exists $w'' \in R(W_i, S')$ such that $w'' \notin R(W_i, S)$. Since $w'' \in R(W_i, S')$, there exists a $W_i - \{w''\}$ path, say Q', in $G - S'$. And since $w'' \notin R(W_i, S)$, S must intersect the path Q'. Let $u' \in V(Q') \cap S$. Thus, $u' \in S$ but $u' \notin S'$. But then $u' \notin (S \setminus S_1)$ as $S \setminus S_1 \subseteq S'$. Therefore, $u' \in S_1$, which implies that $u' \notin R(W_i, S_1)$. Now, note that $u' \in R(W_i, S')$, as $u' \in V(Q') \subseteq R(W_i, S')$. Therefore, $u' \in R(W_i, S_1')$ as well, because $S' \supseteq S_1'$, and therefore, $R(W_i, S') \subseteq R(W_i, S_1')$. We thus have $u' \in R(W_i, S_1')$ but $u' \notin R(W_i, S_1)$, which contradicts Claim 2. □

Claim 3, along with the fact that $R(W_i, S)$ is bipartite, immediately implies that $R(W_i, S')$ is bipartite for every $i \in [p] \setminus \{1\}$. This completes the proof of Lemma 7. □

Consider an instance $(G, \{W_1, W_2, \ldots, W_p\}, k)$ of BMC FOR SETS. Lemma 7 tells us that if $(G, \{W_1, W_2, \ldots, W_t\}, k)$ is a yes-instance of BMC FOR SETS, then this instance has a solution that contains a well-domination maximal $W_1 - W'$ separator. And the following result due to Lokshtanov and Ramanujan [16] says that we can find a sufficiently small set of vertices that intersects the well-domination maximal separator that we are looking for.

Lemma 8 [16,19]. *There is an algorithm that, given a graph G, disjoint vertex subsets $X, Y \subseteq V(G)$ and a non-negative integer k as input, runs in time $2^{\mathcal{O}(k^3)} n^{\mathcal{O}(1)}$, and either returns a set $J(X, Y, k) \subseteq V(G) \setminus (X \cup Y)$ of size $2^{\mathcal{O}(k^3)}$ such that $J(X, Y, k)$ intersects all well-domination maximal X-Y separators of size at most k, or correctly reports that no well-domination maximal X-Y separator of size at most k exists.*

4.1 Algorithm for BMC For Sets

We design a branching algorithm for BMC For Sets using Lemmas 7 and 8, and thus prove Theorem 2. Before proceeding to the algorithm, we introduce the following reduction rule, which removes the W_is that are already unreachable from all the other W_js and $R_G(W_i)$s induce bipartite subgraphs.

Reduction Rule 1. *Let* $(G, \{W_1, \ldots, W_p\}, k)$ *be an instance of* BMC For Sets. *If there exists* $i \in [p]$ *such that* $\lambda_G(W_i, W_j) = 0$ *for every* $j \in [p] \setminus \{i\}$ *and* $R_G(W_i)$ *is bipartite, then delete* $R_G(W_i)$ *from* G.

We are now ready to describe our algorithm, which we call Algo-BMCS.

Description of the Algorithm Algo-BMCS. We are given an instance of BMC For Sets $(G, \{W_1, \ldots, W_p\}, k)$ as input, where G is an n-vertex graph and k is a non-negative integer. We do as follows.
Step 1. Proceed only if $k \geq 0$. If $\lambda_G(W_i, W_j) = 0$ for every distinct $i, j \in [p]$ and $R(W_i)$ is bipartite for every $i \in [p]$, then return that $(G, \{W_1, \ldots, W_p\}, k)$ is a yes-instance and terminate.
Step 2. Apply Reduction Rule 1 exhaustively.
Step 3. Compute $\lambda_G(W_1, W')$, where $W' = \bigcup_{i=2}^{p} W_i$. If $\lambda_G(W_1, W') = 0$, then for each k_1, where $0 \leq k_1 \leq k$, use the algorithm in Lemma 2, to check if $\text{oct}(G[R(W_1)]) \leq k_1$. Proceed only if $\text{oct}(G[R(W_1)]) \leq k_1$ for some $k_1 \leq k$. Let $k' = \min\{k_1 \mid \text{oct}(G[R(W_1)]) \leq k_1\}$. Call Algo-BMCS on the instance $(G - R(W_1), \{W_2, W_3, \ldots, W_p\}, k - k')$.
Step 4. Proceed only if $\lambda_G(W_1, W') > 0$, where $W' = \bigcup_{i=2}^{p} W_i$. Call the algorithm in Lemma 8 on the input (G, W_1, W', k). Proceed only if the algorithm in Lemma 8 returns a set $J(W_1, W', k) \subseteq V(G) \setminus (W_1 \cup W')$ of size at most $2^{\mathcal{O}(k^3)}$. For each $L \subseteq J(W_1, W', k)$ such that $|L| \leq k$, do as follows: Construct the instance $(G - L, \{W_1, W_2, \ldots, W_p\}, k - |L|)$ of BMC For Sets, and call Algo-BMCS on this instance.
Step 5. Return that $(G, \{W_1, \ldots, W_p\}, k)$ is a no-instance, and terminate.

5 Conclusion

We show that MOCT admits an FPT algorithm with runtime $(2\ell + k)^{2\ell+k} 2^{\mathcal{O}(k^5)} n^{\mathcal{O}(1)}$. An immediate question is to what extent can the dependence of the runtime on ℓ be improved. For example, can it be shown that MOCT admits an algorithm with runtime $2^{\mathcal{O}(\ell)} 2^{k^{\mathcal{O}(1)}} n^{\mathcal{O}(1)}$ so that as long as the number of arcs ℓ is at most $\log n$, the algorithm runs in time $2^{k^{\mathcal{O}(1)}} n^{\mathcal{O}(1)}$? Notice that any improvement in our algorithm for BMC For Sets will immediately imply a faster algorithm for MOCT. So, it would be interesting to see if the $2^{\mathcal{O}(k^5)}$ factor in the algorithm for BMC For Sets can be improved.

References

1. Bonsma, P., Lokshtanov, D.: Feedback vertex set in mixed graphs. In: Nešetřil, J., Ossona de Mendez, P. (eds.) Algorithms and Data Structures, pp. 122–133. Springer, Heidelberg (2011). https://doi.org/10.1007/978-3-642-27875-4
2. Cao, Y., Chen, J., Liu, Y.: On feedback vertex set: new measure and new structures. Algorithmica **73**(1), 63–86 (2015)
3. Chen, J., Fomin, F.V., Liu, Y., Lu, S., Villanger, Y.: Improved algorithms for feedback vertex set problems. J. Comput. Syst. Sci. **74**(7), 1188–1198 (2008)
4. Chen, J., Liu, Y., Lu, S.: An improved parameterized algorithm for the minimum node multiway cut problem. Algorithmica **55**(1), 1–13 (2009)
5. Chen, J., Liu, Y., Lu, S., O'sullivan, B., Razgon, I.: A fixed-parameter algorithm for the directed feedback vertex set problem. In: Proceedings of the Fortieth Annual ACM Symposium on Theory of Computing, pp. 177–186 (2008)
6. Cygan, M., et al.: Parameterized Algorithms. Springer, Cham (2015). https://doi.org/10.1007/978-3-319-21275-3
7. Cygan, M., Nederlof, J., Pilipczuk, M., Pilipczuk, M., van Rooij, J.M.M., Wojtaszczyk, J.O.: Solving connectivity problems parameterized by treewidth in single exponential time. In: IEEE 52nd Annual Symposium on Foundations of Computer Science, FOCS, pp. 150–159. IEEE Computer Society (2011)
8. Cygan, M., Pilipczuk, M., Pilipczuk, M., Wojtaszczyk, J.O.: On multiway cut parameterized above lower bounds. ACM Trans. Comput. Theory (TOCT) **5**(1), 1–11 (2013)
9. Guo, J., Gramm, J., Hüffner, F., Niedermeier, R., Wernicke, S.: Compression-based fixed-parameter algorithms for feedback vertex set and edge bipartization. J. Comput. Syst. Sci. **72**(8), 1386–1396 (2006)
10. Iwata, Y., Oka, K., Yoshida, Y.: Linear-time FPT algorithms via network flow. In: Proceedings of the Twenty-Fifth Annual ACM-SIAM Symposium on Discrete Algorithms, pp. 1749–1761. SIAM (2014)
11. Iwata, Y., Wahlström, M., Yoshida, Y.: Half-integrality, LP-branching, and FPT algorithms. SIAM J. Comput. **45**(4), 1377–1411 (2016)
12. Kawarabayashi, K.i., Reed, B.: An (almost) linear time algorithm for odd cycles transversal. In: Proceedings of the Twenty-First Annual ACM-SIAM Symposium on Discrete Algorithms, pp. 365–378. SIAM (2010)
13. Kociumaka, T., Pilipczuk, M.: Faster deterministic feedback vertex set. Inf. Process. Lett. **114**(10), 556–560 (2014)
14. Li, J., Nederlof, J.: Detecting feedback vertex sets of size k in $O^*(2.7^k)$ time. In: Proceedings of the Fourteenth Annual ACM-SIAM Symposium on Discrete Algorithms SODA, pp. 971–989. SIAM (2020)
15. Lokshtanov, D., Narayanaswamy, N., Raman, V., Ramanujan, M., Saurabh, S.: Faster parameterized algorithms using linear programming. ACM Trans. Algorithms (TALG) **11**(2), 1–31 (2014)
16. Lokshtanov, D., Ramanujan, M.: Parameterized tractability of multiway cut with parity constraints. In: Montanari, U., Rolim, J.D.P., Welzl, E. (eds.) International Colloquium on Automata, Languages, and Programming. LNCS, pp. 750–761. Springer, Heidelberg (2012). https://doi.org/10.1007/3-540-45022-X
17. Lokshtanov, D., Ramanujan, M., Saurab, S., Zehavi, M.: Parameterized complexity and approximability of directed odd cycle transversal. In: Proceedings of the Fourteenth Annual ACM-SIAM Symposium on Discrete Algorithms, pp. 2181–2200. SIAM (2020)

18. Marx, D.: Parameterized graph separation problems. Theoret. Comput. Sci. **351**(3), 394–406 (2006)
19. Ramanujan, M.: Parameterized graph separation problems: new techniques and algorithms. Ph.D. thesis, The Institute of Mathematical Sciences, HBNI (2013)
20. Reed, B., Smith, K., Vetta, A.: Finding odd cycle transversals. Oper. Res. Lett. **32**(4), 299–301 (2004)

Preventing Small (s, t)-Cuts by Protecting Edges

Niels Grüttemeier, Christian Komusiewicz, Nils Morawietz([✉]), and Frank Sommer

Fachbereich Mathematik und Informatik, Philipps-Universität Marburg, Marburg, Germany
{niegru,komusiewicz,morawietz,fsommer}@informatik.uni-marburg.de

Abstract. We introduce and study Weighted Min (s,t)-Cut Prevention, where we are given a graph $G=(V,E)$ with vertices s and t and an edge cost function and the aim is to choose an edge set D of total cost at most d such that G has no (s,t)-edge cut of capacity at most a that is disjoint from D. We show that Weighted Min (s,t)-Cut Prevention is NP-hard even on subcubcic graphs when all edges have capacity and cost one and provide a comprehensive study of the parameterized complexity of the problem. We show, for example W[1]-hardness with respect to d and an FPT algorithm for a.

1 Introduction

Network interdiction is a large class of optimization problems with direct applications in operations research [5,6,16–18]. In these problems one player wants to achieve a certain goal (for example finding a short path between two given vertices s and t), and another player wants to modify the network to prevent this. Given the enormous importance of the max-flow/min-cut problem it comes as no surprise that two-player games where an attacker wants to decrease the maximum (s,t)-flow of a network by deleting edges have been considered [5,17]. We study an inverse problem: an attacker wants to find an (s,t)-cut of capacity at most a and a defender wants to protect edges in order to increase the capacity of any minimum (s,t)-cut in G to at least $a+1$. The formal problem definition reads as follows.

Weighted Min (s,t)-Cut Prevention (WMCP)
Input: A graph $G=(V,E)$, two vertices $s,t \in V$, an edge cost function $c: E \rightarrow \mathbb{N}$, a capacity function $\omega: E \rightarrow \mathbb{N}$, and integers d and a.
Question: Is there a set $D \subseteq E$ with $c(D) := \sum_{e \in D} c(e) \leq d$ such that for every (s,t)-cut $A \subseteq (E \setminus D)$ in G we have $\omega(A) := \sum_{e \in A} \omega(e) > a$?

The special case where we have only unit capacities and unit costs is referred to as Min (s,t)-Cut Prevention (MCP). A related problem called Minimum

N. Morawietz—Supported by the DFG, project OPERAH, KO 3669/5-1.

Ł. Kowalik et al. (Eds.): WG 2021, LNCS 12911, pp. 143–155, 2021.
https://doi.org/10.1007/978-3-030-86838-3_11

Table 1. Parameter overview for WMCP and MCP. We write NP-h if the problem is NP-hard even if the corresponding parameter is a constant.

	a	d	Δ	$d+\Delta$	vc	pw + fvs
WMCP	FPT	W[1]-h	NP-h	W[1]-h if $\Delta = 3$	Weakly NP-h Theorem 7	Weakly NP-h Theorem 7
					W[1]-h	W[1]-h
	Theorem 5	Lemma 3	Theorem 1	Theorem 1	Theorem 8	Theorem 8
MCP	FPT	W[1]-h	NP-h	FPT	FPT	W[1]-h
	Theorem 5	Lemma 3	Theorem 3	Theorem 3	Theorem 6	Theorem 9

st-Cut INTERDICTION has been studied recently [1]; in this problem the graph is directed and an interdictor may freely choose the amount of increase in edge capacities. In our formulation, the defender may only decide to fully protect an edge or to leave it unprocted. To the best of our knowledge, this formulation of WMCP has not been considered so far. We study the classical complexity of WMCP and its parameterized complexity with respect to a, d, and important structural parameterizations of the input graph G.

Related Work. Many interdiction problems have been studied from a (parameterized) complexity perspective: In MATCHING INTERDICTION [18], one wants to remove vertices or edges to decrease the weight of a maximum-weight matching. In the MOST VITAL EDGES IN MST problem, one aims to remove edges to decrease the weight of any maximum spanning tree. In SHORTEST-PATH INTERDICTION [12], also known as SHORTEST PATH MOST VITAL EDGES [3,9] and MINIMUM LENGTH-BOUNDED CUT [2], one wants to remove edges to increase the length of a shortest (s,t)-path above a certain threshold. All of these problems are NP-hard and the study of their classical and parameterized complexity has received a lot of attention [3,9,10,18].

Our Results. An overview of our results is given in Table 1. We show that WMCP and MCP are NP-hard even on subcubic graphs. This motivates a parameterized complexity study with respect to the natural parameters defender budged d and attacker budget a and with respect to structural parameters of the input graph G. Here, we consider the structural parameters treewidth $\text{tw}(G)$ and vertex cover number $\text{vc}(G)$ of G as well as pathwidth $\text{pw}(G)$ and feedback vertex set number $\text{fvs}(G)$ of G. Our main results are as follows. MCP and WMCP are W[1]-hard with respect to the defender budget d and FPT with respect to the attacker budget a. MCP and WMCP are W[1]-hard with respect to the combined parameter pathwidth of G plus feedback vertex set number of G and thus also W[1]-hard with respect to the treewidth of G. The hardness for these parameters motivates a study of the vertex cover number $\text{vc}(G)$. We show that MCP is FPT with respect to $\text{vc}(G)$, whereas WMCP is weakly NP-hard even for $\text{vc}(G) = 2$ and W[1]-hard with respect to $\text{vc}(G)$ even when all capacities and costs are encoded in unary. Finally, we provide a polynomial kernel for WMCP parameterized by $\mathrm{vc}(G) + a$ and complement this result by showing that MCP and WMCP do not admit polynomial kernels with respect to the large combined parameter $d + a + \mathrm{tw}(G) + \mathrm{lp}(G) + \Delta(G)$ where $\mathrm{lp}(G)$ denotes the length of a longest path in G and $\Delta(G)$ denotes the maximum degree. Overall, our results give a comprehensive complexity overview of WMCP and MCP.

The proofs of statements marked with a (*) are deferred to a full version[1].

2 Preliminaries

For integers i and j with $i \leq j$, we define $[i, j] := \{k \in \mathbb{N} \mid i \leq k \leq j\}$.

An (undirected) graph $G = (V, E)$ consists of a set of vertices V and a set of edges $E \subseteq \binom{V}{2}$. Throughout this work, let $n := |V|$, $m := |E|$, and $|G| := |V| + |E|$. For vertex sets $S \subseteq V$ and $T \subseteq V$ we denote with $E_G(S, T) := \{\{s, t\} \in E \mid s \in S, t \in T\}$ the edges between S and T. Moreover, we define $E_G(S) := E_G(S, S)$ and $E_G(v, S) := E_G(\{v\}, S)$ for $v \in V$. For a vertex set $S \subseteq V$ we denote with $G[S] := (S, E_G(S))$ the *induced subgraph of S in G*. Moreover, for an edge set $D \subseteq E$, let $G - D := (V, E \setminus D)$ the graph obtained by deleting D. For a vertex $v \in V$, we denote with $N_G(v) := \{w \in V \mid \{v, w\} \in E\}$ the *open neighborhood* of v in G. Analogously, for a vertex set $S \subseteq V$, we define $N_G(S) := \bigcup_{v \in S} N_G(v) \setminus S$. If G is clear from the context, we may omit the subscript. A sequence of distinct vertices $P = (v_0, \ldots, v_k)$ is a *path* or (v_0, v_k)-*path* of length k in G if $\{v_{i-1}, v_i\} \in E$ for all $i \in [1, k]$. Let s and t be distinct vertices of V. An edge set $A \subseteq E$ is an (s, t) (edge)-cut in G if there is no (s, t)-path in $G - A$. A graph $G = (V, E)$ is *connected* if there is an (a, b)-path in G for each pair of distinct vertices $a, b \in V$.

Two instances I and I' of the same decision problem L are equivalent if I is a yes-instance of L if and only if I' is a yes-instance of L. A *reduction rule* for a decision problem L is an algorithm A that transforms any instance I of L into another instance $A(I)$ of L. We call A *safe*, if for each instance I of L, I and $A(I)$ are equivalent instances of L. A reduction rule A is exhaustively applied for an instance I if $A(I) = I$.

For details on parameterized complexity and the definitions of all graph parameters considered in this work, we refer to the standard monograph [7].

Let $I = (G = (V, E), s, t, c, \omega, d, a)$ be an instance of any of the above problems (in the case of MCP, $c(e) := \omega(e) := 1$ for all $e \in E$). We call an edge set $D \subseteq E$ a *solution* of I if every (s, t)-cut $A \subseteq E \setminus D$ has capacity at least $a + 1$ according to ω. A solution D of I is called a *minimum solution* of I, if there is no solution D' of I with $c(D') < c(D)$.

Basic Observations. We assume that G is connected and that $c(e) \leq d + 1$ and $\omega(e) \leq a + 1$ for each edge $e \in E$, as otherwise we can decrease these weights accordingly. Furthermore, we can assume that $d \leq c(E)$ and $a \leq \omega(E)$.

[1] https://arxiv.org/abs/2107.04482.

Fact 1. *Let $G = (V, E)$ be a graph, let $\omega : E \to \mathbb{N}$ be a capacity function, and let $D \subseteq E$. Then, in $n^{\mathcal{O}(1)}$ time we can compute an (s,t)-cut $A \subseteq E \setminus D$ with $\omega(A) \leq a$ or report that no such (s,t)-cut exists.*

Lemma 1 (*). *Let $I = (G = (V, E), s, t, c, \omega, d, a)$ be an instance of* WMCP. *Then, one can compute in $(n + a + d)^{\mathcal{O}(1)}$ time an equivalent instance $I' = (G', s', t', d, a)$ of* MCP.

The next definition will be a useful tool in several proofs in this work.

Definition 1. *Let $I = (G = (V, E), s, t, c, \omega, d, a)$ be an instance of* WMCP, *and let $e = \{u, w\} \in E$. The* merge of u and w in I *is the instance I' obtained from I by removing u and w from G and adding a new vertex $v_{\{u,w\}}$ which is adjacent to $N(\{u, w\})$. The cost and capacity for each edge in $E \cap E'$ are set to the corresponding cost and capacity in E, and for each $x \in N(\{u, w\})$,*

- $c'(\{v_{\{u,w\}}, x\}) = \min\{c(e') \mid e' \in E(x, \{u, w\})\}$, *and*
- $\omega'(\{v_{\{u,w\}}, x\}) = \sum_{e' \in E(x,\{u,w\})} \omega(e')$.

Rule 1. *If G contains an edge $e^* = \{u^*, w^*\} \in E$ which is not contained in any inclusion-minimal (s,t)-cut of capacity at most a in G, then merge u^* and w^*.*

Lemma 2. (*). *Rule 1 is safe and can be applied exhaustively in $n^{\mathcal{O}(1)}$ time.*

3 NP-hardness and Parameterization by the Defender Budget d

In this section we prove that MCP is NP-hard and we analyze parameterization by d and $\Delta(G)$. In particular, we provide a complexity dichotomy for $\Delta(G)$.

Lemma 3. WMCP *is* NP*-complete and* W[1]*-hard when parameterized by d even if G is bipartite, $\omega(e) = 1$, and $c(e) \in \mathcal{O}(|G|)$ for all $e \in E$.*

Proof. We describe a parameterized reduction from a variant of INDEPENDENT SET which is known to be W[1]-hard when parameterized by k [7,8].

REGULAR-INDEPENDENT SET
Input: An r-regular graph $G = (V, E)$ for some integer r and an integer k.
Question: Is there an independent set $S \subseteq V$ of size at least k in G?

Let $I := (G = (V, E), k)$ be an instance of REGULAR-INDEPENDENT SET. We describe how to construct an instance $I' := (G' = (V', E'), s, t, c, \omega, d, a)$ of WMCP in polynomial time such that I is a yes-instance of REGULAR-INDEPENDENT SET if and only if I' is a yes-instance of WMCP.

We start with an empty graph G' and add all vertices of V to G'. For each vertex $v \in V$ we also add an additional vertex v'. Furthermore, for each edge $e \in E$ we add a vertex w_e, and two new vertices s and t to G'. Moreover, we add the edges $\{s, v\}$, $\{v, v'\}$ and $\{v', t\}$ to G' for each vertex $v \in V$. Next, we add the edges $\{u, w_e\}, \{v, w_e\}$, and $\{w_e, t\}$ to G' for each edge $e = \{u, v\} \in E$. Now, we set $\omega(e') := 1$ for all $e' \in E'$. Furthermore, for each $e' \in E'$, we set $c(e') := 1$ if $s \in e'$ and $c(e') := k + 1$ otherwise. Finally, we set $d := k$ and $a := n+kr-1$ where $n := |V|$. This completes the construction of I'. Observe that G' is bipartite with one partite set being $\{t\} \cup V$. Note that only the edges incident with s can be protected, since all other edges have cost exactly $d + 1$.

Next, we show that I is a yes-instance of REGULAR-INDEPENDENT SET if and only if I' is a yes-instance of WMCP.

($\Rightarrow$) Let $S \subseteq V$ be an independent set of G of size exactly $k = d$. We set $D' := \{\{s, v\} \mid v \in S\}$. Note that D' has cost exactly d. It remains to show that D' is a solution of I'. To this end, we provide $a + 1$ many paths whose edge sets may only intersect in D'.

Note that for each vertex $v \in V \setminus S$ we have a path (s, v, v', t). These are $n-k$ many. Next, consider a vertex $v \in S$. Observe that (s, v, v', t) and $\{(s, v, w_e, t) \mid e \in E, v \in e\}$ are $r + 1$ paths only sharing the edge $\{s, v\} \in D'$. Since $|S| = k$ and G is r-regular, these are $kr + k$ many paths. Moreover, since S is an independent set no two vertices $u, v \in S$ have a common neighbor w_e in G' for $e = \{u, v\}$. Hence, there are $n - k + kr + k = n + kr = a + 1$ many (s, t)-paths in G' whose edge sets only intersect in D'.

($\Leftarrow$) Suppose that I' is a yes-instance of WMCP. Let D' be a solution with cost at most d of I'. Recall that $c(e) = d + 1$ for each edge $e' \in E'$ with $s \notin e'$. Hence, $D' \subseteq \{\{s, v\} \mid v \in V\}$. If $|D'| < d$, then we add exactly $d - |D'|$ many edges of the form $\{s, v\}$ which are not already contained in D' to D'. Note that D' remains a solution of I'. Thus, in the following we can assume that $|D'| = d = k$. Let $S := \{v \mid \{s, v\} \in D'\}$. We prove that S is an independent set in G.

Assume towards a contradiction that S is no independent set in G and let e^* be an edge of $G[S]$. In the following, we construct an (s, t)-cut $A \subsetneq (E' \setminus D')$ in G' of size at most a. Let $A_{V \setminus S} := \{\{s, v\} \mid v \notin S\}$, $A_S := \{\{v, v'\} \mid v \in S\}$, and $A_E := \{\{v, w_e\} \in E' \mid v \in S, e \neq e^*\}$. We show that $A := A_{V \setminus S} \cup A_S \cup A_E \cup \{\{w_{e^*}, t\}\}$ is an (s, t)-cut of size at most a in G'. Note that $|A_{V \setminus S}| + |A_S| = n$. Moreover, since $|S| = k$ and each vertex $v \in V$ has degree exactly r in G, $|A_E| \leq kr - 2$. Hence, A has capacity at most $n + kr - 1 = a$ since $\omega(e') = 1$ for each $e' \in E'$. It remains to show that A is an (s, t)-cut in G'. Let $G^* := G' - A$. Note that $N_{G^*}(s) = S$ and $N_{G^*}(v) = \{s\}$ for each $v \in S \setminus e^*$. Moreover, note that $N_{G^*}(v) = \{s, w_{e^*}\}$ for each $v \in e^*$ and $N_{G^*}(w_{e^*}) = e^*$. Hence, A is an (s, t)-cut in G' with capacity at most a. A contradiction.

Consequently, S is an independent set of size k in G and, therefore, I is a yes-instance of REGULAR-INDEPENDENT SET. □

By applying Lemma 1, we can extend the hardness results to MCP.

Corollary 1. MCP *is* NP*-complete and* W[1]*-hard when parameterized by d, even on bipartite graphs.*

Next, we provide a complexity dichotomy for the classical complexity with respect to the maximum degree of the graph.

Theorem 1 (*). WMCP *can be solved in polynomial time on graphs of maximum degree two.* WMCP *is* NP-*hard and* W[1]-*hard when parameterized by* d *even on subcubic graphs and even if* $c(e) = 1$ *and* $\omega(e) \in \mathcal{O}(|G|)$ *for all* $e \in E$.

We provide a simple search tree algorithm for $a+d$. If the graph has an (s,t)-cut of capacity at most a we branch on the at most a possibilities to protect one of the edges of this (s,t)-cut. The depth of this search tree is bounded by d since each choice decreases the defender budget by at least one.

Theorem 2 (*).WMCP *can be solved in* $a^d \cdot n^{\mathcal{O}(1)}$ *time.*

Next, we strengthen the NP-hardness of WMCP on subcubic graphs to MCP. Furthermore, we show that a is bounded by a function only depending on $d + \Delta(G)$ implying fixed-parameter tractability for $d + \Delta(G)$.

Theorem 3 (*). MCP *is* NP-*complete even on subcubic graphs. Furthermore,* MCP *can be solved in* $((d/2+1) \cdot \Delta(G))^d \cdot n^{\mathcal{O}(1)}$ *time.*

4 Parameterization by the Attacker Budget

In this section, we show that WMCP admits an FPT-algorithm for the parameter a. To this end, we first provide an algorithm with a running time of $a^{f(\text{tw}(G))} \cdot n$ for some computable function f, where $\text{tw}(G)$ denotes the treewidth of the graph. Afterwards, we show that for every instance of WMCP we can obtain an equivalent instance I' of WMCP in polynomial time, where every edge is contained in an inclusion-minimal (s,t)-cut of size at most a in I'. Due to previous results [11,15], the graph of I' then has treewidth at most $g(a)$ for some computable function g. In combination with the algorithm for a and $\text{tw}(G)$, we thus obtain the stated FPT-algorithm for the parameter a.

The algorithm with a running time of $a^{f(\text{tw}(G))} \cdot n$ relies on dynamic programming over a tree decomposition. Essentially, what the attacker can achieve in the current subgraph is to disconnect specific parts of the bag and thus obtain a cheap partition. Roughly speaking, the algorithm computes the minimum cost for an edge set D such that each choice of the attacker to obtain any partition disjoint from D is expensive. Hence, before we describe the algorithm, we first introduce some notations for partitions.

Let X be a set. We denote with $B(X)$ the collection of all partitions of X. Let $P \in B(X)$ be a partition of X and let $v \in X$. Then, we define with $P - v := \{R \setminus \{v\} \mid R \in P\} \setminus \{\emptyset\}$ the partition of $X \setminus \{v\}$ after removing v from P. Analogously, for every $w \notin X$ we define $P + w := \{P' \in B(X \cup \{w\}) \mid P' - w = P\}$. Note that $B(X \setminus \{v\}) = \{P - v \mid P \in B(X)\}$ and $B(X \cup \{w\}) = \{P + w \mid P \in B(X)\}$. Moreover, we denote with $P(v)$ the unique set of P containing v for a partition P of X and an element $v \in X$.

Let $(\mathcal{T} := (\mathcal{V}, \mathcal{A}, r), \beta)$ be a tree decomposition of a graph G. For a node $x \in \mathcal{V}$, we define with V_x the union of all bags $\beta(y)$, where y is reachable from x in $\mathcal{T}$, $G_x := G[V_x]$, and $E_x := E_G(V_x)$.

Let P be a partition of $\beta(x)$, then we call an edge set $A \subseteq E_x$ a *partition-cut for P in G_x* if v and w are in different connected components in $G_x - A$ for every pair of distinct vertices $\{v, w\}$ of $\beta(x)$ with $P(v) \neq P(w)$. Note that all edges between distinct sets of P are contained in every partition-cut for P in G_x.

Theorem 4. *Let* $\text{tw}(G)$ *denote the treewidth of* G*. Then,* WMCP *can be solved in* $a^{\text{tw}(G)^{\mathcal{O}(\text{tw}(G))}} \cdot n + m$ *time.*

Proof. Let $I = (G = (V, E), s, t, c, \omega, d, a)$ be an instance of WMCP. In the following, we assume that there is no edge $\{s, t\} \in E$ since if $c(\{s, t\}) \leq d$, then $\{\{s, t\}\}$ is a valid solution with cost at most d and, thus, I is a trivial yes-instance of WMCP. Otherwise, this edge is contained in every (s, t)-cut and, thus, we can simply remove the edge from the graph and reduce a by $\omega(\{s, t\})$.

We describe a dynamic programming algorithm on a tree decomposition. First, we compute a nice tree decomposition $(\mathcal{T} = (\mathcal{V}, \mathcal{A}, r), \beta')$ of $G - \{s, t\}$ with $|\mathcal{V}| \leq 4n$ such that the bag of the root and the bag of each leaf is the empty set in $\text{tw}^{\mathcal{O}(\text{tw}^3)} \cdot n + m$ time [4,14]. Next, we set $\beta(x) := \beta'(x) \cup \{s, t\}$ for each $x \in \mathcal{V}$. Note that $(\mathcal{T}, \beta)$ is a tree decomposition of width at most $\text{tw} + 2$ for G. Recall that for a node $x \in \mathcal{V}$, the vertex set V_x is the union of all bags $\beta(y)$, where y is reachable from x in $\mathcal{T}$, $G_x := G[V_x]$, and $E_x := E_G(V_x)$.

The dynamic programming table T has entries of type $T[x, f_x, D_x]$ with $x \in \mathcal{V}$, $f_x : B(\beta(x)) \to [0, a+1]$, and $D_x \subseteq E(\beta(x))$. Each entry stores the minimal cost of an edge set $D \subseteq E_x$ with $D_x := D \cap E(\beta(x))$ such that for every $P \in B(\beta(x))$ the capacity of every partition-cut $A \subseteq E_x \setminus D$ of P in G_x is at least $f_x(P)$.

For each entry of T, we will sketch the proof of the correctness of its recurrence; the formal correctness proof is omitted.

We start to fill the table T by setting for each leaf node ℓ of $\mathcal{T}$:

$$T[\ell, f_\ell, \emptyset] := \begin{cases} 0 & \text{if } f_\ell(\{\{s\}, \{t\}\}) = f_\ell(\{\{s, t\}\}) = 0, \\ \infty & \text{otherwise.} \end{cases}.$$

Recall that $\beta(\ell) = \{s, t\}$ and that we assumed that there is no edge between s and t in G. Hence, G_ℓ contains no edges and, thus, the empty set is a partition-cut for both $\{\{s\}, \{t\}\}$ and $\{\{s, t\}\}$, and has capacity zero.

To compute the remaining entries $T[x, f_x, D_x]$, we distinguish between the three types of non-leaf nodes.

Forget Node: Let x be a forget node with child node y and let v be the unique vertex in $\beta(y) \setminus \beta(x)$. Then we compute the table entries for x by:

$$T[x, f_x, D_x] := \min_{E_v \subseteq E(v, \beta(x))} T[y, f_y, D_x \cup E_v]$$

where $f_y(P) := f_x(P - v)$ for each $P \in B(\beta(y))$.

The idea behind the definition of $f_y(P)$ is that every partition cut for P in G_y must be as expensive as the partition cut of the unique partition of $\beta(x)$ that agrees with P on $\beta(x)$. By the fact that $G_x = G_y$, it follows that for each partition $P \in B(\beta(x))$, an edge set $A \subseteq E_x$ is a partition-cut for P in G_x if and only if A is also a partition-cut for some $P' \in P + v$ in G_x. Since we are looking for the minimal costs of an edge set $D \subseteq E_x$ such that every partition-cut disjoint from D for $P - v$ in G_x has capacity at least $f_x(P - v)$, it is thus necessary and sufficient that every partition-cut for P in G_y has capacity at least $f_x(P - v)$.

Introduce Node: Let x be an introduce node with child node y and let v be the unique vertex in $\beta(x) \setminus \beta(y)$. Then we compute the table entries for x by:

$$T[x, f_x, D_x] := T[y, f_y, D_x \cap E(\beta(y))] + c(D_x \setminus E(\beta(y)))$$

where $f_y(P) := \max(\{0\} \cup \{f_x(P') - \omega(A_{P'}) \mid P' \in (P + v), D_x \cap A_{P'} = \emptyset\})$ for each $P \in B(\beta(y))$ and $A_{P'} := E(v, \beta(y) \setminus P'(v))$.

The idea behind the definition of $f_y(P)$ is that, since every partition in $P+v$ agrees with P in $\beta(y)$, every partition cut for P in G_y must be sufficiently large to ensure that every partition cut for any partition in $P + v$ is as least as expensive as desired. Since we are looking for the minimum cost of an edge set $D \subseteq E_x$ which intersects with $E(\beta(x))$ in exactly the set D_x, the cost of D is exactly $c(D \cap E(\beta(y))) + c(D_x \setminus E(\beta(y)))$. Let $P' \in B(\beta(x))$. Note that $A_{P'}$ is a subset of every partition-cut for P' in G_x. Hence, if $D_x \cap A_{P'} = \emptyset$, then $f_y(P' - v)$ has to be at least $f_x(P') - \omega(A_{P'})$. Otherwise, if $D_x \cap A_{P'} \neq \emptyset$, then there is no partition-cut for P' in G_x disjoint from D.

Join Node: Let x be a join node with child nodes y and z. Then we compute the table entries for x by:

$$T[x, f_x, D_x] := \min_{f_y : B(\beta(y)) \to [0, a+1]} T[y, f_y, D_x] + T[z, f_z, D_x] - c(D_x)$$

where the mapping f_z is given by

$$f_z(P) := \max\left(0, \min\left(a + 1, f_x(P) - f_y(P) + \omega(E(\beta(x)) \setminus E(P))\right)\right)$$

with $E(P) := \cup_{R \in P} E(R)$ for each $P \in B(\beta(z))$.

The idea behind the definition of $f_z(P)$ is that the no partition cut for P in G_z is more expensive than the sum of any combination of partition cuts for P in G_y and G_z minus the capacity of the cut-edges in the current bag. Recall that we are looking for the minimum cost of an edge set $D \subseteq E_x$ such that for each partition $P \in B(\beta(x))$, every partition-cut for P in G_x disjoint from D has capacity at least $f_x(P)$. Since $E_y \cap E_z = E(\beta(x))$ it follows that the cost of D is $c(S_y) + c(S_z) - c(D_x)$, where $S_y := E_y \cap D$ and $S_z := E_z \cap D$. Moreover, note that for every partition $P \in B(\beta(x))$, every partition-cut $A_\alpha \subseteq E_\alpha$ for P in G_α has to contain all edges of $E(\beta(x)) \setminus E(P)$, where $\alpha \in \{x, y, z\}$. Thus, we have to guarantee that $f_y(P) + f_z(P) - \omega(E(\beta(x)) \setminus E(P)) \geq f_x(P)$, $f_y(P) > f_x(P)$, or $f_z(P) > f_x(P)$.

Then, there is a solution D of cost at most d of I if and only if $T[r, f_r, \emptyset] \leq d$, where r is the root of $\mathcal{T}$, $f_r(\{\{s,t\}\}) = 0$ and $f_r(\{\{s\},\{t\}\}) = a + 1$. Moreover, the corresponding set D can be found via traceback. The analysis of the running time is deferred to the full version. □

Next, we show that we can use Theorem 4 to obtain an FPT-algorithm for WMCP when parameterized by a. To this end, we first obtain the following corollary which follows from a result of Gutin et al. [11, Lemma 12].

Corollary 2. *Let $G = (V, E)$ be a graph, let s and t be distinct vertices of G, and let a be an integer. If every edge $e \in E$ is contained in an inclusion-minimal (s,t)-cut of size at most a, then $\text{tw}(G) \leq g(a)$ for some computable function g.*

Hence, to obtain an FPT-algorithm for WMCP with the parameter a, we only have to find an equivalent instance in polynomial time where each edge is contained in some inclusion-minimal (s,t)-cut of size at most a. Since each edge in an instance of WMCP has capacity at least one, by applying Rule 1 exhaustively we obtain an equivalent instance of WMCP where each edge is contained in some inclusion-minimal (s,t)-cut of *size* at most a. Hence, we obtain the following by combining Lemma 2, Corollary 2, and Theorem 4.

Theorem 5. WMCP *is* FPT *when parameterized by a.*

5 Parameterization by Vertex Cover Number

We investigate the parameterization by the vertex cover number $\text{vc}(G)$. Observing that for MCP the number of protected edges d is at most $2\text{vc}(G)$ in nontrivial instances, eventually leads to the following FPT result.

Theorem 6 (*). MCP *can be solved in $2^{\mathcal{O}(\text{vc}(G)^2)} \cdot n^{\mathcal{O}(1)}$ time.*

Theorem 4 implies that WMCP can be solved in pseudopolynomial time on graphs with a constant treewidth and therefore on graphs with a constant vertex cover number. With the next two theorems we show that significant improvements of this result are presumably impossible.

Theorem 7 (*). WMCP *is weakly* NP*-hard even if* $\text{vc}(G)$ *is two.*

Theorem 8. WMCP *is* W[1]*-hard when parameterized by* $\text{vc}(G)$ *even if* $c(e) + \omega(e) \in n^{\mathcal{O}(1)}$ *and G is a biclique.*

Proof. We describe a parameterized reduction from Bin Packing which is W[1]-hard when parameterized by k even if the size of each item is polynomial in the input size [13].

Bin Packing
Input: A set U of items, a size-function $f : U \to \mathbb{N}$, and integers B and k.
Question: Is there a k-partition $(U_1, \ldots, U_k)$ of U with $\sum_{u \in U_i} f(u) = B$ for all $i \in [1, k]$?

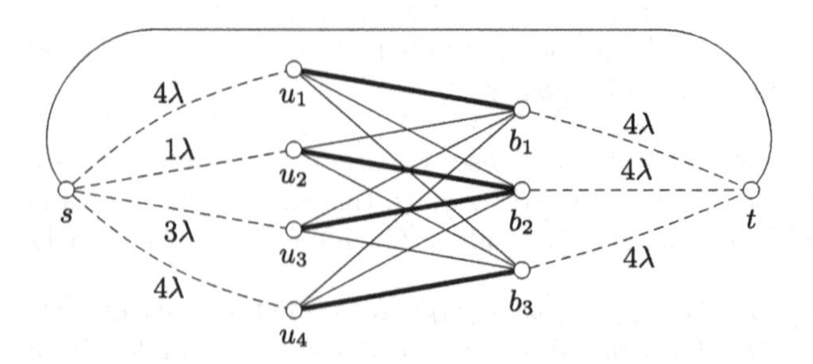

Fig. 1. An example of the construction from the proof of Theorem 8 for a Bin Packing instance with $f(u_1) = f(u_4) = 4$, $f(u_2) = 1$, $f(u_3) = 3$, $B = 4$, and $k = 3$. The thick edges represent a minimum solution D. The edge-labels represent all edge capacities that are bigger than one. Observe that every (s,t)-cut avoiding D contains dashed edges that have a capacity sum of at least 12λ.

Let $I := (U, f, B, k)$ be an instance of Bin Packing where the size of each item is polynomial in the input size. We can assume without loss of generality that $\sum_{u \in U} f(u) = Bk$, as, otherwise, I is a trivial no-instance of Bin Packing. Moreover, we assume $B \geq k$ since instances with $B < k$ can be solved in $f(k) \cdot |I|^{\mathcal{O}(1)}$ time. We construct an equivalent instance $I' := (G = (V, E), s, t, c, \omega, d, a)$ of WMCP where G has a vertex cover of size $k+1$. The graph G is a biclique with bipartition $(\{s\} \cup \mathcal{B}, \{t\} \cup U)$ where $\mathcal{B} := \{b_1, \ldots, b_k\}$. We set $d := |U|$, and

$$c(e) := \begin{cases} 1 & \text{if } e \in \{\{u, b\} \mid u \in U, b \in \mathcal{B}\}, \text{ and} \\ d+1 & \text{otherwise.} \end{cases}$$

Let $\lambda := 2B \cdot |U|$, we set

$$\omega(e) := \begin{cases} \lambda \cdot f(u) & \text{if } e = \{s, u\} \text{ with } u \in U, \\ \lambda \cdot B & \text{if } e = \{t, b\} \text{ with } b \in \mathcal{B}, \text{ and} \\ 1 & \text{otherwise.} \end{cases}$$

Finally, we set $a := |U| \cdot k + \lambda(Bk - 1)$. This completes the construction of I'. Figure 1 shows an example of the construction. Note that $\{s\} \cup \mathcal{B}$ is a vertex cover of G of size $k+1$. It remains to show that I is a yes-instance of Bin Packing if and only if I' is a yes-instance of WMCP.

($\Rightarrow$) Suppose that I is a yes-instance of Bin Packing. Then, there is a k-partition $(U_1, \ldots, U_k)$ of U, such that $\sum_{u \in U_i} f(u) = B$ for all $i \in [1, k]$. We set $D := \{\{u, b_i\} \mid i \in [1, k], u \in U_i\}$. Note that $c(D) = d$. We next show that D is a solution.

Let $A \subseteq E \setminus D$ be an (s,t)-cut in G and let $i \in [1, k]$. Since for each $u \in U_i$, D contains the edge $\{u, b_i\}$, the (s,t)-path $P_u := (s, u, b_i, t)$ can only be cut if $\{s, u\} \in A$ or $\{b_i, t\} \in A$. Consequently, $\{\{s, u\} \mid u \in U_i\} \subseteq A$ or $\{b_i, t\} \in A$. Recall that $\sum_{u \in U_i} f(u) = B$. Hence, $\sum_{u \in U_i} \omega(\{s, u\}) = \sum_{u \in U_i} \lambda f(u) = \lambda B = \omega(\{b_i, t\})$. Since P_u and P_w are edge-disjoint if u and w are in distinct parts of the k-partition, we obtain that $\omega(A) \geq k\lambda B > a$ and, thus, I' is a yes-instance of WMCP.

($\Leftarrow$) Suppose that I' is a yes-instance of WMCP. Then, there is a solution $D \subseteq E$ with $c(D) \leq d$. By construction, $D \subseteq E(U, \mathcal{B})$, since all other edges have cost $d+1$.

Note that for each $u \in U$, there is some $b \in \mathcal{B}$, such that $\{u, b\} \in D$, as, otherwise $A := \{\{s,t\}\} \cup \{\{s,u'\} \mid u' \in U \setminus \{u\}\} \cup \{\{u,b'\} \mid b' \in \mathcal{B}\}\}$ is an (s,t)-cut in G with capacity $\lambda(Bk - f(u)) + k + 1 < a$. Since $|D| \leq d$, we obtain that for each $u \in U$, there is exactly one $b \in \mathcal{B}$, such that $\{u, b\} \in D$.

We set $U_i := \{u \in U \mid \{u, b_i\} \in D\}$ for all $i \in [1,k]$. By the above, we obtain that $(U_1, \ldots, U_k)$ is a k-partition of U. We show that $\sum_{u \in U_i} f(u) = B$ for all $i \in [1,k]$.

Assume towards a contradiction that $\sum_{u \in U_i} f(u) \neq B$ for some $i \in [1,k]$. This is the case if and only if there is some $j \in [1,k]$ with $\sum_{u \in U_j} f(u) < B$. We set $A := \{\{s,t\}\} \cup \{\{s,u\} \mid u \in U_j\} \cup \{\{b,t\} \mid b \in \mathcal{B} \setminus \{b_j\}\} \cup (E(U,\mathcal{B}) \setminus D)$. Note that $\omega(A) = 1 + \lambda(\sum_{u \in U_j} f(u)) + \lambda B(k-1) + |U| \cdot (k-1) \leq \lambda(B-1) + \lambda B(k-1) + |U| \cdot k = \lambda(Bk-1) + |U| \cdot k = a$, since $\sum_{u \in U_j} f(u) < B$. It remains to show that A is an (s,t)-cut in G. Observe that $N_{G-A}(t) = b_j$. Since $N_{G-A}(b_j) = \{t\} \cup U_j$ and $N_{G-A}(u) = \{b_j\}$ for each $u \in U_j$, we conclude that A is indeed an (s,t)-cut in G. This contradicts the fact that there is no (s,t)-cut disjoint to D in G of capacity at most a. As a consequence, $\sum_{u \in U_i} f(u) = B$ for all $i \in [1,k]$ and, thus, I is a yes-instance of Bin Packing. □

We use Theorem 8 to show W[1]-hardness of MCP when parameterized by $\mathrm{pw}(G) + \mathrm{fvs}(G)$ and thus also when parameterized by $\mathrm{tw}(G)$. As a consequence the parameter $\mathrm{vc}(G)$ in the running time stated in Theorem 6 can presumably not be replaced by $\mathrm{pw}(G) + \mathrm{fvs}(G)$.

Theorem 9 (*). MCP *is* W[1]*-hard when parameterized by* $\mathrm{pw}(G) + \mathrm{fvs}(G)$.

6 On Problem Kernelization

On the positive side, we show that WMCP admits a polynomial kernel when parameterized by $\mathrm{vc}(G) + a$. The main tool for this kernelization is the merge of vertices according to Definition 1.

Let $J := (G = (V, E), s, t, c, \omega, d, a)$ be an instance of WMCP. We first provide two simple reduction rules that remove degree-one vertices.

Rule 2. *If s has exactly one neighbor w and $\omega(\{s,w\}) \leq a$, then delete s, set $s := w$, and decrease d by $c(\{s,w\})$. Analogously, if t has exactly on neighbor v and $\omega(\{t,v\}) \leq a$, then delete t, set $t := v$, and decrease d by $c(\{t,v\})$.*

Rule 3. *If there exists a degree-one vertex $v \notin \{s,t\}$, then delete v.*

The next reduction rule is the main idea behind the problem kernelization.

Rule 4. *If there are vertices $u, v \in V$ such that a minimum (u,v)-cut has capacity at least $a+1$, then merge u and v.*

The proof of the safeness of the Rules 2–4 is deferred to the full version. We now assume that J is reduced regarding Rules 2–4. Now we show the following.

Theorem 10 (*). WMCP *admits a polynomial problem kernel with* $2\mathrm{vc}(G) \cdot a$ *edges when parameterized by* $\mathrm{vc}(G) + a$.

Corollary 3 (*). MCP *admits a polynomial kernel with* $4\mathrm{vc}(G) \cdot a^2$ *edges.*

On the negative side, we provide an OR-composition to exclude a polynomial kernel for the combination of almost all considered parameters.

Theorem 11 (*). *Both* MCP *and* WMCP *do not admit a polynomial kernel when parameterized by* $d + a + \mathrm{lp}(G) + \Delta(G) + \mathrm{td}(G)$, *unless* $\mathrm{NP} \subseteq \mathrm{coNP/poly}$, *where* $\mathrm{td}(G)$ *denotes the tree-depth of* G.

References

1. Abdolahzadeh, A., Aman, M., Tayyebi, J.: Minimum *st*-cut interdiction problem. Comput. Ind. Eng. **148**, 106708 (2020)
2. Baier, G., et al.: Length-bounded cuts and flows. ACM Trans. Algorithms **7**(1), 4:1–4:27 (2010)
3. Bazgan, C., Fluschnik, T., Nichterlein, A., Niedermeier, R., Stahlberg, M.: A more fine-grained complexity analysis of finding the most vital edges for undirected shortest paths. Networks **73**(1), 23–37 (2019)
4. Bodlaender, H.L.: A linear-time algorithm for finding tree-decompositions of small treewidth. SIAM J. Comput. **25**(6), 1305–1317 (1996)
5. Chestnut, S.R., Zenklusen, R.: Hardness and approximation for network flow interdiction. Networks **69**(4), 378–387 (2017)
6. Cormican, K.J., Morton, D.P., Wood, R.K.: Stochastic network interdiction. Oper. Res. **46**(2), 184–197 (1998)
7. Cygan, M., et al.: Parameterized Algorithms. Springer, Cham (2015). https://doi.org/10.1007/978-3-319-21275-3
8. Downey, R.G., Fellows, M.R.: Fundamentals of Parameterized Complexity. TCS, Springer, London (2013). https://doi.org/10.1007/978-1-4471-5559-1
9. Fluschnik, T., Hermelin, D., Nichterlein, A., Niedermeier, R.: Fractals for kernelization lower bounds. SIAM J. Discret. Math. **32**(1), 656–681 (2018)
10. Guo, J., Shrestha, Y.R.: Parameterized complexity of edge interdiction problems. In: Cai, Z., Zelikovsky, A., Bourgeois, A. (eds.) COCOON 2014. LNCS, vol. 8591, pp. 166–178. Springer, Cham (2014). https://doi.org/10.1007/978-3-319-08783-2_15
11. Gutin, G.Z., Jones, M., Sheng, B.: Parameterized complexity of the *k*-arc Chinese postman problem. J. Comput. Syst. Sci. **84**, 107–119 (2017)
12. Israeli, E., Wood, R.K.: Shortest-path network interdiction. Networks **40**(2), 97–111 (2002)
13. Jansen, K., Kratsch, S., Marx, D., Schlotter, I.: Bin packing with fixed number of bins revisited. J. Comput. Syst. Sci. **79**(1), 39–49 (2013)

14. Kloks, T. (ed.): Treewidth, Computations and Approximations. LNCS, vol. 842. Springer, Heidelberg (1994). https://doi.org/10.1007/BFb0045375
15. Marx, D., O'Sullivan, B., Razgon, I.: Finding small separators in linear time via treewidth reduction. ACM Trans. Algorithm **9**(4), 30:1–30:35 (2013)
16. Smith, J.C., Prince, M., Geunes, J.: Modern network interdiction problems and algorithms. In: Pardalos, P.M., Du, D.-Z., Graham, R.L. (eds.) Handbook of Combinatorial Optimization, pp. 1949–1987. Springer, New York (2013). https://doi.org/10.1007/978-1-4419-7997-1_61
17. Wood, R.K.: Deterministic network interdiction. Math. Comput. Model. **17**(2), 1–18 (1993)
18. Zenklusen, R.: Matching interdiction. Discret. Appl. Math. **158**(15), 1676–1690 (2010)

Completion to Chordal Distance-Hereditary Graphs: A Quartic Vertex-Kernel

Christophe Crespelle[1], Benjamin Gras[2,3(✉)], and Anthony Perez[3]

[1] UCBL, DANTE/INRIA, LIP UMR CNRS 5668, Université de Lyon, Lyon, France
[2] Universität Trier, Fachbereich IV, Informatikwissenschaften, 54296 Trier, Germany
benjamin.gras@univ-orleans.fr
[3] Univ. Orléans, INSA Centre Val de Loire, LIFO EA 4022, 45067 Orléans, France

Abstract. Given a class of graphs $\mathcal{G}$ and a graph $G = (V, E)$, the aim of the $\mathcal{G}$-COMPLETION problem is to find a set of at most k non-edges whose addition in G results in a graph that belongs to $\mathcal{G}$. Completion to chordal or to natural subclasses of chordal graphs cover a broad range of classical NP-Complete problems, that have been extensively studied. When $\mathcal{G}$ coincides with the class of chordal graphs, the problem is the well-known MINIMUM FILL-IN problem. Other notable examples include completion to proper interval, threshold or trivially perfect graphs. Aforementioned problems are known to admit polynomial kernels, and it has been conjectured that completion to subclasses of chordal graphs further characterized by a finite number of forbidden induced subgraphs admits polynomial kernels. We investigate this line of research by considering completion to an important subclass of chordal graphs, namely chordal distance-hereditary graphs. Chordal distance-hereditary graphs are a natural generalization of trivially perfect graphs and have been extensively studied from the structural viewpoint. However, to the best of our knowledge, completion to chordal distance-hereditary graphs has not received attention so far. We thus initiate the first algorithmic study of this problem, and prove its NP-Completeness and that it admits a kernel with $O(k^4)$ vertices. To that aim, we rely on several known characterizations of chordal distance-hereditary graphs. In particular, such graphs admit a tree-like decomposition, so-called *clique laminar tree*. Unlike all aforementioned subclasses of chordal graphs, this decomposition does not correspond to a partition of the vertex set at hand. To circumvent this, we propose an approach based on the notion of clique (minimal) separator decomposition and a new characterization of chordal distance-hereditary graphs that might be of independent interest.

Keywords: Parameterized complexity · Kernelization algorithms · Chordal graphs · Distance-hereditary graphs

1 Introduction

Given a class of graphs $\mathcal{G}$ and a graph $G = (V, E)$, the aim of the parameterized $\mathcal{G}$-COMPLETION problem is to find a set of at most k non-edges whose addition

Ł. Kowalik et al. (Eds.): WG 2021, LNCS 12911, pp. 156–168, 2021.
https://doi.org/10.1007/978-3-030-86838-3_12

in G results in a graph that belongs to $\mathcal{G}$. Completion problems cover a broad range of NP-Complete problems [15,18,19,28,31,36] and have been extensively studied in the last decades. When $\mathcal{G}$ coincides with the class of chordal graphs, this is the well-known MINIMUM FILL-IN problem [18,36]. MINIMUM FILL-IN has been tackled from the parameterized complexity viewpoint by Kaplan et al. [24] who gave both parameterized and kernelization algorithms. Following this line of research, many completion problems towards subclasses of chordal graphs have been studied [4–6,13,14,21]. The motivation for the study of such completion problems mainly comes from practical applications, covering a wide range of fields such as bioinformatics, database management or artificial intelligence (see for instance [6]). Notable examples include completion to 3-leaf power, threshold or trivially perfect graphs, which are known to admit polynomial kernels [4,14,21]. In this work we consider an important subclass of chordal graphs, namely chordal distance-hereditary graphs [8,25]. These graphs do not contain any induced cycle of length at least 4 (or *hole*) and are moreover distance-hereditary: the distances in every connected induced subgraph are the same as in the original graph. Such graphs are a natural generalization of chordal cographs (*i.e.* trivially perfect graphs) and contain all the aforementioned classes. Moreover, they are known to admit a laminar structure [35], *i.e.* a tree-like decomposition that captures the subset relation on nonempty intersections of maximal cliques. Chordal distance-hereditary graphs have been extensively investigated from the structural viewpoint [2,7,9,22,23,25,32,33]. However, to the best of our knowledge, completion to chordal distance-hereditary graphs has not received any attention so far. We thus initiate the algorithmic study of this problem, mainly from the parameterized complexity viewpoint.

Parameterized Complexity. A parameterized problem Π is a problem whose input is a pair (I, k), where $k \in \mathbb{N}$ is called *parameter*. A parameterized problem Π is *fixed-parameter tractable* whenever any instance (I, k) of Π can be decided in time $f(k) \times p(|I|)$, where f is a computable function and p is a polynomial in the input size. A kernelization algorithm (*kernel* for short) for a parameterized problem Π is an algorithm that given any instance (I, k) of Π outputs in polynomial time an equivalent instance $(I', g(k))$ such that $|I'| \leqslant h(k)$ and $g(k) \leqslant h(k)$ for some function h. A kernel is said to be *polynomial* whenever h is a polynomial. It is well known that a parameterized problem Π is FPT if and only if it admits a kernel (see e.g. [17]). Formally, we consider the following problem:

CHORDAL DISTANCE-HEREDITARY COMPLETION
- **Input**: A graph $G = (V, E)$, $k \in \mathbb{N}$
- **Question**: Does there exist a set $F \subseteq (V \times V)$ of size at most k such that $H = (V, E \cup F)$ is chordal distance-hereditary?

Chordal distance-hereditary graphs are also known as *ptolemaic* graphs in the literature [25]. For the sake of readability, we will henceforth mainly refer to such graphs as *ptolemaic graphs* and consider the PTOLEMAIC COMPLETION problem.

Related Work. Cai [10] provided a dichotomy result for parameterized complexity of more general graph modification problems. Whenever the target graph class $\mathcal{G}$ can be characterized by finitely many *obstructions* (*i.e.* forbidden induced subgraphs), the corresponding modification problem (including $\mathcal{G}$-COMPLETION) is fixed-parameter tractable. While several polynomial kernels are known for graph modification problems [4,5,12,14,20,21], there exist such problems that do not admit polynomial kernels [11,20,26]. When $\mathcal{G}$ is characterized by a single obstruction, several recent results towards a dichotomy have been obtained [1,11,30]. We refer the reader to [12,17,27] for recent surveys on the subject. Regarding (sub)classes of chordal graphs, Kaplan et al. [24] considered completion problems to chordal and proper interval graphs, providing a quadratic vertex- kernel for the former problem. Guo [21] provided several kernelization algorithms for completion problems towards split, threshold and trivially perfect graphs. A recent result of Drange and Pilipczuk extended the latter to the TRIVIALLY PERFECT EDITING problem [14]. Other examples of such kernels are the ones for 3-LEAF POWER COMPLETION [4] and PROPER INTERVAL COMPLETION [5]. Bessy and Perez [5] conjectured that completion problems to subclasses of chordal graphs further characterized by a finite number of obstructions admit polynomial kernels. In all the aforementioned problems, the kernelization algorithms rely on the finite set of obstructions and on tree-like decompositions of the graph classes at hand that provide a *partition* of the vertex set which is exploited by reduction rules. However, the laminar structure of ptolemaic graphs is defined on intersections of maximal cliques and hence does not provide such a partition. This implies that standard techniques (such as the notion of *branches* [4,5]) cannot be applied directly.

Our Results. We prove that PTOLEMAIC COMPLETION is NP-Complete and admits a kernel with $O(k^4)$ vertices. Our method is inspired by the notion of clique (minimal) separator decomposition introduced by Tarjan [34]. This allows us to detect and reduce parts of the instance that are properly connected to the rest of the graph. This process can actually be reproduced on most previously mentioned kernelization algorithms for completion problems to subclasses of chordal graphs. This might bring new insights towards the design of kernelization algorithms for completion problems to other subclasses of chordal graphs.

Outline. We begin with preliminary definitions and results about ptolemaic graphs (Sect. 2). We then provide structural properties and a new decomposition theorem for such graphs (Sect. 3). We next describe the main structures that will be used (Sect. 4). Finally, we give our set of reduction rules (Sect. 5) and we conclude by bounding the size of a reduced instance (Sect. 6).

2 Preliminaries

We consider simple undirected graphs $G = (V, E)$ where V denotes the *vertex set* and E the *edge set* of G. We will sometimes use $V(G)$ and $E(G)$ to clarify the context. Given a vertex $u \in V$, $N_G(u)$ denotes the (open) *neighborhood* of u in G, that is $N_G(u) = \{v \in V : uv \in E\}$. The distance between two vertices u and v is the length of a shortest path from u to v. We similarly define $N_G^i(u)$, $i \geqslant 2$, as the set of vertices at distance at most i from u in G. The closed neighborhood of u is defined as $N_G[u] = N_G(u) \cup \{u\}$. Two vertices u and v are *true twins* whenever $N_G[u] = N_G[v]$ and *false twins* $N_G(u) = N_G(v)$ ($uv \notin E$ in this case). Given a subset $S \subseteq V$ of vertices, $N_G(S)$ is the set $\cup_{v \in S}(N_G(v) \setminus S)$. Similarly, $N_G^i(S)$, $i \geqslant 2$, is the set $\cup_{v \in S}(N_G^i(v) \setminus S)$. We moreover consider the closed neighborhoods $N_G[S]$ and $N_G^i[S]$ as natural extensions of the previous definitions. The *frontier* of S is defined as $\delta_G(S) = \{v \in S : N(v) \cap (V \setminus S) \neq \emptyset\}$. We omit the mention to graph G whenever the context is clear. The subgraph $G[S] = (S, E_S)$ *induced* by S is defined as $E_S = \{uv \in E : u \in S, v \in S\}$. For the sake of readability, given a subset $S \subseteq V$ we define $G \setminus S$ as $G[V \setminus S]$. A subset of vertices $C \subseteq V$ is a *connected component* of G if $G[C]$ is a maximal connected subgraph of G. We will sometimes refer to $G[C]$ as C. A *semi-split* of G is a subset $C \subseteq V$ such that the edges between $\delta_G(C)$ and $V \setminus C$ induce a non-empty complete bipartite graph. Let $G = (V, E)$ be a connected graph. A set $S \subseteq V$ is a *separator* of G if $G \setminus S$ has more connected components than G. Given two vertices u and v of G, the separator S is a uv-separator if u and v lie in distinct connected components of $G \setminus S$. Moreover, S is a *minimal* uv-separator if no proper subset of S is a uv-separator. Finally, a separator S is *minimal* if there exists a pair $\{u, v\}$ such that S is a minimal uv-separator.

Ptolemaic Graphs. We use a forbidden induced subgraph characterization of ptolemaic graphs [23] as well as a tree decomposition defined on intersections of maximal cliques [29,35]. Hereafter, the gem is the graph on 5 vertices with an induced $P_4 = \{p_1, p_2, p_3, p_4\}$ and a universal vertex t.

Theorem 1 [2,23]. *The following conditions are equivalent:*

(i) G is chordal distance-hereditary (or ptolemaic*).*
(ii) G does not contain any hole nor gem as an induced subgraph.
(iii) Given two maximal cliques P, Q of G such that $P \cap Q \neq \emptyset$, $P \cap Q$ separates $P \setminus Q$ and $Q \setminus P$ in G.
(iv) G can be obtained from a single vertex by repeating the following operations: adding a degree-one vertex, a true twin to some vertex u or a false twin to some vertex v, in which case $N_G(v)$ must be a clique.

An instance $(G = (V, E), k)$ of Ptolemaic Completion is a YES-instance whenever there exists a set $F \subseteq (V \times V)$ of size at most k such that $H = (V, E \cup F)$ is ptolemaic. The set F is called a *k-completion* of G (into a ptolemaic graph), and we will denote the resulting graph $H = G + F$. A *completion* refers to any set $F \subseteq (V \times V)$ such that $H = G + F$ is ptolemaic. Moreover, an *optimal completion* is a minimum-sized completion of G. A vertex is *affected* by a completion F whenever it is contained in some pair of F.

Lemma 1. Ptolemaic Completion *is NP-Complete.*

Clique Laminar Tree of Ptolemaic Graphs. We now describe a canonical tree representation for ptolemaic graphs due to Uehara and Uno [35].

Definition 1 ((Strong) Laminar family). *Let U be a universe and $\mathcal{F} \subseteq 2^{|U|}$ a family of subsets of U. The family $\mathcal{F}$ is* laminar *if and only if for all $A, B \in \mathcal{F}$, A and B are either disjoint or comparable by inclusion,* i.e. *either $A \cap B = \emptyset$, $A \subseteq B$ or $B \subseteq A$ holds. We say that $\mathcal{F}$ is a* strong *laminar family whenever there exists a set $A \in \mathcal{F}$ that contains all other sets of $\mathcal{F}$.*

Let U be a universe of n elements, and $\mathcal{F} \subseteq 2^{|U|}$ a family of subsets of U. Let $\vec{D}_{\mathcal{F}} = (V, A)$ be a directed graph where V contains a vertex x for every set $X \in \mathcal{F}$ and there is an arc from x to y in A if and only if $Y \subsetneq X$ (sets corresponding to y and x, respectively) and there does not exist $Z \in \mathcal{F}$ such that $Y \subsetneq Z \subsetneq X$. The digraph $\vec{D}_{\mathcal{F}}$ is called the *transitive reduction digraph* of $\mathcal{F}$. We use $D_{\mathcal{F}}$ to denote the underlying undirected graph. Moreover, if $\mathcal{F}$ is defined as a collection of subsets of vertices of a graph $G = (V, E)$, we refer to the vertices of $\vec{D}_{\mathcal{F}}$ as *bags* in order to avoid confusion with vertices of G. The notation t will henceforth denote a bag, while V_t will stand for the vertices of G contained in the set of $\mathcal{F}$ corresponding to t. Given a ptolemaic graph $G = (V, E)$, let $\mathcal{M}(G)$ be the set of maximal cliques of G and $\mathcal{C}(G)$ be the set of nonempty intersections of some maximal cliques of G. Notice in particular that $\mathcal{C}(G)$ contains the set $\mathcal{M}(G)$. Let $\mathcal{L}(G) = \mathcal{C}(G) \setminus \mathcal{M}(G)$ be the set of all nonempty intersections of at least two distinct maximal cliques of G. Any family $\mathcal{F}$ of sets of $\mathcal{L}(G)$ that are contained in a same maximal clique of $\mathcal{M}(G)$ is a laminar family [29,35]. This leads to the following characterization.

Theorem 2 [29,35]. *A graph $G = (V, E)$ is ptolemaic if and only if $D_{\mathcal{C}(G)}$ is a tree.*

We say that $\vec{D}_{\mathcal{C}(G)}$ is the *clique laminar tree* of G (see Fig. 1), it may also refer to $D_{\mathcal{C}(G)}$, the notation will always clarify this point. For the sake of readability, we will use $\vec{T}_G$ (resp. T_G) to denote $\vec{D}_{\mathcal{C}(G)}$ (resp. $D_{\mathcal{C}(G)}$).

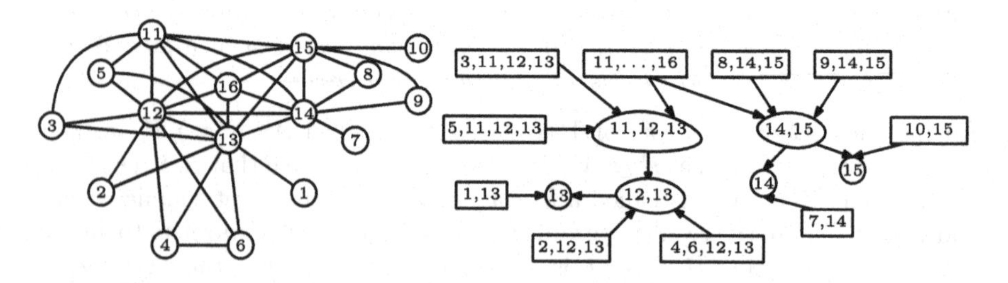

Fig. 1. A ptolemaic graph together with its laminar tree [35].

3 Structural Properties of Ptolemaic Graphs

In this section, we provide a complete characterization of ptolemaic graphs in terms of laminar family and semi-splits (Theorem 3). This result was partially known to exist [16,29,35], but we present a stronger statement. The following observation will be useful when considering decomposition of ptolemaic graphs.

Observation 1. *Let $G = (V, E)$ be a chordal graph, $S \subseteq V$ a clique of G and $\{C_1, \cdots, C_\ell\}$, with $\ell \geq 1$, the set of connected components of $G \setminus S$. Then for every $1 \leqslant i \leqslant \ell$, $G[\delta(C_i)]$ is connected.*

As we shall see Theorem 3, ptolemaic graphs can be characterized in terms of decomposition around complete subgraphs. To that aim, given a clique $S \subseteq V$, one needs to consider the set of neighborhoods of each connected component $G \setminus S$ w.r.t. S. This leads to the following definition.

Definition 2 (Footprint and trace). *Let $G = (V, E)$ be a graph, $S \subseteq V$ a clique of G and $\{C_1, \ldots, C_\ell\}$ the set of connected components of $G \setminus S$. The* footprint *of C_i, $1 \leqslant i \leqslant \ell$, is defined as the family of sets:*

$$\Phi_S^G(C_i) = \{N(v) \cap S : \ v \in \delta(C_i)\} \cup \{N(\delta(C_i)) \cap S\}$$

The set $N(\delta(C_i)) \cap S$ is called the trace *of component C_i, and denoted $\tau_S^G(C_i)$.*

The main difference between these two notions is that the *trace* of a given component represents the neighborhood of the whole component in S, while the *footprint* considers in addition the neighborhoods of every vertex of the frontier of the component. In both notations we omit the reference to the clique S or the graph G whenever the context is clear. Notice that $\{\tau(C_i)\} = \Phi(C_i)$ whenever C_i is a semi-split. We illustrate Definition 2 on Fig. 1 with $S = \{12, 13, 14, 15, 16\}$. Notice that the only components of $G \setminus S$ containing at least two vertices are $C_1 = \{3, 5, 11\}$ and $C_2 = \{4, 6\}$. Hence for any component C containing one vertex, we have $\Phi_S^G(C) = \{\tau_C^G(S)\}$. A similar observation holds for component C_2 since vertices 4 and 6 are true twins in G. Finally, we have $\tau_S^G(C_1) = N(11) \cap S = \{12, 13, 14, 15, 16\}$ while $\Phi_S^G(C_1) = \{\{12, 13\}, \{12, 13, 14, 15, 16\}\}$.

Definition 3 (Overlap and $\oslash$ notation). *Let U be a universe and A, B two subsets of U. The sets A and B* overlap*, denoted $A \oslash B$, whenever A and B are neither disjoint nor comparable by inclusion. Two families $\mathcal{A}, \mathcal{B} \subseteq 2^{|U|}$ of subsets of U overlap, denoted $\mathcal{A} \oslash \mathcal{B}$, if there exist $A \in \mathcal{A}$ and $B \in \mathcal{B}$ such that $A \oslash B$.*

For the sake of readability, we will abuse this notation for a set A and a family $\mathcal{B}$, that is $A \oslash \mathcal{B}$ rather than $\{A\} \oslash \mathcal{B}$.

Theorem 3. *Let $G = (V, E)$ be a graph, $S \subseteq V$ a clique of G and $\mathcal{C}$ the set of connected components of $G \setminus S$. The graph G is ptolemaic if and only if there is an order $\{C_1, \ldots, C_\ell\}$ on the components of $\mathcal{C}$ such that:*

(i) $G[S \cup C_i]$ *is ptolemaic for every* $1 \leqslant i \leqslant \ell$
(ii) $\bigcup_{1 \leqslant i \leqslant \ell} \Phi_S(C_i)$ *is a laminar family*
(iii) C_i *is a semi-split of* G *for every* $1 \leqslant i \leqslant \ell - 1$
(iv) if C_ℓ *is not a semi-split then:*
- *there exists* $v \in C_\ell$ *such that* $N(v) \cap S = \tau(C_\ell) = S$ *and*
- *no set* $\Gamma \in \Phi(C_\ell)$ *satisfies* $\Gamma \subsetneq \tau(C_i)$, $1 \leqslant i \leqslant \ell - 1$.

Corollary 1. *Let* $G = (V, E)$ *be a graph, and* $S \subseteq V$ *a clique minimal separator or a maximal clique of* G. *Let* $\mathcal{C} = \{C_1, \dots, C_\ell\}$ *be the set of connected components of* $G \setminus S$. *The graph* G *is ptolemaic iff the conditions (i) and (ii) of Theorem 3 hold, and condition (iii) of Theorem 3 holds for every* $1 \leqslant i \leqslant \ell$ *(i.e. condition (iv) of Theorem 3 does not occur).*

4 Decomposing the Instance

We first give some known and new results about completions into chordal distance-hereditary graphs. The soundness of the following result comes from the fact that such graphs are hereditary and closed under true twin addition (Theorem 1 (iv)).

Lemma 2 [4]. *Let* $\mathcal{G}$ *be an hereditary class of graphs closed under true twin addition. For every graph* $G = (V, E)$, *there exists an optimal completion* F *into a graph of* $\mathcal{G}$ *such that for any two maximal sets of true twins* M *and* M' *either* $(M \times M') \subseteq F$ *or* $(M \times M') \cap F = \emptyset$.

Lemma 3. *Let* G *be a graph and* S *a clique of* $G = (V, E)$ *such that any connected component* C *of* $G \setminus S$ *is a semi-split. Then, there exists an optimal completion* F *of* G, *with* $H = G + F$, *such that every connected component* C' *of* $H \setminus S$ *is a semi-split.*

Corollary 2. *Let* $G = (V, E)$ *be a graph and* $S \subseteq V$ *a clique of* G. *Let* $S' \subseteq S$ *be such that for every connected component* C *of* $G \setminus S$ *the following holds:*

(i) S' *and* $\Phi_S(C)$ *do not overlap and*
(ii) if C *is not a semi-split, then no set* $\Gamma \in \Phi_S(C)$ *satisfies* $\Gamma \subsetneq S'$.

Then there exists an optimal completion F *of* G *such that for any connected component* C' *of* $H \setminus S$, *with* $H = G + F$, *the following conditions hold:*

(i) S' *and* $\Phi_S^H(C')$ *do not overlap and*
(ii) if C' *is not a semi-split, then no set* $\Gamma \in \Phi_S^H(C')$ *satisfies* $\Gamma \subsetneq S'$.

4.1 Clams and Tentacles

We now introduce the main structures that will be considered by our kernelization algorithm. We consider parts of the graph that are *properly* connected to the rest of the graph (in terms of overlap, Definition 3).

Definition 4 (Clam). *Let $G = (V, E)$ be a graph, $S \subseteq V$ a clique separator of G and $\mathcal{C} = \{C_1, \ldots, C_p\}$, with $p \geq 1$, a maximal collection of connected components of $G \setminus S$ such that:*

(i) $G[S \cup C_i]$ is ptolemaic, $1 \leqslant i \leqslant p$
(ii) C_i is a semi-split of G, $1 \leqslant i \leqslant p$
(iii) $\tau(C_i) = \tau(C_j)$ for every $1 \leqslant i < j \leqslant p$

*The collection $\mathcal{C}$ is called an S-*clam *of G.*

Let $G = (V, E)$ be a graph, $S \subseteq V$ a clique separator of G and $\mathcal{C} = \{C_1, \ldots, C_p\}$ an S-clam of G. We let $V(\mathcal{C}) = \cup_{i=1}^{p} C_i$, and $\delta(\mathcal{C}) = \cup_{i=1}^{p} \delta(C_i)$. The set $\delta(\mathcal{C})$ is called the *frontier* of the S-clam $\mathcal{C}$. We first prove that there always exists an optimal completion that affects only (and uniformly) the frontier of a given clam.

Lemma 4. *Let $G = (V, E)$ be a graph, $S \subseteq V$ a clique separator of G and $\mathcal{C} = \{C_1, \ldots, C_p\}$ an S-clam of G. Any inclusion-minimal completion F of G into a ptolemaic graph satisfies the two following properties:*

(i) if F contains a pair $\{u, v\}$ where $u \in V(\mathcal{C})$ then $u \in \delta(\mathcal{C})$ and $v \in V \setminus V(\mathcal{C})$
(ii) if F contains a pair $\{u, v\}$, where $u \in \delta(\mathcal{C})$ and $v \in V \setminus V(\mathcal{C})$, then F contains all pairs $\{u', v\}$ with $u' \in \delta(\mathcal{C})$.

We now give a reduction rule to deal with *clams*. This is needed to define properly the other structures considered by our kernelization algorithm. As we shall see, using Lemma 4 on clique *minimal* separators will ease the polynomial-time application of our reduction rules.

Rule 1 (Clams). *Let $S \subseteq V$ be a clique minimal separator of G and $\mathcal{C} = \{C_1, \ldots, C_p\}$ an S-clam of G. Replace $V(\mathcal{C})$ by a clique C_S of size* $\min(k + 1, |\delta(\mathcal{C})|)$ *with edges $(C_S, N(\delta(\mathcal{C})) \cap S)$.*

Lemma 5. *Rule 1 is sound.*

In the remaining of this section, we assume that the instance at hand is reduced under Rule 1. In particular, this means that any clam of the given instance contains exactly one connected component. In order to avoid confusion, we henceforth refer to such clams as *tentacles*.

Definition 5 (Tentacle). *Let $(G = (V, E), k)$ be an instance of* Ptolemaic Completion *reduced under Rule 1 and $S \subseteq V$ be a clique separator of G. Let $\mathcal{C}$ be an S-clam. Since G is reduced under Rule 1, $\mathcal{C}$ contains exactly one connected component C. This component is called an S-*tentacle *of G.*

We now refine the notion of tentacles by considering different *types* of such structures according to hypotheses of Corollary 2.

Definition 6. *Let $(G = (V, E), k)$ be an instance of* PTOLEMAIC COMPLETION *reduced under Rule 1, $S \subseteq V$ a clique separator of G and C an S-tentacle of G. If there does not exist a component C' of $G \setminus (S \cup C)$ such that $\tau(C) \oslash \tau(C')$ then C is a:*

(i) ***type-α S-tentacle*** *if for every component C'' of $G \setminus (S \cup C)$ that is not a semi-split, no set $\Gamma \in \Phi(C'')$ satisfies $\Gamma \subsetneq \tau(C)$.*
(ii) ***type-β S-tentacle*** *if there exists a component C'' of $G \setminus (S \cup C)$ that is not a semi-split and a set $\Gamma \in \Phi(C'')$ such that $\Gamma \subsetneq \tau(C)$.*

Otherwise C is a ***type-γ S-tentacle****.*

5 Reducing the Instance

We now provide the set of reduction rules that will constitute our kernelization algorithm. For the sake of readability, we assume in the remaining of this section that we are given an instance $(G = (V, E), k)$ of PTOLEMAIC COMPLETION.

Rule 2. *Let $C \subseteq V$ be a subset of vertices such that $G[C]$ is a ptolemaic connected component of G. Remove C from G.*

Rule 3. *Let $\{X_1, \ldots, X_p\}$, $p > k$, be a set of distinct induced cycles of length 4 having a non-edge $\{u, v\}$ in common. Add the pair $\{u, v\}$ to the solution and decrease k by 1.*

The soundness of the following rule comes from Lemma 2 since ptolemaic graphs are hereditary and closed under true twin addition (Theorem 1 (iv)).

Rule 4. *Let $C \subseteq V$ be a maximal set of true twins of G such that $|C| > k + 1$. Remove $|C| - (k + 1)$ arbitrary vertices in C from G.*

Definition 7 (Gem-breaker). *Let $G = (V, E)$ be a graph and $X = \{t, p_1, p_2, p_3, p_4\}$ an induced gem of G. Each of the two pairs $\{p_1, p_3\}$ and $\{p_2, p_4\}$ is called a* gem-breaker *of X.*

Definition 8 (Gem-sunflower). *A collection $\mathcal{X} = \{X_1, \ldots, X_p\}$ of induced gems with $X_i \subseteq V$, $1 \leqslant i \leqslant p$, is a* gem-sunflower *if $p > k$ and:*

(i) there exist $u, v \in \cap_{i=1}^{p} X_i$ such that $\{u, v\}$ is a gem-breaker of each X_i, and
(ii) X_i and X_j do not share a gem-breaker $\{u', v'\} \neq \{u, v\}$, $1 \leqslant i < j \leqslant p$.

The gem-breaker $\{u, v\}$ is called the center *of $\mathcal{X}$.*

Rule 5. *Let $\mathcal{X} = \{X_1, \ldots, X_p\}$ be a gem-sunflower. Add the center of $\mathcal{X}$ to the solution and decrease k by 1.*

Lemma 6. *Rules 2 to 5 are sound and can be applied in polynomial time.*

5.1 Fishing and Eating the Seafood

We now turn our attention to reduction rules that consider structures defined in Sect. 4.1. In the remaining of this section we assume that we are given an instance $(G = (V, E), k)$ of Ptolemaic Completion reduced under Rule 1.

Rule 6 (Type-α tentacles). *Let $S \subsetneq V$ be a clique minimal separator of G, and $C \in G \setminus S$ be a type-α S-tentacle of G. Remove C from G.*

Rule 7 (Type-β tentacles). *Let $S \subseteq V$ be a clique minimal separator of G and $\mathcal{C} = \{C_1, \ldots, C_p\}$, $p \geqslant 2k+1$, be a set of type-β S-tentacles. If there exist a non semi-split component C' of $G \setminus S$ and a set $\Gamma \in \Phi(C')$ with $\Gamma \subsetneq \tau(C_i)$ for every $1 \leqslant i \leqslant p$, then add all pairs of $(\delta(C') \times \tau(C')) \setminus E$ to the solution and decrease k accordingly.*

Rule 8. *Let $S \subseteq V$ be a clique minimal separator of G such that:*

1. *S separates G into exactly two connected components C_1 and C_2 and*
2. *$G[N^2_G[S]]$ is ptolemaic and*
3. *$\forall i \in \{1, 2\}$, there exist a clique minimal separator $S_i \subseteq C_i \setminus N^2_G(S)$ of $G[C_i]$ and a connected component C'_i of $G[C_i] \setminus S_i$ such that $N^2_G(S) \cap C_i \subseteq C'_i$ and C'_i is an S_i-clam of $G[C_i]$.*

Then, remove S from G.

Lemma 7. *Rules 6 to 8 are sound.*

We now state that above reductions rules can be applied in polynomial time. We rely on the following result.

Theorem 4 [3,34]. *Given a graph $G = (V, E)$, a clique minimal separator decomposition can be obtained in $O(nm)$ time. Moreover, all clique minimal separators are used in the decomposition.*

We would like to mention that our objective here is to prove that all reduction rules can be applied in polynomial time. In particular, we do not give the explicit running time of our algorithms nor try to optimize them.

Lemma 8. *Rules 1, 6, 7 and 8 can be applied in polynomial time.*

6 Bounding the Size of Reduced Instances

We are now ready to bound the size of reduced instances $(G = (V, E), k)$ of Ptolemaic Completion. We need the following results.

Lemma 9. *Let $(G = (V, E), k)$ be a YES-instance of* Ptolemaic Completion *reduced under Rule 1 and Rules 5 to 7. Let $S \subsetneq V$ be a clique separator of G and $\mathcal{C} = \{C_1, \ldots, C_\ell\}$ the connected components of $G \setminus S$. Then $|\mathcal{C}| = O(k^2)$.*

Lemma 10. *Let $G = (V, E)$ be a connected ptolemaic graph without true twins. Let $\vec{T}_G$ be its clique laminar tree with $\left|V(\vec{T}_G)\right| = p$. Then $|V(G)| \leqslant p$.*

Theorem 5. Ptolemaic Completion *admits a kernel with $O(k^4)$ vertices.*

Proof (sketch). We say that an instance $(G = (V, E), k)$ of Ptolemaic Completion is *reduced* whenever none of the described reduction rules can be applied to G. Let $(G = (V, E), k)$ be a reduced YES-instance of Ptolemaic Completion, and F a k-completion of G. Let $\{C_1, \ldots, C_c\}$ denote the connected components of G. We work on the ptolemaic graph $H = G + F$ with connected components $\{H_1, \ldots, H_c\}$. Since G is reduced under Rule 2, we know that $c \leqslant k$. We assume that $F = \cup_{i=1}^{c} F_i$ with $H_i = G[C_i] + F_i$ and $|F_i| = k_i$ for $1 \leqslant i \leqslant c$. By Theorem 2, let $\vec{T}_H$ be the clique laminar forest of H, and $\vec{T}_{H_i}$ be the clique laminar tree of H_i, $1 \leqslant i \leqslant c$. Let $\mathcal{F}_i$ be the set of all *filled bags* of $\vec{T}_{H_i}$, that is $\mathcal{F}_i = \{t \in V(\vec{T}_{H_i}) : \exists \{u, v\} \in F, \{u, v\} \subseteq V_t\}$ and $\mathcal{F} = \cup_{i=1}^{c} \mathcal{F}_i$.

Claim 1. For every $1 \leqslant i \leqslant c$, $|\mathcal{F}_i| \leqslant k_i \cdot (6k - 1)$. Hence $|\mathcal{F}| \leqslant 6k^2 - k$.

Tentacles. Let T_i denote a minimal tree spanning vertices of $\mathcal{F}_i$ in T_{H_i}, $1 \leqslant i \leqslant c$. Let $\mathcal{S}_i$ be the set of maximal subtrees of $T_{H_i} \setminus V(T_i)$. We will count the bags of $\mathcal{S}_i$ together with $\mathcal{A}_i$, the set of bags of $T_i \setminus \mathcal{F}_i$ containing at least one affected vertex.

Claim 2. $|V(\mathcal{S}_i) \cup \mathcal{A}_i| = O(k_i \cdot k^2)$ with $V(\mathcal{S}_i)$ the set of bags of subtrees in $\mathcal{S}_i$.

Let $\mathcal{R}_i^{\geqslant 3}$ be the set of bags of $T_i \setminus \mathcal{F}_i$ having degree at least 3 in T_i. One can see that $\left|\mathcal{R}_i^{\geqslant 3}\right| \leqslant k_i$. Finally, let P be a path of $T_i \setminus (\mathcal{F}_i \cup \mathcal{R}_i^{\geq 3} \cup \mathcal{A}_i)$. Since G is reduced under Rule 8, one can see that $|V(P)| \leqslant 15$ where $V(P)$ denotes the bags of P. To conclude the proof, notice that by construction, any bag of $\vec{T}_H$ is either in $\mathcal{F}$, or in the set $V_i = V(T_i) \setminus \mathcal{F}_i$ of bags of the minimal tree T_i spanning $\mathcal{F}_i$ or in a maximal subtree of $T_{H_i} \setminus V(T_i)$ for some $1 \leqslant i \leqslant c$. We thus obtain:

$$\left|V(\vec{T}_H)\right| = \left|\mathcal{F} \bigcup (\cup_{i=1}^{c} V(\mathcal{S}_i)) \bigcup (\cup_{i=1}^{c} V_i)\right| \leqslant O(k^2) + \sum_{i=1}^{c} O(k_i \cdot k^2)$$
$$+ \sum_{i=1}^{c} (k_i + O(k_i \cdot k^2) + 15k_i)$$

In turn, since $\sum_{i=1}^{c} k_i = k$, we have $\left|V(\vec{T}_H)\right| \in O(k^3)$. By Lemma 10 and Rule 4, we thus conclude that G contains $O(k^4)$ vertices. Since all reduction rules can be applied in polynomial time (Lemmata 6 and 8), the result follows. □

Acknowledgements. This work has received funding from the European Union's Horizon 2020 research and innovation program under the Marie Skłodowska-Curie grant agreement No 749022. Benjamin Gras is partially supported by the French National Research Agency, ANR project GraphEn (ANR-15-CE40-0009).

References

1. Aravind, N., Sandeep, R., Sivadasan, N.: Dichotomy results on the hardness of H-free edge modification problems. SIAM J. Discrete Math. **31**(1), 542–561 (2017)
2. Bandelt, H.J., Mulder, H.M.: Distance-hereditary graphs. J. Comb. Theory Ser. B **41**(2), 182–208 (1986)
3. Berry, A., Pogorelcnik, R., Simonet, G.: An introduction to clique minimal separator decomposition. Algorithms **3**(2), 197–215 (2010)
4. Bessy, S., Paul, C., Perez, A.: Polynomial kernels for 3-leaf power graph modification problems. Discrete Appl. Math. **158**(16), 1732–1744 (2010). https://doi.org/10.1016/j.dam.2010.07.002
5. Bessy, S., Perez, A.: Polynomial kernels for proper interval completion and related problems. Inf. Comput. **231**, 89–108 (2013)
6. Bliznets, I., Cygan, M., Komosa, P., Pilipczuk, M., Mach, L.: Lower bounds for the parameterized complexity of minimum fill-in and other completion problems. ACM Trans. Algorithm (TALG) **16**(2), 1–31 (2020)
7. Brandstädt, A., Hundt, C.: Ptolemaic graphs and interval graphs are leaf powers. In: Laber, E.S., Bornstein, C., Nogueira, L.T., Faria, L. (eds.) LATIN 2008. LNCS, vol. 4957, pp. 479–491. Springer, Heidelberg (2008). https://doi.org/10.1007/978-3-540-78773-0_42
8. Brandstädt, A., Le, V.B., Spinrad, J.P.: Graph classes: a survey. SIAM (1999)
9. Brešar, B., Changat, M., Klavžar, S., Kovše, M., Mathews, J., Mathews, A.: Cover-incomparability graphs of posets. Order **25**(4), 335–347 (2008)
10. Cai, L.: Fixed-parameter tractability of graph modification problems for hereditary properties. Inf. Process. Lett. **58**(4), 171–176 (1996)
11. Cai, L., Cai, Y.: Incompressibility of H-free edge modification problems. Algorithmica **71**(3), 731–757 (2015)
12. Crespelle, C., Drange, P.G., Fomin, F.V., Golovach, P.A.: A survey of parameterized algorithms and the complexity of edge modification. CoRR abs/2001.06867 (2020). https://arxiv.org/abs/2001.06867
13. Drange, P.G., Dregi, M.S., Lokshtanov, D., Sullivan, B.D.: On the threshold of intractability. In: Bansal, N., Finocchi, I. (eds.) ESA 2015. LNCS, vol. 9294, pp. 411–423. Springer, Heidelberg (2015). https://doi.org/10.1007/978-3-662-48350-3_35
14. Drange, P.G., Pilipczuk, M.: A polynomial kernel for trivially perfect editing. Algorithmica **80**(12), 3481–3524 (2018)
15. El-Mallah, E.S., Colbourn, C.J.: The complexity of some edge deletion problems. IEEE Trans. Circ. Syst. **35**(3), 354–362 (1988)
16. Fagin, R.: Degrees of acyclicity for hypergraphs and relational database schemes. J. ACM (JACM) **30**(3), 514–550 (1983)
17. Fomin, F.V., Lokshtanov, D., Saurabh, S., Zehavi, M.: Kernelization: Theory of Parameterized Preprocessing. Cambridge University Press, New York (2019)
18. Garey, M.R., Johnson, D.S.: Computers and Intractability, vol. 29. WH freeman, New York (2002)
19. Golumbic, M.C., Kaplan, H., Shamir, R.: On the complexity of DNA physical mapping. Adv. Appl. Math. **15**(3), 251–261 (1994)
20. Guillemot, S., Havet, F., Paul, C., Perez, A.: On the (non-)existence of polynomial kernels for P_l-free edge modification problems. Algorithmica **65**(4), 900–926 (2013). https://doi.org/10.1007/s00453-012-9619-5

21. Guo, J.: Problem Kernels for NP-complete edge deletion problems: split and related graphs. In: Tokuyama, T. (ed.) ISAAC 2007. LNCS, vol. 4835, pp. 915–926. Springer, Heidelberg (2007). https://doi.org/10.1007/978-3-540-77120-3_79
22. Howorka, E.: On metric properties of certain clique graphs. J. Comb. Theory Ser. B **27**(1), 67–74 (1979)
23. Howorka, E.: A characterization of ptolemaic graphs. J. Graph Theory **5**(3), 323–331 (1981)
24. Kaplan, H., Shamir, R., Tarjan, R.E.: Tractability of parameterized completion problems on chordal, strongly chordal, and proper interval graphs. SIAM J. Comput. **28**(5), 1906–1922 (1999)
25. Kay, D.C., Chartrand, G.: A characterization of certain ptolemaic graphs. Can. J. Math. **17**, 342–346 (1965)
26. Kratsch, S., Wahlström, M.: Two edge modification problems without polynomial kernels. Discrete Optim. **10**(3), 193–199 (2013)
27. Liu, Y., Wang, J., Guo, J.: An overview of kernelization algorithms for graph modification problems. Tsinghua Sci. Tech. **19**(4), 346–357 (2014)
28. Mancini, F.: Graph modification problems related to graph classes. Ph.D. degree dissertation, University of Bergen Norway 2 (2008)
29. Markenzon, L., Waga, C.F.E.M.: New results on ptolemaic graphs. Discrete Appl. Math. **196**, 135–140 (2015)
30. Marx, D., Sandeep, R.B.: Incompressibility of H-free edge modification problems: Towards a dichotomy. In: Grandoni, F., Herman, G., Sanders, P. (eds.) 28th Annual European Symposium on Algorithms, ESA 2020, September 7–9, 2020, Pisa, Italy (Virtual Conference). LIPIcs, vol. 173, pp. 72:1–72:25. Schloss Dagstuhl - Leibniz-Zentrum für Informatik (2020). https://doi.org/10.4230/LIPIcs.ESA.2020.72
31. Natanzon, A., Shamir, R., Sharan, R.: Complexity classification of some edge modification problems. Discrete Appl. Math. **113**(1), 109–128 (2001)
32. Nieminen, J.: The center and the distance center of a ptolemaic graph. Oper. Res. Lett. **7**(2), 91–94 (1988)
33. Takahara, Y., Teramoto, S., Uehara, R.: Longest path problems on ptolemaic graphs. IEICE Trans. Inf. Syst. **91**(2), 170–177 (2008)
34. Tarjan, R.E.: Decomposition by clique separators. Discrete Math. **55**, 221–232 (1985)
35. Uehara, R., Uno, Y.: Laminar structure of ptolemaic graphs with applications. Discrete Appl. Math. **157**(7), 1533–1543 (2009)
36. Yannakakis, M.: Computing the minimum fill-in is NP-complete. SIAM J. Algebraic Discrete Methods **2**(1), 77–79 (1981)

A Heuristic Approach to the Treedepth Decomposition Problem for Large Graphs

Sylwester Swat(✉) and Marta Kasprzak

Institute of Computing Science, Poznan University of Technology, Piotrowo 2, 60-965 Poznan, Poland
{sylwester.swat,marta.kasprzak}@put.poznan.pl

Abstract. In this article, we describe algorithms and techniques used in the method ExTREEm for the treedepth decomposition problem. ExTREEm won the heuristic track of the 5th Parameterized Algorithms and Computational Experiments Challenge (PACE 2020). It searches for a minimum-height treedepth decomposition of a graph via computing graph separators. Among concepts that are incorporated into the approach, we can distinguish a new objective function for evaluating separators, preprocessing based on finding treedepth decompositions in cactus subgraphs and on identification of graphlets, five algorithms for finding separators, a separator minimization method for a refinement of found separators, and a refinement of an obtained treedepth decomposition by merging techniques of tree rotations. This approach enables us to quickly obtain low-depth decompositions of very large graphs.

Keywords: Treedepth decomposition · Elimination tree · Separator

1 Introduction

In this paper, we provide a description of algorithms of the method ExTREEm, which is a heuristic approach to the treedepth decomposition problem. The goal of the problem is to find a treedepth decomposition for a given graph with a height as small as possible. A treedepth decomposition of a connected graph $G = (V, E)$ is a rooted tree $T = (V, E_T)$ such that every edge of G connects a pair of nodes that have an ancestor-descendant relationship in T. A treedepth of a connected graph is a minimum possible height of its treedepth decomposition. There are many equivalent notions to treedepth. The most commonly used ones include the notion of elimination tree of a graph and corresponding elimination height [14], ordered coloring, vertex ranking [10] and centered coloring [13].

There are many fields where the treedepth decomposition is applicable. One of them is parallel factorization of sparse matrices using the Cholesky factorization method [7]. Elimination trees are also used in routing algorithms such as the Customizable Contraction Hierarchies algorithm [5], where finding good balanced separators and nested dissection orders is of utmost practical importance, especially when operating on graphs with millions of vertices. Such routing algorithms are used in many navigating systems, as well as in the field of computer

Ł. Kowalik et al. (Eds.): WG 2021, LNCS 12911, pp. 169–181, 2021.
https://doi.org/10.1007/978-3-030-86838-3_13

games, where shortest paths in a given map graph need to be computed many times each second. The treedepth notion is also relevant for theoretical reasons, e.g. in the design of fixed-parameter-tractable algorithms [8].

It was shown that the decision variant of the treedepth problem is NP-complete. There are classes of graphs for which the treedepth problem is solvable in polynomial time; e.g., it is possible for trees [12] and interval graphs [1]. In general, the construction of a minimum height elimination tree for a large graph in reasonable time seems to be very unlikely. Therefore, instead of exact algorithms, heuristics are used to create good decompositions. ExTREEm was designed for such optimization variant of the problem. It enables us to quickly find good treedepth decompositions even for very large graphs.

2 Preliminaries

Before we proceed to the description of algorithms, let us fix some natural notations and definitions. For a given connected graph $G = (V, E)$, we denote by $T(G)$ its treedepth decomposition, by $h(T)$ we denote height of tree T, and by $root(T)$ we denote the root of T. We denote by $G \setminus S$ an induced graph $G[V \setminus S]$, and by $C(G, S)$ the set of connected components of $G \setminus S$. For $a \in V$ we define $N(a) = \{v \in V : \{a, v\} \in E\}$, and for $A \subset V$ we take $N(A) = \bigcup_{v \in A} N(v)$. To indicate that a neighborhood is considered in graph H (and not in G), we use notations $N_H(a)$ and $N_H(A)$, respectively. We denote by $size_n(G)$ (or $|V|$) the number of nodes in G and by $size_e(G)$ (or $|E|$) the number of edges in G. For $a, b \in V$ we denote by $d(a, b)$ the distance between nodes a and b. For sets $A, B \subset V$ we denote by $d(A, B)$ the distance between sets A and B, $d(A, B) = \min\{d(a, b) : a \in A, b \in B\}$. A subset $S \subset V$ such that $G \setminus S$ is disconnected is called a *separator*. If additionally, for every $C \in C(G, S)$, the condition $size(C) \leq b \cdot size(G)$ holds, where $size(C)$ is either $size_n(C)$ or $size_e(C)$, then we say that separator S is *b-balanced*. By *balanced* we mean b-balanced, where b is a fixed parameter. Given two sets $A, B \subset V$ we denote $G_{A,B} = (A \cup B, \{\{a, b\} \subset E : a \in A, b \in B\})$.

3 Algorithms

In ExTREEm we search for a treedepth decomposition using a nested dissection approach[1]. The algorithm works in iterations, each iteration is executed independently of the others and with modified parameters. We refer to those iterations as *main iterations*. In each main iteration we apply some preprocessing to a given graph G. Then, we use a set of different heuristics to create a set of separators of G. Each of five best (according to a certain criterion) candidates is further refined. After selecting the best one after the refinement, we recursively obtain treedepth decompositions for components in $C(G, S)$. Finally, we merge

[1] ExTREEm is available at https://doi.org/10.5281/zenodo.3873126.

separator S and the decompositions we found into an elimination tree $T(G)$. At the end we apply some additional improvements to $T(G)$, trying to further minimize height of the treedepth decomposition.

3.1 Separator Evaluation

There are a few commonly used objective functions used to evaluate separators (see e.g. [3,9]). We propose a new approach to evaluate separators, based on the estimated height of the whole elimination tree.

To assess the quality of a separator S, we need to know values $mn(G, S)$ and $me(G, S)$ denoting, respectively, the maximum number of nodes and the maximum number of edges of a graph from $C(G, S)$. Now, let us define

$$score_n(S) = |S| \cdot \frac{1 - \beta^{\lceil -\frac{\log |V|}{\log \beta} \rceil}}{1 - \beta}, \quad \text{where } \beta = \frac{mn(G, S)}{|V|}.$$

We analogously define $score_e(S)$ by taking $\beta = \frac{me(G,S)}{|E|}$. Objectives $score_n$ and $score_e$ are used to quickly estimate a total height of the elimination tree, assuming that all subgraphs considered in recursive calls will have roughly the same ratio $\frac{|S|}{size_n(G)}$, respectively $\frac{|S|}{size_e(G)}$. Now, we can define the final objective function used to evaluate the quality of separators:

$$score(S) = \theta \cdot score_n(S) + (1 - \theta) \cdot score_e(S),$$

where $\theta \in [0, 1]$ is a parameter.

For given two balanced (b-balanced) or two unbalanced separators S_1 and S_2, we say that S_1 is better than S_2 if $score(S_1) < score(S_2)$. Additionally, a balanced separator is always better than an unbalanced one. The balance parameter b is modified in every main iteration. Greater values are used to enable the objective *score* to find tiny separators that disjoin relatively small subgraphs from the rest of the graph (as it happens, e.g., in road graphs, see [9]), whereas smaller values are used mainly to find separators with better balance at the topmost levels of the decomposition.

More details about objective functions $score_n(S)$ and $score_e(S)$ can be found in Appendix A.

3.2 Preprocessing

The preprocessing phase works in two steps. The first step consists in detecting some cactus-subgraphs of a given graph G. This is achieved by repetitively performing a vertex contraction operation (see [5]) on a node of degree at most 2, unless its neighbors are already connected by an edge. For each such cactus component, we find a treedepth decomposition using recursively the Articulation Point Separator Creator method (see Sect. 3.3). By G_1 we denote graph G after the first preprocessing step.

The second step of the preprocessing has two substeps. In the first one we find an independent set I_3 of an induced graph $G_1[\{v \in V : deg(v) = 3\}]$ and perform vertex contraction on each $v \in I_3$. In the second substep, we find a maximum independent set I_4 of nodes that are "center nodes" of induced subgraphs of G_1 isomorphic to some graphlet from the set $\{G_i : i \in \{22, 24, 26, 27, 28, 29\}\}$ (according to the numeration from Fig. 2 of [19]). All nodes in the set I_4 have degree 4 and the neighborhood of each of those nodes induces a connected subgraph. We proceed with set I_4 in the same way as with I_3. By G_2 we denote graph G after the second preprocessing step. After finding recursively a decomposition of G_2, each node $v \in I_4$ is attached to the deepest of its neighbors (in G_1), then we analogously proceed with set I_3.

Let us now examine how the preprocessing influences the height of a final treedepth decomposition $T(G)$. For each removed cactus subgraph, we attach the root of its corresponding decomposition to the lowest of its neighbors in G. This way, for each cactus C, the value $h(T(G_1))$ after attaching $T(C)$ increases by exactly $\max\{0, h(T(C)) - h(T(G_1)) + \max_{v \in N(C) \setminus C} d_T(v, root(G_1))\}$. Since the treedepth decomposition of a cactus graph is of size $O(\log |V|)$, for most cactus subgraphs it simply does not cause any increase. Attaching nodes from I_3 can increase $h(T(G_2))$ by at most one. The same holds for I_4. Hence, we have $h(T(G_1)) \leq h(T(G_2)) + 2$.

Finding decompositions of detected cacti is done in $O(|V| \cdot \log |V|$ time, while the second preprocessing step works in time $O(|E| + |V| \cdot \log |V|)$. The overall complexity of the preprocessing phase is $O(|E| + |V| \cdot \log |V|)$.

3.3 Separator Creation

After the preprocessing, we generate a set of separator candidates using several heuristics. From that set we select five best ones (with respect to their value of the objective *score*) that are further subjected to a refinement process called *minimization*. As a final separator we take the best one after the minimization.

Articulation Point Separator Creator. In this method, we find separators that contain only articulation points (cut vertices) of the graph G. At the beginning, we find a set A of articulation points of G using Tarjan's algorithm [4] for finding biconnected components. We want to find, for each articulation point $v \in A$, values $mn(G, \{v\})$ and $me(G, \{v\})$. To do this, during the depth-first search we additionally keep track of the number of visited nodes and edges. Whenever we are processing node v and backtracking from node u such that u and v do not belong to the same biconnected component, we are able to count the number of nodes and edges in the component $C \in C(G, \{v\})$ that contains u. Hence, for each $a \in A$, we can find values $size_n$ and $size_e$ for every component in $C(G, \{a\})$. It is now easy to obtain values $mn(G, \{v\})$ and $me(G, \{v\})$.

This method of creating separators is used mainly during the preprocessing phase, where its complexity is $O(|V| \cdot \log |V|)$ for creating a treedepth decomposition for each of the processed cactus-subgraphs.

BFS Separator Creator. Here, we create separators basing on the known fact that set $D_l(B) = \{v \in V : d(v, u) = l, u \in B\}$, for a given set $B \subset V$, is a separator in graph G. We propose a modification to this algorithm that makes it useful in practice, especially in the context of the BFS Separator Minimizer (see Sect. 3.4), where the algorithm's running time is crucial.

Given a set $B \subset V$, we run the standard breadth-first search with source-nodes set B. Let $L = \max_{v \in V \setminus B} d(B, \{v\})$ and let $G_i = G[V \setminus (\bigcup_{j=0}^{i-1} D_j(B))]$, for $1 \leq i \leq L$. We divide nodes in $D_i(B)$ into blocks, two nodes belong to the same block if they belong to the same connected component of G_i. Now, for each such block X we find the minimum vertex cover of a graph $G_{X, D_{i+1} \cap N_{G_i}(X)}$ using Kőnig's theorem and algorithm of Hopcroft and Karp for finding maximum matching in a bipartite graph (see [15]). In order to perform the whole procedure quickly, we process sets $D_i(B)$ in the reverse order (from L to 1), dynamically keeping track of the $size_n$ and $size_e$ values of components in graph G_i. We consider all found vertex covers and all blocks as different separator candidates.

We run the described procedure multiple times, for small random subsets $B \subset V$. During each iteration we solve multiple instances of finding a vertex cover of a bipartite graph. By observing that each edge can occur in at most one such bipartite graph, we obtain the bound $O(|E| \cdot |V|^{\frac{1}{2}})$ on the running time.

Component Expansion Separator Creator. The method described in this subsection is an improvement of another known approach. It consists in fixing some initial set of nodes B, then iteratively expanding set B by adding to it a node from $N(B) \setminus B$. We select each time a node with the tightest connection to B. In case of a tie, a node with fewer neighbors outside B is preferred. We store a sequence $ord = (v_0, v_1, \dots v_n)$ of nodes added to B. We call that sequence an *expansion order*. Creating an expansion order for a given initial set B works in time $O(|E| \cdot \log |V|)$, as the information about node candidates is updated using a binary heap.

Let us denote $P_i = \bigcup_{j=0}^{i} \{v_i\}$. For a given expansion order $(v_0, v_1, \dots v_n)$, we consider separators of the form $S_i = \{v_j \in P_i : |N(v_j) \setminus P_i| > 0\}$. In order to do this quickly, we process nodes from ord in the reverse order and dynamically keep track of all necessary information required to calculate sizes of components in graphs $G \setminus P_i$. By processing the nodes from ord in the original order, we find those information for components in graphs $G[P_i]$. We are therefore able to quickly find values $mn(G, S_i)$ and $me(G, S_i)$ for all $0 < i < n$.

It often happens that found separators are not minimal. To avoid those situations, we want to rearrange nodes in ord in such a way that, when iterating over i from 1 to $n-1$, if $|C(G, P_{i-1})| < |C(G, P_i)|$ then all nodes from smaller components will occur in ord before nodes that are in larger components.

To obtain time complexity better than $O(|E| \cdot |V|)$, we create an auxiliary graph H. We initially set $H = (V, \emptyset)$, then we process nodes v_i in the reverse order, dynamically keeping track of the number of nodes and edges in graphs $G \setminus P_i$.

For each i we consider the set of representatives of connected components in $G \backslash P_i$ in which v_i has a neighbor, then we sort those representatives by ascending order of the corresponding values $size_e$. Finally, considering representatives r in this order, we add a directed edge (v_i, v_j) to graph H, where $j > i$ is the smallest integer such that v_j and r belong to the same connected component in $G \backslash P_i$. After processing all nodes, we run a depth-first search on H, starting from v_0 and processing neighbors in the order of adding them to H. We obtain the rearranged sequence ord by listing nodes in the order in which they were visited during the traversal. We observed that the use of optimized orders for generating separators S_i almost always results in a huge decrease of the value $score(S_i)$.

To create the auxiliary graph H, we need to keep track of component sizes and their representatives. We additionally need, for each $0 < i < n$, to sort a set of designated representatives. Fortunately, for given index i only one of found representatives can occur again for another index $j < i$, as the components are merged. Hence, creating a rearranged, optimized order takes time $O(|E| \cdot \log |V|)$. The overall running time thus remains $O(|E| \cdot \log |V|)$.

FlowCutter Separator Creator. We use our own implementation of a slightly modified version of the FlowCutter algorithm (see [9]). At the beginning, we create a set L, by initializing it with a random node and iteratively adding to L a random node v that lies furthest to L. We stop adding nodes when $|L| = 50$.

As the initial source node and target node for the FlowCutter iteration we take a random pair of nodes from L. Additionally, we enable expansion of the larger of the source-reachable and target-reachable sets only if both grow to size at least $\frac{|V|}{10}$. When the final cut is found, we consider four different expansion orders based on the order of adding graph nodes to sources and targets. For each order we find separators using the Component Expansion Separator Creator. We also consider as a separator candidate a vertex cover of a bipartite graph $G_{X,Y}$, where X and Y denote final sets of sources and targets.

It is necessary to mention here, that we do not use FlowCutter Separator Creator if $score(S)$ is large for the the best separator found by other methods. In those cases graphs seem not to have balanced separators of small size and the algorithm's running time $O(c \cdot |E|)$, where c is the size of the most balanced cut, is too expensive.

Flow Separator Creator. We consider sets of the form $B = N(N(N(u)))$ and $E = N(N(N(v)))$, where u, v are randomly selected nodes. We find a maximum set of node-disjoint paths that begin in B and end in E. This is done by running a unit-flow algorithm with unit capacity constraints imposed on nodes. As a separator we consider the union of all paths and refine that separator using Greedy Minimizer (see Sect. 3.4). It is worth noting that separators created using this method are much worse than those from FlowCutter Separator Creator, but this algorithm works pessimistically in time $O(|E| \cdot \min(|V|^{\frac{2}{3}}, |E|^{\frac{1}{2}}))$, has much smaller constant factor and works well in the context of Flow Minimizer (see Sect. 3.4), where sets B and E are not created for random u and v.

3.4 Separator Minimization

After creating separator candidates, we proceed to the refinement step. For each candidate S we try to minimize the value of $score(S)$ using iteratively the following methods, most of which are based on a usage of separator creators with specific initial settings:

1. Vertex Cover Minimizer - in this minimization technique we find a vertex cover of a bipartite graph $G_{S,N(S)\cap C_d}$, where C_d is a subset containing a minimal number of largest components of $C(G,S)$ whose total sum is at least $\frac{|V \setminus S|}{4}$. The algorithm works in time $O(|E| \cdot |V|^{\frac{1}{2}})$.
2. BFS Minimizer and Component Expansion Minimizer - we find separators using BFS Separator Creator and Component Expansion Separator Creator, respectively, with the initial source set containing all nodes from S.
3. Greedy Minimizer - we greedily remove nodes from S, each time selecting a node v which minimizes $size_e(C)$, where C is a component obtained by merging v and its adjacent components from $C(G,S)$. It is done by operating on an auxiliary weighted bipartite graph with bipartition (A,B), where set A represents nodes in S, nodes in set B represent connected components of $C(G,S)$, and weights represent the number of edges between corresponding node and component. By using a binary heap to quickly update size values and removing lazily nodes from the auxiliary graph when a node from S is removed, we achieve running time $O(|E| + |S|^2 \cdot \log |V|)$.
4. FlowCutter Minimizer and Flow Minimizer - we find separators using Flow-Cutter Separator Creator and Flow Separator Creator, respectively. In the first variant, as the initial set of sources we take any subset $B \subset C_d$ with size $|B| = \frac{|C_d|}{2}$ and with the property that there does not exist any $v \notin B$ with $d(\{v\},S) > d(B,S)$ (C_d is taken in the same way as in Vertex Cover Minimizer). We analogously create the initial set of targets, but we take nodes from $V \setminus (S \cup C_d)$. In the second variant, we consider initial sets of the form $\{v \in X : d(\{v\},S) = t\}$, where X is C_d or $V \setminus (S \cup C_d)$, respectively, and t is a small value (usually between 2 and 4). Algorithms work in time $O(|E| + |S| \cdot |V|)$, but with greatly reduced constant factor.

3.5 Attaching Subtrees

After finding decompositions for all components in $C(G,S) = \{C_1, \dots, C_k\}$, we need to merge the results to obtain $T(G)$. To do that, we sort all trees $T(C_i)$ by their depths in the nonincreasing order. Then, we create a sequence S' (starting with $S' = \emptyset$) by iteratively adding to S' nodes from $(S \cap N(C_i)) \setminus S'$. As the tree $T(G)$ we initially take the tree (path) represented by sequence S', with the root set to its first element. We attach each tree $T(C_i)$ to the last node from sequence S' that occurs in $N(C_i)$. By using counting sort for sorting heights of trees and considering only edges with an end in S, the whole procedure works in time $O(|V| + M)$, where M is the number of edges in a graph $G[S \cup N(S)]$.

3.6 Tree Improvements

In the literature, there were proposed few techniques aiming at the reducing height of the elimination tree and basing on tree rotations. In this section we describe two structural improvements. The first one is a variation of a rotation-based algorithm described in [11] that enables us to implement it very easily. The second algorithm is, to the best of our knowledge, a new approach to the reduction of the height of an elimination tree via tree rotations.

Let us fix some additional notations. In this section, by T we mean tree $T(G)$. For $v \in V$ we denote by T_v the subtree of T with root in v. By $depth_T(v)$ we denote the depth of node v in tree T. By the block of $v \in V$ in a tree $T(G)$ we mean a maximal path in $T(G)$ which contains v, such that each node on that path, apart from the deepest one, has at most one son.

Block Pivots. Let us fix any block B of tree $T(G)$ and let S be a path from $root(T)$ to the topmost node in B. We consider S as a separator in graph G. Treedepth decompositions $T_i = T(C_i)$ are already constructed for each component $C_i \in C(G, S)$, $1 \leq i \leq k$, they form connected components of $T \setminus S$. We now rearrange nodes in S using algorithm 3.5 for separator S and trees T_i. We repeat the algorithm for several blocks B on the longest root-leaf path.

Hall-Set Pivots. The algorithm based on block pivots always needs to rearrange order of a given, contiguous sequence of initial nodes on a root-leaf path in tree T. In the following method, we propose a rotation-based technique without that constraint, which works exceptionally well for graphs that do not contain balanced separators of small size.

Let us fix any block B in $T(G)$ that lies on a longest root-leaf path P and let v be the topmost node in that block. Let $U(v)$ be a set of nodes on path P from the root to the parent of v and $D(v)$ be a set containing the remaining nodes on path P. We now consider a maximum matching M in a bipartite graph $G_{U(v),N(U(v))\cap R}$, where $R = T_v \setminus D(v)$. If matching M does not saturate set $U(v)$, then there exists in graph G a set $H_M \subset U(v)$ with property $|H_M| > |N(H_M) \cap R|$. In our algorithm, from all sets H_M violating Hall's condition for the existence of a matching saturating set $U(v)$, we select the one with maximum size. Set H_M contains all nodes $u \in U(v)$ to which there exists in $G_{U(v),N(U(v))\cap R}$ an M-alternating path starting in an unsaturated node from $U(v)$. We now remove from tree T_v all nodes belonging to the set $N(H_M)$. For each node s with $par(s) \in N(H_M) \cap T_v$ we set $par(s)$ to the deepest ancestor of s in $P \setminus (H_M \cup N(H_M))$. Let us now consider set $S = U(v) \cup N(H_M)$ and a set of treedepth decompositions $T(C_i)$ for $C_i \in T \setminus S$. Using Algorithm 3.5, we obtain a new treedepth decomposition $T'(G)$.

In the transformation, we removed from path P at least $|H_M|$ nodes and there are at most $|N(H_M) \cap R| < |H_M|$ new nodes in T' that became ancestors of a node from T_v. It follows that for each node $w \in T_v$ we have $depth_{T'}(w) < depth_T(w)$. Let us mention here, that it not necessarily means that

$h(T') < h(T)$, as for some other node $l \in T$ we may have $depth_{T'}(l) \geq h(T)$. From our observations, however, the Hall-set pivot technique works very well on graphs for which the ratio $\frac{h(T(G))}{|V|}$ is large.

Running the algorithm for a single block B is dominated by the factor of finding a maximum matching in a graph $G_{U(v),N(U(v))\cap R}$, which takes time $O(|E| \cdot |h(T)|^{\frac{1}{2}})$. The selection of the block B is not unimportant. We therefore consider all blocks B on path P in a leaf-to-root manner. If, at some moment, for the i-th considered block B_i we obtain a tree T' with $h(T') < h(T)$, we terminate the algorithm and run it again for T'. If, however, valid set H_M does not exist or $h(T') \geq h(T)$, we check the next block B_{i+1} above B_i. We can now initialize next matching M_{i+1} with all edges from previous matching M_i that have an endpoint in $H_{M_i} \setminus B_{i+1}$. Such initialization resulted in a considerable performance improvement. It is also worth mentioning that when a node $p \in U$ becomes saturated in a matching M_i, then it becomes saturated in all further matchings M_j $(j > i)$, until it leaves U. We therefore terminate processing blocks as soon as the whole set U is saturated.

4 Results

A thorough comparison of ExTREEm and other methods was made within the heuristic track of the contest PACE 2020: 5th Parameterized Algorithms and Computational Experiments Challenge. There, 55 heuristic methods were submitted and tested on 200 instances differing in size and properties. ExTREEm won this contest. Here, we focus on a subset of these instances, large graphs. More information about the contest and short descriptions of several solvers can be found in [20].

Other Methods. To the comparison we selected four other best heuristics from the contest[2]. They are: FlowCutter [16] (2nd in the contest), Sallow [18] (3rd), Tweed-Plus [17] (4th), and Fluid [2] (5th).

FlowCutter is based on the previously mentioned algorithm of the same name from 2016 for finding separators of a graph [9]. Present version of FlowCutter is additionally supported by two approaches: iterative node contraction and label propagation. Sallow is another method that incorporates the ideas of FlowCutter from 2016, supplemented with greedy heuristics. The order in which vertices are processed in greedy heuristics is based on different criteria and is updated on the fly. Tweed-Plus creates and next improves an elimination tree with two known methods: nested dissection and the minimum-degree ordering algorithm. In each of its phases, if a graph is small enough, the computations are repeated many times with different results due to randomisation. Fluid realizes four separate strategies and selects as a result the best found solution. Two strategies iteratively select a vertex with a best score, two others iteratively search for separators and remove them. The score-based strategies compute an elimination

[2] https://pacechallenge.org/2020/results/#heuristic-track.

order, later used in the tree construction, many times with randomness involved. The separator-based strategies use a greedy approach or the asynchronous fluid communities algorithm.

Data Set. From among public instances of PACE 2020 for the heuristic track[3], we chose graphs of more than 1000 vertices. They represent a wide range of graph kinds; the most important groups according to their origin are biological and social networks, road graphs and graphs generated according to different rules. These 66 instances include from 1013 to 1.32 million vertices and have the average vertex degree from 2 to 148, their presumable heights of the minimum treedepth decompositions vary from 3 to several thousands. All the instances are simple connected graphs.

Comparison. For the purposes of the comparison, the smallest value of the criterion function obtained for a graph in the contest by any of the 55 heuristics is assumed as the optimum value for this graph. We partitioned the data set into groups from the point of view of a few parameters: graph order, optimum tree height, average vertex degree, fraction of vertices with degree at least three. This way it is easy to notice how compared methods deal with smaller groups of similar instances. The groups are presented in Table 1.

Table 1. Partition of the set of 66 instances into groups by different parameters. In a row, the cardinality of a subset of instances is followed by the parameter by which the subset has been determined and the range of its values.

Group of instances	Cardinality	Parameter	Range of values
A	22	Graph order	$\langle 1000; 7000)$
B	24		$\langle 7000; 50000)$
C	20		$\langle 50000; 1.32$ mln$)$
D	23	Optimum tree height	$\langle 1; 100)$
E	20		$\langle 100; 300)$
F	23		$\langle 300; 78500)$
G	24	Average vertex degree	$\langle 2; 3)$
H	22		$\langle 3; 8)$
I	20		$\langle 8; 150)$
J	21	Fraction of V with degree ≥ 3	$\langle 13\%; 77\%)$
K	24		$\langle 77\%; 88\%)$
L	21		$\langle 88\%; 100\%\rangle$

During the PACE 2020 challenge, every heuristic had the limit of 30 min for returning a solution for an instance. Thus, the results are comparable in this

[3] https://pacechallenge.org/files/pace2020-heur-public.tgz.

sense. Figure 1 presents the average quality of solutions obtained by the five best methods for the particular groups of instances and for all 66 instances. The quality is calculated as a ratio of the assumed optimum value of the criterion function to the value returned by a given method.

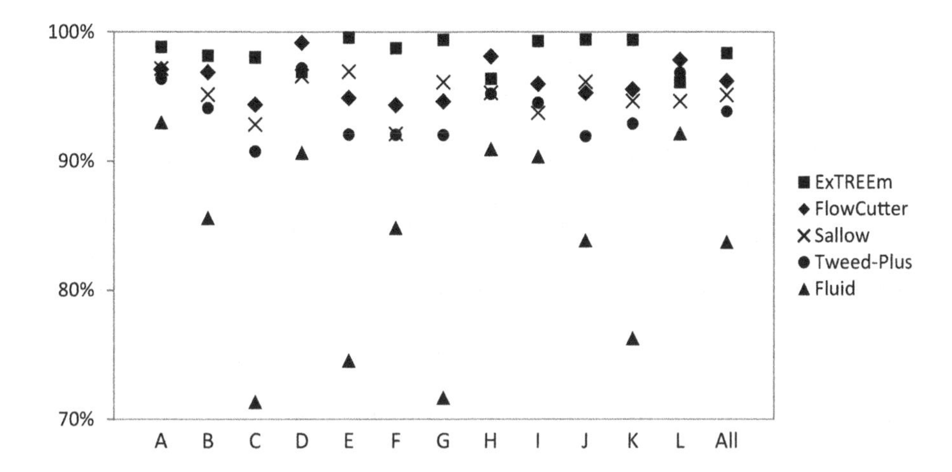

Fig. 1. The comparison of the PACE 2020 results for particular groups of large instances. Every marker stands for an average result of a given heuristic for the data set. Y axis shows how close to the optimum the results are.

ExTREEm generated solutions of the best quality (on average) for 9 out of 12 groups of instances A–L and for all large instances as well. For all the instances, ExTREEm achieved the ratio 98.34%, FlowCutter took the second rank with 96.18%, and Sallow was the next with 95.11%. The partitioning of the data set on the basis of graph order led to the same hierarchy. Groups D, H, and L were best solved by FlowCutter, ExTREEm took the second (H) or the third position (D, L). It means that a little harder for our method, in comparison to the others, were instances with a resulting tree of a small height or instances with a small fraction of vertices having only one or two neighbors.

5 Conclusions

We proposed the new method for the treedepth decomposition problem, which proved its efficiency in a wide comparison with top and current other algorithms. As ExTREEm very well solves large graphs, it may be useful in practical applications involving wide and complex networks, for example in industry of computer games or navigating systems. On the other hand, the method also deals well with smaller graphs. The public instances of PACE 2020 with less than 1000 vertices were solved by ExTREEm with the quality equal to 99%. Therefore, the applicability of ExTREEm is even wider.

Appendix A

Let us define

$$score_n(S) = |S| \cdot \frac{1 - \beta^{\lceil -\frac{\log |V|}{\log \beta} \rceil}}{1 - \beta}, \quad \text{where } \beta = \frac{mn(G,S)}{|V|},$$

$$score_e(S) = |S| \cdot \frac{1 - \gamma^{\lceil -\frac{\log |E|}{\log \gamma} \rceil}}{1 - \gamma}, \quad \text{where } \gamma = \frac{me(G,S)}{|E|}.$$

We now show why the proposed objective functions $score_n(S)$ and $score_e(S)$ estimate the height of a treedepth decomposition. We specify only the case of $score_n(S)$, arguments for $score_e(S)$ are analogous.

Let S be a separator of a graph $G = (V, E)$, $\alpha = \frac{|S|}{|V|}$, $\beta = \frac{mn(G,S)}{|V|}$, and let $EH(G, \alpha, \beta)$ be a function that estimates the height of a sought decomposition $T(G)$. It is calculated on the basis of values α and β, that is on information that we can obtain knowing only graph G and separator S. For each $C \in C(G, S)$ the decomposition $T(C)$ will be attached to some node from the set $S \cap N(C)$ (see Sect. 3.5), therefore we use the following estimation:

$$EH(G, \alpha, \beta) \leq |S| + \max_{C \in C(G,S)} h(T(C)) = \alpha \cdot |V| + \max_{C \in C(G,S)} h(T(C))$$

Assuming that in each recursive call the values of parameters α and β do not change, we can replace $h(T(C))$ with $EH(C, \alpha, \beta)$ to obtain the following assessment:

$$\begin{aligned} EH(G, \alpha, \beta) &\leq \alpha \cdot |V| + \max_{C \in C(G,S)} h(T(C)) \\ &\approx \alpha \cdot |V| + \max_{C \in C(G,S)} EH(C, \alpha, \beta) \\ &\approx \alpha \cdot |V| + \alpha \cdot \beta \cdot |V| + \max_{C' \in C(C,S')} EH(C', \alpha, \beta) \\ &\approx \alpha \cdot |V| + \alpha \cdot \beta \cdot |V| + \alpha \cdot \beta^2 \cdot |V| + \dots \\ &\approx \sum_{i=0}^{\lceil \log_{\beta^{-1}} |V| \rceil} \alpha \cdot |V| \cdot \beta^i = \alpha \cdot |V| \cdot \sum_{i=0}^{\lceil -\frac{\log |V|}{\log \beta} \rceil} \beta^i \\ &\approx |S| \cdot \frac{1 - \beta^{\lceil -\frac{\log |V|}{\log \beta} \rceil}}{1 - \beta} \end{aligned}$$

Let us note here that the formulas for the objective functions can be further simplified via the estimation $\beta^{\lceil \log_{\beta^{-1}} |V| \rceil} \approx \beta^{\log_{\beta^{-1}} |V|} = \frac{1}{|V|}$. We found, however, test cases where the replacement made a difference to the evaluation of separators.

References

1. Aspvall, B., Heggernes, P.: Finding minimum height elimination trees for interval graphs in polynomial time. BIT Numer. Math. **34**, 484–509 (1994)
2. Bannach, M., Berndt, S., Schuster, M., Wienobst, M.: Solver description of Fluid. In: 5th Parameterized Algorithms and Computational Experiments Challenge, PACE 2020 (2020). https://doi.org/10.5281/zenodo.3871709
3. Castillo-Garcia, N., Fraire-Huacuja, H., Flores, J., Rangel, R., Gonzalez Barbosa, J., Carpio Valadez, J.: Comparative study on constructive heuristics for the vertex separation problem. Stud. Comput. Intell. **601**, 465–474 (2015)
4. Cormen, T.H., Leiserson, C.E., Rivest, R.L., Stein, C.: Introduction to Algorithms, 2nd edn. MIT Press, Cambridge (2001)
5. Dibbelt, J., Strasser, B., Wagner, D.: Customizable contraction hierarchies. In: Gudmundsson, J., Katajainen, J. (eds.) SEA 2014. LNCS, vol. 8504, pp. 271–282. Springer, Cham (2014). https://doi.org/10.1007/978-3-319-07959-2_23
6. Galler, B.A., Fisher, M.J.: An improved equivalence algorithm. Commun. ACM **7**, 301–303 (1964). 364099.364331
7. Gupta, A.: Sparse direct methods. In: Padua, D. (ed.) Encyclopedia of Parallel Computing, pp. 1877–1886. Springer, Boston (2011). https://doi.org/10.1007/978-0-387-09766-4_507
8. Gutin, G., Jones, M., Wahlstrom, M.: The mixed Chinese postman problem parameterized by pathwidth and treedepth. SIAM J. Discret. Math. **30**, 2177–2205 (2016)
9. Hamann, M., Strasser, B.: Graph bisection with pareto-optimization. In: Proceedings of the Meeting on Algorithm Engineering and Experiments, ALENEX 2016, Arlington, USA, pp. 90–102 (2016.) https://doi.org/10.1137/1.9781611974317.8
10. Karpas, I., Neiman, O., Smorodinsky, S.: On vertex rankings of graphs and its relatives. Discret. Math. **338**, 1460–1467 (2015)
11. Liu, J.W.: Reordering sparse matrices for parallel elimination. Parallel Comput. **11**, 73–91 (1989)
12. Manne, F.: An algorithm for computing an elimination tree of minimum height for a tree. Preprint at ResearchGate (1998)
13. Nesetril, J., de Mendez, P.O.: Sparsity: Graphs, Structures, and Algorithms. Springer, Heidelberg (2012). https://doi.org/10.1007/978-3-642-27875-4
14. Pieck, J.: Formele definitie van een e-tree. Technische Hogeschool Eindhoven, memorandum 80–06 (1980)
15. Schrijver, A.: A course in combinatorial optimization. Preprint at ResearchGate (2003)
16. Strasser, B.: FlowCutter. In: 5th Parameterized Algorithms and Computational Experiments Challenge, PACE 2020 (2020). https://github.com/ben-strasser/flow-cutter-pace20
17. Trimble, J.: Tweed: a heuristic solver for treedepth. In: 5th Parameterized Algorithms and Computational Experiments Challenge, PACE 2020 (2020). https://doi.org/10.5281/zenodo.3881441
18. Wrochna, M.: Sallow - a heuristic algorithm for treedepth decompositions. In: 5th Parameterized Algorithms and Computational Experiments Challenge, PACE 2020, preprint arXiv:2006.07050 (2020). https://doi.org/10.5281/zenodo.3870565
19. Yaveroglu, O., Fitzhugh, S., Kurant, M., Markopoulou, A., Butts, C., Przulj, N.: Ergm.graphlets: a package for ERG modeling based on graphlet statistics. J. Stat. Softw. **65** (2014)
20. Proceedings of the 15th International Symposium on Parameterized and Exact Computation, IPEC 2020, Leibniz International Proceedings in Informatics 180 (2020)

The Perfect Matching Cut Problem Revisited

Van Bang Le[1(✉)] and Jan Arne Telle[2]

[1] Institut für Informatik, Universität Rostock, Rostock, Germany
van-bang.le@uni-rostock.de
[2] Department of Informatics, University of Bergen, 5020 Bergen, Norway
Jan.Arne.Telle@uib.no

Abstract. In a graph, a perfect matching cut is an edge cut that is a perfect matching. PERFECT MATCHING CUT (PMC) is the problem of deciding whether a given graph has a perfect matching cut, and is known to be NP-complete. We revisit the problem and show that PMC remains NP-complete when restricted to bipartite graphs of maximum degree 3 and arbitrarily large girth. Complementing this hardness result, we give two graph classes in which PMC is polynomial time solvable. The first one includes claw-free graphs and graphs without an induced path on five vertices, the second one properly contains all chordal graphs. Assuming the Exponential Time Hypothesis, we show there is no $O^*(2^{o(n)})$-time algorithm for PMC even when restricted to n-vertex bipartite graphs, and also show that PMC can be solved in $O^*(1.2721^n)$ time by means of an exact branching algorithm.

1 Introduction

In a graph $G = (V, E)$, a *cut* is a partition $V = X \cup Y$ of the vertex set into disjoint, non-empty sets X and Y. The set of all edges in G having an endvertex in X and the other endvertex in Y, written $E(X, Y)$, is called the *edge cut* of the cut (X, Y). A *matching cut* is an edge cut that is a (possibly empty) matching. Another way to define matching cuts is as follows; see [8,12]: a cut (X, Y) is a matching cut if and only if each vertex in X has at most one neighbor in Y and each vertex in Y has at most one neighbor in X. MATCHING CUT (MC) is the problem of deciding if a given graph admits a matching cut and this problem has received much attention lately; see [7,10] for recent results.

An interesting special case, where the edge cut $E(X, Y)$ is a *perfect matching*, was considered in [13]. The authors proved that PMC, the problem of deciding if a given graph admits an edge cut that is a perfect matching, is NP-complete. A perfect matching cut (X, Y) can be described as a (σ, ρ) 2-partitioning problem [22], as every vertex in X must have exactly one neighbor in Y and every vertex in Y must have exactly one neighbor in X. By results of [6,22,23] it can therefore be solved in FPT time when parameterized by treewidth or cliquewidth (to mention only the two most famous width parameters) and in XP time when parameterized by mim-width (maximum induced matching-width) of a given decomposition of

Ł. Kowalik et al. (Eds.): WG 2021, LNCS 12911, pp. 182–194, 2021.
https://doi.org/10.1007/978-3-030-86838-3_14

the graph. For several classes of graphs, like interval and permutation, a decomposition of bounded mim-width can be computed in polynomial-time [3], thus the problem is polynomial on such classes.

In this paper, we revisit the PMC problem. Our results are:

- While MC is polynomial time solvable when restricted to graphs of maximum degree 3 and its computational complexity is still open for graphs with large girth, we prove that PMC is NP-complete in the class of bipartite graphs of maximum degree 3 and arbitrarily large girth. Further, we show that PMC cannot be solved in $O^*(2^{o(n)})$ time for n-vertex bipartite graphs and cannot be solved in $O^*(2^{o(\sqrt{n})})$ time for bipartite graphs with maximum degree 3 and arbitrarily girth.
- We provide the first exact algorithm to solve PMC on n-vertex graphs, of runtime $O^*(1.2721^n)$. Note that the fastest algorithm for MC has runtime $O^*(1.3280^n)$ and is based on the current-fastest algorithm for 3-SAT [17].
- We give two graph classes of unbounded mim-width in which PMC is solvable in polynomial time. The first class contains all claw-free graphs and graphs without an induced path on 5 vertices, the second class contains all chordal graphs.

Related Work. The computational complexity of MC was first considered by Chvátal in [8], who proved that MC is NP-complete for graphs with maximum degree 4 and polynomial time solvable for graphs with maximum degree at most 3. Hardness results were obtained for further restricted graph classes such as bipartite graphs, planar graphs and graphs of bounded diameter (see [4,19,21]). Further graph classes in which MC is polynomial time solvable were identified, such as graphs of bounded tree-width, claw-free, hole-free and Ore-graphs (see [4,7,21]). FPT algorithms and kernelization for MC with respect to various parameters has been discussed in [1,2,10,11,17,18]. The current-best exact algorithm solving MC has a running time of $O^*(1.3280^n)$ where n is the vertex number of the input graph [17]. Faster exact algorithms can be obtained for the case when the minimum degree is large [7]. The recent paper [10] addresses enumeration aspects of matching cuts.

Very recently, a related notion has been discussed in [5]. In this paper, the authors consider perfect matchings $M \subseteq E$ of a graph $G = (V, E)$ such that $G \setminus M = (V, E \setminus M)$ is disconnected, which they call perfect matching-cuts. To avoid confusion, we call such a perfect matching a *disconnected perfect matching.* Note that, by definition, every perfect matching cut is a disconnected perfect matching but a disconnected perfect matching need not be a perfect matching cut. Indeed, all perfect matchings of the cycle on $4k + 2$ vertices are disconnected perfect matchings and none of them is a perfect matching cut. In [5], the authors showed, among others, that recognizing graphs having a disconnected perfect matching is NP-complete even when restricted to graphs with maximum degree 4, and left open the case of maximum degree 3. It is not clear whether our hardness result on degree-3 graphs can be modified to obtain a hardness result of recognizing degree-3 graphs having a disconnected perfect matching.

Notation and Terminology. Let $G = (V, E)$ be a graph with vertex set $V(G) = V$ and edge set $E(G) = E$. The neighborhood of a vertex v in G, denoted by $N_G(v)$, is the set of all vertices in G adjacent to v; if the context is clear, we simply write $N(v)$. Let $\deg(v) := |N(v)|$ be the degree of the vertex v, and $N[v] := N(v) \cup \{v\}$ be the closed neighborhood of v. For a subset $F \subseteq V$, $G[F]$ is the subgraph of G induced by F, and $G - F$ stands for $G[V \setminus F]$. We write $N_F(v)$ and $N_F[v]$ for $N(v) \cap F$ and $N[v] \cap F$, respectively, and call the vertices in $N(v) \cap F$ the *F-neighbors* of v. The *girth* of G is the length of a shortest cycle in G, assuming G contains a cycle. The path on n vertices is denoted by P_n, the complete bipartite graph with one color class of size p and the other of size q is denoted by $K_{p,q}$; $K_{1,3}$ is also called a *claw*.

When an algorithm branches on the current instance of size n into r subproblems of sizes at most $n - t_1, n - t_2, \ldots, n - t_r$, then $(t_1, t_2, \ldots, t_r)$ is called the *branching vector* of this branching, and the unique positive root of $x^n - x^{n-t_1} - x^{n-t_2} - \cdots - x^{n-t_r} = 0$, denoted by $\tau(t_1, t_2, \ldots, t_r)$, is called its *branching factor*. The running time of a branching algorithm is $O^*(\alpha^n)$, where $\alpha = \max_i \alpha_i$ and α_i is the branching factor of branching rule i, and the maximum is taken over all branching rules. Throughout the paper we use the O^* notation which suppresses polynomial factors. We refer to [9] for more details on exact branching algorithms.

Algorithmic lower bounds in this paper are conditional, based on the Exponential Time Hypothesis (ETH) [14]. The ETH states that there is no $O^*(2^{o(n)})$-time algorithm for 3-SAT where n is the variable number of the input 3-CNF formula. It is known that the hard case for 3-SAT already consists of formulas with $O(n)$ clauses [15]. Thus, assuming ETH, there is no $O^*(2^{o(m)})$-time algorithm for 3-SAT where m is the clause number of the input formula.

Observe that a graph has a perfect matching cut if and only if each of its connected components has a perfect matching cut. Thus, we may assume that all graphs in this paper are connected.

Due to space restriction, most proofs are given in the full version [20].

2 Hardness Results

In this section, we give two polynomial time reductions from POSITIVE NAE 3-SAT to PMC. Recall that an instance for POSITIVE NAE 3-SAT is a 3-CNF formula $F = C_1 \wedge C_2 \wedge \cdots \wedge C_m$ over n variables $x_1, x_2, \ldots, x_n$, in which each clause C_j consists of three distinct variables. The problem asks whether there is a truth assignment of the variables such that every clause in F has one true and one false variable. Such an assignment is called *nae assignment.*

It is well-known that there is a polynomial reduction from 3-SAT to POSITIVE NAE 3-SAT where the variable number of the reduced formula is linear in the clause number of the original formula. Hence, the ETH implies that there is no subexponential time algorithm for POSITIVE NAE 3-SAT in the number of variables.

Theorem 1. *Assuming ETH,* PMC *cannot be solved in subexponential time in the vertex number, even when restricted to bipartite graphs.*

Proof. We give a polynomial reduction from POSITIVE NAE 3-SAT to PMC restricted to bipartite graphs. Given a 3-CNF formula F, construct a graph G as follows.

For each clause $C_j = \{c_{j1}, c_{j2}, c_{j3}\}$, let $G(C_j)$ be the cube with *clause vertices* labeled c_{j1}, c_{j2}, c_{j3}, respectively, as depicted in Fig. 1. For each variable x_i, we introduce a *variable vertex* x_i and a dummy vertex x_i' adjacent only to x_i. Finally, we connect a variable vertex x_i to a clause vertex in $G(C_j)$ if and only if C_j contains the variable x_i, i.e., $x_i = c_{jk}$ for some $k \in \{1, 2, 3\}$.

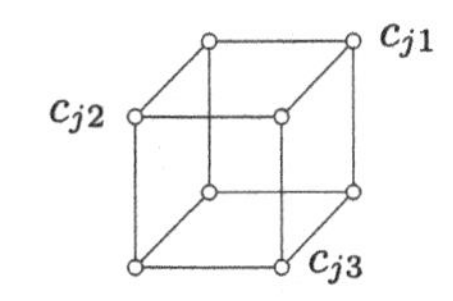

Fig. 1. The graph $G(C_j)$.

Observe that G is bipartite and has the following property: no perfect matching M of G (in particular, no perfect matching cut) contains an edge between a clause vertex and a variable vertex. Thus, for every perfect matching cut $M = E(X, Y)$ of G, the restriction $M_j = E(X_j, Y_j)$ on $G(C_j)$ is a perfect matching cut of $G(C_j)$. Moreover, $G(C_j)$ has the following property: it has exactly three perfect matching cuts, and in any perfect matching cut of $G(C_j)$ not all clause vertices belong to the same part. Conversely, any bipartition of C_j can be extended (in a unique way) to a perfect matching cut M_j of $G(C_j)$. See also Fig. 2.

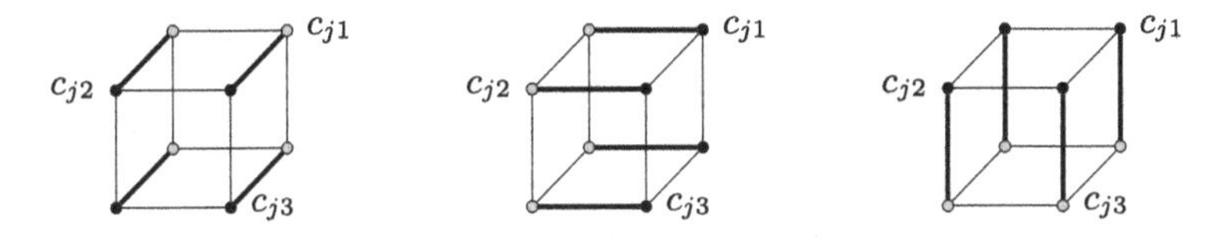

Fig. 2. The three perfect matching cuts of $G(C_j)$; black vertices in X, gray vertices in Y.

We are now ready to see that F has a nae assignment if and only if G has a perfect matching cut: First, if there is a nae assignment for F then put all true variable vertices into X, all false variable vertices into Y, and extend X and Y (in a unique way) to a perfect matching cut of G; note that x_i' and x_i have to belong to different parts. Second, if (X, Y) is a perfect matching cut of G then defining x_i be `true` if $x_i \in X$ and `false` if $x_i \in Y$ we obtain a nae assignment for F.

Observe that G has $N = O(n+m)$ vertices. Hence the reduction implies that, assuming ETH, PMC has no subexponential time algorithm in vertex number N, even when restricted to bipartite graphs. □

We now describe how to avoid vertices of degree 4 and larger (the clause and variable vertices) in the previous reduction to obtain a bipartite graph with maximum degree 3 and large girth.

Theorem 2. *Let $g > 0$ be a given integer.* PMC *remains* NP*-complete when restricted to bipartite graphs of maximum degree three and girth at least g.*

Proof. Let $h \geq 0$ be a fixed integer, which will be more concrete later.

Clause gadget: we subdivide every edge of the cube with $4h+4$ new vertices, fix a vertex c_j of degree 3 and label the three neighbors of c_j with c_{j1}, c_{j2} and c_{j3}, respectively. We denote the obtained graph again by $G(C_j)$ and call the labeled vertices the *clause vertices*.

Variable gadget: for each variable x_i we introduce m *variable vertices* x_i^j one for each clause C_j, $1 \leq j \leq m$, as follows. (We assume that the formula F consists of $m \geq 3$ clauses.) First, take a cycle with m vertices $x_i^1, x_i^2, \ldots, x_i^m$ and edges $x_i^1 x_i^2, x_i^2 x_i^3, \ldots, x_i^{m-1} x_i^m$ and $x_i^1 x_i^m$. Then subdivide every edge with $4h+3$ new vertices to obtain the graph $G(x_i)$. Thus, $G(x_i)$ is a cycle on $4m(h+1)$ vertices.

Now the graph G is obtained by connecting the variable vertex x_i^j in $G(x_i)$ to a clause vertex in $G(C_j)$ by an edge whenever x_i appears in clause C_j, i.e., $x_i = c_{jk}$ for some $k \in \{1, 2, 3\}$. It follows from construction, that G is bipartite, has maximum degree 3 and girth at least $\min\{4m(h+1), 8(h+2)\}$. As in the proof of Theorem 1, we can argue that F has a nae assignment if and only if G has a perfect matching cut. Finally, given $g > 0$, let $h \geq 0$ be an integer at least $\max\{\frac{g}{4m} - 1, \frac{g}{8} - 2\}$. Then G has girth at least $\min\{4m(h+1), 8(h+2)\} \geq g$. The details are given in the full version. □

Note that the graph G in the proof of Theorem 2 has $N = O(m+nm)$ vertices, where n and m are the variable number and clause number, respectively, of the formula F. Since we may assume that F has $m = O(n)$ clauses, G has $N = O(n^2)$ vertices. Hence we obtain the following.

Theorem 3. *Assuming ETH, there is no $O^*(2^{o(\sqrt{n})})$-time algorithm for* PMC *even when restricted to n-vertex bipartite graphs with maximum degree 3 and arbitrarily large girth.*

3 An Exact Exponential Algorithm

Recall that, assuming ETH, there is no $O^*(2^{o(n)})$-time algorithm for PMC on n-vertex (bipartite) graphs. The main result in this section is an algorithm solving PMC in $O^*(1.2721^n)$ time.

Our algorithm follows the idea of known branching algorithms for MC [7,17,18]. We adapt basic reduction rules for matching cuts to perfect matching cuts, and add new reduction and branching rules for perfect matching cuts.

If the input graph $G = (V, E)$ has a perfect matching cut (X, Y), then some edge has an endvertex a in X and the other endvertex b in Y. The branching algorithm will be executed for all possible edges $ab \in E$, hence $O(m)$ times. To do this set $A := \{a\}$, $B := \{b\}$, and $F := V \setminus \{a, b\}$ and call the branching algorithm. At each stage of the algorithm, A and B will be extended or it will be determined that there is no perfect matching cut *separating* A and B, that is a perfect matching cut (X, Y) with $A \subseteq X$ and $B \subseteq Y$. We describe our algorithm by a list of reduction and branching rules given in preference order, i.e., in an execution of the algorithm on any instance of a subproblem one always applies the first rule applicable to the instance, which could be a reduction or a branching rule. A reduction rule produces one subproblem while a branching rule results in at least two subproblems, with different extensions of A and B. Note that G has a perfect matching cut that separates A from B if and only if in at least one recursive branch, extensions A' of A and B' of B are obtained such that G has a perfect matching cut that separates A' from B'. Typically a rule assigns one or more free vertices, vertices of F, either to A or to B and removes them from F, that is, we always have $F = V \setminus (A \cup B)$.

Reduction Rule 1

- *If a vertex in A has two B-neighbors, or a vertex in B has two A-neighbors then STOP: "G has no matching cut separating A, B".*
- *If $v \in F$, $|N(v) \cap A| \geq 2$ and $|N(v) \cap B| \geq 2$ then STOP: "G has no matching cut separating A, B".*
- *If there is an edge xy in G such that $x \in A$ and $y \in B$ and $N(x) \cap N(y) \cap F \neq \emptyset$ then STOP: "G has no matching cut separating A, B".*
- *If a vertex in A and a vertex in B have three or more common neighbors in F then STOP: "G has no matching cut separating A, B".*
- *If a vertex in A (respectively in B) has no neighbor in $B \cup F$ (respectively in $A \cup F$) then STOP: "G has no* perfect *matching cut separating A, B".*
- *If there are $x \in A$ and $y \in B$ such that $N(x) \cap F = N(y) \cap F = \{v\}$ then STOP: "G has no* perfect *matching cut separating A, B".*

Reduction Rule 2

- *If $v \in F$ has at least 2 A-neighbors (respectively B-neighbors) then $A := A \cup \{v\}$ (respectively $B := B \cup \{v\}$).*
- *If $v \in F$ with $|N(v) \cap N(z) \cap F| \geq 3$ for some $z \in A$ (respectively $z \in B$) then $A := A \cup \{v\} \cup (N(v) \cap N(z) \cap F)$ (respectively $B := B \cup \{v\} \cup (N(v) \cap N(z) \cap F)$).*

Reduction Rule 3. *If $x \in A$ (respectively $y \in B$) has two adjacent F-neighbors u, v then $A := A \cup \{u, v\}$ (respectively $B := B \cup \{u, v\}$).*

Reduction Rule 4. *If there is an edge xy in G such that $x \in A$ and $y \in B$ then add $N(x) \cap F$ to A, and add $N(y) \cap F$ to B.*

Reduction Rule 5 below is given in [17] and remains correct for perfect matching cuts.

Reduction Rule 5. *If there are vertices $u, v \in F$ such that $N(u) = N(v) = \{x, y\}$ with $x \in A, y \in B$, then $A := A \cup \{u\}$, $B := B \cup \{v\}$.*

The remaining reduction rules work for perfect matching cuts but not for matching cuts in general.

Reduction Rule 6. *If $x \in A$ (respectively $y \in B$) has exactly one neighbor $v \in F$ then $B := B \cup \{v\}$ (respectively $A := A \cup \{v\}$).*

Reduction Rule 7. *Let $z \in A$ (respectively $z \in B$) and let $v \in N(z) \cap F$.*

- *If* $\deg(v) = 1$ *then* $B := B \cup \{v\}$ *(respectively* $A := A \cup \{v\}$*).*
- *If* $\deg(v) = 2$ *and* $w \in F$ *is other neighbor of* v *then* $B := B \cup \{w\}$ *(respectively* $A := A \cup \{w\}$*).*

Reduction Rule 8. *Let $x \in A$ and $y \in B$ with $|N(x) \cap N(y) \cap F| = 2$. If $|N(x) \cap F| \geq 3$ or $|N(y) \cap F| \geq 3$ then $A := A \cup N(x) \setminus N(y)$, $B := B \cup N(y) \setminus N(x)$.*

We now describe the branching rules. All branching rules are based on the fact: if some vertex in A has no neighbor in B, it must have a neighbor in F that must go to Y, and if some vertex in B has no neighbor in A, it must have a neighbor in F that must go to X.

To determine the branching vectors which correspond to our branching rules, we set the size of an instance (G, A, B) as its number of free vertices, i.e., $|V(G)| - |A| - |B|$. Vertices in $A \cup B$ having exactly two neighbors in F will be covered by the first four branching rules.

Branching Rule 1. *Let $x \in A$ and $y \in B$ with $N(x) \cap N(y) \cap F = \{u, v\}$. By Reduction Rule 8, $N(x) \cap F = N(y) \cap F = \{u, v\}$. We branch into two subproblems.*

- *First, add $N[u] \cap F$ to A. Then $N[v] \cap F$ has to be added to B.*
- *Second, add $N[u] \cap F$ to B. Then $N[v] \cap F$ has to be added to A.*

The branching factor of Branching Rule 1 is at most $\tau(3, 3) < 1.2560$.

Branching Rule 2. *Let $x \in A$ with $N(x) \cap F = \{u, v\}$ and $N(u) \cap B = \{y_1\}$, $N(v) \cap B = \{y_2\}$. We branch into 2 subproblems.*

- *First, add u to B. Then v has to be added to A and $N_2 := N(y_2) \cap F \setminus \{v\}$ has to be added to B.*
- *Second, add v to B. Then u has to be added to A and $N_1 := N(y_1) \cap F \setminus \{u\}$ has to be added to B.*

Symmetrically for $y \in B$ with $N(y) \cap F = \{u, v\}$ and $N(u) \cap A = \{x_1\}$, $N(v) \cap A = \{x_2\}$.

The branching factor is at most $\tau(3, 3) < 1.2560$.

Branching Rule 3. *Let $x \in A$ with $N(x) \cap F = \{u, v\}$ and $N(u) \cap B = \emptyset$, $N(v) \cap B = \{y\}$. We branch into two subproblems.*

- *First, add u to B. Then v has to be added to A and $N := N(u) \cap F$ has to be added to B.*
- *Second, add v to B. Then u has to be added to A.*

Symmetrically for $y \in B$ with $N(y) \cap F = \{u, v\}$ and $N(u) \cap A = \emptyset$, and $N(v) \cap A = \{x\}$.

The branching factor of Branching Rule 3 is at most $\tau(4, 2) < 1.2721$.

Branching Rule 4. *Let $x \in A$ with $N(x) \cap F = \{u_1, u_2, \ldots, u_r\}$, $r \geq 2$, and $N(u_i) \cap B = \emptyset$, $1 \leq i \leq r$. We branch into r subproblems. For each $1 \leq i \leq r$, the instance of the i-th subproblem is obtained by adding u_i to B. Then $N(x) \cap F \setminus \{u_i\}$ has to be added to A and $N_i := N(u_i) \cap F$ has to be added to B.*
Symmetrically for $y \in B$ with $N(y) \cap F = \{v_1, v_2, \ldots, v_r\}$ and v_i has no neighbor in A, $1 \leq i \leq r$.

The branching factor of Branching Rule 4 is at most $\tau(r+2, r+2, \ldots, r+2) = \sqrt[r+2]{r} < 1.2600$.

Branching Rules 1 and 4 together with the remaining branching rules cover vertices in $A \cup B$ having at least three neighbors in F. Branching Rule 5 deals with the case $z \in A$ (respectively $z \in B$) in which at least two vertices in $N(z) \cap F$ have neighbors in B (respectively in A).

Branching Rule 5. *Let $x \in A$ with $N(x) \cap F = \{u_1, \ldots, u_p, v_1, v_2, \ldots, v_q\}$, $p \geq 0$, $q \geq 2$, such that $N(u_i) \cap B = \emptyset$, $1 \leq i \leq p$ and $N(v_j) \cap B = \{y_j\}$, $1 \leq j \leq q$. We branch into $r = p + q$ subproblems.*

- *For each $1 \leq i \leq p$, the instance of the i-th subproblem is obtained by adding u_i to B. Then $N(x) \cap F \setminus \{u_i\}$ has to be added to A and all $N_j := N(y_j) \cap F \setminus \{v_j\}$, $1 \leq j \leq q$, have to be added to B.*
- *For each $1 \leq j \leq q$, the instance of the $p+j$-th subproblem is obtained by adding v_j to B. Then $N(x) \cap F \setminus \{v_j\}$ has to be added to A and all $N_k := N(y_j) \cap F \setminus \{v_j\}$, $1 \leq k \leq q$, $k \neq j$, have to be added to B.*

Symmetrically for $y \in B$ with $N(y) \cap F = \{u_1, \ldots, u_p, v_1, v_2, \ldots, v_q\}$, $p \geq 0$, $q \geq 2$ such that $N(u_i) \cap A = \emptyset$, $1 \leq i \leq p$ and $N(v_j) \cap A = \{x_j\}$, $1 \leq j \leq q$.

The branching factor is at most $\tau(r+2q, \ldots, r+2q, r+2(q-1), \ldots, r+2(q-1)) < \tau(r+2, \ldots, r+2) = \sqrt[r+2]{r} < 1.2600$.

The last two branching rules deal with the case $z \in A$ (respectively $z \in B$) in which exactly one vertex in $N(z) \cap F$ has a unique neighbor in B (respectively in A).

Branching Rule 6. *Let $x \in A$ with $N(x) \cap F = \{u_1, u_2, \ldots, u_r, v\}$, $r \geq 2$, such that $N(u_i) \cap B = \emptyset$, $1 \leq i \leq r$, and $N(v) \cap B = \{y\}$. Write $N(y) \cap F \setminus \{v\} = \{v_1, \ldots, v_s\}$, $s \geq 2$. Assume that some u_i has two neighbors in $\{v_1, \ldots, v_s\}$. We branch into 2 subproblems.*

- *First, add v to A. Then $\{v_1, \ldots, v_s\}$ and u_i have to be added to B, and $\{u_1, \ldots, u_r\} \setminus \{u_i\}$ has to be added to A.*
- *Second, add v to B. Then $\{u_1, \ldots, u_r\}$ has to be added to A.*

Symmetrically for $y \in B$ with $N(y) \cap F = \{u_1, u_2, \ldots, u_r, v\}$ such that $N(u_i) \cap A = \emptyset$, $1 \leq i \leq r$, and $N(v) \cap A = \{x\}$ and some u_i has two neighbors in $N(x) \cap F \setminus \{v\}$.

The branching vector of Branching Rule 6 is $\tau(r+s+1, r+1) \leq \tau(5, 3) < 1.1939$.

Branching Rule 7. *Let $x \in A$ with $N(x) \cap F = \{u_1, u_2, \ldots, u_r, v\}$, $r \geq 2$, such that $N(u_i) \cap B = \emptyset$, $1 \leq i \leq r$, and $N(v) \cap B = \{y\}$. Write $N(y) \cap F \setminus \{v\} = \{v_1, \ldots, v_s\}$, $s \geq 2$. We branch into $r + s$ subproblems.*

- *For each $1 \leq i \leq r$, the instance of the i-th subproblem is obtained by adding u_i to B. Then $\{u_1, \ldots, u_r\} \setminus \{u_i\}$ and v have to be added to A, $N_i := N(u_i) \cap F$ and $\{v_1, \ldots, v_s\}$ have to be added to B.*
- *For each $1 \leq j \leq s$, the instance of the $r + j$-th subproblem is obtained by adding v_j to A. Then $\{v_1, \ldots, v_s\} \setminus \{v_j\}$ and v have to be added to B, $M_j := N(v_j) \cap F$ and $\{u_1, \ldots, u_r\}$ have to be added to A.*

Symmetrically for $y \in B$ with $N(y) \cap F = \{u_1, u_2, \ldots, u_r, v\}$ such that $N(u_i) \cap A = \emptyset$, $1 \leq i \leq r$, and $N(v) \cap A = \{x\}$.

The branching factor is at most $\tau(r+s+2, \ldots, r+s+2) = \sqrt[r+s+2]{r+s} < 1.2600$.

The description of all branching rules is completed. Among all branching rules, Branching Rule 3 has the largest branching factor of 1.2721. Consequently, the running time of our algorithm is $O^*(1.2721^n)$.

If all reduction and branching rules are not longer applicable, then no vertex in $A \cup B$ has a neighbor in F. Hence, by connectedness of G, $F = \emptyset$. Therefore, G has a perfect matching cut separating A and B if and only if (A, B) is a perfect matching cut. In summary, we obtain:

Theorem 4. *There is an algorithm for* PMC *running in $O^*(1.2721^n)$ time.*

4 Two Polynomial Solvable Cases

In this section, we provide two graph classes in which PMC is solvable in polynomial time. Both classes are well motivated by the hardness results.

Excluding a (Small) Tree of Maximum Degree Three. Let H be a fixed graph. A graph G is H-free if G contains no induced subgraph isomorphic to H. Theorem 2 implies that PMC remains NP-complete on H-free graphs whenever H has a vertex of degree larger than three or has a cycle. This suggests studying the computational complexity of PMC restricted to H-free graphs for a fixed forest H with maximum degree at most three.

As the first step in this direction, we show that PMC is solvable in polynomial time for H-free graphs, where H is the tree T with 6 vertices obtained from the claw $K_{1,3}$ by subdividing two edges each with one new vertex; see Fig. 3. In particular, PMC is polynomial time solvable for $K_{1,3}$-free graphs but hard for $K_{1,4}$-free graphs (by Theorem 2).

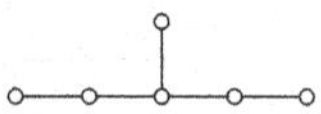

Fig. 3. The tree T.

Given a connected T-free graph $G = (V, E)$, our algorithm works as follows. Fix an edge $ab \in E$ and decide if G has a perfect matching cut $M = E(X, Y)$ separating $A = \{a\}$ and $B = \{b\}$. We use the notations and reduction rules from Sect. 3. In addition, we need one new reduction rule; recall that $F = V \setminus (A \cup B)$. This additional reduction rule is correct for matching cuts in general and is already used in [7].

Reduction Rule 9. *If there are vertices $u, v \in F$ with a common neighbor in A (respectively in B) and $|N(u) \cap N(v) \cap F| \geq 2$, then $A := A \cup \{u, v\}$ (respectively $B := B \cup \{u, v\}$).*

Now, we apply the Reduction Rules 1–9 exhaustively. If $F = V \setminus (A \cup B)$ is empty, then G has a perfect matching cut separating A and B if and only if (A, B) is a perfect matching cut of G.

In case $F \neq \emptyset$ it can be shown that G has no perfect matching cut separating A and B, or G contains the tree T as an induced subgraph. So, after at most $|E|$ rounds, each for a candidate $ab \in E$ and in polynomial time, our algorithm will find out whether G has a perfect matching cut at all. In summary, we obtain:

Theorem 5. PMC *is solvable in polynomial time for T-free graphs.*

Interval, Chordal and Pseudo-chordal Graphs. Recall that a graph has girth at least g if and only if it has no induced cycles of length less than g. Thus, Theorem 2 implies that PMC remains hard when restricted to graphs without short induced cycles. This suggests studying PMC restricted to graphs without long induced cycles, i.e., k-chordal graphs. Here, given an integer $k \geq 3$, a graph is *k-chordal* if it has no induced cycles of length larger than k; the 3-chordal graphs are known as chordal graphs.

We will show that PMC can be solved in polynomial time when restricted to what we call pseudo-chordal graphs, that contain the class of 3-chordal graphs and thus known to have unbounded mim-width [16].

We begin with a concise characterization of interval graphs having perfect matching cuts, to yield a polynomial-time algorithm deciding if an interval graph has a perfect matching cut which is much simpler than what we get by the mim-width approach [6].

Fact 1. *Let G have a vertex set $U \subseteq V(G)$ such that $G[U]$ is connected with every edge of $G[U]$ belonging to a triangle. Then if (X, Y) is a perfect matching cut of G we must have $U \subseteq X$ or $U \subseteq Y$.*

This since otherwise we must have a triangle K and two vertices u, v with $u \in K \cap X$ and $v \in K \cap Y$ having a common neighbor in K so this cannot be a perfect matching cut.

If an interval graph G has a cycle then it has a 3-clique. By Fact 1 these 3 vertices would have to belong to the same side of the cut, and each would need to have a unique neighbor on the other side of the cut. But then those 3 neighbors would form an asteroidal triple, contradicting that G was an interval graph. Thus an interval graph which is not a tree does not have a perfect matching cut. A tree T is an interval graph if and only if it does not have the subdivided claw as a subgraph. It is not hard to verify the following.

Fact 2. *An interval graph has a perfect matching cut if and only if it is a caterpillar with basic path $x_1, \ldots, x_k$ such that any x_i for $1 < i < k$ has either zero or one leaf, and any maximal sub-path of $x_1, \ldots, x_k$ with zero leaves contains an even number of vertices.*

We will show a polynomial-time algorithm for what we call pseudo-chordal graphs. The maximal 2-connected subgraphs of a graph are called its blocks, and a block is non-trivial if it contains at least 3 vertices.

A graph is *pseudo-chordal* if, for every non-trivial block B, every edge of B belongs to a triangle. Note that chordal graphs are pseudo-chordal, but pseudo-chordal graphs may contain induced cycles of any length.

Theorem 6. *There is a polynomial-time algorithm deciding if a pseudo-chordal graph G has a perfect matching cut.*

Proof. We first compute the blocks of G and let D be the subgraph of G formed by the edges of non-trivial blocks of G. Let $D_1, D_2, \ldots, D_k$ be the connected components of D. Note that by collapsing each D_i into a supernode we can treat the graph G as having a tree structure T (related to the block structure) with one node for each $v \in V(G) \setminus V(D)$, and a supernode for each D_i. See Fig. 4. Note that since G is pseudo-chordal then by Fact 1 all the vertices in a fixed supernode D_i must be on the same side in any perfect matching cut of G. Our algorithm will pick a root R of T and proceed by bottom-up dynamic programming on the rooted tree T. The details are given in the full version. □

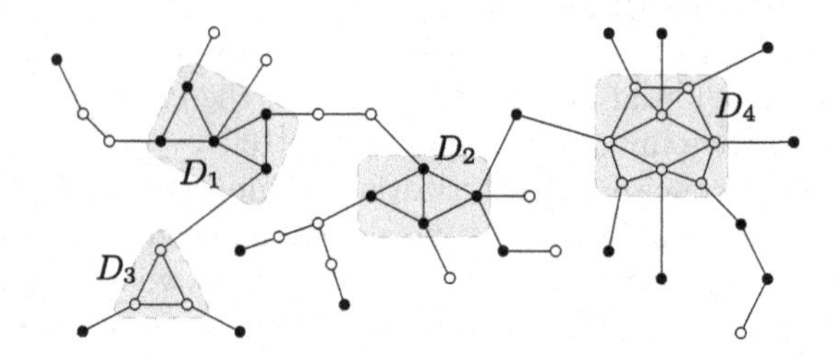

Fig. 4. A pseudo-chordal graph and perfect matching cut given by (X, Y) with X being black vertices. Note the tree structure composed of (i) those vertices that do not belong to a clique of size 3 and (ii) the four supernodes D_1, D_2, D_3, D_4.

5 Conclusion

We have shown that PMC remains NP-complete when restricted to bipartite graphs of maximum degree 3 and arbitrarily large girth. This implies that PMC remains NP-complete when restricted to H-free graphs where H is any fixed graph having a vertex of degree at least 4 or a cycle. This suggests the following problem for further research: Let F be a fixed forest with maximum degree at most 3. What is the computational complexity of PMC restricted to F-free graphs? We have proved a first polynomial case where F is a certain 6-vertex tree, including claw-free graphs and graphs without an induced 5-path. Our hardness result also suggests studying PMC restricted to graphs without long induced cycles: What is the computational complexity of PMC on k-chordal graphs? It follows from our results that PMC is polynomially solvable for 3-chordal graphs.

We have also given an exact branching algorithm for PMC running in $O^*(1.2721^n)$ time. It is natural to ask whether the running time of the branching algorithm can be improved. Finally, as for matching cuts, also for perfect matching cuts it would be interesting to study counting and enumeration as well as FPT and kernelization algorithms.

References

1. Aravind, N.R., Kalyanasundaram, S., Kare, A.S.: On structural parameterizations of the matching cut problem. In: Gao, X., Du, H., Han, M. (eds.) COCOA 2017. LNCS, vol. 10628, pp. 475–482. Springer, Cham (2017). https://doi.org/10.1007/978-3-319-71147-8_34
2. Aravind, N.R., Saxena, R.: An FPT algorithm for matching cut. CoRR, abs/2101.06998 (2021). https://arxiv.org/abs/2101.06998. arxiv:2101.06998
3. Belmonte, R., Vatshelle, M.: Graph classes with structured neighborhoods and algorithmic applications. Theor. Comput. Sci. **511**, 54–65 (2013). https://doi.org/10.1016/j.tcs.2013.01.011
4. Bonsma, P.S.: The complexity of the matching-cut problem for planar graphs and other graph classes. J. Graph Theory **62**(2), 109–126 (2009). https://doi.org/10.1002/jgt.20390
5. Bouquet, V., Picouleau, C.: The complexity of the perfect matching-cut problem. CoRR, abs/2011.03318 (2020). https://arxiv.org/abs/2011.03318. arXiv:2011.03318
6. Bui-Xuan, B.-M., Telle, J.A., Vatshelle, M.: Fast dynamic programming for locally checkable vertex subset and vertex partitioning problems. Theor. Comput. Sci. **511**, 66–76 (2013). https://doi.org/10.1016/j.tcs.2013.01.009
7. Chen, C.-Y., Hsieh, S.-Y., Le, H.-O., Le, V.B., Peng, S.-L.: Matching cut in graphs with large minimum degree. Algorithmica **83**(5), 1238–1255 (2021). https://doi.org/10.1007/s00453-020-00782-8
8. Chvátal, V.: Recognizing decomposable graphs. J. Graph Theory **8**(1), 51–53 (1984). https://doi.org/10.1002/jgt.3190080106
9. Fomin, F.V., Kratsch, D.: Exact Exponential Algorithms. Springer, Heidelberg (2010). https://www.springer.com/gp/book/9783642165320
10. Golovach, P.A., Komusiewicz, C., Kratsch, D., Le, V.B.: Refined notions of parameterized enumeration kernels with applications to matching cut enumeration. In: 38th International Symposium on Theoretical Aspects of Computer Science, STACS 2021, volume 187 of LIPIcs, pp. 37:1–37:18 (2021). https://doi.org/10.4230/LIPIcs.STACS.2021.37
11. Gomes, G.C.M., Sau, I.: Finding cuts of bounded degree: complexity, FPT and exact algorithms, and kernelization. In: 14th International Symposium on Parameterized and Exact Computation, IPEC 2019, pp. 19:1–19:15 (2019). https://doi.org/10.4230/LIPIcs.IPEC.2019.19
12. Graham, R.L.: On primitive graphs and optimal vertex assignments. Ann. N. Y. Acad. Sci. **175**(1), 170–186 (1970)
13. Heggernes, P., Telle, J.A.: Partitioning graphs into generalized dominating sets. Nord. J. Comput. **5**(2), 128–142 (1998)
14. Impagliazzo, R., Paturi, R.: On the complexity of k-sat. J. Comput. Syst. Sci. **62**(2), 367–375 (2001). https://doi.org/10.1006/jcss.2000.1727

15. Impagliazzo, R., Paturi, R., Zane, F.: Which problems have strongly exponential complexity? J. Comput. Syst. Sci. **63**(4), 512–530 (2001). https://doi.org/10.1006/jcss.2001.1774
16. Kang, D.Y., Kwon, O., Strømme, T.J.F., Telle, J.A.: A width parameter useful for chordal and co-comparability graphs. Theor. Comput. Sci. **704**, 1–17 (2017). https://doi.org/10.1016/j.tcs.2017.09.006
17. Komusiewicz, C., Kratsch, D., Le, V.B.: Matching cut: kernelization, single-exponential time FPT, and exact exponential algorithms. Discret. Appl. Math. **283**, 44–58 (2020). https://doi.org/10.1016/j.dam.2019.12.010
18. Kratsch, D., Le, V.B.: Algorithms solving the matching cut problem. Theor. Comput. Sci. **609**, 328–335 (2016). https://doi.org/10.1016/j.tcs.2015.10.016
19. Le, H.-O., Le, V.B.: A complexity dichotomy for matching cut in (bipartite) graphs of fixed diameter. Theor. Comput. Sci. **770**, 69–78 (2019). https://doi.org/10.1016/j.tcs.2018.10.029
20. Le, V.B., Telle, J.A.: The perfect matching cut problem revisited. CoRR, abs/2107.06399 (2021). https://arxiv.org/abs/2107.06399. arXiv:2107.06399
21. Moshi, A.M.: Matching cutsets in graphs. J. Graph Theory **13**(5), 527–536 (1989). https://doi.org/10.1002/jgt.3190130502
22. Telle, J.A., Proskurowski, A.: Algorithms for vertex partitioning problems on partial k-trees. SIAM J. Discret. Math. **10**(4), 529–550 (1997). https://doi.org/10.1137/S0895480194275825
23. Vatshelle, M.: New width parameters of graphs. Ph.D. thesis, University of Bergen, Norway (2012). https://bora.uib.no/bora-xmlui/handle/1956/6166

The Complexity of Gerrymandering over Graphs: Paths and Trees

Matthias Bentert, Tomohiro Koana(✉), and Rolf Niedermeier

Faculty IV, Algorithmics and Computational Complexity, TU Berlin, Berlin, Germany
{matthias.bentert,tomohiro.koana,rolf.niedermeier}@tu-berlin.de

Abstract. Roughly speaking, gerrymandering is the systematic manipulation of the boundaries of electoral districts to make a specific (political) party win as many districts as possible. While typically studied from a geographical point of view, addressing social network structures, the investigation of gerrymandering over graphs was recently initiated by Cohen-Zemach et al. [AAMAS 2018]. Settling three open questions of Ito et al. [AAMAS 2019, TCS 2021], we classify the computational complexity of the NP-hard problem GERRYMANDERING OVER GRAPHS when restricted to paths and trees. Our results, which are mostly of negative nature (that is, worst-case hardness), in particular yield two complexity dichotomies for trees. For instance, the problem is polynomial-time solvable for two parties but becomes weakly NP-hard for three. Moreover, we show that the problem remains NP-hard even when the input graph is a path.

1 Introduction

How to influence an election? One answer to this is gerrymandering [4,8,14]. Gerrymandering is the systematic manipulation of the boundaries of electoral districts in favor of a particular party. It has been studied in the political sciences for decades [13]. In recent years, various models of gerrymandering were investigated from an algorithmic and computational perspective. For instance, Lewenberg et al. [11] and Eiben et al. [6] studied the (parameterized) computational complexity of gerrymandering assuming that the voters are points in a two-dimensional space and the task is to place k polling stations where each voter is assigned to the polling station closest to her. Cohen-Zemach et al. [5] introduced a version of gerrymandering over graphs (which may be seen as models of social networks) where the question is whether a given candidate can win at least ℓ districts. This leads to the question whether there is a partition of

This work was initiated at the research retreat of the Algorithmics and Computational Complexity group held in September 2020 in Zinnowitz, Germany. A full version of this work is available on arXiv [2].
T. Koana—Supported by the Deutsche Forschungsgemeinschaft (DFG), project FPT-inP, NI 369/19.

Ł. Kowalik et al. (Eds.): WG 2021, LNCS 12911, pp. 195–206, 2021.
https://doi.org/10.1007/978-3-030-86838-3_15

the graph into k connected subgraphs such that at least ℓ of these are won by a designated candidate; herein, k and ℓ are part of the input of the computational problem. Cohen-Zemach showed that this version is NP-complete even when restricted to planar graphs. Following up on their pioneering work, Ito et al. [10] performed a refined complexity analysis, particularly taking into account the special graph structures of cliques, paths, and trees. Indeed, their formal model is slightly different from the one of Cohen-Zemach et al. [5] and their work will be our main point of reference. Notably, both studies focus on the perhaps simplest voting rule, Plurality. In parallel and independently of our work, Gupta et al. [9] developed algorithms and complexity results for gerrymandering over graphs, where their main focus was on paths.

We mention in passing that earlier work also studied the special case of gerrymandering on grid graphs. More specifically, Apollonio et al. [1] analyzed gerrymandering in grid graphs where each district in the solution has to be of (roughly) the same size. Focusing on two candidates (equivalently, two parties), they evaluated the margin of votes between the two equally supported candidates. Later, Borodin et al. [3] also considered gerrymandering on grid graphs with two parties (expressed by colors red and blue), but here each vertex represents a polling station and thus is partially "red" and partially "blue" colored. They provided a worst-case analysis for a two-party situation in terms of the total fraction of votes the party responsible for the gerrymandering process gets. They also confirmed their findings with experiments.

To formally define our central computational problem, we continue with a few definitions. For a vertex-colored graph and for each color r, let S_r be the set of r-colored vertices. A vertex-weighted graph is *q-colored* if for each color r it holds that $\sum_{v \in S_q} w(v) \geq \sum_{v \in S_r} w(v)$. A vertex-weighted graph is *uniquely q-colored* if $\sum_{v \in S_q} w(v) > \sum_{v \in S_r} w(v)$ for each color $r \neq q$. Analogously, we say that a set of vertices is q-colored if they induce a connected q-colored graph. Thus, we arrive at the central problem of this work, going back to Ito et al. [10].

GERRYMANDERING OVER GRAPHS

Input: An undirected, connected graph $G = (V, E)$, a weight function $w\colon V \to \mathbb{N}$, a set C of colors, a target color $p \in C$, a coloring function $\mathrm{col}\colon V \to C$, and an integer k.

Question: Can V be partitioned into exactly k subsets $\mathcal{V} = \{V_1, \ldots, V_k\}$ such that each $V_i \in \mathcal{V}$, $i \in [k]$, induces a connected subgraph in G and the number of uniquely p-colored induced subgraphs exceeds the number of r-colored induced subgraphs for each $r \in C \setminus \{p\}$?

Figure 1 presents a simple example of GERRYMANDERING OVER GRAPHS. We remark that all our results except for Theorem 1 (that is, the NP-hardness on paths) also transfer to the slightly different model of Cohen-Zemach et al. [5].[1]

[1] In fact, we conjecture that the gerrymandering problem of Cohen-Zemach et al. [5] is polynomial-time solvable on paths.

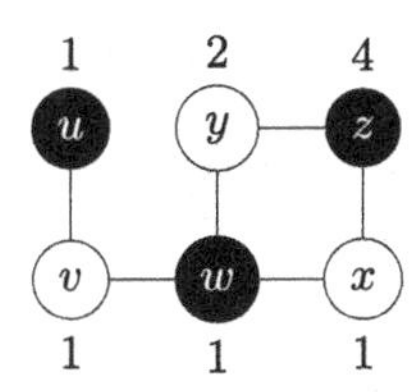

Fig. 1. An example input instance for GERRYMANDERING OVER GRAPHS with two colors (black and white) where black is the target color and where the numbers next to the vertices illustrate the vertex weights. For $k = 2$, a solution for this instance is $\mathcal{V} = \{\{u, v, w\}, \{x, y, z\}\}$ as each of these two parts induces a uniquely black-colored connected subgraph.

We also use an equivalent interpretation of solutions for GERRYMANDERING OVER GRAPHS. Since each part $V_i \in \mathcal{V}$ has to induce a connected subgraph, in the spirit of edge deletion problems from algorithmic graph theory, we also represent solutions by a set of edges such that removing these yields the disjoint union of subgraphs induced by each part $V_i \in \mathcal{V}$. In Fig. 1, removing the edges $\{w, x\}$ and $\{w, y\}$ yields a solution for $k = 2$.

Finally, regarding notation, for a color q we use $w_q(v) := w(v)$ if v is of color q and $w_q(v) = 0$ if v has another color. Further, we use $w(V') := \sum_{v \in V'} w(v)$ and $w_q(V') := \sum_{v \in V'} w_q(v)$.

Known and New Results. As mentioned before, we essentially build upon the work of Ito et al. [10], in particular studying exactly the same computational problem. We only focus on the case of path and tree graphs as input, whereas they additionally studied cliques. For cliques, they showed NP-hardness already for $k = 2$ and two colors. On the positive side, they provided for cliques a pseudo-polynomial-time algorithm for $k = 2$ and a polynomial-time algorithm for each fixed $k \geq 3$. Moving to paths and trees, besides some positive algorithmic and hardness results, Ito et al. [10] particularly left three open problems:

1. Existence of a polynomial-time algorithm for paths when $|C|$ is not fixed.
2. Existence of a polynomial-time algorithm for trees when $|C|$ is a constant.
3. Existence of a polynomial-time algorithm for trees of diameter exactly three.

Indeed, they called the first two questions the "main open problems" of their paper. We settle all three questions, the first two in the negative by showing NP-hardness. See Table 1 for an overview on some old and our new results. Notably, our new results (partially together with the previous results of Ito et al. [10]) reveal two sharp complexity dichotomies for trees. For up to two colors, the problem is polynomial-time solvable, whereas it gets NP-hard with three or more colors. Moreover, it is polynomial-time solvable for trees with diameter at most three but NP-hard for trees with diameter at least four.

Gupta et al. [9] also studied the model of Ito et al. [10] and showed (parameterized) exponential-time algorithms for general graphs. Moreover, they proved

Table 1. Results overview. The diameter of a graph is denoted by diam.

	Restriction	Complexity	Reference
Paths	No	NP-hard	Theorem 1
	Constant $\lvert C\rvert$	Polynomial time	[10, Theorem 6]
Trees	Constant $\lvert C\rvert$	Pseudo-polynomial time	[10, Theorem 7]
	$\lvert C\rvert = 2$	Polynomial time	Proposition 1
	$\lvert C\rvert \geq 3$	Weakly NP-hard	Theorem 2
	diam $= 2$	Polynomial time	[10, Theorem 5]
	diam $= 3$	Polynomial time	Proposition 2
	diam ≥ 4	NP-hard	[10, Theorem 3]

(independently from us) that Gerrymandering over Graphs remains NP-hard on paths. Notably, their corresponding reduction does not result in unit weights (as our reduction does).

2 NP-Hardness on Paths

Ito et al. [10] showed that Gerrymandering over Graphs on paths can be solved in polynomial time for fixed $|C|$. They left open the question of polynomial-time solvability on paths when $|C|$ is unbounded. Negatively answering their question, we show that Gerrymandering over Graphs remains NP-hard on paths.

Theorem 1. Gerrymandering over Graphs *restricted to paths is NP-hard even if all vertices have unit weight.*

Proof. We reduce from Clique on regular graphs, an NP-hard problem [12]. Let (G, ℓ) be an instance of Clique, where G is d-regular for some integer d, and ℓ is the sought solution size. We first construct an equivalent instance $\mathcal{I}$ of Gerrymandering over Graphs where the graph consists of disjoint paths. We then modify the construction to obtain an instance $\mathcal{J}$ on one connected path.

All vertices in the following constructions have weight one. Let n and m be the number of vertices and edges in G, respectively, and let $N := 4n^2$. We introduce a path P_v on $4N - 1$ vertices for each vertex $v \in V$ and a path P'_e on four vertices for each edge $e \in E$. Moreover, we introduce an independent set S of $2N - (n - \ell) + 1$ vertices. We denote by $G' = (V', E')$ the disjoint union of all P_v for $v \in V$, all P'_e for $e \in E$, and S. Note that G' has $z := 2N + \ell + m + 1$ connected components.

We introduce colors p, q, r, a unique color c_v for each $v \in V$. Additionally, we introduce a set D of colors, where for each color $c \in D$, there will be only a single vertex of color c in the resulting graph. The target color is p. We color $N + 1$ vertices of S with color p and $N - (n - \ell)$ vertices of S with color q. For each vertex $v \in V$, we color the vertices in P_v as follows.

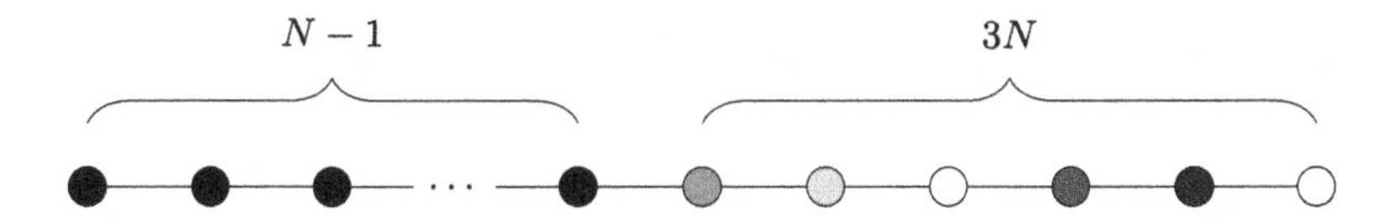

Fig. 2. An example of a gadget for a vertex v. Black and white vertices represent q-colored and c_v-colored vertices, respectively, and each other vertex has a distinct color.

- The first $N-1$ vertices receive color q,
- for each $i \in [N]$, the $(N-1+3i)$-th vertex receives color c_v, and
- each remaining vertex receives a (distinct) color from D.

An illustration of the path P_v is shown in Fig. 2. For each edge $e = \{u, v\} \in E$, we color the two inner vertices of P'_e with color r and the endpoints with colors c_u and c_v, respectively. Finally, we set $k := (n-\ell) \cdot 3N + d\ell + \binom{\ell}{2} + z$. We call the resulting instance $\mathcal{I}$.

First, we show that if G contains a clique K of size ℓ, then the constructed instance $\mathcal{I}$ is a yes-instance. We will specify the set E'' of exactly $k-z$ edges such that the connected components of $G'' = (V', E' \setminus E'')$ correspond to a solution. Note that each removal of an edge increases the number of connected components by exactly one.

- For each vertex $v \in V \setminus K$, the edge set E'' contains all $3N$ edges in P_v that are not between two q-colored vertices. There are $(n-\ell) \cdot 3N$ such edges.
- For each vertex $v \in K$ and each edge $e = \{u, v\}$, the edge set E'' contains the edge incident to the c_v-colored vertex in P'_e. There are $d\ell$ such edges as each vertex in the input graph has d neighbors.
- For each edge e where both endpoints are contained in K, the edge set E'' contains the edge between the two inner (r-colored) vertices in P'_e. There are $\binom{\ell}{2}$ such edges.

Thus, E'' contains $(n-\ell) \cdot 3N + d\ell + \binom{\ell}{2} = k - z$ edges in total, leaving k connected components in the graph G''.

Next, we examine the color of each connected component in G''. First, note that there are $N+1$ connected components that are uniquely p-colored. We now show that for each color c other than p there are at most N connected components which are c-colored.

- For color q, observe that there are $N-(n-\ell)$ isolated vertices of color q in S and for each vertex $v \in V \setminus K$ there is exactly one q-colored connected component contained in P_v and for every vertex $v \in K$ there is no q-colored connected component in P_v. Hence, there are $N-(n-\ell)+(n-\ell) = N$ connected components that are q-colored.
- For color r, note that there are $2m < N$ vertices of color r. Thus, there are less than N connected components that are r-colored.
- For each color c_v with $v \in V \setminus K$, there are N connected components in P_v that are c_v-colored. All other vertices of color c_v are contained in P'_e for some $e \in E$ and those belong to r-colored component by construction. Hence, there are N connected components that are c_v-colored.

- For each color c_v with $v \in K$, the whole path P_v remains one connected component which is c_v-colored. All other vertices of color c_v are contained in P'_e for some $e \in E$ and since $N > m$, there are at most N connected components that are c_v-colored.

Thus, if G contains a clique of size ℓ, then the constructed instance $\mathcal{I}$ is a yes-instance.

Conversely, we show that if $\mathcal{I}$ has a solution $\mathcal{V}$, then there is a clique of size ℓ in G. Let E'' be a set of exactly $k-z$ edges in G' such that the connected components of $G'' = (V', E' \setminus E'')$ correspond to $\mathcal{V}$. Let J be the set of vertices $v \in V$ such that P_v contains an edge of E'' and let $K := V \setminus J$. For each vertex $v \in J$, let n_v^q and n_v^c be the number of connected components of $P_v - E''$ which are q-colored and c_v-colored, respectively. Our goal is to show that K forms a clique of size ℓ in G. To this end, we derive an upper bound on the size of E'' in terms of n_v^q, n_v^c, and $|J|$:

1. For each vertex $v \in J$, there are at most $n_v^q - 1$ edges in P_v of E'' whose endpoints have color q. Since S contains $N+1$ isolated vertices of color p and $N-(n-\ell)$ isolated vertices of color q, it holds that $\sum_{v \in J} n_v^q \leq n - \ell$. Thus, E'' contains at most $\sum_{v \in J}(n_v^q - 1) = n - \ell - |J|$ edges in P_v both of whose endpoints have color q.
2. For each vertex $v \in J$, the edge set E'' contains at most $3n_v^c$ edges in P_v where at least one endpoint does not have color q.
3. For each vertex $v \in J$, the edge set E'' contains at most $N - n_v^c$ edges incident to a vertex of color c_v in a P'_e for some edge $e \in E$.
4. For each vertex $v \in K$, there are exactly d edges incident to a vertex of color c_v that are contained in a P'_e for some edge $e \in E$. Thus, E'' contains at most $d \cdot |K| = d \cdot (n - |J|)$ such edges.
5. Finally, we consider edges between inner vertices of P'_e for $e \in E$. Observe that if such an edge $e = \{u, v\}$ is contained in E'', then G'' has one c_u-colored component and one c_v-colored component. Thus, $|E''|$ contains at most

$$\binom{|K|}{2} + \left(\sum_{v \in J} N - n_v^c\right) = \binom{n - |J|}{2} + \sum_{v \in J} N - n_v^c$$

such edges.

Summing over these edges yields that E'' contains at most

$$\begin{aligned}
&(n - \ell - |J|) + \left(\sum_{v \in J} 3n_v^c\right) + \left(\sum_{v \in J} N - n_v^c\right) \\
&+ d \cdot (n - |J|) + \binom{n - |J|}{2} + \left(\sum_{v \in J} N - n_v^c\right) \\
&\leq (n - \ell - |J|) + 3N \cdot |J| + d \cdot (n - |J|) + \binom{n - |J|}{2}
\end{aligned}$$

edges. Here, the inequality is due to the fact that $n_v^c \leq N$. Thus, $|E''| \leq f(|J|)$, where

$$f(x) := (n - \ell - x) + 3N \cdot x + d \cdot (n - x) + \binom{n-x}{2}.$$

Next, we show that $|J| \leq n - \ell$. Recall that G has $N+1$ isolated vertices of color p and $N - (n - \ell)$ isolated vertices of color q. Since the path P_v contains at least one q-colored part for every vertex $v \in J$, we obtain $|J| \leq n - \ell$.

Notice that $f(x)$ is monotonically increasing for $x \geq 0$ and that from this follows that $k - z = |E''| \leq f(|J|) \leq f(n - \ell)$. Note that $f(n - \ell) = k - z$ by the definition of f. Consequently, we have $f(|J|) = f(n - \ell)$ and hence $|J| = n - \ell$. Finally, note that for any solution where $|J| = n - \ell$, we cannot remove any edges between two vertices of color q (as this would result in at least $N+1$ connected components that are q-colored). Hence, $n_v^q = 1$ for each $v \in V$ and thus summing up all edges in E'' (except for those between two vertices of color q) yields

$$|E''| \leq \left(\sum_{v \in J} 2N + n_v^c\right) + d\ell + \binom{\ell}{2} \leq (n - 3\ell) \cdot 3N + d\ell + \binom{\ell}{2} = k - z.$$

Here the second inequality follows from the fact $n_v^c \leq N$ for each vertex $v \in J$. Since E'' has to contain $k - z$ edges, we obtain $n_v^c = N$ for each vertex $v \in J$. Hence, there are exactly $\binom{\ell}{2}$ edges in E'' between two vertices of color r in P_e' for edges $e \in E$. Note that for each such edge $e \in E$ it has to hold that both endpoints of e are in K as otherwise there are $N+1$ connected components in G'' of color c_v (where $v \in J$ is an endpoint of e). Thus, there are ℓ vertices in K that share $\binom{\ell}{2}$ edges between them, that is, K induces a clique of size ℓ.

Finally, we show how to construct an equivalent instance $\mathcal{J}$ of GERRYMANDERING OVER GRAPHS on one path. To do so, we simply connect G' with paths on $M := 4nN + 3m$ vertices. More precisely, we do the following. We fix an arbitrary ordering of the connected components of G'. For every consecutive connected components P and P', we introduce an path Q of M vertices each with a distinct color from D. We then add an edges between the last vertex of P and the first vertex of Q and an edge between the last vertex of Q and the first vertex of P'. Let H be the resulting graph. Recall that G' consists of z paths. So H contains $(z-1)M$ additional vertices. To conclude the construction of $\mathcal{J}$, we set the partition size $k' := k + (z-1)M$.

We now proceed to show the equivalence between $\mathcal{I}$ and $\mathcal{J}$. If $\mathcal{I}$ is a yes-instance, then we obtain a solution of $\mathcal{J}$ by removing the edges added to H. That is, the partition $\mathcal{V}' = \mathcal{V} \cup \{\{v\} \mid v \in V_H\}$ is a solution of $\mathcal{J}$, where $\mathcal{V}$ is a solution of $\mathcal{I}$ and V_H is the set of vertices introduced in the construction of H. Conversely, suppose that $\mathcal{J}$ admits a solution $\mathcal{V}'$. Let F be the set of $k'-1 = k+(z-1)M-1$ edges of H such that $\mathcal{V}'$ is the connected components of the graph obtained by deleting F from H. Since M is sufficiently large, F contains at least one edge of each path of M vertices between the connected components of G'. It follows that any pair of vertices that are in different connected components in G' belong to different parts of $\mathcal{V}'$. We can then show that F contains at most $k - z$ edges of

G' as described above. In fact, F contains exactly $k-z$ edges of G' and all the edges introduced when constructing H: Since we added $(z-1)(M+1)$ edges, F contains $|F|-(z-1)(M+1)=k-z$ edges of G'. Consequently, $\mathcal{V}'$ contains exactly k subsets consisting of vertices in G', which form a solution of $\mathcal{I}$. □

In the above reduction, we use an unbounded number of colors. This appears to be inevitable since GERRYMANDERING OVER GRAPHS is polynomial-time solvable on paths for any constant $|C|$ [10]. We wonder whether there are other graph classes for which GERRYMANDERING OVER GRAPHS can be solved in polynomial time when $|C|$ is constant. Caterpillars form a possible candidate.

3 Complexity on Trees

In this section, we investigate trees. We first address the special case of three colors (NP-hard), then two colors (polynomial-time solvable), and finally we show polynomial-time solvability for diameter-three trees.

Ito et al. [10] developed a pseudo-polynomial-time algorithm for GERRYMANDERING OVER GRAPHS on trees for constant $|C|$, which led them to ask whether it is also polynomial-time solvable for fixed $|C|$. Notably, Gupta et al. [9] asked whether a generalized version of GERRYMANDERING OVER GRAPHS on trees is fixed-parameter tractable with respect to $|C|$. We show that GERRYMANDERING OVER GRAPHS on trees is weakly NP-hard even if $|C|=3$, answering both aforementioned questions in the negative. Afterwards, we will show the polynomial-time solvability for $|C|=2$, yielding a tight classification.

Theorem 2. GERRYMANDERING OVER GRAPHS *restricted to trees is weakly NP-hard even if* $|C|=3$.

Proof. We reduce from PARTITION, which is known to be weakly NP-hard [7]. Given a multi-set A of n non-negative integers $a_1, a_2, \ldots, a_n$, the task is to find a subset $B \subseteq A$ of exactly $n/2$ integers whose sum is $s/2$, where $s := \sum_{a \in A} a$. We can assume that s is a multiple of n (otherwise we multiply each element of A by n). Let $N := s+1$ and let M be some natural number greater than $N \cdot 2^n(n+1)+s/2+1$. For the construction, we use a set $C=\{p,q,r\}$ of three colors, where p is the target color. We start with a star with a center vertex z and a set L of $n/2$ leaves. We color each vertex in the star with color p. We assign the weights $w(z) := Mn+s/2+1$ to the center z and $w(\ell) := 1$ for each leaf $\ell \in L$. For each $a_i \in A$, we do the following.

- We introduce two vertices x_i^q and y_i^q of color q and two vertices x_i^r and y_i^r of color r. Let $X_i := \{x_i^q, x_i^r\}$, $Y_i := \{y_i^q, y_i^r\}$, and $Z_i := X_i \cup Y_i$.
- We add four edges $\{z, x_i^q\}$, $\{z, y_i^q\}$, $\{x_i^q, x_i^r\}$, and $\{y_i^q, y_i^r\}$.
- We define the weights for each vertex in Z_i as

$$w(x_i^q) := M + N \cdot 2^i + a_i, \qquad w(x_i^r) := M - N \cdot 2^i,$$

$$w(y_i^r) := M + N \cdot 2^i - a_i + \frac{2s}{n}, \text{ and} \qquad w(y_i^q) := M - N \cdot 2^i.$$

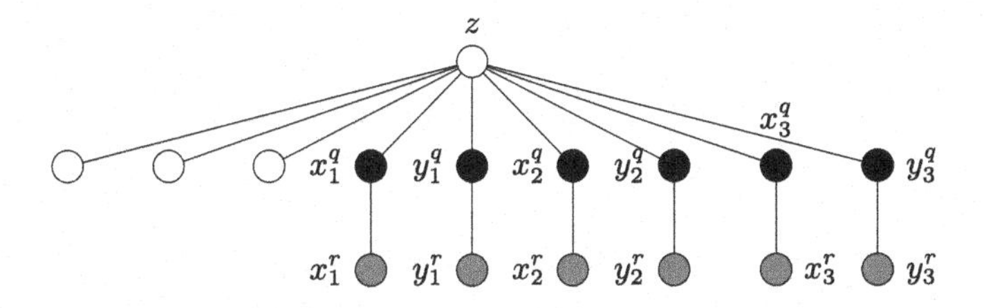

Fig. 3. An illustration of the construction in the proof of Theorem 2. White represents vertices of color p, black represents vertices of color q, and gray represents vertices of color r.

Observe that the weights are integral since s is divisible by n. In addition, observe that X_i is q-colored and that Y_i is r-colored.

The constructed graph $G = (V, E)$ is illustrated in Fig. 3. It is clearly a tree.

To conclude the construction of the GERRYMANDERING OVER GRAPHS instance, we set $k := 3n/2 + 1$.

We next show that the construction is correct. Suppose that there is a subset $I \subseteq [n]$ of exactly $n/2$ indices such that $\sum_{i \in I} a_i = s/2$. Then, the partition

$$\mathcal{V} = \{V'\} \cup \{\{\ell\} \mid \ell \in L\} \cup \{X_i \mid i \in [n] \setminus I\} \cup \{Y_i \mid i \in [n] \setminus I\},$$

where $V' := \{z\} \cup \{v \in X_i \cup Y_i \mid i \in I\}$ is a solution for the constructed instance of GERRYMANDERING OVER GRAPHS. First, observe that V' is p-colored as $w_p(V') = Mn + s/2 + 1$ and $w_q(V') = w_r(V') = Mn + s/2$. Second, observe that the singleton $\{\ell\}$ is p-colored for each leaf $\ell \in L$, and hence $\mathcal{V}$ has $n/2 + 1$ parts which are p-colored. Since X_i is q-colored and Y_i is r-colored for each $i \in [n]$, exactly $n - |I| = n/2$ subsets of $\mathcal{V}$ are q-colored and exactly $n - |I| = n/2$ subsets of $\mathcal{V}$ are r-colored. Thus, $\mathcal{V}$ is indeed a solution.

Conversely, suppose that there is a solution $\mathcal{V}$. We show that the PARTITION instance is a yes-instance. Note that there are at least $k/|C| = (3n/2+1)/3 > n/2$ parts in $\mathcal{V}$ which are uniquely p-colored. Since there are exactly $n/2+1$ vertices of color p, each vertex of color p is contained in a distinct part in $\mathcal{V}$. In particular, this means that $\{\ell\} \in \mathcal{V}$ for each leaf $\ell \in L$.

Let $V_z \in \mathcal{V}$ denote the subset containing the center z, and let n_q and n_r denote the number of vertices of color q and r in V_z, respectively. As each vertex of color q or r has weight at least $M - N \cdot 2^n$, we have $w_q(V_z) \geq (M - N \cdot 2^n) \cdot n_q$ and $w_r(V_z) \geq (M - N \cdot 2^n) \cdot n_r$. Since V_z is uniquely p-colored, we have

$$\max\{w_q(V_z), w_r(V_z)\} < w_p(V_z) = w(z) = Mn + s/2 + 1 \text{ and}$$
$$\max\{n_q, n_r\} < \frac{Mn + s/2 + 1}{M - N \cdot 2^n} = n + \frac{N \cdot 2^n n + s/2 + 1}{M - N \cdot 2^n} < n + 1.$$

Here, the last inequality follows since $M > N \cdot 2^n(n+1) + s/2 + 1$. Thus, it holds that $|V_z|$ contains at most $n_q + n_r + 1 \leq 2n + 1$ vertices.

Let $\mathcal{V}' := \mathcal{V} \setminus (\{V_z\} \cup \{\{\ell\} \mid \ell \in L\})$ be the collection of parts of $\mathcal{V}$ not containing any vertices of color p. Notice that $|\mathcal{V}| = k = 3n/2 + 1$ and that $|\mathcal{V}'| = |\mathcal{V}| - (n/2+1) = n$. Now, consider some $V' \in \mathcal{V}'$. We have $V' \subseteq X_i$ or $V' \subseteq Y_i$ for some $i \in [n]$ by construction. Since $|X_i| = |Y_i| = 2$ for all $i \in [n]$, we have $|V'| \leq 2$ and thus $\left|\bigcup_{V' \in \mathcal{V}'} V'\right| \leq 2n$. Moreover, since there are $n/2 + 1 + 4n = 9n/2 + 1$ vertices in G, we have $\left|\bigcup_{V' \in \mathcal{V}'} V'\right| = |V| - |V_z| - |L| \geq 2n$. Hence, $\left|\bigcup_{V' \in \mathcal{V}'} V'\right| = 2n$ and thus, for each part $V' \in \mathcal{V}'$, it holds that $|V'| = 2$ yielding $V' = X_i$ or $V' = Y_i$ for some $i \in [n]$. Let $I_x := \{i \in [n] \mid X_i \in \mathcal{V}'\}$ and $I_y := \{i \in [n] \mid Y_i \in \mathcal{V}'\}$. Since all X_i are q-colored and all Y_i are r-colored, we have $|I_x| \leq n/2$ and $|I_y| \leq n/2$. Then, since $|I_x| + |I_y| = |\mathcal{V}'| = n$, we obtain $|I_x| = |I_y| = n/2$.

Let $J_x := [n] \setminus I_x$ and $J_y := [n] \setminus I_y$. The total weights of vertices of color q and r in V_z are

$$w_q(V_z) = \sum_{j \in J_x} w(x_j^q) + \sum_{j \in J_y} w(y_j^q) = Mn + \sum_{j \in J_x} a_j + N\left(\sum_{j \in J_x} 2^i - \sum_{j \in J_y} 2^j\right) \text{ and} \tag{1}$$

$$w_r(V_z) = \sum_{j \in J_x} w(x_j^r) + \sum_{j \in J_y} w(y_j^r) = Mn + s - \sum_{j \in J_y} a_j + N\left(\sum_{j \in J_y} 2^j - \sum_{j \in J_x} 2^j\right), \tag{2}$$

respectively. Now, assume for the sake of contradiction that $J_x \neq J_y$. Then, there exists an index $j_{\max} := \max\{(J_x \setminus J_y) \cup (J_y \setminus J_x)\}$. If $j_{\max} \in J_x$, then each element in $J_y \setminus J_x$ is smaller than $j_{\max}$, and hence

$$\sum_{j \in J_x} 2^j - \sum_{j \in J_y} 2^j = \sum_{j \in J_x \setminus J_y} 2^j - \sum_{j \in J_y \setminus J_x} 2^i \geq 2^{j_{\max}} - \sum_{j \in [j_{\max}-1]} 2^j = 2. \tag{3}$$

Combining Inequalities 1 and 3 yields

$$w_q(V_z) \geq Mn + \sum_{j \in J_x} a_j + 2N \geq Mn + 2N > w_p(V_z),$$

which is a contradiction to V_z being uniquely p-colored. We analogously obtain a contradiction for $j_{\max} \in J_y$ and thus it holds that $J_x = J_y$. Observe that for $J := J_x = J_y$ Inequality 1 implies $w_q(V_z) = Mn + \sum_{j \in J} a_j$ and Inequality 2 implies $w_r(V_z) = Mn + s - \sum_{j \in J} a_j$.

Since $w_q(V_z) < w_p(V_z)$ and $w_r(V_z) < w_p(V_z)$, we obtain

$$w_q(V_z) = w_r(V_z) = Mn + s/2$$

and thus $\sum_{j \in J} a_j = s/2$. Consequently, J is a solution to the original instance of PARTITION. □

We continue with a complexity analysis for the case $|C| = 2$. Note that GERRYMANDERING OVER GRAPHS on trees is pseudo-polynomial-time solvable for any constant $|C|$ (and thereby for $|C| = 2$) [10]. To complement this result and also Theorem 2, we next show that for $|C| = 2$ there is a polynomial-time algorithm for trees, adapting a pseudo-polynomial-time algorithm of Ito et al. [10, Theorem 7]. We thus obtain a dichotomy with respect to $|C|$. As in Ito et al. [10, Theorem 7], we employ a dynamic programming approach. More specifically, we store the maximum winning margin of the target color p over the other color in the district containing v for each vertex v, each possible number $k' \leq k$ of districts in the subtree rooted in v, and each possible number $\ell' \leq k'$ of districts won by p in this subtree. The key difference from the algorithm of Ito et al. [10] is that we only store the *maximum* winning margin. Due to lack of space, the proof of this and the next proposition are deferred to the full version [2].

Proposition 1. *For $|C| = 2$,* GERRYMANDERING OVER GRAPHS *restricted to trees can be solved in $O(n^3)$ time.*

Finally, we bridge the gap for trees of fixed diameter by generalizing the known polynomial-time algorithm for trees of diameter two [10] to trees of diameter three. NP-hardness on trees of diameter four was shown by Ito et al. [10].

Proposition 2. *For diameter-three trees,* GERRYMANDERING OVER GRAPHS *can be solved in $O(|C|^2 \cdot n^5)$ time.*

The key observation is that a tree of diameter three can be obtained from two stars by adding an edge between their centers. Our algorithm then adapts a polynomial-time algorithm for stars [10]. Proposition 2 yields a complexity dichotomy for trees with respect to the diameter parameter. Clearly, our polynomial-time solvability is mainly of classification nature; it remains a future task to lower the degree in the polynomial of the running time.

4 Conclusion

Answering open questions of Ito et al. [10] and Gupta et al. [9] in the negative, we presented an NP-hardness result on paths and a weak NP-hardness result on trees. Now, one may claim that the computational complexity of GERRYMANDERING OVER GRAPHS restricted to paths and trees is well-understood. The results indicate that, through the lens of worst-case complexity analysis, GERRYMANDERING OVER GRAPHS is extremely hard. Indeed, from our and previous findings, one can also deduce negative results in terms of parameterized complexity analysis, that is NP-hardness for constant values of each (single) of the following graph parameters: vertex cover number, maximum leaf number, and vertex deletion number to cliques. In parameterized complexity theory, these are among the "weakest" parameters.

As previous work, we focused on the Plurality voting rule, leaving open to study GERRYMANDERING OVER GRAPHS also for other voting rules. Moreover, we focused on theoretical results. Since worst-case intractability is clearly no shield against susceptibility of real-world instances to gerrymandering, following the example of Cohen-Zemach et al. [5] it may be promising to investigate empirical issues.

References

1. Apollonio, N., Becker, R.I., Lari, I., Ricca, F., Simeone, B.: Bicolored graph partitioning, or: gerrymandering at its worst. Discret. Appl. Math. **157**(17), 3601–3614 (2009)
2. Bentert, M., Koana, T., Niedermeier, R.: The complexity of gerrymandering over graphs: paths and trees. CoRR abs/2102.08905 (2021)
3. Borodin, A., Lev, O., Shah, N., Strangway, T.: Big city vs. the great outdoors: voter distribution and how it affects gerrymandering. In: Proceedings of the 27th International Joint Conference on Artificial Intelligence (IJCAI 2018), pp. 98–104. ijcai.org (2018)
4. Chatterjee, T., DasGupta, B., Palmieri, L., Al-Qurashi, Z., Sidiropoulos, A.: On theoretical and empirical algorithmic analysis of the efficiency gap measure in partisan gerrymandering. J. Comb. Optim. **40**(2), 512–546 (2020). https://doi.org/10.1007/s10878-020-00589-x
5. Cohen-Zemach, A., Lewenberg, Y., Rosenschein, J.S.: Gerrymandering over graphs. In: Proceedings of the 17th International Conference on Autonomous Agents and MultiAgent Systems (AAMAS 2018), pp. 274–282. International Foundation for Autonomous Agents and Multiagent Systems (2018)
6. Eiben, E., Fomin, F.V., Panolan, F., Simonov, K.: Manipulating districts to win elections: fine-grained complexity. In: Proceedings of the 34th AAAI Conference on Artificial Intelligence (AAAI 2020), pp. 1902–1909. AAAI Press (2020)
7. Garey, M.R., Johnson, D.S.: Computers and Intractability: A Guide to the Theory of NP-Completeness. W. H. Freeman, San Francisco (1979)
8. Guest, O., Kanayet, F.J., Love, B.C.: Gerrymandering and computational redistricting. J. Comput. Soc. Sci. **2**(2), 119–131 (2019). https://doi.org/10.1007/s42001-019-00053-9
9. Gupta, S., Jain, P., Panolan, F., Roy, S., Saurabh, S.: Gerrymandering on graphs: computational complexity and parameterized algorithms. CoRR abs/2102.09889 (2021). 14th International Symposium on Algorithmic Game Theory, SAGT (accepted)
10. Ito, T., Kamiyama, N., Kobayashi, Y., Okamoto, Y.: Algorithms for gerrymandering over graphs. Theoret. Comput. Sci. **868**, 30–45 (2021)
11. Lewenberg, Y., Lev, O., Rosenschein, J.S.: Divide and conquer: using geographic manipulation to win district-based elections. In: Proceedings of the 16th International Conference on Autonomous Agents and Multiagent Systems (AAMAS 2017), pp. 624–632. ACM (2017)
12. Mathieson, L., Szeider, S.: Editing graphs to satisfy degree constraints: a parameterized approach. J. Comput. Syst. Sci. **78**(1), 179–191 (2012)
13. Schuck, P.H.: The thickest thicket: partisan gerrymandering and judical regulation of politics. Columbia Law Rev. **87**(7), 1325–1384 (1987)
14. Tasnádi, A.: The political districting problem: a survey. Soc. Econ. **33**(3), 543–554 (2011)

Feedback Vertex Set on Hamiltonian Graphs

Dario Cavallaro and Till Fluschnik(✉)

Faculty IV, Algorithmics and Computational Complexity,
Technische Universität Berlin, Berlin, Germany
cavallaro@campus.tu-berlin.de, till.fluschnik@tu-berlin.de

Abstract. We study the computational complexity of FEEDBACK VERTEX SET on subclasses of Hamiltonian graphs. In particular, we consider Hamiltonian graphs that are regular or are planar and regular. Moreover, we study the less known class of p-Hamiltonian-ordered graphs, which are graphs that admit for any p-tuple of vertices a Hamiltonian cycle visiting them in the order given by the tuple. We prove that FEEDBACK VERTEX SET remains NP-hard in these restricted cases, even if a Hamiltonian cycle is additionally given as part of the input.

Keywords: Planar graphs · Regular graphs · Connected graphs · Ordered graphs · Hamiltonian-ordered graphs

1 Introduction

Hamiltonian graphs are graphs admitting a cycle that visits every vertex (exactly once). We study the computational complexity of the following classic NP-complete [9] problem on subclasses of Hamiltonian graphs.

Problem 1. FEEDBACK VERTEX SET (FVS)
Input: An undirected graph $G = (V, E)$ and an integer $k \in \mathbb{N}_0$.
Question: Is there $U \subseteq V$ with $|U| \leq k$ such that $G - U$ is acyclic?

We additionally restrict Hamiltonian graphs to be planar (can be drawn on the two-dimensional plane with no two edges crossing except at their endpoints) or regular (every vertex has the same degree). In particular, we study the classes of 4-regular planar Hamiltonian graphs and of 5-regular planar Hamiltonian graphs (recall that there is no 6-regular planar graph). Moreover, we consider the class of p-Hamiltonian-ordered graphs [13]. These Hamiltonian graphs admit for each p-tuple $(x_1, \dots, x_p)$ of vertices a Hamiltonian cycle that visits the vertices $x_1, \dots, x_p$ in this order. The class of p-Hamiltonian-ordered graphs form a subclass of p-ordered Hamiltonian graphs, the latter being Hamiltonian graphs that for any p-tuple $(x_1, \dots, x_p)$ admit a cycle that visits $x_1, \dots, x_p$ in this order. The class of p-ordered Hamiltonian graphs form a subclass of $(p-1)$-connected

T. Fluschnik—Supported by DFG, project TORE (NI 369/18).

Ł. Kowalik et al. (Eds.): WG 2021, LNCS 12911, pp. 207–218, 2021.
https://doi.org/10.1007/978-3-030-86838-3_16

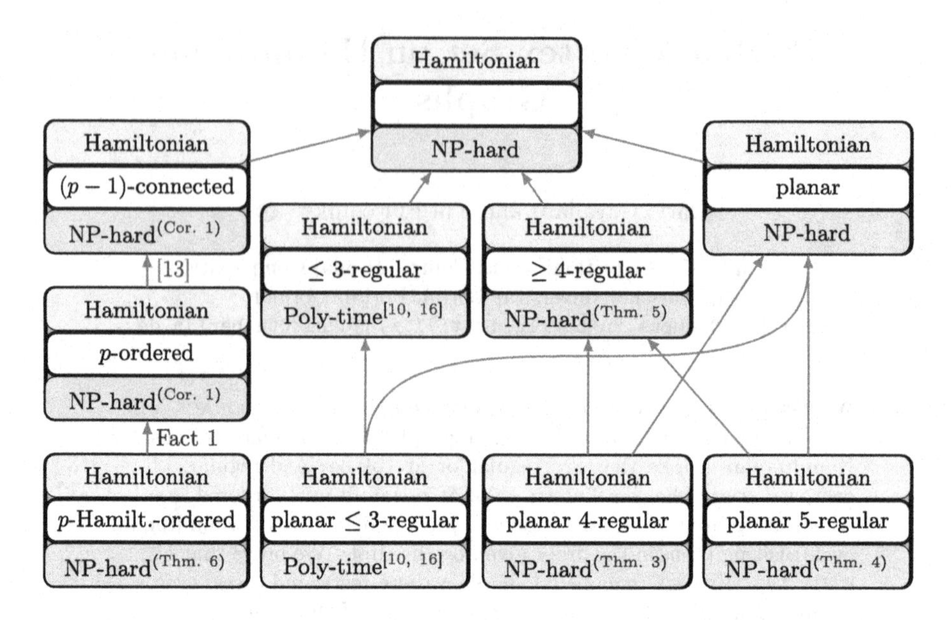

Fig. 1. Overview of our results. In each box, the lowest level describes the computational complexity (NP-hard versus polynomial-time) of FEEDBACK VERTEX SET on the graph class described by the two upper layers. An arrow from a box A to a box B describes that A's graph class is included in B's graph class. All shown NP-hardness results hold true even if a Hamiltonian cycle is provided as part of the input.

Hamiltonian graphs (graphs in which every pair of vertices is connected via $(p-1)$ internally vertex-disjoint paths). Finally, for FEEDBACK VERTEX SET on these subclasses of Hamiltonian graphs, we also study the more restricted case when a Hamiltonian cycle is additionally provided in the input (recall that computing a Hamiltonian cycle is NP-complete in general [9]).

Related Work. INDEPENDENT SET remains NP-complete on 3- and 4-regular Hamiltonian graphs [7], which enabled to prove NP-hardness for a temporal graph problem with two layers [8]. 3-COLORING remains NP-complete on 4- and 5-regular planar graphs [3], and on 4-regular Hamiltonian graphs [6]. FEEDBACK VERTEX SET remains NP-complete on planar graphs of maximum degree four [15] and on line graphs of planar cubic bipartite graphs [12], and is polynomial-time solvable on maximum degree-three [16] and 3-regular graphs [10], chordal graphs, permutation graphs, split graphs [5].

Our Contributions. Figure 1 gives an overview of our results. We prove that FEEDBACK VERTEX SET is NP-hard on 4- and 5-regular planar Hamiltonian graphs as well as on p-regular Hamiltonian graphs for every $p \geq 4$. Moreover, we prove that FEEDBACK VERTEX SET is NP-hard on p-Hamiltonian-ordered graphs for every $p \geq 3$, which implies NP-hardness on p-ordered Hamiltonian graphs and further NP-hardness on $(p-1)$-connected Hamiltonian graphs. Finally, all our NP-hardness results still hold true if a Hamiltonian cycle is additionally provided as part of the input.

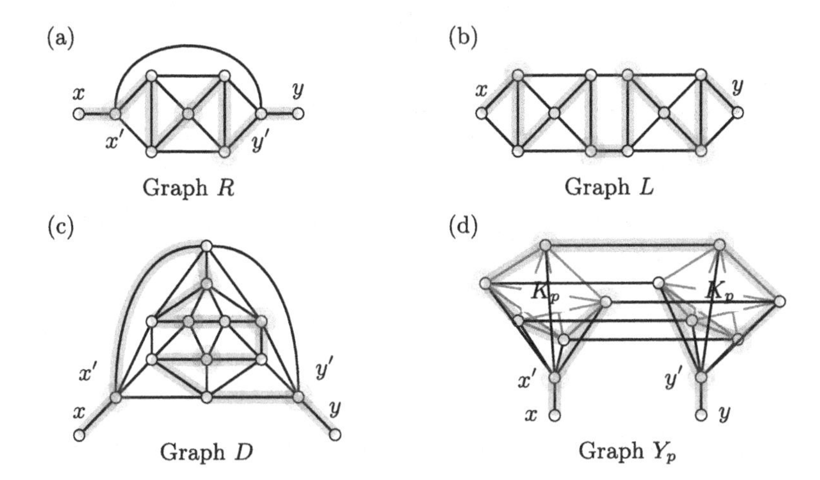

Fig. 2. Our graph tool box with (a) the graph R, (b) the graph L, (c) the graph D, and (d) the graph Y_p. For each graph, a Hamiltonian path (blue) as well as a minimum feedback vertex set (orange) are depicted. (Color figure online)

2 Preliminaries

We denote by $\mathbb{N}$ and $\mathbb{N}_0$ the natural numbers excluding and including zero, respectively. We use basic notations from graph theory [1,4]. Details and proofs (marked with ★) are deferred to a full version.

Graph Theory. For two graphs G, H, we denote by $G*H$ the graph with vertex set $V(G) \cup V(H)$ and edge set $E(G) \cup E(H) \cup \{\{v, w\} \mid v \in V(G), w \in V(H)\}$. We denote by K_n the complete graph on $n \in \mathbb{N}$ vertices. We denote by C_n the cycle on $n \in \mathbb{N}$ vertices. The neighborhood $N_G(v)$ of a vertex $v \in V$ in G is the vertex set $\{w \in V \mid \{v, w\} \in E\}$. Let v, w be two distinct vertices in $G = (V, E)$. The graph obtained by *identifying* v *with* w has vertex set $(V \setminus \{v, w\}) \cup \{vw\}$, where vw is a new vertex, and edge set $(E \setminus \{e \in E \mid \{v, w\} \cap e \neq \emptyset\}) \cup \{\{vw, x\} \mid x \in (N_G(v) \cup N_G(w)) \setminus \{v, w\}\}$.

Hamiltonian Graphs and Subclasses. A graph G is p-ordered if for every sequence $v_1, \ldots, v_p$ of distinct vertices of G there exists a cycle C in G that encounters the vertices $v_1, \ldots, v_p$ in this order. A graph G is called p-Hamiltonian-ordered if for every sequence $v_1, \ldots, v_p$ of distinct vertices of G there exists a Hamiltonian cycle C that encounters the vertices $v_1, \ldots, v_p$ in this order. Clearly:

Fact 1. Every p-Hamiltonian-ordered graph is p-ordered Hamiltonian.

Graph Tool Box. We use several graphs as gadgets for our NP-hardness reductions that we collect in this "graph tool box" (see Fig. 2).

The Graph R (see Fig. 2(a)): Let R denote the graph obtained from a $C_4 * K_1$ by adding two vertices x' and y' and making each adjacent with exactly two neighboring vertices of degree three such that all vertices except for x' and y' have degree four. Add vertex x and make it adjacent with x', and add vertex y and make it adjacent with y'. Finally, make x' adjacent with y'.

We have the following simple yet useful observation on a $C_4 * K_1$:

Observation 1. *The graph $C_4 * K_1$ admits no feedback vertex set of size one yet one of size two.*

The Graph L (see Fig. 2*(b)):* Let L denote the graph obtained as follows. Take two disjoint $C_4 * K_1$s. Let $\{v, w\}$ be an edge of one $C_4 * K_1$ with both v, w being of degree three, and $\{v', w'\}$ analogously from the other $C_4 * K_1$. Make v adjacent with v' and w adjacent with w'. Let $\{x', x''\}$ and $\{y', y''\}$ denote the two edges with vertices of degree three. Add a vertex x and make it adjacent with x', x'', and add a vertex y and make it adjacent with y', y''.

The Graph D (see Fig. 2*(c)):* Let D denote the graph obtained as follows. Take a C_6, say with vertex set $\{c_0, \ldots, c_5\}$ and edge set $\{\{c_i, c_{i+1 \bmod 6}\} \mid i \in \{0, \ldots, 5\}\}$. Add a K_3, say with vertex set $\{v_1, v_2, v_3\}$. Make v_1 adjacent with c_0, c_1, c_5, v_2 adjacent with c_1, c_2, c_3, and v_3 adjacent with c_3, c_4, c_5. Add a vertex z and make it adjacent with c_2, c_3, c_4. Add vertices x and x', and make x' adjacent with x, z, c_0, c_1, c_2. Finally, add vertices y and y', and make y' adjacent with y, z, c_0, c_4, c_5.

The Graphs Y_p (see Fig. 2*(d)):* Let $p \in \mathbb{N}$ with $p \geq 3$. Let Y_p denote the graph obtained as follows. Take two disjoint $A := K_p$ and $B := K_p$. Add a perfect matching between the vertices of A and B. Next, add two vertices x, x' and make x' adjacent to all vertices in $V(A) \cup \{x\}$. Finally, add two vertices y, y' and make y' adjacent to all vertices in $V(B) \cup \{y\}$.

Definition 1 (Insertion). *Let G be a graph and $u, v \in V(G)$. An H-insertion at u, v with $H \in \{R, L, D\} \cup \bigcup_{p \geq 3}\{Y_p\}$ results in the graph obtained from G by adding a copy of H to G and identifying x with u and y with v.*

3 Planar Regular Hamiltonian Graphs

In this section, we prove that FEEDBACK VERTEX SET remains NP-hard on 4-regular planar Hamiltonian graphs and on 5-regular planar Hamiltonian graphs, in both cases even if a Hamiltonian cycle is provided. We first prove that FVS is NP-hard on 4-regular planar graphs (Sect. 3.1), then make the graph Hamiltonian (Sect. 3.2), and finally make the graph 5-regular (Sect. 3.3).

3.1 4-Regular Planar

FEEDBACK VERTEX SET is NP-hard even on connected planar graphs of maximum degree four [14,15]. We strengthen this with the following.

Theorem 1 (★). FEEDBACK VERTEX SET *is* NP-*hard on connected 4-regular planar graphs.*

We point out that a result of Munaro [12] implies that FVS remains NP-hard on 4-regular planar graphs. We give an alternative proof.

We can delete degree-zero and -one vertices from a graph, and obtain an equivalent instance. Next, we deal first with degree-two vertices to prove that FVS is NP-hard on planar graphs of minimum degree three and maximum degree four (Proposition 1), and then deal with the remaining vertices of degree three.

Degree-Two Vertices. We start by making each degree-two vertex a degree-four vertex to obtain the following.

Proposition 1 (★). FEEDBACK VERTEX SET *is* NP-*hard on connected planar graphs of minimum degree three and maximum degree four.*

To prove Proposition 1, we will perform an R-insertion at v, v on each degree-two vertex v. We have the following crucial observation on R.

Lemma 1 (★). *Graph R is planar, admits a Hamiltonian x-y path, and has no feedback vertex set of size at most two, yet one of size three containing x' or y', but none of size three containing x or y.*

An immediate consequence of Lemma 1 is the following.

Observation 2. *Let $\mathcal{I} = (G, k)$ be an instance of* FEEDBACK VERTEX SET *and let $u, v \in V(G)$. Let G' be the graph obtained from an R-insertion at u, v and let $k' := k + 3$. Then $\mathcal{I}$ is a* `yes`*-instance if and only if (G', k') is a* `yes`*-instance of* FEEDBACK VERTEX SET.

We are set to prove Proposition 1 (★).

Degree-Three Vertices. Next, we deal with degree-three vertices. We will employ the following specific straight-line embedding of our graph on a grid.

Theorem 2 ([2]). *For any planar graph with n vertices one can compute in $\mathcal{O}(n)$ time a straight-line embedding on the $(2n - 4)$ by $(n - 2)$ grid.*

We start with an embedding. Let $p(v) = (i, j) \in \{1, \ldots, 2n - 4\} \times \{1, \ldots, n - 2\}$ be the coordinate of vertex v in the grid-embedding. We aim for connecting the remaining degree-three vertices in a pairwise manner (note that there is an even number of these). To this end, we construct chains of Rs connecting two degree-three vertices. To ensure polynomial running time and planarity of the construction, we need to identify the pairs of degree-three vertices which we want to connect such that the "R-chains" are pairwise non-crossing and vertex-disjoint. To this end, we apply a "left-to-right bottom-to-top" approach as follows (see Fig. 3 for an illustration). We iterate over vertices from left to right

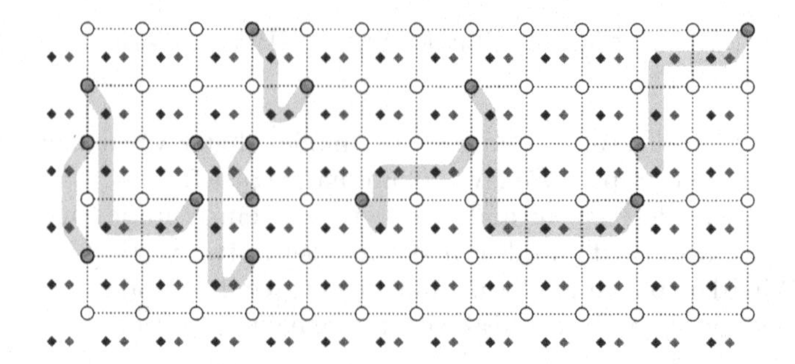

Fig. 3. Illustration to how we connect pairs of degree-three vertices. Round vertices correspond to the vertices in our graph, where filled round vertices are the vertices we want to connect. Diamond-shaped vertices correspond to the points in the grid shifted by $1/3 \pm \varepsilon$ horizontally and $1/2$ vertically. Thick lines depict the pairwise connections.

by coordinates, that is, by $(i,\cdot)$ for increasing i. Thereby, for each i, we iterate over (i,j) with increasing j. Once two vertices of degree three are discovered, we connect them in a "down-first out-most"-manner with an "R-chain". As we thereby possibly introduce edge crossings, we need to *dissolve* them as follows.

Definition 2. *Let G be a graph embedded in the two-dimensional plane such that at most two edges cross at one point. Let e_1, e_2 be two edges crossing at point (i,j).* Dissolving *the crossing is doing the following: subdivide edge e_1 (denote the vertex v_1) and edge e_2 (denote the vertex v_2), identify v_1 with v_2 (denote the vertex v), and embed v at point (i,j).*

The way we dissolve crossings immediately gives the following.

Observation 3. *Every vertex resulting from a dissolution has degree four and dissolving all edge-crossings yields a planar graph.*

We will add and embed edges between disjoint pairs of degree-three vertices, dissolve each newly formed crossing, and replace edges introduced by the dissolution by R-insertions on its endpoints. Formally:

Definition 3. *Let $0 < \varepsilon < 1/3$. R-connecting vertex v with v', where $p(v) = (i,j)$ and $p(v') = (i',j')$, is doing the following:*

1. *Add and embed a new edge $f = \{v,v'\}$ as follows:*
   if $i = i'$: *It goes from (i,j) to $(i - \frac{1}{3} - \varepsilon, j + \frac{1}{2})$ to $(i' - \frac{1}{3} - \varepsilon, j' - \frac{1}{2})$ and finally to (i',j').*
   if $i \neq i'$: *It goes from (i,j) to $(i + \frac{1}{3} - \varepsilon, j - \frac{1}{2})$ to $(i + \frac{1}{3} - \varepsilon, j' - \frac{1}{2})$ to $(i' - \frac{1}{3} + \varepsilon, j' - \frac{1}{2})$ and finally to (i',j').*
2. *Dissolve every crossing. and let $f_1, \ldots, f_\ell$ denote the edges in which edge f is dissolved.*
3. *Replace each edge f_i, $i \in \{1,\ldots,\ell\}$, with an R-insertion on its endpoints.*

Algorithm 1: Computing an equivalent instance (G', k') with G' being 4-regular planar from (G, k) with G being of minimum degree three and maximum degree four embedded with straight-lines on the $(2n-4 \times n-2)$-grid, where n denotes the number of vertices of G.

```
1  d ← 0; a ← ∅; G' ← G;
2  for x from 1 to 2n − 4 do                                   // x-coordinates
3      for y from 1 to n − 2 do                                // y-coordinates
4          if p^{-1}(x, y) = v is a degree-three vertex then
5              if a = ∅ then
6                  a ← v;
7              else
8                  R-connect a with v in G' (denote the obtained graph again
                   by G') and let d' denote the number of R-insertions;
9                  d ← d + d'; a ← ∅;
10 return (G', k + 3d)
```

As a technical remark, we choose ε in Definition 3 such that no existing slope is resampled. Note that in our embedding (Theorem 2) the number of slopes is finite. Thus, we can R-connect any two vertices in polynomial time.

We employ Algorithm 1 to construct our instance (G', k') (see Fig. 4 for an illustration). The following invariant immediately holds for Algorithm 1 by our "left-to-right bottom-to-top" approach in an embedding given by Theorem 2.

Observation 4. *When Algorithm 1 detects two degree-three vertices v (first) and v' (second) with $p(v) = (i, j)$ and $p(v') = (i', j')$ to connect, then there is no degree-three vertex w, $p(w) = (x, y)$, with*

if $i = i'$**:** *$x = i$ and $j < y < j'$;*
if $i \neq i'$**:** *(i) $i < x < i'$, (ii) $x = i$ and $y > j$, or (iii) $x = i'$ and $y < j'$.*

It follows that every two R-connections will be non-crossing and vertex-disjoint.

Proposition 2 (★). *Let $\mathcal{I} = (G, k)$ be an instance of* FEEDBACK VERTEX SET *with G being planar and of minimum degree three and maximum degree four. Then one can compute an equivalent instance $\mathcal{I}' = (G', k')$ with $k' \in \mathcal{O}(|V(G)|^2)$ and G' having $\mathcal{O}(|V(G)|^2)$ vertices and edges, and being 4-regular planar.*

We are set to prove Theorem 1 (★).

3.2 4-Regular Planar Hamiltonian

In Sect. 3.1, we proved that FEEDBACK VERTEX SET is NP-hard on 4-regular planar graphs. We next give a polynomial-time many-one reduction to an equivalent instance with a 4-regular planar Hamiltonian graph.

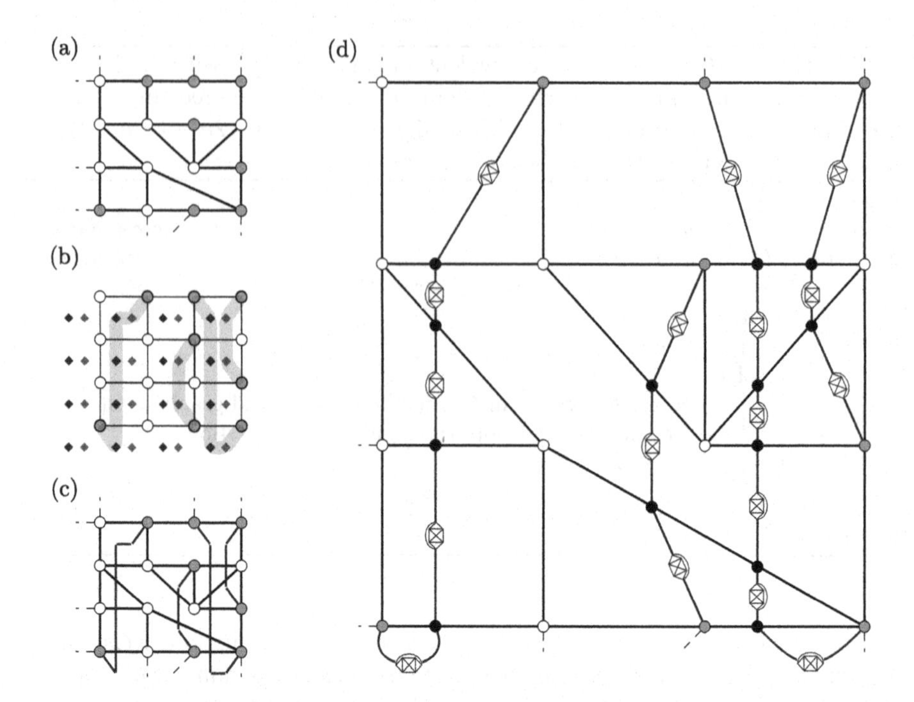

Fig. 4. Illustration to the proof of Proposition 2. (a) An example graph G (excerpt) embedded on a grid where filled vertices correspond to vertices to connect (vertices of degree three). (b) The connecting paths constructed in the grid. (c) Embedding the edges in G. (d) Dissolving the edge-crossings and replacing edges by R-insertions.

Theorem 3 (★). Feedback Vertex Set *on 4-regular planar Hamiltonian graphs is* NP*-hard, even if a Hamiltonian cycle is provided.*

We will follow the idea of Fleischner and Sabidussi [6]: We first compute a *2-factor*, that is, a spanning 2-regular subgraph, in polynomial time, and then iteratively connect cycles from the 2-factor by L-insertions to obtain a Hamiltonian cycle. Note that L contains two vertex-disjoint $C_4 * K_1$s, hence admits no feedback vertex set of size three and no feedback vertex set of size four containing x or y (Observation 1). Yet, L admits feedback vertex sets each of size four disconnecting x and y (see Fig. 2).

Lemma 2. *Graph L is planar, admits a Hamiltonian x-y path, has no feedback vertex set of size three, but one of size four disconnecting x and y, and none of size four containing one of x or y.*

Lemma 2 immediately implies the following.

Observation 5. *Let $\mathcal{I} = (G, k)$ be an instance of* Feedback Vertex Set *and let $u, v \in V(G)$. Let G' be the graph obtained from G by an L-insertion at u, v and let $k' := k + 4$. Then $\mathcal{I}$ is a* **yes***-instance if and only if (G', k') is a* **yes***-instance of* Feedback Vertex Set.

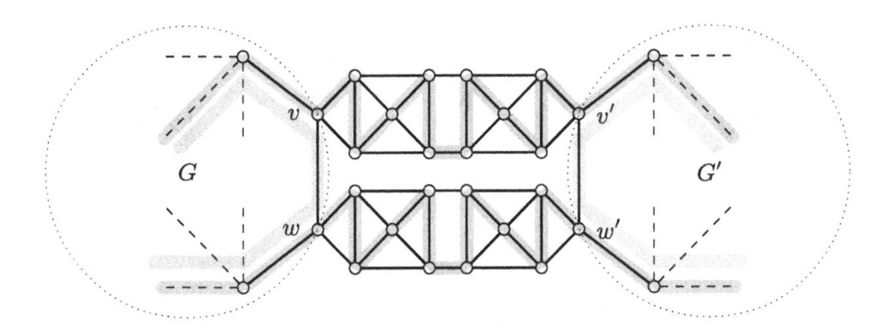

Fig. 5. Illustration to the proof of Lemma 3. Hamiltonian cycles C and C' of G (after doubly subdividing an edge to obtain v, w) and G' (after doubly subdividing an edge to obtain v', w') are depicted on the left and right in magenta and cyan, respectively. Hamiltonian cycle C^* obtained after L-insertions at v, v' and w, w' is depicted in blue. (Color figure online)

We represent a 2-factor by its components $Q = \{Q_1, \ldots, Q_q\}$, where Q_i is a cycle for every $i \in \{1, \ldots, q\}$. Every 4-regular graph admits a 2-factor computable in polynomial time [6,11]. We first compute a 2-factor $Q = \{Q_1, \ldots, Q_q\}$ of G, and then iteratively make L-insertions at vertices obtained from subdivisions to merge cycles from Q until $|Q| = 1$. We defer the details and rest of the proof to a full version (★).

3.3 5-Regular Planar Hamiltonian

In this section, we prove that Feedback Vertex Set is also NP-hard on 5-regular planar Hamiltonian graphs with provided Hamiltonian cycle. To this end, we start from a 4-regular planar Hamiltonian graph and then make it 5-regular by a D-insertion at every second edge of a Hamiltonian cycle. Hence, we need to ensure the 4-regular Hamiltonian graph to have an even number of vertices. To this end, we take two disjoint copies of the input graph and connect them via two L-insertions at vertices obtained from subdivisions (see Fig. 5 for an illustration).

Lemma 3 (★). Feedback Vertex Set *is* NP*-hard on 4-regular planar Hamiltonian graphs with an even number of vertices, even if a Hamiltonian cycle is provided.*

Using Lemma 3, we will prove next the following main result of this section.

Theorem 4 (★). Feedback Vertex Set *is* NP*-hard on 5-regular planar Hamiltonian graphs even if a Hamiltonian cycle is given.*

We will perform a series of D-insertions (see Fig. 6 for an illustration). Hence, we discuss the following in advance.

Lemma 4 (★). *Graph D is planar, admits a Hamiltonian x-y path, and has no feedback vertex set of size at most five yet one of size six containing x' or y' but none of size six containing x or y.*

We are set to prove Theorem 4 (★).

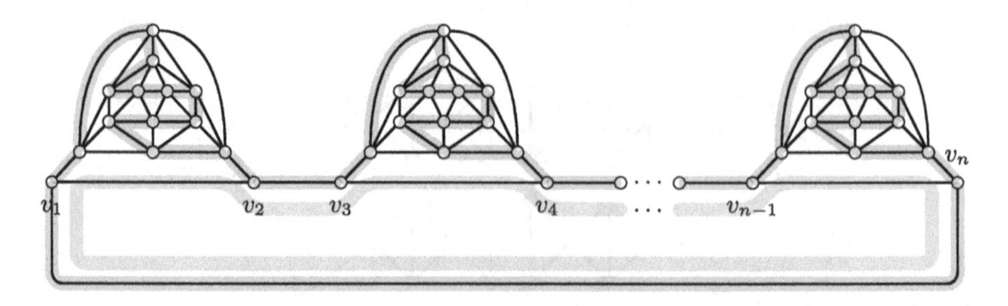

Fig. 6. Illustration to the proof of Theorem 4. The magenta path depicts the Hamiltonian cycle before the D-insertions, and the blue path depicts the Hamiltonian cycle after the D-insertions. (Color figure online)

4 Regular Hamiltonian Graphs

In this section, we prove that FEEDBACK VERTEX SET remains NP-hard on p-regular Hamiltonian graphs for *every* $p \geq 4$.

Theorem 5 (★). *For every $p \geq 4$,* FEEDBACK VERTEX SET *on p-regular Hamiltonian graphs is* NP*-hard, even if a Hamiltonian cycle is provided.*

We have seen that FEEDBACK VERTEX SET is NP-hard on 5-regular graphs, even if a Hamiltonian cycle C is provided. Every 5-regular graph has an even number of vertices. Thus, we find a perfect matching M on C. We will make a Y_p-insertion on every pair of vertices from M.

Lemma 5 (★). *Graph Y_p admits a Hamiltonian path with endpoints x and y and admits no feedback vertex set of size at most $2(p+1)-5$ yet one of size $2(p+1)-4$ containing x' or y'.*

From Lemma 5 we immediately get the following.

Observation 6. *Let $\mathcal{I} = (G, k)$ be an instance of* FEEDBACK VERTEX SET *and let $v, w \in V(G)$ be two distinct vertices. Let G' be the graph obtained from G by the Y_p-insertion at v, w, and let $k' := k + 2(p+1) - 4$. Then, $\mathcal{I}$ is a* `yes`*-instance if and only if (G', k') is a* `yes`*-instance of* FEEDBACK VERTEX SET.

We are set to prove Theorem 5 (★).

5 Ordered and Connected Hamiltonian Graphs

For every cycle, and for every three vertices on it, we can shift the start of the cycle and its orientation to encounter the three vertices in any order. Thus, we get the following.

Observation 7. *Every Hamiltonian graph is 3-Hamiltonian-ordered.*

We start from here and prove inductively the following.

Theorem 6 (★). *For every $p \geq 3$,* Feedback Vertex Set *on p-Hamiltonian-ordered graphs is* NP*-hard, even if a Hamiltonian cycle is provided.*

Construction 1. Let $\mathcal{I} = (G, k, C)$ with $G = (V, E)$ be an input instance of Feedback Vertex Set with Hamiltonian cycle C of G. Let $n := |V|$. We construct an instance $\mathcal{I}' := (G', k', C')$ with $k' := 3n+k$ as follows. Let $H := K_{3n}$. Construct $G' := G * H$. Moreover, add two new vertices x, y to G' and make x adjacent to all vertices in $V(H) \cup \{y\}$ and y adjacent to all vertices in $V(H) \cup \{x\}$. Extend C through $V(H) \cup \{x, y\}$ to obtain C'. $\diamond$

Observation 8 (★). *Let instance $\mathcal{I}' = (G', k', C')$ be obtained from an input instance $\mathcal{I} = (G, k, C)$ of* Feedback Vertex Set *using Construction 1. If G is p-Hamiltonian-ordered for $3 \leq p \leq n$, then G' is $(p+1)$-Hamiltonian-ordered.*

Our proof of Observation 8 employs the following.

Fact 2 ([13])**.** Let $G = (V, E)$ be a graph with $|V| \geq 3$ and let $p \in \{3, \ldots, n\}$. If $\deg(v) + \deg(w) \geq |V| + 2p - 6$ for every non-adjacent $v, w \in V$, then G is p-Hamiltonian-ordered.

Lemma 6 (★). *Let $\mathcal{I}' = (G', k', C')$ be the instance obtained from an input instance $\mathcal{I} = (G, k, C)$ of* Feedback Vertex Set *using Construction 1. Then, $\mathcal{I}$ is a* `yes`*-instance if and only if $\mathcal{I}'$ is a* `yes`*-instance.*

We are set to prove Theorem 6 (★).

Recall that every p-Hamiltonian-ordered graph is trivially also p-ordered Hamiltonian. In addition, the following holds true.

Fact 3 ([13])**.** If a graph is p-ordered for any $p \geq 3$, then it is $(p-1)$-connected.

Hence, we get the following.

Corollary 1. Feedback Vertex Set *is* NP*-hard on p-ordered and $(p-1)$-connected Hamiltonian graphs, $p \geq 3$, even if a Hamiltonian cycle is provided.*

6 Conclusion

Feedback Vertex Set remains NP-hard on quite restricted cases even if the graph is additionally Hamiltonian and a Hamiltonian cycle is provided. Which problems are NP-hard on Hamiltonian graphs and become non-trivially tractable if a Hamiltonian cycle is provided? Which problems become non-trivially tractable on p-Hamiltonian-ordered graphs? As to the class of p-Hamiltonian-ordered graphs, we are not aware of a computational complexity study on this class next to ours. Further, it is interesting to study Feedback Vertex Set on the intersections of the classes of regular graphs, planar graphs, p-Hamiltonian-ordered graphs, and p-ordered Hamiltonian graphs.

References

1. Balakrishnan, R., Ranganathan, K.: A Textbook of Graph Theory. Springer, New York (2012). https://doi.org/10.1007/978-1-4614-4529-6
2. Chrobak, M., Payne, T.H.: A linear-time algorithm for drawing a planar graph on a grid. Inf. Process. Lett. **54**(4), 241–246 (1995). https://doi.org/10.1016/0020-0190(95)00020-D
3. Dailey, D.P.: Uniqueness of colorability and colorability of planar 4-regular graphs are NP-complete. Discrete Math. **30**(3), 289–293 (1980). https://doi.org/10.1016/0012-365X(80)90236-8. https://www.sciencedirect.com/science/article/pii/0012365X80902368
4. Diestel, R.: Graph Theory, Graduate Texts in Mathematics, vol. 173, 4th edn. Springer, Heidelberg (2010). https://doi.org/10.1007/978-3-662-53622-3
5. Festa, P., Pardalos, P.M., Resende, M.G.C.: Feedback set problems. In: Du, D., Pardalos, P.M. (eds.) Handbook of Combinatorial Optimization, pp. 209–258. Springer, New York (1999). https://doi.org/10.1007/978-1-4757-3023-4_4
6. Fleischner, H., Sabidussi, G.: 3-colorability of 4-regular Hamiltonian graphs. J. Graph Theory **42**(2), 125–140 (2003). https://doi.org/10.1002/jgt.10079
7. Fleischner, H., Sabidussi, G., Sarvanov, V.I.: Maximum independent sets in 3- and 4-regular Hamiltonian graphs. Discrete Math. **310**(20), 2742–2749 (2010). https://doi.org/10.1016/j.disc.2010.05.028
8. Fluschnik, T., Niedermeier, R., Rohm, V., Zschoche, P.: Multistage vertex cover. In: Proceedings of 14th IPEC. LIPIcs, vol. 148, pp. 14:1–14:14. Schloss Dagstuhl - Leibniz-Zentrum für Informatik (2019). https://doi.org/10.4230/LIPIcs.IPEC.2019.14
9. Karp, R.M.: Reducibility among combinatorial problems. In: Proceedings of a Symposium on the Complexity of Computer Computations, pp. 85–103. The IBM Research Symposia Series, Plenum Press, New York (1972). https://doi.org/10.1007/978-1-4684-2001-2_9
10. Li, D., Liu, Y.: A polynomial algorithm for finding the minimum feedback vertex set of a 3-regular simple graph 1. Acta Mathematica Scientia **19**(4), 375–381 (1999)
11. Mulder, H.M.: Julius Petersen's theory of regular graphs. Discrete Math. **100**(1–3), 157–175 (1992). https://doi.org/10.1016/0012-365X(92)90639-W
12. Munaro, A.: On line graphs of subcubic triangle-free graphs. Discrete Math. **340**(6), 1210–1226 (2017). https://doi.org/10.1016/j.disc.2017.01.006
13. Ng, L., Schultz, M.: k-ordered Hamiltonian graphs. J. Graph Theory **24**(1), 45–57 (1997). https://doi.org/10.1002/(SICI)1097-0118(199701)24:1⟨45::AID-JGT6⟩3.0.CO;2-J
14. Speckenmeyer, E.: Untersuchungen zum Feedback Vertex Set Problem in ungerichteten Graphen. Ph.D. thesis, Paderborn (1983)
15. Speckenmeyer, E.: On feedback vertex sets and nonseparating independent sets in cubic graphs. J. Graph Theory **12**(3), 405–412 (1988). https://doi.org/10.1002/jgt.3190120311
16. Ueno, S., Kajitani, Y., Gotoh, S.: On the nonseparating independent set problem and feedback set problem for graphs with no vertex degree exceeding three. Discrete Math. **72**(1–3), 355–360 (1988). https://doi.org/10.1016/0012-365X(88)90226-9

Towards Classifying the Polynomial-Time Solvability of Temporal Betweenness Centrality

Maciej Rymar[1], Hendrik Molter[1,2(✉)], André Nichterlein[1], and Rolf Niedermeier[1]

[1] TU Berlin, Algorithmics and Computational Complexity, Berlin, Germany
{m.rymar,andre.nichterlein,rolf.niedermeier}@tu-berlin.de
[2] Department of Industrial Engineering and Management, Ben-Gurion University of the Negev, Beer-Sheva, Israel
molterh@post.bgu.ac.il

Abstract. In static graphs, the betweenness centrality of a graph vertex measures how many times this vertex is part of a shortest path between any two graph vertices. Betweenness centrality is efficiently computable and it is a fundamental tool in network science. Continuing and extending previous work, we study the efficient computability of betweenness centrality in *temporal* graphs (graphs with fixed vertex set but time-varying arc sets). Unlike in the static case, there are numerous natural notions of being a "shortest" temporal path (walk). Depending on which notion is used, it was already observed that the problem is #P-hard in some cases while polynomial-time solvable in others. In this conceptual work, we contribute towards classifying what a "shortest path (walk) concept" has to fulfill in order to gain polynomial-time computability of temporal betweenness centrality.

Keywords: Temporal graphs · Temporal paths and walks · Network science · Network centrality measures · Counting complexity

1 Introduction

Network science is a central pillar of data science. It relies on spotting and analyzing important network (graph) properties. Betweenness centrality, introduced by Freeman [14] and made a practical tool of high relevance by Brandes [9], is a key instrument in this area, in particular in the context of social network analysis. Informally, the *betweenness centrality* of a graph vertex v correlates to the probability that v is visited by a randomly chosen shortest path. With the advent of investigating dynamically changing network structures and, thus, the

M. Rymar—Partially supported by the DFG, project MATE (NI 369/17).
H. Molter—Supported by the DFG, project MATE (NI 369/17), and by the ISF, grant No. 1070/20. Main part of this work was done while affiliated with TU Berlin.

Ł. Kowalik et al. (Eds.): WG 2021, LNCS 12911, pp. 219–231, 2021.
https://doi.org/10.1007/978-3-030-86838-3_17

growing interest in temporal graphs, studying concepts of temporal betweenness centrality and their algorithmic complexity became very popular over the recent years [1,3,11,16,17,19,24–27].

The temporal graphs we are considering have fixed vertex set and edge set(s) that change over discrete time steps. A temporal path in such a graph has to respect time, that is, the path has to traverse edges at non-decreasing time steps. The study of *temporal* betweenness centrality is significantly richer than in static graphs since in temporal graphs the term "shortest path" may have numerous different but natural interpretations. Indeed, the shortest transfer from a start vertex to a target vertex may even be a walk (and not just a path) and there is also an intensive study on shortest-path and shortest-walk computations in temporal graphs [8,10,12,28]. We refrain from going into the details here but refer to our predecessor work [11] for a more extensive discussion. What is important to note, however, is that the complexity of temporal betweenness centrality computation, which essentially boils down to a counting problem, crucially depends on the concept used. More specifically, the complexity may vary from polynomial-time solvable (with different polynomial degrees) to #P-hard. To systematically investigate this issue and to develop a better understanding of when one has to expect such a huge jump in computational complexity is the main motivation of our work.

The by far closest reference point for our work is a previous paper from our group [11]. It also surveys the literature roughly till the year 2020. Since then, Simard et al. [24] studied a continuous-time scenario and betweenness based on shortest paths, while we focus on discrete time. Our work has a significantly stronger conceptual objective than Buß et al. [11] had. So our classification results comprise the results there. They are based on coining the concept of *prefix-compatible* temporal walks. These walks can be counted in polynomial time and thus the corresponding temporal betweenness centrality value can be computed in polynomial time. To this end, we provide simple (still tunable) polynomial-time algorithms that apply to a whole class of temporal betweenness centrality problems. Moreover, we indicate that slightly relaxing from prefix-compatibility typically already yields #P-hardness. Due to space constraints, detailed proofs (marked with a $\star$) are deferred to a full version [22].

2 Preliminaries

The fundamental concept we use in this work is a *temporal graphs.* A directed *temporal graph* $\mathcal{G}$ is a triple $(V, \mathcal{E}, T)$ such that V is a set of vertices, $\mathcal{E} \subseteq \{(u, v, t) \mid u, v \in V, u \neq v, t \in [T]\}$ is a set of time arcs, and $T \in \mathbb{N}$, where $[T] := \{1, \dots, T\}$ is a set of time steps; see Fig. 1 for an illustration.

Throughout this work, let $n := |V|$ and $M := |\mathcal{E}|$. We call $V \times [T]$ the set of (possible) *vertex appearances.* We consider directed temporal graphs as temporal paths and walks are implicitly directed because of the ascending time labels. We call a time arc $e = (v, w, t)$ also the *transition* from v to w at time step t. We call v the starting point and w the endpoint of the transition. Using this, we can now define temporal walks and temporal paths; see Fig. 1 for an illustration.

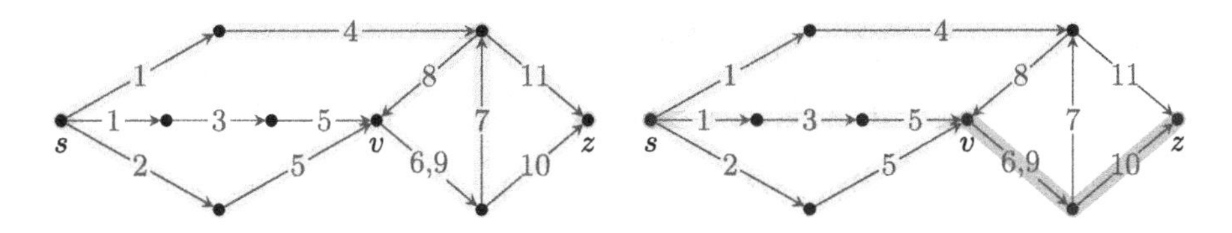

Fig. 1. Our running example for a temporal graph $\mathcal{G}$ with 9 vertices and 13 time-arcs (the outgoing arc from vertex v denotes two time-arcs at time steps 6 and 9). The number(s) on the arcs denote the time steps. *Left:* highlighted are the unique shortest s-z-path (top path in green) in $\mathcal{G}$ and a fastest s-z-walk (bottom walk in blue, v and its successor appear twice in the walk). *Right:* three foremost s-z-paths in $\mathcal{G}$ are highlighted ($\mathcal{G}$ has two more foremost s-z-walks; for visibility not highlighted). (Color figure online)

Definition 1 (Temporal Walk). *A* temporal walk W *is an ordered sequence of transitions* $(e_1, \ldots, e_k) \in \mathcal{E}^k$ *such that for each* $i \in [k-1]$ *the endpoint of* e_i *is the starting point of* e_{i+1} *and* $t_i \leq t_{i+1}$*, where* t_i *and* t_{i+1} *are the time labels of transitions* e_i *and* e_{i+1}*, respectively. The length of* W *is* $\text{length}(W) := k$.

Let $W = (e_1, \ldots, e_k)$ be a temporal walk. We call W a *temporal s-z-walk* if $e_1 = (s, v, t)$ and $e_k = (w, z, t')$ for some v, w and some t, t' and we call W a *temporal s-(z, t')-walk* if $e_1 = (s, v, t)$ and $e_k = (w, z, t')$ for some v, w and some t.

A temporal walk may visit the same vertex more than once. In contrast to that, a temporal *path* visits each vertex at most once. This is analogous to static graphs. In contrast to the static setting, there are several canonical notions of "optimal" temporal walks. The three most important ones [10] are *shortest* temporal walks, which are temporal walks between two vertices that uses a minimum number of time arcs, *fastest* temporal walks, which are temporal walks between two vertices with a minimum difference between the time steps of the first and last transition used by the walk, and *foremost* temporal walks, which are temporal walks between two vertices with a minimum time step on their last transition (see Fig. 1 for an illustration). The corresponding optimal temporal paths are defined analogously. We remark that in the case of "shortest", every shortest temporal walk is in fact a temporal path, similarly to the static case.[1] However, for "fastest" and "foremost" temporal walks this is generally not the case.

For readability, we use the notation $v \xrightarrow{t} w$ instead of the triple (v, w, t). If a temporal walk W contains the transition $v \xrightarrow{t} w$, then we say that W *visits* (or *goes through*) vertex appearance (w, t) (in Fig. 1 (left) the blue walk visits $(v, 5)$ and $(v, 8)$). Let $P = v_0 \xrightarrow{t_1} v_1 \xrightarrow{t_2} \ldots \xrightarrow{t_{\ell-1}} v_{\ell-1} \xrightarrow{t_\ell} v_\ell$ be a temporal path. For any $0 \leq i < j \leq \ell$, we write (if $i = 0$, then we define the following for $t_0 = 0$):

[1] In fact, all optimal temporal path concepts (we are aware of) where path counting and computing the betweenness centrality can be done in polynomial time have this property, ensuring that optimal walks are indeed paths.

$$P[(v_i, t_i), (v_j, t_j)] := v_i \overset{t_{i+1}}{\to} v_{i+1} \overset{t_{i+2}}{\to} \dots \overset{t_j}{\to} v_j,$$
$$P[\bullet, (v_j, t_j)] := v_0 \overset{t_1}{\to} v_1 \overset{t_2}{\to} \dots \overset{t_j}{\to} v_j,$$
$$P[(v_i, t_i), \bullet] := v_i \overset{t_{i+1}}{\to} v_{i+1} \overset{t_{i+2}}{\to} \dots \overset{t_\ell}{\to} v_\ell.$$

We use an analogous notation for temporal walks. Note that for a non-strict temporal walk W the above notation may not be well-defined since the same vertex appearance (v, t) may appear more than once in W. In this case we define $W[(v_i, t_i), (v_j, t_j)]$ to be the subwalk of W from the first appearance of $u \overset{t_i}{\to} v_i$ for any u to the last appearance of $u \overset{t_j}{\to} v_j$ for any u. The cases of $P[\bullet, (v_j, t_j)]$ and $P[(v_i, t_i), \bullet]$ are handled analogously. Furthermore, we use $\oplus$ to denote *concatenations* of temporal walks, that is, let W, W' be two temporal walks such that W ends in v and W' starts in v. Let t be the time label on the last transition of W and t' the label on the first transition of W'. Then if $t' \geq t$, we denote with $W \oplus W'$ the concatenation of W and W'.

Given a temporal graph $\mathcal{G}$, we denote by walks($\mathcal{G}$) the set of all temporal walks in $\mathcal{G}$. Subsequently, we will need to consider the successors (or dually, the predecessors) of each vertex appearance on temporal walks in some $\mathcal{W} \subseteq$ walks($\mathcal{G}$).

Definition 2 (Direct predecessor set, direct successor set). *Let $\mathcal{G} = (V, \mathcal{E}, T)$ be a temporal graph and $\mathcal{W} \subseteq$ walks($\mathcal{G}$) be a subset of its temporal walks. Fix a source vertex $s \in V$. Let $\mathcal{W}_s \subseteq \mathcal{W}$ be the set of temporal walks in $\mathcal{W}$ that start in s. Now let $(w, t') \in V \times [T]$ be any vertex appearance. Then* $\text{Pre}_s^{\mathcal{W}}(w, t')$ *is the set of all direct predecessors of (w, t') on temporal walks in $\mathcal{W}_s$. Formally,*

$$\text{Pre}_s^{\mathcal{W}}(w, t') := \left\{(v, t) \in V \times [T] \mid \exists W \in \mathcal{W}_s \colon u \overset{t}{\to} v \overset{t'}{\to} w \in W\right\}$$
$$\cup \left\{(s, 0) \mid \exists W \in \mathcal{W}_s \colon s \overset{t'}{\to} w \in W\right\}.$$

The set $\text{Succ}_s^{\mathcal{W}}(v, t)$ *of successors of a vertex appearance (v, t) is the "inverse" of the predecessor relation, formally,*

$$\text{Succ}_s^{\mathcal{W}}(v, t) := \left\{(w, t') \mid (v, t) \in \text{Pre}_s^{\mathcal{W}}(w, t')\right\}.$$

Clearly, the direct predecessor sets induce a relation over vertex appearances. We use this to define the following directed graph. We remark that this graph is similar to a so-called static expansion [18,28,29] that is tailored to a specific source vertex.

Definition 3 (Predecessor graph). *Let $\mathcal{G} = (V, \mathcal{E}, T)$ be a temporal graph and $\mathcal{W} \subseteq$ walks($\mathcal{G}$) be a subset of its temporal walks. Fix a source vertex $s \in V$. Then $G_s^{\text{Pre}} := (U, A)$ is the* predecessor graph (of s, with respect to $\mathcal{W}$), *where*

$$U := \{(s, 0)\} \cup \{(w, t') \mid \text{Pre}_s^{\mathcal{W}}(w, t') \neq \emptyset\}, \text{ and}$$
$$A := \{((v, t), (w, t')) \mid (v, t) \in \text{Pre}_s^{\mathcal{W}}(w, t')\}.$$

We next introduce some notation and definitions for temporal walk counting and temporal betweenness.

Definition 4. *Let $\mathcal{G} = (V, \mathcal{E}, T)$ be a temporal graph and $\mathcal{W} \subseteq$ walks($\mathcal{G}$) be a subset of its temporal walks. Let $s, v, z \in V$ and $t \in [T]$. Then,*

- *$\sigma_{sz}^{\mathcal{W}}$ is the number of temporal s-z-walks in $\mathcal{W}$ that start in s and end in z.*
- *$\sigma_{sz}^{\mathcal{W}}(v)$ is the number of temporal s-z-walks in $\mathcal{W}$ that go through the vertex v. Furthermore, $\sigma_{sz}^{\mathcal{W}}(z) = \sigma_{sz}^{\mathcal{W}}(s) = \sigma_{sz}^{\mathcal{W}}$ and $\sigma_{ss}^{\mathcal{W}}(s) = \sigma_{ss}^{\mathcal{W}}$;*
- *$\sigma_{sz}^{\mathcal{W}}(v, t)$ is the number of temporal s-z-walks in $\mathcal{W}$ that go through the vertex appearance (v, t). Furthermore, $\sigma_{sz}^{\mathcal{W}}(s, 0) = \sigma_{sz}^{\mathcal{W}}(s) = \sigma_{sz}^{\mathcal{W}}$ and $\sigma_{sz}^{\mathcal{W}}(s, t') = 0$ for all $t' \in [T]$.*

Based on this definition we can define the notions of dependency of vertices on other vertices, similar to how it was done by Brandes [9] in the static case. We remark that the notions for the temporal setting introduced in the following are very similar to the ones used by Buß et al. [11]. We give them again here for completeness and since we adapt them for general sets of temporal walks.

Definition 5 (Pair dependency, cumulative dependency). *Let $\mathcal{G}$ be a temporal graph and $\mathcal{W} \subseteq$ walks($\mathcal{G}$) be a subset of its temporal walks. Then,*

$$\delta_{sz}^{\mathcal{W}}(v) := \begin{cases} 0, & \text{if } \sigma_{sz}^{\mathcal{W}} = 0, \\ \frac{\sigma_{sz}^{\mathcal{W}}(v)}{\sigma_{sz}^{\mathcal{W}}}, & \text{otherwise;} \end{cases} \qquad \delta_{s\bullet}^{\mathcal{W}}(v) := \sum_{z \in V} \delta_{sz}^{\mathcal{W}}(v)$$

are the pair dependency *of s and z on v and the* cumulative dependency *of s on v, respectively.*

In other words, $\delta_{sz}^{\mathcal{W}}(v)$ is the fraction of temporal s-z-walks that go through v. Intuitively, the higher this fraction is, the more important v is to the connectivity of s and z in the graph. Furthermore, $\delta_{s\bullet}^{\mathcal{W}}(v)$ is the cumulative dependency of s on v for all possible destinations. These notions can be used to define temporal betweenness centrality, which intuitively captures how *all* other vertices depend on v for their connectivity.

Definition 6 (Temporal betweenness centrality). *Let $\mathcal{G}$ be a temporal graph and $\mathcal{W} \subseteq$ walks($\mathcal{G}$) be a subset of its temporal walks. Then, for any vertex $v \in V$,*

$$C_B^{\mathcal{W}}(v) := \sum_{s \neq v \neq z} \delta_{sz}^{\mathcal{W}}(v) \qquad \text{and} \qquad \hat{C}_B^{\mathcal{W}}(v) := \sum_{s,z \in V} \delta_{sz}^{\mathcal{W}}(v)$$

are the temporal betweenness centrality *$C_B^{\mathcal{W}}(v)$ of v and* total temporal betweenness centrality *$\hat{C}_B^{\mathcal{W}}(v)$ of v (with respect to $\mathcal{W}$).*

If the set of walks $\mathcal{W}$ in question is clear from the context, then we omit the $\mathcal{W}$. The main reason behind mainly using total temporal betweenness centrality instead of the standard temporal betweenness in the following is that it simplifies some of our proofs as it works well with our definition of cumulative dependency:

Observation 1. *For any vertex $v \in V$, $\hat{C}_B^{\mathcal{W}}(v) = \sum_{s \in V} \delta_{s\bullet}^{\mathcal{W}}(v)$.*

Due to space constraints, we defer some further concepts to a full version.

3 Prefix-Compatibility

In this section, our goal is to find an easy-to-understand-and-use property for optimality concepts for temporal walks and paths that is sufficient for polynomial-time solvability of (1) counting optimal temporal walks and (2) computing the temporal betweenness with respect to that optimality concept for temporal walks. We call this property "prefix-compatibility". Intuitively, a class of temporal walks is prefix-compatible if prefixes of optimal temporal walks are also optimal ("prefix-optimality") and prefixes of optimal temporal walks can be exchanged ("prefix-exchangeability"). To formally define optimality concepts for temporal walks, we use cost functions.

Definition 7 (Cost function). *Let $\mathcal{W}$ be the set of all temporal walks in a temporal graph $\mathcal{G}$. A function of the form $c : \mathcal{W} \to \mathbb{R} \cup \{\infty\}$ is a* cost function.

We remark that for this work we assume that the cost function can be computed in constant time, which turns out to be a valid assumption for many optimality concepts. When considering cost functions that need polynomial time to be evaluated, this polynomial factor would form an extra multiplicative term in our running time results.

Let c be a cost function and let $\mathcal{G} = (V, \mathcal{E}, T)$ be a temporal graph. Fix a source $s \in V$. Then, for every vertex appearance $(v, t) \in V \times [T]$ we define $c_s^*(v, t)$ to be the minimum value of c assumed over all temporal s-(v, t)-walks. That is, we have $c_s^*(v, t) = \min_{s\text{-}(v,t)\text{-walk } W}\{c(W)\}$. If the minimum is not defined or there is no temporal s-(v, t)-walk, then let $c_s^*(v, t) := \infty$. Similarly, we define the optimal c-values for the vertices $v \in V$ as $c_s^*(v) := \min_{t \in [T]}\{c_s^*(v, t)\}$. We call a temporal s-(v, t)-walk W *c-optimal* if we have $c(W) = c_s^*(v, t) < \infty$. Similarly, we call a temporal s-v-walk W *c-optimal* if we have $c(W) = c_s^*(v) < \infty$. Observe that this notion of c-optimal walks is very general and allows to capture essentially all natural walk concepts, see Fig. 2 for some examples.

We can now define the set of walks in a temporal graph that is optimal with respect to some cost function c in a straightforward way:

Definition 8 (Induced set of optimal temporal walks). *Let c be a cost function and let $\mathcal{G} = (V, \mathcal{E}, T)$ be a temporal graph. For $s, z \in V$, let $\mathcal{W}_{sz}$ be the set of all temporal s-z-walks in $\mathcal{G}$. Then*

$$\mathcal{W}^{(c)} := \bigcup_{s,z \in V} \{W \in \mathcal{W}_{sz} \mid c(W) < \infty \land c(W) = c_s^*(z)\}$$

is the induced set of optimal temporal walks (of c).

From now on, we introduce the two properties for cost functions that we need to obtain prefix-compatibility; in Fig. 3 we illustrate that (restless) fastest paths do not satisfy one of these. We start with "prefix-optimality", which intuitively states that prefixes of optimal temporal walks are also optimal.

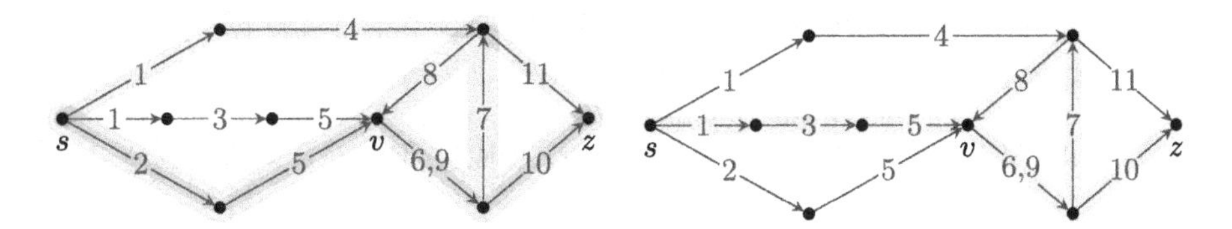

Fig. 2. Examples for c-optimal walks for various c. *Left:* the top (green) s-z-path is the shortest s-z-walk, that is, $c(W)$ is the number of time arcs in the walk W. The blue walk and the red path are the two fastest s-z-walks in the graph; here $c(W)$ is the difference of the time steps of the first and last time arc in the walk W. *Right:* highlighted is the only *2-restless* s-z-walk, that is, the difference between the time steps of two consecutive time arcs is at most two [8,12]. This could be encoded in c as follows: for a walk $W = (e_1, \ldots, e_k)$ we have $c(W) = 1$ if $t(e_i) + 2 \geq t(e_{i+1})$ for all $i \in [k-1]$ and $c(W) = \infty$ otherwise. Notably, in general it is NP-hard to decide whether such a 2-restless s-z-*path* exists [12], but for walks even the optimization variants (shortest, fastest, ...) are polynomial-time solvable (Color figure online) [8].

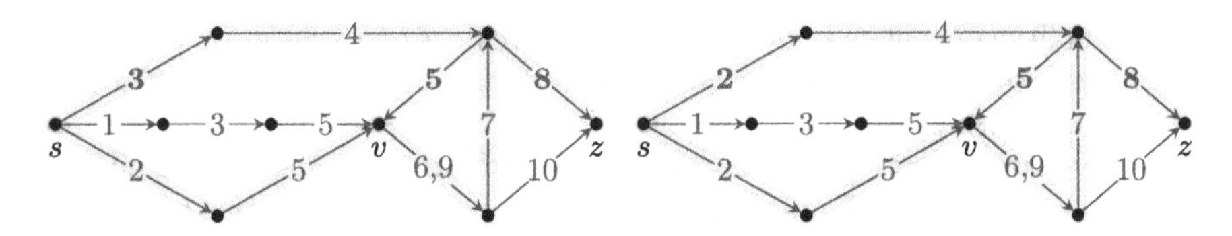

Fig. 3. Counter examples showing that (restless) fastest paths neither satisfy prefix-optimality nor prefix-exchangeability. Inhere, three time steps (indicated by bold numbers) are updated in our standard temporal graph. *Left:* the blue path (starting at time step 2) is the unique fastest 3-restless (cf. Fig. 2) s-z-path with travel time $8 - 2 = 6$; it is not prefix-optimal as the red path is a faster s-$(v,5)$-path (travel time $5 - 2 = 3$ vs. $5 - 3 = 2$). *Right:* the blue path (starting at time step 2) is the fastest s-z-path with travel time $8 - 2 = 6$; it is not prefix-exchangeable as the red path is also a fastest s-$(v,5)$-path but the corresponding prefix cannot be replaced with the red path as the resulting walk would not be a path. (Color figure online)

Definition 9 (Prefix-optimality). *Let c be a cost function and let $\mathcal{W}^{(c)}$ be a set of optimal temporal walks in a temporal graph $\mathcal{G}$ that is induced by c. Then, c is* prefix-optimal *if for every temporal walk $W \in \mathcal{W}^{(c)}$ and for every vertex appearance $(v,t) \in W$ it holds that $c(W[\bullet,(v,t)]) = c_s^*(v,t)$.*

Note that we do not require that the prefixes be optimal temporal walks *to a vertex*, so the temporal walk $c(W[\bullet,(v,t)])$ is not required to be in the induced set of optimal temporal walks $\mathcal{W}^{(c)}$. If c is clear from the context and there is no danger of confusion, then we drop the superscript (c).

The second property we introduce is "prefix-exchangeability". It intuitively states that a prefix of an optimal temporal walk can be exchanged by certain other temporal walks.

Definition 10 (Prefix-exchangeability). *Let c be a cost function and let $\mathcal{W}^{(c)}$ be a set of optimal temporal walks in a temporal graph $\mathcal{G}$ that is induced* *by* c*. Then,* c *is* prefix-exchangeable *if for every vertex appearance* $(v,t) \in V \times [T]$ *such that there exist* $s, z \in V$ *for which there is a temporal* s*-*z*-walk* $W \in \mathcal{W}^{(c)}$ *going through* (v,t) *and for every temporal* s*-*(v,t)*-walk* W' *with* $c(W') = c_s^*(v,t)$ *it holds that* $W' \oplus W[(v,t), \bullet] \in \mathcal{W}^{(c)}$*, that is,* $c(W' \oplus W[(v,t), \bullet]) = c_s^*(z)$*.*

In other words, if there is an optimal temporal s-z-walk W going through (v,t), then all c-optimal temporal s-(v,t)-walks can be substituted for the first part of W to get a c-optimal temporal s-z-walk.

It is convenient to combine the two main properties into one.

Definition 11 (Prefix-compatibility). *Let* c *be a cost function. Then* c *is* prefix-compatible *if it is both prefix-optimal and prefix-exchangeable.*

3.1 Examples of Prefix-Compatible Cost Functions

We exemplarily show that five well-known optimality concepts for temporal paths and walks [8,10,28] can be expressed by prefix-compatible cost functions. The optimality concepts we consider here are *foremost temporal walks* and *shortest temporal paths.* We further consider *shortest fastest temporal paths*, which are temporal paths that are shortest among all fastest temporal paths and *shortest restless temporal walks* [8], which are the shortest temporal walks among all temporal walks where the difference of the time labels of two consecutive transitions is bounded by some constant. Lastly, we consider *strict prefix-foremost temporal paths* [28]. A temporal path is strict prefix-foremost if it is foremost and every prefix is also foremost and all transitions have increasing time labels.

Proposition 1 ($\star$). *The cost functions describing the following optimality concepts are prefix-compatible:*

- *foremost temporal walk,*
- *shortest temporal path,*
- *shortest fastest temporal path,*
- *shortest restless temporal walks, and*
- *strict prefix-foremost temporal path.*

We remark that e.g. "shortest fastest" temporal path have to the best of our knowledge not been considered yet for temporal betweenness computation while being a very natural optimality criterion. Furthermore, many other natural optimality concepts can be shown to be prefix-compatible in a similar way to our examples.

Since we aim for a very general framework, it is not surprising that our running times for temporal betweenness computation can be improved for specific optimality concepts by tailored algorithms. As we will show in Theorems 2 and 3, we can count c-optimal temporal walks and compute the temporal betweenness centrality with respect to optimal walks for prefix-compatible cost functions c in $O(n^2MT^2)$ time. However, for example for strict prefix-foremost paths and shortest temporal paths, the corresponding temporal betweenness computation can be done in $O(nM \log M)$ and $O(n^3T^2)$ time, respectively [11]. Also the space requirements can be improved for specific optimality concepts [11]. We remark that the well-known techniques for static graphs [6,9,13] can mostly be transferred to the temporal setting in straightforward ways.

3.2 Necessity of Prefix-Optimality and Prefix-Exchangeability

We now briefly motivate why we need a cost function c to be both prefix-optimal and prefix-exchangeable in order to be able to count c-optimal temporal walks in polynomial time. We do this by giving examples of cost functions which have only one of the two properties and where the corresponding problem of counting c-optimal temporal walks is #P-hard. Note that this implies that the corresponding temporal betweenness computation problem is also #P-hard [11]. We remark that this does not imply that prefix-optimality and prefix-exchangeability cannot be replaced by some weaker requirements.

First, if we leave out prefix-exchangeability, then we obtain hardness.

Proposition 2 ($\star$). *There exists a prefix-optimal cost function c for which counting the number of c-optimal temporal walks is #P-hard.*

Second, if we leave out prefix-optimality, then we also obtain hardness.

Proposition 3 ($\star$). *There exists a prefix-exchangeable cost function c for which counting the number of c-optimal temporal walks is #P-hard.*

4 Counting Walks

In this section, complementing the hardness shown in Sect. 3.2, we extend classic algorithms for path and walk counting to our setting. This will provide a polynomial-time algorithm for counting optimal walks with respect to a prefix-compatible cost function.

The general idea is roughly as follows: First, compute the static predecessor graph $G_s^{\text{Pre}}(c)$ with respect to c (see Definition 3) using a slightly modified version of the classic Bellman-Ford algorithm [6,13]. Second, count the walks in this static graph $G_s^{\text{Pre}}(c)$ with known approaches; the results correspond to the number of c-optimal walks in the temporal input graph.

We start with statements explaining the connection between $G_s^{\text{Pre}}(c)$ and the number of walks in the temporal input graph. Here, an important corner case is that there might be infinitely many c-optimal walks.

Definition 12 (Finiteness). *Let c be a cost function for a temporal graph $\mathcal{G}$. Then, c is* finite *on $\mathcal{G}$ if the induced set $\mathcal{W}^{(c)}$ of optimal temporal walks of c has finite cardinality.*

As stated next, finiteness of the cost function c coincides with the predecessor graph $G_s^{\text{Pre}}(c)$ containing directed cycles and is, thus, easy to detect.

Lemma 1 ($\star$). *Let c be a prefix-compatible cost function. Then c is finite if and only if the predecessor graph $G_s^{\text{Pre}}(c)$ is acyclic.*

Assuming that we can efficiently compute $G_s^{\mathrm{Pre}}(c)$, Lemma 1 allows us to detect and deal wit the cases of infinitely many c-optimal walks. Moreover, if c is finite, then $G_s^{\mathrm{Pre}}(c)$ is a DAG and, hence, counting walks is easy. Hence, we arrive at the following statement.

Lemma 2. *Let c be a prefix-compatible cost function and $G_s^{\mathrm{Pre}}(c)$ a predecessor graph for a temporal graph $\mathcal{G}$ and a vertex s in $\mathcal{G}$. Given $G_s^{\mathrm{Pre}}(c)$, the number of c-optimal temporal walks between from s to any vertex v and any vertex appearance (v,t) in $\mathcal{G}$ can be computed in $O(|G_s^{\mathrm{Pre}}(c)|)$ time.*

Proof (Sketch). First, run Kosaraju's algorithm [2] on the graph to compute in linear time the strongly connected components (SCCs) in $G_s^{\mathrm{Pre}}(c)$ while also keeping track of their size. Mark every SCC of size > 1 with ∞. Then, run a BFS over the SCCs starting in all nodes marked ∞ and label all the nodes reached during the BFS with ∞. Now, for every vertex appearance (v,t) belonging to an SCC marked with ∞, we set the number of temporal s-(v,t)-walks as ∞ and then remove it from $G_s^{\mathrm{Pre}}(c)$. The correctness of this step follows from the proof of Lemma 1.

Let G' be the remaining graph. Clearly, G' is a DAG. We next show that counting paths to a vertex in G' will exactly correspond to counting c-optimal temporal walks to a vertex appearance corresponding to that vertex:

We shall prove the statement above by induction on the vertices of G', taken in the topological ordering. First, $(s,0)$ must clearly be the first vertex in that ordering. Obviously, there is only one c-optimal path to $(s,0)$, so the computed value will be correct here.

Now, consider a vertex v_i corresponding to some appearance (v,t). By definition, all its direct predecessors v_j come before v_i in the topological ordering in the graph. Since $v_j \in \mathrm{Pre}_s^{\mathcal{W}}(v_i)$, by prefix-exchangeability, every c-optimal walk to v_j can be extended to a c-optimal walk to v_i. Conversely, by prefix-optimality, we are also not missing any c-optimal walks to (v,t). Hence, the computed number of paths to v_i will also be correct.

To compute the number of c-optimal temporal walks *to a vertex* $v \in V$ we can first find $c_s^*(v) = \min_{t\in[T]} c_s^*(v,t)$, and then compute $\sigma_{sv}^{\mathcal{W}} = \sum_{t|c_s^*(v,t)=c_s^*(v)} \sigma_{s(v,t)}^{\mathcal{W}}$. This path counting in a DAG is clearly doable in time linear in the size of $G_s^{\mathrm{Pre}}(c)$. □

To employ Lemma 2, we need to compute $G_s^{\mathrm{Pre}}(c)$. To this end, we run a slight variation of the classical Bellman-Ford algorithm. This leads to the following lemma. Recall that we assume here that c can be evaluated in $O(1)$ time.

Lemma 3 ($\star$). *Let c be a prefix-compatible cost function for a temporal graph $\mathcal{G}$. Let s be a vertex in $\mathcal{G}$. Then the predecessor graph $G_s^{\mathrm{Pre}}(c)$ can be computed in $O(nMT^2)$ time.*

Applying Lemmas 2 and 3 starting from each vertex yields the following.

Theorem 2 (Walk counting). *Let c be a prefix-compatible cost function for a temporal graph $\mathcal{G} = (V, \mathcal{E}, T)$. Then the number of c-optimal temporal walks from each vertex $s \in V$ to any vertex appearance (v, t) can be computed in $O(n^2MT^2)$ time.*

5 Computing Temporal Betweenness

In this section, we discuss how to compute temporal betweenness centrality efficiently for optimal temporal walks defined by a prefix-compatible cost function. In order to do this, we adapt the machinery of showing a recursive relation of the temporal dependencies [11] to our generalized setting. Due to space constraints, we refer to a full version for details. Together with the fact that we can compute the walk-counts in polynomial time (Theorem 2), we can use a Brandes-like [9] approach to compute the temporal betweenness values in polynomial time.

Lemma 4 ($\star$). *Let c be a finite prefix-compatible cost function for a temporal graph $\mathcal{G} = (V, \mathcal{E}, T)$. Given $G_s^{\text{Pre}}(c)$ for each $s \in V$, the temporal betweenness centrality of all vertices in $\mathcal{G}$ can be computed in $O(\sum_{s \in V} |G_s^{\text{Pre}}(c)| + nM)$ time.*

Using the running time bound from Lemma 3 together with Lemma 4, we immediately get our main result of this work.

Theorem 3 (General betweenness computation). *Let c be a finite prefix-compatible cost function. Then the betweenness centrality of all vertices can be computed in $O(n^2MT^2)$ time.*

Combining Proposition 1 and Theorem 3 yields the following result.

Corollary 1. *The betweenness centrality of all vertices in a temporal graph can be computed in $O(n^2MT^2)$ time with respect to:*

- *foremost temporal walk,*
- *shortest temporal path,*
- *shortest fastest temporal path,*
- *shortest restless temporal walks, and*
- *strict prefix-foremost temporal path.*

We remark that, while foremost temporal walk, shortest temporal path, and strict prefix-foremost temporal path were known from previous work [1,11,19], this is a new classification for shortest fastest temporal paths and shortest restless temporal walks.

6 Conclusion

The very nature of this work is conceptual. It goes without saying that to achieve improved efficiency, exploiting specific properties of the various temporal path and walk concepts may clearly allow for further improved polynomial running times. As to future research, we wonder whether our concept of prefix-compatibility may finally lead to a full characterization of polynomial-time computable temporal betweenness centrality values. As to the computationally hard cases (but not only them), for high efficiency in practice, one might also explore the possibilities of efficient data reductions or approximation algorithms. This proved useful in the static graphs case, with respect to data reduction [5,7,20,23] as well as with respect to approximation [4,15,21].

References

1. Afrasiabi Rad, A., Flocchini, P., Gaudet, J.: Computation and analysis of temporal betweenness in a knowledge mobilization network. Comput. Soc. Netw. **4**(1), 5 (2017)
2. Aho, A.V., Hopcroft, J.E., Ullman, J.D.: Data Structures and Algorithms. Addison-Wesley (1983)
3. Alsayed, A., Higham, D.J.: Betweenness in time dependent networks. Chaos Solitons Fractals **72**, 35–48 (2015)
4. Bader, D.A., Kintali, S., Madduri, K., Mihail, M.: Approximating betweenness centrality. In: Bonato, A., Chung, F.R.K. (eds.) WAW 2007. LNCS, vol. 4863, pp. 124–137. Springer, Heidelberg (2007). https://doi.org/10.1007/978-3-540-77004-6_10
5. Baglioni, M., Geraci, F., Pellegrini, M., Lastres, E.: Fast exact computation of betweenness centrality in social networks. In: Proceedings of the 4th International Conference on Advances in Social Networks Analysis and Mining (ASONAM 2012), pp. 450–456. IEEE Computer Society (2012)
6. Bellman, R.: On a routing problem. Q. Appl. Math. **16**(1), 87–90 (1958)
7. Bentert, M., Dittmann, A., Kellerhals, L., Nichterlein, A., Niedermeier, R.: An adaptive version of Brandes' algorithm for betweenness centrality. J. Graph Algorithms Appl. **24**(3), 483–522 (2020)
8. Bentert, M., Himmel, A.-S., Nichterlein, A., Niedermeier, R.: Efficient computation of optimal temporal walks under waiting-time constraints. Appl. Netw. Sci. **5**(1), 73 (2020)
9. Brandes, U.: A faster algorithm for betweenness centrality. J. Math. Sociol. **25**(2), 163–177 (2001)
10. Bui-Xuan, B.-M., Ferreira, A., Jarry, A.: Computing shortest, fastest, and foremost journeys in dynamic networks. Int. J. Found. Comput. Sci. **14**(02), 267–285 (2003)
11. Buß, S., Molter, H., Niedermeier, R., Rymar, M.: Algorithmic aspects of temporal betweenness. In: Proceedings of the 26th ACM SIGKDD International Conference on Knowledge Discovery & Data Mining (KDD 2020), pp. 2084–2092. Association for Computing Machinery (2020)
12. Casteigts, A., Himmel, A.-S., Molter, H., Zschoche, P.: The computational complexity of finding temporal paths under waiting time constraints. In: Proceedings of the 31st International Symposium on Algorithms and Computation (ISAAC 2020), LIPIcs, vol. 181, pages 30:1–30:18. Schloss Dagstuhl-Leibniz-Zentrum für Informatik (2020)
13. Ford Jr., L.R.: Network flow theory. Technical report, Rand Corp Santa Monica CA (1956)
14. Freeman, L.C.: A set of measures of centrality based on betweenness. Sociometry **40**(1), 35–41 (1977)

15. Geisberger, R., Sanders, P., Schultes, D.: Better approximation of betweenness centrality. In: Proceedings of the 10th Meeting on Algorithm Engineering & Expermiments (ALENEX 2008), pp. 90–100. SIAM (2008)
16. Gunturi, V.M., Shekhar, S., Joseph, K., Carley, K.M.: Scalable computational techniques for centrality metrics on temporally detailed social network. Mach. Learn. **106**(8), 1133–1169 (2017)
17. Habiba, H., Tantipathananandh, C., Berger-Wolf, T.Y.: Betweenness centrality measure in dynamic networks. Technical Report 19, Department of Computer Science, University of Illinois at Chicago, Chicago, DIMACS Technical Report (2007)
18. Kempe, D., Kleinberg, J., Kumar, A.: Connectivity and inference problems for temporal networks. J. Comput. Syst. Sci. **64**(4), 820–842 (2002)
19. Kim, H., Anderson, R.: Temporal node centrality in complex networks. Phys. Rev. E **85**(2), 026107 (2012)
20. Puzis, R., Elovici, Y., Zilberman, P., Dolev, S., Brandes, U.: Topology manipulations for speeding betweenness centrality computation. J. Complex Netw. **3**(1), 84–112 (2015)
21. Riondato, M., Kornaropoulos, E.M.: Fast approximation of betweenness centrality through sampling. Data Min. Knowl. Discov. **30**(2), 438–475 (2016)
22. Rymar, M., Molter, H., Nichterlein, A., Niedermeier, R.: Towards classifying the polynomial-time solvability of temporal betweenness centrality. CoRR, abs/2105.13055 (2021). https://arxiv.org/abs/2105.13055
23. Sariyüce, A.E., Kaya, K., Saule, E., Çatalyürek, Ü.V.: Graph manipulations for fast centrality computation. ACM Trans. Knowl. Discovery Data **11**(3):26:1–26:25 (2017)
24. Simard, F., Magnien, C., Latapy, M.: Computing betweenness centrality in link streams. CoRR, abs/2102.06543 (2021). https://arxiv.org/abs/2102.06543
25. Tang, J., Musolesi, M., Mascolo, C., Latora, V., Nicosia, V.: Analysing information flows and key mediators through temporal centrality metrics. In: Proceedings of the 3rd Workshop on Social Network Systems (SNS 2010). Association for Computing Machinery (2010)
26. Tsalouchidou, I., Baeza-Yates, R., Bonchi, F., Liao, K., Sellis, T.: Temporal betweenness centrality in dynamic graphs. Int. J. Data Sci. Anal. **9**(3), 257–272 (2020)
27. Williams, M.J., Musolesi, M.: Spatio-temporal networks: reachability, centrality and robustness. Roy. Soc. Open Sci. **3**(6) (2016)
28. Wu, H., Cheng, J., Ke, Y., Huang, S., Huang, Y., Wu, H.: Efficient algorithms for temporal path computation. IEEE Trans. Knowl. Data Eng. **28**(11), 2927–2942 (2016)
29. Zschoche, P., Fluschnik, T., Molter, H., Niedermeier, R.: The complexity of finding separators in temporal graphs. J. Comput. Syst. Sci. **107**, 72–92 (2020)

The Dynamic Complexity of Acyclic Hypergraph Homomorphisms

Nils Vortmeier[1(✉)] and Ioannis Kokkinis[2,3]

[1] University of Zurich, Zurich, Switzerland
nils.vortmeier@uzh.ch
[2] National Technical University of Athens, Athens, Greece
[3] University of the Aegean, Mytilene, Greece
ikokkinis@aegean.gr

Abstract. Finding a homomorphism from some hypergraph $\mathcal{Q}$ (or some relational structure) to another hypergraph $\mathcal{D}$ is a fundamental problem in computer science. We show that an answer to this problem can be maintained under single-edge changes of $\mathcal{Q}$, as long as it stays acyclic, in the DynFO framework of Patnaik and Immerman that uses updates expressed in first-order logic. If additionally also changes of $\mathcal{D}$ are allowed, we show that it is unlikely that existence of homomorphisms can be maintained in DynFO.

Keywords: Dynamic complexity · Conjunctive queries · Hypergraph homomorphisms

1 Introduction

Many important computational problems can be phrased as the question "is there a homomorphism from $\mathcal{Q}$ to $\mathcal{D}$?", where $\mathcal{Q}$ and $\mathcal{D}$ are hypergraphs, or more generally, relational structures. Examples include evaluation and minimisation of conjunctive queries [4] and solving constraint satisfaction problems, see [10].

The problem HOM – is there a homomorphism from $\mathcal{Q}$ to $\mathcal{D}$? – is NP-complete in its general form. In the static setting it is well understood which restrictions on $\mathcal{Q}$ or $\mathcal{D}$ render the problem tractable [5,14,16]. A particular restriction of great importance in databases is to demand that $\mathcal{Q}$ is *acyclic* [1]. This restriction of HOM, we call it the Acyclic Hypergraph Homomorphism problem AHH, can be solved in polynomial time by Yannakakis' algorithm [25] and is complete for the complexity class LOGCFL [12], the class of problems that can be reduced in logarithmic space to a context-free language.

We are interested in a *dynamic* setting where the input of a problem is subject to changes. The complexity-theoretic framework DynFO for such a dynamic setting was introduced by Patnaik and Immerman [20] and it is closely related to a setting of Dong, Su and Topor [8]. In this setting, a relational input structure is subject to a sequence of changes, which are usually insertions of single tuples into a relation, or deletions of single tuples from a relation. After each change,

Ł. Kowalik et al. (Eds.): WG 2021, LNCS 12911, pp. 232–244, 2021.
https://doi.org/10.1007/978-3-030-86838-3_18

additionally stored auxiliary relations are updated as specified by first-order *update formulas.* The class DynFO contains all problems for which the update formulas can maintain the answer for the changing input.

With few exceptions, for example in parts of [19], research in the DynFO framework takes a *data complexity* viewpoint: all context-free languages [11] and all problems definable in monadic second-order logic MSO [7] are in DynFO if the context-free language or the MSO-definable problem is fixed and not part of the input. Every fixed conjunctive query is trivially in DynFO, as such a query can be expressed in first-order logic and updates defined by first-order formulas can just compute the result from scratch after every change; however, there are also non-trivial maintenance results for fixed conjunctive queries for subclasses of DynFO [11,26]. The complexity results for HOM and AHH of [12,25] are however from a *combined complexity* perspective: both $\mathcal{Q}$ and $\mathcal{D}$ are part of the input.

Contributions. In this paper we study the combined complexity of AHH in the dynamic setting. As inputs we allow hypergraphs and general relational structures over some fixed schema τ.

As our main positive result, we show that AHH(τ) is in DynFO for every schema τ, if $\mathcal{Q}$ is subject to insertions and deletions of hyperedges but stays acyclic, and $\mathcal{D}$ may initially be arbitrary but is not changed afterwards. A main building block for this result is a proof that a *join tree* for $\mathcal{Q}$ can be maintained in DynFO in such a way that after a single change to $\mathcal{Q}$ the maintained join tree only changes by a constant number of edges. We show that given a join tree for $\mathcal{Q}$ we can maintain the answer to AHH(τ) under changes of single edges of the join tree. The main result follows by compositionality properties of DynFO.

We also give a hardness result for the case that also $\mathcal{D}$ is subject to changes. If AHH(τ) is in DynFO for every schema τ under changes of $\mathcal{Q}$ and $\mathcal{D}$, then all LOGCFL-problems are in (a variant of) DynFO, which we believe not to be the case. So, this result is a strong indicator that maintenance under changes of $\mathcal{D}$ is not possible in DynFO. Note that this result does not follow immediately from the fact that AHH is LOGCFL-complete: the NL-complete problem of reachability in directed graphs is in DynFO [6] as well as a PTIME-complete problem [20], and these results do not imply that all NL-problems and even all PTIME-problems are in DynFO, as this class is not known to be closed under the usual classes of reductions.

Further Related Work. In databases, Incremental View Maintenance is concerned with updating the result of a database query after a change of the input, see [15] for an overview. Koch [17] shows that a set of queries that include conjunctive queries can be maintained incrementally by low-complexity updates. A system for maintaining the result of Datalog-like queries under changes of the data and the queries is described in [13].

Organisation. We introduce preliminaries and the DynFO framework in Sect. 2. Section 3 contains the maintenance result for AHH under changes of $\mathcal{Q}$, the hardness result for changes of $\mathcal{D}$ is presented in Sect. 4. We conclude in Sect. 5. Proof details omitted due to space constraints can be found in [23].

2 Preliminaries and Setting

We introduce some concepts and notation that we need throughout the paper. See also [21] for an overview of Dynamic Complexity. We assume familiarity with first-order logic FO, and refer to [18] for basics of Finite Model Theory.

A *(purely relational) schema* τ consists of a finite set of relation symbols with a corresponding arity. A *structure* $\mathcal{D}$ over schema τ with finite domain D has, for every k-ary relation symbol $R \in \tau$, a relation $R^{\mathcal{D}} \subseteq D^k$. We assume that all structures come with a linear order $\leq$ on their domain D, which allows us to identify D with $\{1, \dots, n\}$, for $n = |D|$. We also assume that first-order formulas have access to this linear order and to compatible relations $+$ and $\times$ encoding addition and multiplication on $\{1, \dots, n\}$.

The Dynamic Complexity Framework. In the dynamic complexity framework as introduced by Patnaik and Immerman [20], the goal of a *dynamic program* is to answer a standing query to an input structure $\mathcal{I}$ under changes. To do so, the program stores and updates an *auxiliary structure* $\mathcal{A}$, which is over the same domain as $\mathcal{I}$. This structure consists of a set of *auxiliary relations.*

The set of admissible changes to the input structure is specified by a set Δ of *change operations.* We mostly consider the change operations INS_R and DEL_R for a relation R of the input structure. A *change* $\delta(\bar{a})$ consists of a change operation $\delta \in \Delta$ and a tuple $\bar{a}$ over the domain of $\mathcal{I}$. The change $\text{INS}_R(\bar{a})$ inserts the tuple $\bar{a}$ into the relation R and the change $\text{DEL}_R(\bar{a})$ deletes $\bar{a}$ from R.

For every change operation $\delta \in \Delta$ and every auxiliary relation S, a dynamic program has a first-order *update rule* that specifies how S is updated after a change over δ. Such a rule is of the form **on change** $\delta(\bar{p})$ **update** $S(\bar{x})$ **as** $\varphi^S_\delta(\bar{p}; \bar{x})$, where the *update formula* φ^S_δ is a first-order formula over the combined schema of $\mathcal{I}$ and $\mathcal{A}$. After a change $\delta(\bar{a})$ is applied, the relation S is updated to $\{\bar{b} \mid (\mathcal{I}, \mathcal{A}) \models \varphi^S_\delta(\bar{a}; \bar{b})\}$.

We say that a dynamic program $\mathcal{P}$ *maintains* a query Q under changes from Δ if a dedicated auxiliary relation ANS contains the answer to Q for the current input structure after each sequence of changes over Δ. The class DynFO contains all queries that can be maintained by dynamic programs with first-order update rules, starting from initially empty input and auxiliary relations. We also say that Q *can be maintained* in DynFO under Δ changes.

In this paper we are interested in scenarios where only parts of the input are subject to changes. To have a meaningful setting we then have to allow non-empty initial input relations. We then say that a query can be maintained in DynFO *starting from* non-empty inputs. Sometimes we then also allow the auxiliary relations to be initialised within some complexity bound. We say that a query Q is in DynFO *with $\mathcal{C}$ initialisation*, for a complexity class $\mathcal{C}$, if there is a $\mathcal{C}$-algorithm A such that Q can be maintained in DynFO if for an initial input $\mathcal{I}_0$ the initial auxiliary relations are set to the result of A applied to $\mathcal{I}_0$.

The reductions usually used in dynamic complexity are bounded first-order reductions [20]. A reduction f is *bounded* if there is a global constant c such that if the structure $\mathcal{D}'$ is obtained from the structure $\mathcal{D}$ by inserting or deleting one tuple, then $f(\mathcal{D}')$ can be obtained from $f(\mathcal{D})$ by inserting and/or deleting at most c tuples. We will not directly employ these reductions here, but we will use the simple proof idea to show that DynFO is closed under these reductions (see [20]): if a query Q can be maintained by a dynamic program $\mathcal{P}$ under insertions and deletions of single tuples, then there is also a dynamic program that can maintain Q under insertions and deletions of up to c tuples, for any constant c. That dynamic program can be obtained by nesting c copies of the update formulas of $\mathcal{P}$.

Hypergraphs and Homomorphisms. We use the term *hypergraph* in a very broad sense. For this paper, a hypergraph $\mathcal{H}$ is just a relational structure over a purely relational schema $\tau = \{E_1, \dots, E_m\}$, that is, a structure $\mathcal{H} = (\mathcal{V}, E_1, \dots, E_m)$, where the domain $\mathcal{V}$ is a set of nodes and the relations $E_1, \dots, E_m$ are sets of (labelled) hyperedges. This definition implies that the maximal size of any hyperedge, that is, the maximal arity of a relation E_i, is a constant that only depends on τ. Sometimes we ignore the labels and denote $\mathcal{H}$ as a tuple $(\mathcal{V}, \mathcal{E})$, where $\mathcal{E} = E_1 \cup \dots \cup E_m$ is the set of all hyperedges.

A *spanning forest* of an undirected graph $G = (V, E)$ is defined in the usual way. We encode a spanning forest as a structure (V, F, P) where F is the set of spanning edges and P is a ternary relation that describes paths in the spanning forest. A tuple $(s, t, u) \in P$ indicates that (1) s and t are in the same connected component of the spanning forest and (2) the unique path from s to t in the spanning forest is via the node u. Patnaik and Immerman [20] have shown that spanning forests with this encoding can be maintained in DynFO under insertions and deletions of single edges [20, Theorem 4.1].

A *join forest* $J(\mathcal{H})$ of a hypergraph $\mathcal{H} = (\mathcal{V}, E_1, \dots, E_m)$ is a forest whose nodes are the hyperedges of $\mathcal{H}$, such that if two hyperedges e, e' have a node $v \in \mathcal{V}$ in common, then they are in the same connected component of $J(\mathcal{H})$ and all nodes on the unique path from e to e' in $J(\mathcal{H})$ are hyperedges of $\mathcal{H}$ that also include v. We encode a join forest using relations F_{ij} and P_{ijk} with the same intended meaning as for spanning forests, where $i, j, k \in \{1, \dots, m\}$. The arity of F_{ij} is the sum of the arities of E_i and E_j, a tuple $(e, e') \in F_{ij}$ indicates that $J(\mathcal{H})$ has an edge between the hyperedges $e \in E_i$ and $e' \in E_j$. The use of P_{ijk} is analogous.

We define that a hypergraph is *acyclic* if it has a join forest. This definition coincides with the notion of *α-acyclicity* introduced by Fagin [9]. See also [12, Section 2.2] for a detailed discussion of this notion.

A *homomorphism* from a hypergraph $\mathcal{H} = (\mathcal{V}, E_1^{\mathcal{H}}, \dots, E_m^{\mathcal{H}})$ to a hypergraph $\mathcal{H}' = (\mathcal{V}', E_1^{\mathcal{H}'}, \dots, E_m^{\mathcal{H}'})$ is a map $h \colon \mathcal{V} \to \mathcal{V}'$ that preserves the hyperedge relations. So, for all relations E_i and all tuples $(v_1, \dots, v_\ell)$ over $\mathcal{V}$, where ℓ is the arity of E_i, if $(v_1, \dots, v_\ell) \in E_i^{\mathcal{H}}$ is a hyperedge of $\mathcal{H}$, then $(h(v_1), \dots, h(v_\ell)) \in E_i^{\mathcal{H}'}$ is a hyperedge of $\mathcal{H}'$.

The main problem we study is the Acyclic Hypergraph Homomorphism problem AHH(τ), where τ is a fixed schema. It asks, for two given hypergraphs $\mathcal{Q}$ and $\mathcal{D}$ over schema τ (where $\mathcal{Q}$ is acyclic), also called *query hypergraph* and *data hypergraph* respectively, whether there is a homomorphism from $\mathcal{Q}$ to $\mathcal{D}$.

3 Maintenance Under Changes of the Query Hypergraph

The goal of this section is to show that AHH can be maintained under changes of the query hypergraph $\mathcal{Q}$, as long as it stays acyclic. We also show that a DynFO program can recognise that a change would make $\mathcal{Q}$ cyclic. So, we do not need to assume that only changes occur that preserve acyclicity, if we allow the program to "deny" all other changes.

We introduce some notation of [12]. The *weighted hyperedge graph* $\text{WG}(\mathcal{H})$ of a hypergraph $\mathcal{H}$ is the undirected weighted graph $\text{WG}(\mathcal{H}) = (V, E, w)$ whose nodes V are the hyperedges of $\mathcal{H}$ and the set E contains an undirected edge (e, e') if e, e' are different hyperedges of $\mathcal{H}$ that have at least one node in common. The weight $w((e, e'))$ of such an edge is the number of nodes that e and e' have in common.

The *weight* $\text{W}(\mathcal{H})$ of a hypergraph $\mathcal{H}$ is the sum over the degrees of the non-isolated nodes of $\mathcal{H}$, where each degree is decremented by one. So, if for $\mathcal{H} = (\mathcal{V}, E_1, \dots, E_m)$ the set $\mathcal{V}_{\text{NI}} \subseteq \mathcal{V}$ contains all nodes of $\mathcal{H}$ that appear in at least one hyperedge, then $\text{W}(\mathcal{H}) = \sum_{v \in \mathcal{V}_{\text{NI}}} (\deg(v) - 1)$.

The following lemma provides the basis for our approach. It was originally proven in [2], we follow the presentation of [12, Proposition 3.5].

Lemma 1 ([2], see also [12]). *Let $\mathcal{H}$ be a hypergraph.*

(a) The hypergraph $\mathcal{H}$ is acyclic if and only if the weight $\text{W}(\mathcal{H})$ of $\mathcal{H}$ is equal to the weight $\text{W}(\text{MSF}(\text{WG}(\mathcal{H})))$ of a maximal-weight spanning forest of $\text{WG}(\mathcal{H})$.
(b) If $\mathcal{H}$ is acyclic, then $\text{MSF}(\text{WG}(\mathcal{H}))$ is a join forest of $\mathcal{H}$.

Using this lemma, we prove that a dynamic program can maintain acyclicity of hypergraphs, as well as a join forest that only changes moderately when the input hypergraph is changed.

Theorem 2. *Let $\tau = \{E_1, \dots, E_m\}$ be a fixed schema. The following can be maintained in DynFO under insertions and deletions of single hyperedges:*

(a) whether a hypergraph over τ is acyclic, and
(b) a join forest for an acyclic hypergraph $\mathcal{H}$ over τ, as long as $\mathcal{H}$ stays acyclic. Moreover, there is a global constant c_τ such that if $J(\mathcal{H})$ is the maintained join forest for $\mathcal{H}$ and $J(\mathcal{H})'$ is the maintained join forest after a single hyperedge is inserted or deleted, then $J(\mathcal{H})$ and $J(\mathcal{H})'$ differ by at most c_τ edges.

The proof follows the idea that is brought forth by Lemma 1: we show that a maximal-weight spanning forest of $\text{WG}(\mathcal{H})$ and its weight can be maintained.

This weight is compared with the weight of $\mathcal{H}$, which is easy to maintain. If the weights are equal, then $\mathcal{H}$ is acyclic and the spanning forest is a join forest.

Already Patnaik and Immerman [20] describe how a spanning forest of an undirected graph can be maintained under changes of single edges, and their procedure [20, Theorem 4.1] can easily be extended towards maximal-weight spanning forests. However, we face the problem that inserting and deleting hyperedges of $\mathcal{H}$ implies insertions and deletions of nodes of $\text{WG}(\mathcal{H})$. While a spanning forest can easily be maintained in DynFO under node insertions, it is an open problem to maintain a spanning forest under node deletions: if the spanning forest is a star and its center node is deleted, then it seems that a spanning forest of the remaining graph needs to be defined from scratch, which is not possible using FO formulas. We circumvent this problem and show that we can maintain a spanning forest where the degree of every node is bounded by a constant.

Proof Sketch. We show how a maximal-weight spanning forest of $\text{WG}(\mathcal{H})$ and the weight $\text{W}(\mathcal{H})$ can be maintained; the result then follows using Lemma 1.

We start with the weight $\text{W}(\mathcal{H})$. If a hyperedge e is inserted, then the weight of the hypergraph increases by the number of nodes it contains that were not isolated before the insertion. Similarly, if e is deleted, then the weight decreases by the number of nodes it contains that do not become isolated. This update can easily be expressed by first-order formulas.

Now we consider maintaining a maximal-weight spanning forest of $\text{WG}(\mathcal{H})$.

Let a_{MAX} be the maximal arity of a relation in τ. Any hyperedge of $\mathcal{H} = (\mathcal{V}, E_1, \ldots, E_m)$ can only include at most a_{MAX} many nodes and there are at most $r \stackrel{\text{def}}{=} 2^{a_{\text{MAX}}} - 1$ many different non-empty sets of nodes that a fixed hyperedge can have in common with any other hyperedge. We can show how to maintain a maximal-weight spanning forest of $\text{WG}(\mathcal{H})$ where each node has degree at most $2r$. More specifically, for any node e of $\text{WG}(\mathcal{H})$ (which is a hyperedge of $\mathcal{H}$) and each non-empty set A of nodes appearing in e, the maintained spanning forest contains at most two edges $(e, e_1), (e, e_2)$ such that the set of nodes that e has in common with e_1 and e_2, respectively, is exactly A. We call this property *Invariant* $(\star)$. The details are omitted here. □

We now present the main maintenance result of this paper.

Theorem 3. *Let* $\tau = \{E_1, \ldots, E_m\}$ *be a fixed schema. The problem* $\text{AHH}(\tau)$ *can be maintained in* DynFO, *starting from an arbitrary initial hypergraph* $\mathcal{D}$ *and an initially empty hypergraph* $\mathcal{Q}$, *under insertions and deletions of single hyperedges of* $\mathcal{Q}$, *as long as this hypergraph stays acyclic.*

The proof uses the idea of Yannakakis' algorithm [25] for evaluating a conjunctive query. This algorithm processes a join tree for a query $\mathcal{Q}$ in a bottom-up fashion. In a first step, for each node $E_i(\bar{x})$ of the join tree (which is a hyperedge of $\mathcal{Q}$) all assignments $\bar{y}$ for its variables are stored such that $E_i(\bar{y})$ exists in the data hypergraph $\mathcal{D}$. Then, bottom-up, each inner node removes all of its variable assignments that are not consistent with the assignments of its children. So, an assignment $\bar{y}$ for a node $E_i(\bar{x})$ is removed if there is a child $E_j(\bar{x}')$ of $E_i(\bar{x})$ such that no stored assignment $\bar{y}'$ of that child agrees with $\bar{y}$ on the common variables of $\bar{x}$ and $\bar{x}'$. All remaining stored assignments for $E_i(\bar{x})$ can be extended to a homomorphism for the subhypergraph of $\mathcal{Q}$ that consists of the hyperedges that are in the subtree of the join tree rooted at $E_i(\bar{x})$. A homomorphism from $\mathcal{Q}$ to $\mathcal{D}$ exists if after the join tree is processed the root has remaining assignments.

Proof. Let $\mathcal{Q}$ be an acyclic hypergraph over some schema τ and let $\mathcal{D}$ be a hypergraph over the same schema. Also, let $J(\mathcal{Q})$ be a join forest of $\mathcal{Q}$.

We adapt a technique that was used by Gelade, Marquardt and Schwentick [11] to show that regular tree languages can be maintained in a subclass of DynFO. For each triple E_i, E_j, E_k of symbols from τ we maintain an auxiliary relation $H_{ijk}(\bar{r}, \bar{x}_1, \bar{x}_2, \bar{y}_1, \bar{y}_2)$ with the following intended meaning. A tuple $(\bar{r}, \bar{x}_1, \bar{x}_2, \bar{y}_1, \bar{y}_2)$ is in H_{ijk} if

(1) the hyperedges $E_i(\bar{r})$, $E_j(\bar{x}_1)$ and $E_k(\bar{x}_2)$ are present in $\mathcal{Q}$ and in the same connected component C of $J(\mathcal{Q})$,
(2) when we consider $E_i(\bar{r})$ to be the root of C then $E_j(\bar{x}_1)$ is a descendant of $E_i(\bar{r})$ and $E_k(\bar{x}_2)$ is a descendant of $E_j(\bar{x}_1)$, and
(3) if we assume that there is a homomorphism h_2 of the subtree of C rooted at $E_k(\bar{x}_2)$ into $\mathcal{D}$ such that $h_2(\bar{x}_2) = \bar{y}_2$, then it follows that there also is a homomorphism h_1 of the subtree of C rooted at $E_j(\bar{x}_1)$ into $\mathcal{D}$ such that $h_1(\bar{x}_1) = \bar{y}_1$.

Phrased differently, $(\bar{r}, \bar{x}_1, \bar{x}_2, \bar{y}_1, \bar{y}_2) \in H_{ijk}$ means that the hyperedges in $J(\mathcal{Q})$ which, considering $E_i(\bar{r})$ to be the root, are in the subtree of $E_j(\bar{x}_1)$ but not in the subtree of $E_k(\bar{x}_2)$, can be mapped into $\mathcal{D}$ by a homomorphism that maps the elements $\bar{x}_1$ to $\bar{y}_1$ and the elements $\bar{x}_2$ to $\bar{y}_2$.

If $(\bar{r}, \bar{x}_1, \bar{x}_2, \bar{y}_1, \bar{y}_2) \in H_{ijk}$ holds we say that $\bar{y}_1$ is a *valid partial assignment* for $E_j(\bar{x}_1)$ *down to* $(E_k(\bar{x}_2), \bar{y}_2)$.

Notice that from these relations one can first-order define relations $H'_{ij}(\bar{r}, \bar{x}, \bar{y})$ with the intended meaning that $(\bar{r}, \bar{x}, \bar{y}) \in H'_{ij}$ if

(1) the hyperedges $E_i(\bar{r})$ and $E_j(\bar{x})$ are in the same connected component C of $J(\mathcal{Q})$, and
(2) when we consider $E_i(\bar{r})$ to be the root of C then there is a homomorphism h of the subtree of C rooted at $E_j(\bar{x})$ into $\mathcal{D}$ such that $h(\bar{x}) = \bar{y}$.

For this, a first-order formula existentially quantifies a hyperedge $E_k(\bar{x}_2)$ and a tuple $\bar{y}_2$ of elements, and checks that $E_k(\bar{x}_2)$ is a leaf of the component C with root $E_i(\bar{r})$, that the hyperedge $E_k(\bar{y}_2)$ exists in $\mathcal{D}$ and that $(\bar{r}, \bar{x}, \bar{x}_2, \bar{y}, \bar{y}_2) \in H_{ijk}$ holds. Whether a node is a leaf in a join tree can be expressed using the join tree's paths relations P_{ijk}, all other conditions are clearly first-order expressible. We assume in the following that these relations are available. If $(\bar{r}, \bar{x}, \bar{y}) \in H'_{ij}$ holds we say that $\bar{y}$ is a *valid partial assignment* for $E_j(\bar{x})$.

We argue next that if we can maintain these auxiliary relations under insertions and deletions of single edges of the join forest, then the statement of the theorem follows.

Notice that from the auxiliary relations a first-order formula can express whether a homomorphism from $\mathcal{Q}$ to $\mathcal{D}$ exists. To check this, a formula needs to express that for every connected component of $J(\mathcal{Q})$ there is a homomorphism from this component to $\mathcal{D}$. This is the case if for each hyperedge $E_i(\bar{r})$ of $\mathcal{Q}$ there is a tuple $\bar{y}$ such that $(\bar{r}, \bar{r}, \bar{y})$ is in H'_{ii}.

It remains to argue that it is sufficient to maintain the auxiliary relations under changes of single edges of the join forest. From Theorem 2 we know that a join forest $J(\mathcal{Q})$ for $\mathcal{Q}$ can be maintained in DynFO under insertions and deletions of single hyperedges, as long as it stays acyclic. Moreover, after each edge change, the maintained join forest only differs in a constant number of edges from its previous version. If we have a dynamic program that is able to process single edge changes of the join forest, then by nesting its update formulas c times we can obtain a dynamic program $\mathcal{P}'$ that is able to process c edge changes at once. In summary, a dynamic program $\mathcal{P}$ for AHH maintains a join forest as described by Theorem 2 and after every change of a hyperedge it uses $\mathcal{P}'$ to update the auxiliary relations and to decide whether a homomorphism exists.

The relations H_{ijk} can be maintained by first-order formulas under insertions and deletions of single edges of the join forest. The details are omitted here. □

4 Hardness Under Changes of the Data Hypergraph

We have seen in the previous section that one can maintain the existence of homomorphisms in DynFO if only the query hypergraph $\mathcal{Q}$ may change and the data hypergraph $\mathcal{D}$ remains the same. The dynamic program we constructed for the proof of Theorem 3 can not directly cope with changes of $\mathcal{D}$. This is because $\mathcal{Q}$ might contain several hyperedges $E_i(\bar{x}_1), \ldots, E_i(\bar{x}_m)$ over a single relation E_i: if a change of $\mathcal{D}$ occurs, then the number of nodes in the join tree for which we have to take this change into account when updating partial valid assignments is a priori unbounded. If we disallow multiple hyperedges over the same relation in $\mathcal{Q}$, then we can actually allow a change to replace an arbitrary number of $\mathcal{D}$-hyperedges, as long as each change only affects a single relation of $\mathcal{D}$. Such a restriction of $\mathcal{Q}$ translates to *self-join free* acyclic conjunctive queries.

Corollary 4. *Let $\tau = \{E_1, \ldots, E_m\}$ be a fixed schema. As long as $\mathcal{Q}$ remains acyclic and contains at most one hyperedge $E_i(\bar{x})$ for each relation $E_i \in \tau$, the problem* AHH(τ) *can be maintained in* DynFO *under insertions and deletions of single hyperedges of $\mathcal{Q}$ and under arbitrary changes of a single relation of $\mathcal{D}$.*

In the remainder of this section, we will see that if $\mathcal{Q}$ might be an arbitrary acyclic hypergraph, then a maintenance result for AHH(τ) under changes of $\mathcal{D}$ is unlikely, even if in turn $\mathcal{Q}$ is not allowed to change.

Gottlob et al. [12] show that it is LOGCFL-complete to decide whether from a given acyclic hypergraph $\mathcal{Q}$ there is a homomorphism into a hypergraph $\mathcal{D}$. The complexity class LOGCFL contains all problems that can be reduced in logarithmic space to a context-free language. This class is contained in AC^1, contains NL and is equivalent to logspace-uniform SAC^1 [22], the class of problems decidable by logspace-uniform families of semi-unbounded Boolean circuits of polynomial size and logarithmic depth. A semi-unbounded Boolean circuit consists of or-gates with unbounded fan-in and and-gates with fan-in 2. There are no negation gates, but for each input gate x_i there is an additional input gate $\neg x_i$ that carries the negated value of x_i.

In their article, Gottlob et al. [12] show that there is a schema τ such that every SAC^1 problem can be reduced in logarithmic space to $\mathrm{AHH}(\tau)$. We slightly adapt their construction and show that the hardness result also holds for *bounded* logspace reductions. Furthermore, if a reduction f maps an instance x to an instance $f(x)$, then the change to $f(x)$ induced by a change to x is first-order definable.

Theorem 5 (adapted from [12, Theorem 4.8]).

(a) There is a schema τ that contains at most binary relations such that $\mathrm{AHH}(\tau)$ is hard for LOGCFL under logspace reductions.
(b) Let $L \in \mathsf{LOGCFL}$. There is a logspace reduction f_L from L to $\mathrm{AHH}(\tau)$ that satisfies the following properties. Assume that x, x' are instances of L with $|x| = |x'|$ and let $(\mathcal{Q}, \mathcal{D}) = f_L(x)$ and $(\mathcal{Q}', \mathcal{D}') = f_L(x')$. Then:
  (i) $\mathcal{Q} = \mathcal{Q}'$,
  (ii) if x and x' differ only in one bit, then $\mathcal{D}'$ differs from $\mathcal{D}$ by at most c hyperedges, for a global constant c, and
  (iii) $\mathcal{D}'$ is first-order definable from $\mathcal{D}$, x and x'.

Proof. Let L be a problem from LOGCFL. As LOGCFL = logspace-uniform SAC^1, there is a logspace-uniform family $(C_n)_{n \in \mathbb{N}}$ of circuits that decides L, where a circuit C_n has size at most n^k for some $k \in \mathbb{N}$, logarithmic depth in n, and the fan-in of every and-gate is bounded by 2. Without loss of generality, see [12, Lemma 4.6], we can assume that C_n also has the following *normal form*:

(1) the circuit consists of layers of gates, and the gates of layer i receive all their inputs from gates at layer $i - 1$,
(2) all layers either only contain or-gates or only contain and-gates,
(3) the first layer after the inputs consists of or-gates,
(4) if layer i is a layer of or-gates, then layer $i + 1$ only consists of and-gates, and vice versa, and
(5) the output gate is an and-gate.

A circuit of this form accepts its input if and only if a proof tree can be homomorphically mapped into it. A *proof tree* T_n for a circuit C_n in normal form has the same depth as C_n and an and-gate as its root. Each and-gate of the proof tree has two or-gates as its children, and each or-gate has one child, which is a gate labelled with the constant 1 for an or-gate at the lowest layer, and an and-gate for all other or-gates. Note that a proof tree is acyclic.

If there is a homomorphism that maps each constant 1 of the proof tree to an input gate of the circuit that is set to 1, each and-gate of the proof tree to an and-gate of the circuit, and for each and-gate of the proof tree its two children to different or-gates in the circuit, then all gates of the circuit that are in the image of the homomorphism evaluate to 1 for the current input. Therefore, the output gate also evaluates to 1, and the circuit accepts its input. It is also clear that if the circuit accepts its input, then there is a homomorphism from the proof tree into the circuit.

We encode circuits and proof trees over the schema $\tau = \{0, 1, \text{OR}, \text{AND-LEFT}, \text{AND-RIGHT}\}$. Each gate g is encoded by a tuple $\text{ENC}(g)$ of k elements. If g is an and-gate with children g_1, g_2, then this is encoded by tuples $(\text{ENC}(g), \text{ENC}(g_1)) \in \text{AND-LEFT}$ and $(\text{ENC}(g), \text{ENC}(g_2)) \in \text{AND-RIGHT}$. If g is an or-gate and g' is one of its children, then this is encoded by the tuple $(\text{ENC}(g), \text{ENC}(g')) \in \text{OR}$. The relations 0 and 1 are used to encode constants and assignments of input gates in the obvious way.

We use the two relations AND-LEFT, AND-RIGHT to ensure that a homomorphism from a proof tree to a circuit maps the two children of an and-gate to two different or-gates.

From the proof of [12, Theorem 4.8] it follows that from an input x of L with $|x| = n$ the corresponding circuit $C_n(x)$, which results from C_n by assigning constants to its inputs gates as specified by x, and the corresponding proof tree T_n can be computed in logarithmic space. In conclusion, this proves that the function f_L that maps x to $(T_n, C_n(x))$ is a logspace reduction from L to $\text{AHH}(\tau)$, and therefore that $\text{AHH}(\tau)$ is hard for LOGCFL under logspace reductions.

We now proceed to prove part (b) of the theorem statement. Consider two input instances x, x' for L with $|x| = |x'| = n$. Both x and x' are inputs of the circuit C_n, so the same proof tree is constructed for them by f_L, yielding part (b)(i). The only differences in the images of f_L are the assignments of constants to the input gates of C_n. If x and x' only differ in one bit, say, the first bit that is represented by the input gate g_1, then we have $\text{ENC}(g_1) \in 0$ and $\text{ENC}(\neg g_1) \in 1$ for one input, and $\text{ENC}(g_1) \in 1$ and $\text{ENC}(\neg g_1) \in 0$ for the other input. So, the encodings of the circuit only differ by 4 tuples, which implies part (b)(ii). Towards part (b)(iii), we can ensure that these tuples are first-order definable by using an appropriate encoding ENC of the gates, for example by encoding the i-th input gate by the i-th tuple in the lexicographic ordering of k-tuples over the domain. □

Building on the hardness result of Theorem 5, we can show that if $\text{AHH}(\tau)$ can be maintained in DynFO under changes of $\mathcal{D}$, then all LOGCFL-problems are in DynFO if we allow a PTIME initialisation. This would be a breakthrough result, as there are already problems in uniform $\mathsf{AC}^0[2]$ (problems decidable by uniform circuits with polynomial size, constant depth and not-, and-, or- and modulo 2-gates with arbitrary fan-in), a much smaller complexity class, that we do not know how to maintain in DynFO [24].

Theorem 6. *If for arbitrary schema τ the problem $\text{AHH}(\tau)$ can be maintained in* DynFO *under insertions and deletions of single hyperedges from $\mathcal{Q}$ and $\mathcal{D}$, as long as $\mathcal{Q}$ stays acyclic, then every problem $L \in$* LOGCFL *can be maintained in* DynFO *with* PTIME *initialisation under insertions and deletions of single tuples.*

The same even holds under the condition that AHH(τ) can only be maintained under changes of single hyperedges of $\mathcal{D}$, but starting from an arbitrary initial acyclic hypergraph $\mathcal{Q}$, even if a PTIME initialisation of the auxiliary relations is allowed. So, we can take this theorem as a strong indicator that AHH might not be in DynFO under changes of $\mathcal{D}$.

Proof. Let $L \in$ LOGCFL be arbitrary. Let τ be the schema and f_L the reduction guaranteed to exist by Theorem 5 such that f_L is a reduction from L to AHH(τ). Let $\mathcal{P}$ be a dynamic program that maintains AHH(τ) under insertions and deletions of single hyperedges. We construct a dynamic program $\mathcal{P}'$ with PTIME initialisation that maintains L.

For an initially empty input structure $\mathcal{I}$ over a domain of size n, the initialisation first constructs the corresponding SAC^1-circuit $C_n(\mathcal{I})$, with the input bits set as given by $\mathcal{I}$, and the proof tree T_n and stores them in auxiliary relations. This is possible in LOGSPACE $\subseteq$ PTIME. Then, using polynomial time, it simulates $\mathcal{P}$ for a sequence of insertions that lead to $C_n(\mathcal{I})$ and T_n from initially empty hypergraphs and stores the produced auxiliary relations.

When a change of $\mathcal{I}$ occurs, $\mathcal{P}'$ identifies the constantly many changes of $C_n(\mathcal{I})$ that are induced by the change, which is possible in first-order logic thanks to Theorem 5, and simulates $\mathcal{P}$ for these changes. □

5 Conclusion and Further Work

In this paper we studied under which conditions the problem AHH can be maintained in DynFO. Our main result is that this problem is in DynFO under changes of single hyperedges of the query hypergraph $\mathcal{Q}$, on the condition that it remains acyclic. This result directly implies that the result of acyclic Boolean conjunction queries can be maintained in DynFO. As the corresponding dynamic program, see the proof of Theorem 3, also maintains partial assignments of existing homomorphisms, this can straightforwardly be extended also to non-Boolean acyclic conjunctive queries. We have also seen that it is unlikely that AHH is in DynFO under changes of the data hypergraph $\mathcal{D}$.

In the static setting, the homomorphism problem is not only tractable for acyclic hypergraphs $\mathcal{Q}$, but for a larger class of graphs [5] which includes the class of graphs with *bounded treewidth*, see [12]. It is therefore interesting whether our DynFO maintenance result can also be extended to allow for cyclic hypergraphs $\mathcal{Q}$, in particular to allow hypergraphs of treewidth at most k, for some k. Results of this form would probably require an analogous result to Theorem 2, so, that a tree decomposition of some width $f(k)$ can be maintained for every hypergraph with treewidth at most k, and that any change of the hypergraph leads to a maintained tree decomposition that can be obtained from its previous version by a constant number of changes.

Outside the DynFO framework, maintenance of tree decompositions for graphs with treewidth $k = 2$, that is, series-parallel graphs, is considered in [3], but a change of the graph may affect a logarithmic number of nodes of the tree decomposition. Preliminary unpublished results show (using different techniques than [3]) that for graphs with treewidth 2 a tree decomposition can be maintained also in DynFO, but it is so far unclear whether they can be maintained such that only a constant-size part changes after a change of the graph.

Acknowledgements. This project has received funding from the European Union's Horizon 2020 research and innovation programme under grant agreement No. 682588.

References

1. Beeri, C., Fagin, R., Maier, D., Yannakakis, M.: On the desirability of acyclic database schemes. J. ACM (JACM) **30**(3), 479–513 (1983). https://doi.org/10.1145/2402.322389
2. Bernstein, P.A., Goodman, N.: Power of natural semijoins. SIAM J. Comput. **10**(4), 751–771 (1981). https://doi.org/10.1137/0210059
3. Bodlaender, H.L.: Dynamic algorithms for graphs with treewidth 2. In: van Leeuwen, J. (ed.) WG 1993. LNCS, vol. 790, pp. 112–124. Springer, Heidelberg (1994). https://doi.org/10.1007/3-540-57899-4_45
4. Chandra, A.K., Merlin, P.M.: Optimal implementation of conjunctive queries in relational data bases. In: Hopcroft, J.E., Friedman, E.P., Harrison, M.A. (eds.) Proceedings of the 9th Annual ACM Symposium on Theory of Computing, Boulder, Colorado, USA, 4–6 May 1977, pp. 77–90. ACM (1977). https://doi.org/10.1145/800105.803397
5. Dalmau, V., Kolaitis, P.G., Vardi, M.Y.: Constraint satisfaction, bounded treewidth, and finite-variable logics. In: Van Hentenryck, P. (ed.) CP 2002. LNCS, vol. 2470, pp. 310–326. Springer, Heidelberg (2002). https://doi.org/10.1007/3-540-46135-3_21
6. Datta, S., Kulkarni, R., Mukherjee, A., Schwentick, T., Zeume, T.: Reachability is in DynFO. J. ACM **65**(5), 33:1–33:24 (2018). https://doi.org/10.1145/3212685
7. Datta, S., Mukherjee, A., Schwentick, T., Vortmeier, N., Zeume, T.: A strategy for dynamic programs: start over and muddle through. Log. Methods Comput. Sci. **15**(2) (2019). https://doi.org/10.23638/LMCS-15(2:12)2019
8. Dong, G., Su, J., Topor, R.W.: Nonrecursive incremental evaluation of datalog queries. Ann. Math. Artif. Intell. **14**(2–4), 187–223 (1995). https://doi.org/10.1007/BF01530820
9. Fagin, R.: Degrees of acyclicity for hypergraphs and relational database schemes. J. ACM (JACM) **30**(3), 514–550 (1983). https://doi.org/10.1145/2402.322390
10. Feder, T., Vardi, M.Y.: The computational structure of monotone monadic SNP and constraint satisfaction: a study through datalog and group theory. SIAM J. Comput. **28**(1), 57–104 (1998). https://doi.org/10.1137/S0097539794266766
11. Gelade, W., Marquardt, M., Schwentick, T.: The dynamic complexity of formal languages. ACM Trans. Comput. Log. **13**(3), 19:1–19:36 (2012). https://doi.org/10.1145/2287718.2287719
12. Gottlob, G., Leone, N., Scarcello, F.: The complexity of acyclic conjunctive queries. J. ACM **48**(3), 431–498 (2001). https://doi.org/10.1145/382780.382783
13. Green, T.J., Olteanu, D., Washburn, G.: Live programming in the LogicBlox system: a MetaLogiQL approach. Proc. VLDB Endow. **8**(12), 1782–1791 (2015). https://doi.org/10.14778/2824032.2824075

14. Grohe, M.: The complexity of homomorphism and constraint satisfaction problems seen from the other side. J. ACM **54**(1), 1:1–1:24 (2007). https://doi.org/10.1145/1206035.1206036
15. Gupta, A., Mumick, I.S.: Maintenance of Materialized Views: Problems, Techniques, and Applications, pp. 145–157. MIT Press, Cambridge (1999)
16. Hell, P., Nesetril, J.: On the complexity of H-coloring. J. Comb. Theory Ser. B **48**(1), 92–110 (1990). https://doi.org/10.1016/0095-8956(90)90132-J
17. Koch, C.: Incremental query evaluation in a ring of databases. In: Paredaens, J., Gucht, D.V. (eds.) Proceedings of the Twenty-Ninth ACM SIGMOD-SIGACT-SIGART Symposium on Principles of Database Systems, PODS 2010, Indianapolis, Indiana, USA, 6–11 June 2010, pp. 87–98. ACM (2010). https://doi.org/10.1145/1807085.1807100
18. Libkin, L.: Elements of Finite Model Theory. Springer, Heidelberg (2004). https://doi.org/10.1007/978-3-662-07003-1
19. Muñoz, P., Vortmeier, N., Zeume, T.: Dynamic graph queries. In: Martens, W., Zeume, T. (eds.) 19th International Conference on Database Theory (ICDT 2016). Leibniz International Proceedings in Informatics (LIPIcs), vol. 48, pp. 14:1–14:18. Schloss Dagstuhl-Leibniz-Zentrum fuer Informatik, Dagstuhl (2016). https://doi.org/10.4230/LIPIcs.ICDT.2016.14
20. Patnaik, S., Immerman, N.: Dyn-FO: a parallel, dynamic complexity class. J. Comput. Syst. Sci. **55**(2), 199–209 (1997). https://doi.org/10.1006/jcss.1997.1520
21. Schwentick, T., Vortmeier, N., Zeume, T.: Sketches of dynamic complexity. SIGMOD Rec. **49**(2), 18–29 (2020). https://doi.org/10.1145/3442322.3442325
22. Venkateswaran, H.: Properties that characterize LOGCFL. J. Comput. Syst. Sci. **43**(2), 380–404 (1991). https://doi.org/10.1016/0022-0000(91)90020-6
23. Vortmeier, N., Kokkinis, I.: The dynamic complexity of acyclic hypergraph homomorphisms. CoRR abs/2107.06121 (2021). https://arxiv.org/abs/2107.06121
24. Vortmeier, N., Zeume, T.: Dynamic complexity of parity exists queries. In: Fernández, M., Muscholl, A. (eds.) 28th EACSL Annual Conference on Computer Science Logic, CSL 2020. LIPIcs, Barcelona, Spain, 13–16 January 2020, vol. 152, pp. 37:1–37:16. Schloss Dagstuhl - Leibniz-Zentrum für Informatik (2020). https://doi.org/10.4230/LIPIcs.CSL.2020.37
25. Yannakakis, M.: Algorithms for acyclic database schemes. In: Proceedings of the 7th International Conference on Very Large Data Bases, Cannes, France, 9–11 September 1981, pp. 82–94. IEEE Computer Society (1981). https://dl.acm.org/doi/10.5555/1286831.1286840
26. Zeume, T.: The dynamic descriptive complexity of k-clique. Inf. Comput. **256**, 9–22 (2017). https://doi.org/10.1016/j.ic.2017.04.005

Linearizable Special Cases of the Quadratic Shortest Path Problem

Eranda Çela[1], Bettina Klinz[1], Stefan Lendl[1,2], James B. Orlin[3], Gerhard J. Woeginger[4(✉)], and Lasse Wulf[1]

[1] Institute of Discrete Mathematics, TU Graz, Graz, Austria
[2] Institute of Operations und Information Systems, University of Graz, Graz, Austria
[3] Sloan School of Management, M.I.T., Cambridge, USA
[4] Department of Computer Science, RWTH Aachen, Aachen, Germany
woeginger@algo.rwth-aachen.de

Abstract. The quadratic shortest path problem (QSPP) in a directed graph asks for a directed path from a given source vertex to a given sink vertex, so that the sum of the interaction costs over all pairs of arcs on the path is minimized. We consider special cases of the QSPP that are linearizable as a shortest path problem in the sense of Bookhold. If the QSPP on a directed graph is linearizable under all possible choices of the arc interaction costs, the graph is called universally linearizable.

We provide various combinatorial characterizations of universally linearizable graphs that are centered around the structure of source-to-sink paths and around certain forbidden subgraphs. Our characterizations lead to fast and simple recognition algorithms for universally linearizable graphs. Furthermore, we establish the intractability of deciding whether a concrete instance of the QSPP (with a given graph and given arc interaction costs) is linearizable.

1 Introduction

An instance of the *Shortest Path Problem* (SPP) consists of a directed graph $G = (V, A)$ together with a source vertex $s \in V$, a sink vertex $t \in V$, and weights $w(a) \in \mathbb{R}_{\geq 0}$ for the arcs $a \in A$. For a simple directed s-t-path P that traverses the arcs $a_1, a_2, \ldots, a_p$ in that order, its *SPP-weight* is given by

$$\mathrm{SPP}(P, w) := \sum_{i=1}^{p} w(a_i). \tag{1}$$

The *Quadratic Shortest Path Problem* (QSPP) takes as input a directed graph $G = (V, A)$ together with a source $s \in V$ and a sink $t \in V$. Every pair of arcs $a, a' \in A$ comes with an interaction cost $q(a, a') \in \mathbb{R}_{\geq 0}$; we assume throughout that $q(a, a') = q(a', a)$ holds for all $a, a' \in A$. For a simple directed s-t-path P that traverses the arcs $a_1, a_2, \ldots, a_p$ in that order, its *QSPP-cost* is given by

$$\mathrm{QSPP}(P, q) := \sum_{i=1}^{p} \sum_{j=1}^{p} q(a_i, a_j). \tag{2}$$

Ł. Kowalik et al. (Eds.): WG 2021, LNCS 12911, pp. 245–256, 2021.
https://doi.org/10.1007/978-3-030-86838-3_19

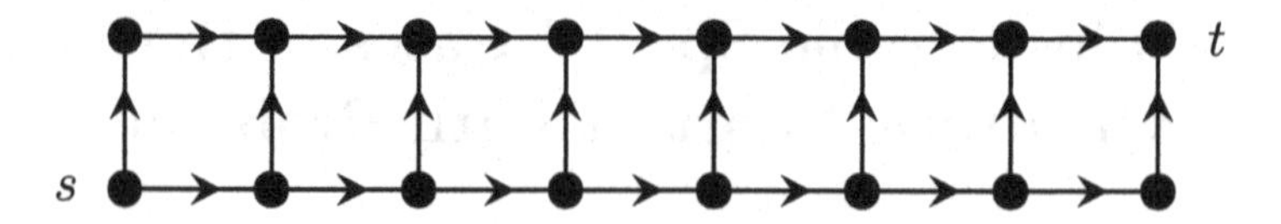

Fig. 1. The grid $G_{2,n}$ consists of two horizontal layers with n vertices. All horizontal arcs are oriented from west to east and all vertical arcs from south to north.

The goal in the SPP and in the QSPP is to determine a simple directed s-t-path P that minimizes the respective objective values (1) and (2). While the SPP is polynomially solvable, the QSPP is NP-hard and extremely difficult to solve; see for instance Rostami et al. [18], and Hu and Sotirov [11–13].

Linearizations. An instance of the QSPP is called *linearizable*, if there exist arc weights $w : A \to \mathbb{R}_{\geq 0}$ for the SPP on the same directed graph G such that

$$\mathrm{QSPP}(P, q) = \mathrm{SPP}(P, w) \qquad \text{for all simple directed } s\text{-}t\text{-paths } P. \tag{3}$$

If we manage to find a linearization of a QSPP instance, we can of course solve the QSPP instance by simply solving the linearized instance of the SPP instead. Hu and Sotirov [11,12] were the first to study linearizations of the QSPP. Among other results, they show in [12] that for an *acyclic* directed graph G with given arc interaction costs, linearizability of the instance can be decided in polynomial time. Furthermore [11] proves that on the grid $G_{2,n}$ the QSPP is *always* linearizable, independently of the concrete arc interaction costs in the instance; see Fig. 1 for an illustration.

Linearizations of non-linear problems form a widely used standard tool in continuous optimization and numerical analysis. On the discrete side, the linearization of hard combinatorial optimization problems by easy combinatorial optimization problems goes back to the seminal work of Bookhold [1], who linearized the NP-hard Quadratic Assignment Problem (QAP) via the polynomially solvable linear assignment problem. Kabadi and Punnen [14,15] gave a polynomial time algorithm for recognizing linearizable instances of the QAP in Koopmans-Beckmann form. Furthermore, [15] presented a purely combinatorial characterization of all linearizable QAP instances with symmetric cost matrices. Further results on linearizations of the QAP have been derived by Erdoğan [5], Erdoğan and Tansel [6,7], and Çela, Deineko and Woeginger [2]. Ćustić and Punnen [3] investigate linearizable instances of the quadratic minimum spanning tree problem, Punnen, Walter and Woods [16] study linearizable instances of the quadratic travelling salesman problem, and De Meijer and Sotirov [4] analyze linearizations of the quadratic cycle cover problem.

Contribution and Organization of This Paper. We call a directed graph G *universally linearizable* (with respect to the QSPP), if every instance of the QSPP on the input graph G is linearizable, for every choice of arc interaction costs q. This concept is motivated by the work of Hu and Sotirov [11], who showed

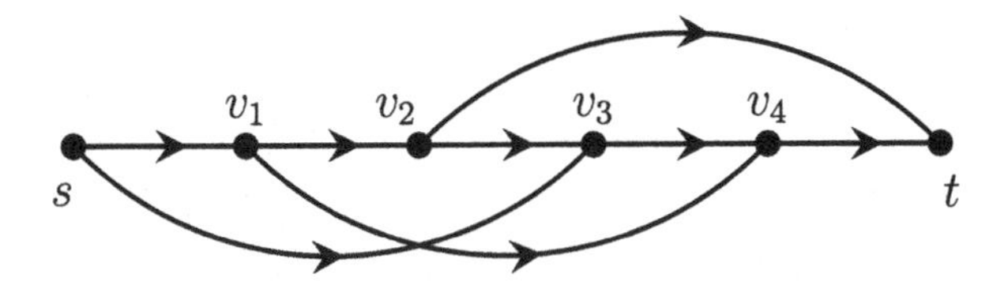

Fig. 2. Another universally linearizable graph.

that the grid $G_{2,n}$ depicted in Fig. 1 is universally linearizable. We provide two further examples, a positive one and a negative one: The graph in Fig. 2 is universally linearizable. The 3×3 grid $G_{3,3}$ (with all horizontal arcs oriented from west to east and all vertical arcs oriented from south to north) is not universally linearizable. The reader may try to settle these statements as a puzzle, or derive them from Theorem 1 in Sect. 3 or from Theorem 2 in Sect. 4.

Which directed graphs are universally linearizable? Section 3 introduces a structural property of the set of s-t-paths that concisely characterizes the universally linearizable graphs. This characterization implies that universally linearizable graphs cannot contain directed cycles (unless these directed cycles do contain some phony type of arcs that do not belong to any s-t-path and that hence are irrelevant for the QSPP). For this reason, Sect. 4 then takes a closer look at *acyclic* graphs, and presents a forbidden subgraph characterization of acyclic universally linearizable graphs that is centered around so-called 12121-subgraphs. Section 5 uses our characterizations to construct fast and simple polynomial time recognition algorithms for universally linearizable graphs, and an even faster and even simpler linear time recognition algorithms for *acyclic* universally linearizable graphs.

Section 6 discusses two closely related decision problems. Problem LINEARIZABLE-QSPP asks whether a given instance of the QSPP (specified by the graph and the arc interaction costs) is linearizable. Problem VALID-LINEARIZATION asks whether a given weight function w forms a linearization of a given instance of the QSPP. Hu and Sotirov [12] have shown that both problems are polynomially solvable in the special case where the underlying directed graph G is acyclic. We show that both problems are coNP-complete in the general case with arbitrary directed graphs that are allowed to contain directed cycles. The coNP-certificate for LINEARIZABLE-QSPP is not straightforward to get.

2 Technical Preliminaries

For a simple directed path P that visits the vertices $v_1, v_2, \ldots, v_p$ in this order and for two indices i and j with $i \leq j$, we denote by $P[v_i, v_j]$ the sub-path of P that starts in vertex v_i and ends in v_j. We often consider vertices v as degenerate paths $P[v, v]$ without arcs, and we consider arcs (u, v) as paths $P[u, v]$ on two vertices. For paths $P = P[v_1, v_2]$ and $Q = Q[v_2, v_3]$, we denote by $P \cdot Q$ (or PQ for short) the path from v_1 to v_3 that results from merging P and Q together at their common vertex v_2. For paths $P = P[v_1, v_2]$ and $Q = Q[v_3, v_4]$ and an arc $(v_2, v_3) \in A$, we denote by $P - Q$ the path $P \cdot (v_2, v_3) \cdot Q$ from v_1 to v_4.

For SPP and QSPP instances, we will always assume that the graph $G = (V, A)$ has a unique source $s \in V$ and a unique sink $t \in V$. By $\mathcal{P}_{st}$ we denote the set of all simple directed s-t-paths in G. If an arc $a \in A$ is not traversed by any path in $\mathcal{P}_{st}$, this arc is irrelevant for SPP and QSPP and in particular is irrelevant for the linearizability of a QSPP instance. Hence we will sometimes assume that every arc in A lies on at least one simple s-t-path, and we will say that graphs with that property are *$\mathcal{P}_{st}$-covered.*

3 The Characterization via Private Arcs

In this section, we provide a combinatorial characterization of universally linearizable directed graphs for the QSPP. For a path $P \in \mathcal{P}_{st}$, we say that an arc $a \in P$ is a *private arc* if no other path $Q \in \mathcal{P}_{st}$ with $Q \neq P$ traverses this arc a.

Theorem 1. *For a digraph $G = (V, A)$, the following statements are equivalent:*

(U) Graph G is universally linearizable.
(P) Every path in $\mathcal{P}_{st}$ possesses a private arc.

Proof. First we show (P) $\Rightarrow$ (U). Let $P_1, \ldots, P_k$ be an enumeration of the paths in $\mathcal{P}_{st}$, and let $a_i \in A$ denote a private arc of path P_i for $i = 1, \ldots, k$. For a QSPP instance on G with arc interaction costs $q : A \times A \to \mathbb{R}_{\geq 0}$, we define arc weights $w : A \to \mathbb{R}_{\geq 0}$ by setting $w(a_i) := \text{QSPP}(P_i, q)$ for $i = 1, \ldots, k$ and $w(a) := 0$ for all remaining arcs $a \in A$. These weights w yield the desired linearization.

Now let us turn to the proof of (U) $\Rightarrow$ (P). Consider some fixed path $P \in \mathcal{P}_{st}$ that traverses the arcs $a_1, a_2, \ldots, a_p$ in that order. For any pair of indices x and y with $1 \leq x \neq y \leq p$, we define a corresponding QSPP instance $I_{x,y}$ by setting the arc interaction costs $q(a_x, a_y) = q(a_y, a_x) = 1$, and setting all other interaction costs to zero. Note that in the resulting instance $I_{x,y}$ of QSPP, the cost of any path $Q \in \mathcal{P}_{st}$ either equals 2 (if Q traverses both arcs a_x and a_y) or equals 0 (if Q skips a_x or a_y). Since the graph G is universally linearizable, the constructed instance $I_{x,y}$ is linearizable as an SPP instance with some weight function $w : A \to \mathbb{R}_{\geq 0}$. Since $\text{QSPP}(P, q) = 2$ and $\text{SPP}(P, w) = 2$, at least one of the arc weights $w(a_i)$ with $1 \leq i \leq p$ is strictly positive; the smallest such index i is denoted by $i(x, y)$.

Now consider an arbitrary path $Q \in \mathcal{P}_{st}$ that traverses the arc $a_{i(x,y)}$. As $\text{SPP}(Q, w) \geq w(a_{i(x,y)}) > 0$ holds in the SPP instance, we conclude that $\text{QSPP}(Q, q) > 0$ and that Q hence traverses both arcs a_x and a_y. Let us summarize our findings so far: For any two indices x and y with $1 \leq x \neq y \leq p$, there exists an index $i(x, y)$, such that every path $Q \in \mathcal{P}_{st}$ that traverses arc $a_{i(x,y)}$ must also traverse the arcs a_x and a_y. In other words, the traversal of arc $a_{i(x,y)}$ enforces the traversal of arc a_x (and the traversal of arc a_y); we denote this situation by the binary relation $a_{i(x,y)} \rightsquigarrow a_x$ (and by $a_{i(x,y)} \rightsquigarrow a_y$). The reflexive and transitive closure of the relation $\rightsquigarrow$ on the arc set $A_P := \{a_1, \ldots, a_p\}$ is denoted by $\rightsquigarrow^*$.

In the last step of the proof, we consider an arc $a_z \in A_P$ that maximizes the number of arcs $a_j \in A_P$ with $a_z \rightsquigarrow^* a_j$. Suppose for the sake of contradiction that there is some arc $a_k \in A_P$ with $a_z \not\rightsquigarrow^* a_k$. But then the existence of arc $a_{i(z,k)}$ contradicts our choice of arc a_z, as all $a_j \in A_P$ with $a_z \rightsquigarrow^* a_j$ satisfy $a_{i(z,k)} \rightsquigarrow a_z \rightsquigarrow^* a_j$, and as furthermore $a_{i(z,k)} \rightsquigarrow a_k$ holds. We conclude that whenever a path $Q \in \mathcal{P}_{st}$ contains the arc a_z, then the path must actually contain all the arcs in A_P and hence coincide with path P. This implies that arc a_z is a private arc for path P, as desired. □

Theorem 1 indicates that universal linearizability imposes heavy constraints on the combinatorial structure of a graph. The following lemma shows that these constraints even prevent the occurrence of directed cycles.

Lemma 1. *Let $G = (V, A)$ be a $\mathcal{P}_{st}$-covered digraph that has a directed cycle. Then G is not universally linearizable and violates the private arc property (P).*

Proof. Let C be a directed cycle in graph G, and let P be a path in $\mathcal{P}_{st}$ that has the largest possible number of arcs in common with the cycle C. Let $y \in V$ denote the last vertex on path P that also belongs to cycle C; note that $y \neq t$. Let $(y, y_Q) \in A$ be the unique arc on C going out of y. As the graph G is $\mathcal{P}_{st}$-covered, there exists another path $Q \in \mathcal{P}_{st}$ that traverses the arc (y, y_Q). Let x be the last vertex on path P (and also the last vertex on path Q) that satisfies

$$P[s, x] \;=\; Q[s, x]. \tag{4}$$

Suppose for the sake of contradiction that $x = y$. Then $P[s, y] = Q[s, y]$, so that path Q contains all the arcs on C that are covered by path P and additionally contains the arc (y, y_Q) on C. As this contradicts our maximizing choice of path P, we conclude $x \neq y$. Furthermore we get that vertex x precedes vertex y on path P as well as on path Q. Let x_P be the successor of x on P, and let x_Q be the successor of x on Q; clearly $x_P \neq x_Q$.

We let z be the last vertex on $P[y, t]$ that also belongs to $Q[x_Q, y]$; this vertex z is well-defined, as vertex y does belong to both paths. We let z_P denote the successor of z on path P; this vertex z_P is well-defined, as $z \in Q[x_Q, y]$ and $y \neq t$ imply $z \neq t$. Similarly, we let z_Q denote the successor of z on path Q; clearly $z_P \neq z_Q$. We introduce the directed path R that is pasted together from three sub-paths of P and Q via the arcs (x, x_Q) and (z, z_P) in the following way:

$$R \;:=\; P[s, x] - Q[x_Q, z] - P[z_P, t] \tag{5}$$

As $P[s, x] = Q[s, x]$ by (4) and as P and Q are simple paths, the first sub-path $P[s, x]$ in R is vertex-disjoint from its second sub-path $Q[x_Q, z]$ and also from its third sub-path $P[z_P, t]$. By our choice of vertex z, also the second and third sub-path of R are vertex-disjoint. All in all, this yields $R \in \mathcal{P}_{st}$. Since path R traverses the arc (x, x_Q), whereas path P traverses the arc (x, x_P), we get

$R \neq P$. Since path R traverses the arc (z, z_P), whereas path Q traverses the arc (z, z_Q), we get $R \neq Q$. As every arc of path $R \in \mathcal{P}_{st}$ is also traversed by path $P \neq R$ or by path $Q \neq R$, path R has no private arc. As graph G violates the private arc property (P), Theorem 1 yields that it is not universally linearizable. □

4 The Characterization via Forbidden Subgraphs

In this section, we analyze certain subgraphs that form obstructions to universal linearizability. A *12121-subgraph* of graph G is a subgraph that is built around the source s, the sink t, four further vertices x_1, x_2, y_1, y_2, and seven simple directed paths P', X_1, X_2, P'', Y_1, Y_2, P''' in G; see Fig. 3 for an illustration. The three paths P', P'', P''' respectively connect vertex s to vertex x_1, vertex x_2 to vertex y_1, and vertex y_2 to vertex t. The two paths X_1 and X_2 each connect x_1 to x_2, and the two paths Y_1 and Y_2 each connect y_1 to y_2. The seven paths have no further vertices in common, and they are arc-disjoint. We require that $x_1 \neq x_2$ and $y_1 \neq y_2$, but we do allow that vertex s coincides with x_1, that vertex x_2 coincides with y_1, and that vertex y_2 coincides with t.

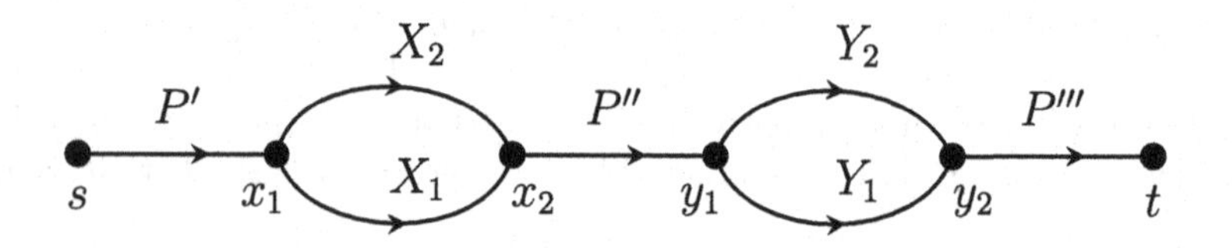

Fig. 3. A 12121-subgraph with the six vertices s, x_1, x_2, y_1, y_2, and t and the seven paths P', X_1, X_2, P'', Y_1, Y_2, and P'''.

Lemma 2. *Let $G = (V, A)$ be a directed graph. If every path in $\mathcal{P}_{st}$ possesses a private arc, then G does not contain any 12121-subgraph.* □

The graph depicted in Fig. 4 demonstrates that the converse of Lemma 2 does not hold true in general. This graph contains only four simple directed s-t-paths: $P_1 = sv_1t$, $P_2 = sv_2t$, $P_3 = sv_1v_2t$, and $P_4 = sv_2v_1t$. It is easily verified that the graph does not contain any 12121-subgraph. Finally, as path P_1 shares its first arc (s, v_1) with P_3 and as it shares its second arc (v_1, t) with P_4, this path does not possess any private arc. Summarizing, the absence of the 12121-subgraph does not guarantee the private arc property (P).

As the trouble-making graph in Fig. 4 is $\mathcal{P}_{st}$-covered and does contain a directed cycle, its violation of the private arc property could also have been concluded from Lemma 1. As it turns out, there are only two types of $\mathcal{P}_{st}$-covered graphs without 12121-subgraph: One type does contain a directed cycle, and is not universally linearizable by Lemma 1. The other type is acyclic, and is universally linearizable by Theorem 1 and the following Lemma 3.

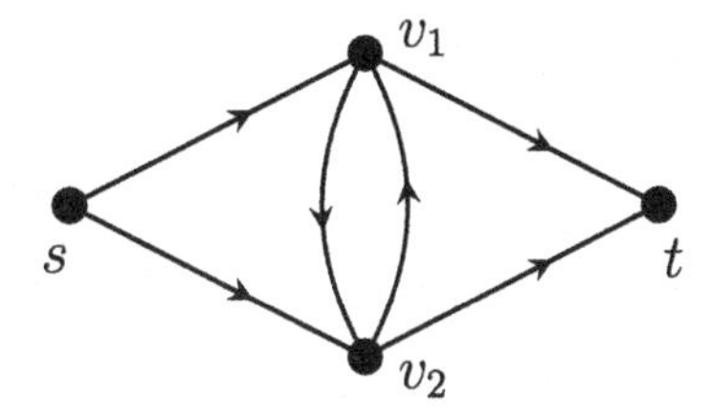

Fig. 4. A graph without 12121-subgraph that violates the private arc property (P).

Lemma 3. *Let $G = (V, A)$ be an acyclic directed graph. If G does not contain any 12121-subgraph, then every path in $\mathcal{P}_{st}$ possesses a private arc.*

Proof. Consider some fixed path $P \in \mathcal{P}_{st}$ that traverses the vertices $v_1, v_2, \ldots, v_p$ in that order (where $v_1 = s$ and $v_p = t$). For integers α and β with $1 \leq \alpha < \beta \leq p$, we say that the closed interval $[\alpha, \beta]$ on the real line is *nice*, if there exists a simple directed path $Q[v_\alpha, v_\beta]$ that leads from vertex v_α to vertex v_β, that does not have any arc in common with path P, and whose inner vertices do not belong to path P. Suppose for the sake of contradiction that there exist two nice intervals $[\alpha, \beta]$ and $[\gamma, \delta]$ with $\beta \leq \gamma$. Since graph G is acyclic, the two underlying paths $Q[v_\alpha, v_\beta]$ and $Q[v_\gamma, v_\delta]$ are arc-disjoint and do not share any inner vertices. Then G contains a 12121-subgraph with $x_1 = v_\alpha$, $x_2 = v_\beta$, $y_1 = v_\gamma$, and $y_2 = v_\delta$; the seven paths in the 12121-subgraph are the five subpaths that result from cutting path P at the vertices $v_\alpha, v_\beta, v_\gamma, v_\delta$ together with the two paths $Q[v_\alpha, v_\beta]$ and $Q[v_\gamma, v_\delta]$. This contradiction yields that any two nice intervals $[\alpha, \beta]$ and $[\gamma, \delta]$ must intersect *properly* in an interval of length at least 1.

Now let $Q \neq P$ be another path in $\mathcal{P}_{st}$ that shares some common vertex or some common arc with path P. As graph G is acyclic, path Q will traverse its common vertices and arcs with P in exactly the same order as P. Let x be the largest index for which the prefix paths $P[s, v_x]$ and $Q[s, v_x]$ coincide, and let y be the smallest index for which the suffix paths $P[v_y, t]$ and $Q[v_y, t]$ coincide. Note that the sub-path $Q[v_x, v_y]$ does only share its start-vertex and end-vertex with P, whereas its inner vertices and its arcs are disjoint from P (as any further overlap or intersection between paths P and Q would yield two nice intervals that do not intersect properly). We associate the nice interval $I(Q) := [x, y]$ with path Q.

Finally let $\mathcal{Q} \subseteq \mathcal{P}_{st}$ denote the set of all paths $Q \in \mathcal{P}_{st}$ that share some arc with path P (if there is no such path Q, then every arc on P is private and we are done). By the above discussion, for any two paths Q_1 and Q_2 in $\mathcal{Q}$ the two associated nice intervals $I(Q_1)$ and $I(Q_2)$ do intersect properly. Then the one-dimensional version of Helly's theorem [10] yields that the intersection of all the intervals $I(Q)$ with $Q \in \mathcal{Q}$ is non-empty and forms an interval $[x^*, y^*]$ of length at least 1. As the arc (v_{x^*}, v_{x^*+1}) lies in every interval $I(Q)$ with $Q \in \mathcal{Q}$, this arc does not belong to any path $Q \in \mathcal{Q}$ and thus constitutes the desired private arc for path P. □

If an acyclic digraph is universally linearizable, every vertex $v \in V - \{s,t\}$ has at least one in-going and at least one out-going arc (as s and t are the unique source and sink). We classify the vertices in $V - \{s,t\}$ into four subsets: vertices in V_{11} have in-degree 1 and out-degree 1; vertices in V_{12} have in-degree 1 and out-degree at least 2; vertices in V_{21} have in-degree at least 2 and out-degree 1; vertices in V_{22} have both in-degree and out-degree at least 2.

Lemma 4. *An acyclic graph $G = (V, A)$ with unique source and unique sink is universally linearizable, if and only if it satisfies the following two conditions:*

(D1) The vertex set V_{22} is empty.
(D2) No directed path connects a vertex in V_{21} to a vertex in V_{12}. □

Lemma 4 leads to yet another, extremely simple combinatorial characterization of universally linearizable acyclic graphs $G = (V, A)$. Let us consider a path $P \in \mathcal{P}_{st}$ on the vertices $v_1, v_2, \ldots, v_p$ in that order (where $v_1 = s$ and $v_p = t$). By condition (D2), on path P every vertex in V_{12} precedes every vertex in V_{21}. Hence by (D1) and (D2), there exists an arc (v_k, v_{k+1}) with $1 \le k \le p-1$, whose removal divides path P into a prefix path on vertices $v_1, \ldots, v_k \in \{s\} \cup V_{11} \cup V_{12}$ and into a suffix path on vertices $v_{k+1}, \ldots, v_p \in \{t\} \cup V_{11} \cup V_{21}$. The first arc with that property on P is denoted $a^*(P)$. It is easily seen that $a^*(P)$ is a private arc for path P. Now let us remove the private arc $a^*(P)$ from every path $P \in \mathcal{P}_{st}$. The remaining graph consists of a component with vertices from $\{s\} \cup V_{11} \cup V_{12}$ that are reachable from the source s, and of another component with vertices from $\{t\} \cup V_{11} \cup V_{21}$ from which the sink t can be reached. This yields that an acyclic universally linearizable graph G is structured into three parts as follows.

- There is an out-tree T^+ whose root is s, and whose inner vertices are the vertices in V_{12} together with some subset of the vertices in V_{11}.
- There is an in-tree T^- whose root is t, and whose inner vertices are the vertices in V_{21} together with the remaining vertices in V_{11}.
- Finally there are the arcs $a^*(P)$ with $P \in \mathcal{P}_{st}$. Each of these arcs connects a vertex in the out-tree T^+ to a vertex in the in-tree T^-.

With this knowledge, it is straightforward to see that the acyclic graph shown in Fig. 2 is indeed universally linearizable: The out-tree T^+ is induced by the three vertices s, v_1, v_2, and the in-tree T^- is induced by the three vertices v_3, v_4, t. The remaining four arcs connect T^+ to T^-, and hence form the private arcs of the paths in $\mathcal{P}_{st}$.

Theorem 2. *For a directed acyclic graph $G = (V, A)$ with unique source and unique sink, the following five statements are pairwise equivalent:*

(U) Graph G is universally linearizable.

(P) Every path in $\mathcal{P}_{st}$ possesses a private arc.

(F) Graph G does not contain any 12121-subgraph.

(D) There is no directed path that leads from a vertex with in-degree at least 2 to a vertex with out-degree at least 2.

(T) Graph G consists of an out-tree rooted at s, an in-tree rooted at t, and of several arcs that connect the out-tree to the in-tree.

Problem: VALID-LINEARIZATION

Instance: A $\mathcal{P}_{st}$-covered directed graph $G = (V, A)$ with $s, t \in V$; arc interaction costs $q : A \times A \to \mathbb{R}_{\geq 0}$; weights $w : A \to \mathbb{R}_{\geq 0}$.

Question: Do the arc weights w form a valid linearization of the given QSPP instance G and q?

Problem: LINEARIZABLE-QSPP

Instance: A $\mathcal{P}_{st}$-covered directed graph $G = (V, A)$ with $s, t \in V$; arc interaction costs $q : A \times A \to \mathbb{R}_{\geq 0}$.

Question: Does there exist a linearization $w : A \to \mathbb{R}_{\geq 0}$ for this QSPP instance?

Fig. 5. The decision problems discussed in Sect. 6.

5 The Recognition of Universally Linearizable Graphs

In this section we present fast, polynomial time recognition algorithms for universally linearizable graphs. Since the algorithmic side of $\mathcal{P}_{st}$-covered graphs is not well-understood, we will not a priori assume that the graphs considered in this section are $\mathcal{P}_{st}$-covered.

Lemma 5. *For a given acyclic directed graph $G = (V, A)$ with unique source and unique sink, it can be decided in linear time $O(|V| + |A|)$ whether G is universally linearizable.* □

Theorem 3. *For a given (not necessarily $\mathcal{P}_{st}$-covered) directed graph $G = (V, A)$ with unique source and unique sink, it can be decided in polynomial time $O(|V| \cdot |A|^2)$ whether G is universally linearizable.* □

6 The Recognition of Linearizable QSPP Instances

Hu and Sotirov [12] have shown that the decision problems VALID-LINEARIZATION and LINEARIZABLE-QSPP in Fig. 5 both are polynomially solvable, if the underlying directed graph G is acyclic. The following (purely combinatorial) proposition can be extracted and deduced from the algorithmic arguments in [12], by combining them with the techniques of our Sect. 4. For subsets $S, T \subseteq A$ of the arc set we denote $q(S, T) = \sum_{s \in S} \sum_{t \in T} q(s, t)$.

Proposition 1. *For an acyclic directed graph $G = (V, A)$ with $s, t \in V$ and for arc interaction costs $q : A \times A \to \mathbb{R}_{\geq 0}$, the following two statements are equivalent.*

(L1) The QSPP instance G and q is linearizable.
(L2) In every 12121-subgraph of G, the subpaths X_1, X_2, Y_1, Y_2 satisfy the equation $q(X_1, Y_1) + q(X_2, Y_2) = q(X_1, Y_2) + q(X_2, Y_1)$. □

We will show that the problems VALID-LINEARIZATION and LINEARIZABLE-QSPP both are coNP-complete in arbitrary (not necessarily acyclic) directed graphs. Note that the combinatorial characterization in Proposition 1 yields a coNP-certificate for the (polynomially solvable) special case of problem LINEARIZABLE-QSPP on acyclic graphs: Any NO-instance contains a 12121-subgraph whose subpaths X_1, X_2, Y_1, Y_2 violate the equation in (L2). Unfortunately this coNP-certificate does not generalize to graphs with cycles, as the following example demonstrates. Consider the graph on six vertices s, t, x_1, x_2, x_3, y in Fig. 6, and define arc interaction costs as follows: The interaction cost of the two arcs (x_1, y) and (y, x_2) is 1; the interaction cost of the two arcs (x_2, y) and (y, x_3) is 1; and the interaction cost of the two arcs (x_3, y) and (y, x_1) is 1; all other arc interaction costs are 0. As the graph does not contain any 12121-subgraph, the QSPP instance trivially satisfies condition (L2) in Proposition 1. However, this instance is not linearizable and hence violates condition (L1): The three paths $P_1 = sx_1yx_2t$, $P_2 = sx_2yx_3t$, $P_3 = sx_3yx_1t$ together traverse each of the twelve arcs exactly once, and the three paths $Q_1 = sx_1yx_3t$, $Q_2 = sx_2yx_1t$, $Q_3 = sx_3yx_2t$ together also traverse each of the twelve arcs exactly once. Therefore any linear weight function $w : A \to \mathbb{R}_{\geq 0}$ will satisfy $\sum_{i=1}^{3} \mathrm{SPP}(P_i, w) = \sum_{i=1}^{3} \mathrm{SPP}(Q_i, w)$, which badly collides with $\sum_{i=1}^{3} \mathrm{QSPP}(P_i, q) = 6$ and $\sum_{i=1}^{3} \mathrm{QSPP}(Q_i, q) = 0$. The following lemma constructs a coNP-certificate for LINEARIZABLE-QSPP based on a different idea.

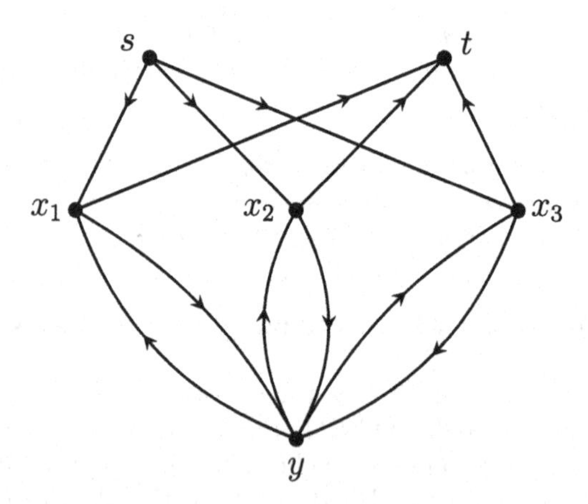

Fig. 6. Proposition 1 does not generalize to graphs with cycles.

Lemma 6. *The problems VALID-LINEARIZATION and LINEARIZABLE-QSPP are contained in the complexity class coNP.*

Proof. Every NO-instance of VALID-LINEARIZATION contains some path P that violates the condition $\text{QSPP}(P, q) = \text{SPP}(P, w)$ in (3). This path P is the coNP-certificate: The path can be described with polynomially many bits, and its violation of (3) is easily verified in polynomial time. Hence VALID-LINEARIZATION is contained in coNP.

Next let us consider a NO-instance of LINEARIZABLE-QSPP. This means that the following system of linear equations and inequalities with real variables w_a for the arcs $a \in A$ is infeasible:

$$\sum_{a \in P} w_a = \text{QSPP}(P, q) \qquad \text{for every path } P \in \mathcal{P}_{st} \tag{6}$$

$$w_a \geq 0 \qquad \text{for every arc } a \in A \tag{7}$$

By Farkas' lemma [8], the infeasibility of (6)–(7) is equivalent to the feasibility of the following dual system with real variables x_P for the paths $P \in \mathcal{P}_{st}$:

$$\sum_{P \in \mathcal{P}_{st}} \text{QSPP}(P, q) \cdot x_P < 0 \tag{8}$$

$$\sum_{P:\, a \in P} x_P \geq 0 \qquad \text{for every arc } a \in A \tag{9}$$

Now fix some feasible solution x^* for (8)–(9), let $\mathcal{P}^+$ denote the set of all paths $P \in \mathcal{P}_{st}$ with $x^*_P \geq 0$, and let $\mathcal{P}^-$ denote the set of all paths $P \in \mathcal{P}_{st}$ with $x^*_P < 0$. We introduce the following additional constraints:

$$\sum_{P \in \mathcal{P}^+} x_P - \sum_{P \in \mathcal{P}^-} x_P \leq 1 \tag{10}$$

$$x_P \geq 0 \qquad \text{for every path } P \in \mathcal{P}^+ \tag{11}$$

$$x_P \leq 0 \qquad \text{for every path } P \in \mathcal{P}^- \tag{12}$$

As the constraints (10)–(12) enforce $|x_P| \leq 1$ for every path P, the underlying feasible region is bounded. Furthermore it is easy to see that the feasibility of the system (8)–(9) implies the feasibility of the system (8)–(12). In particular, the system (8)–(12) possesses a feasible solution x^{**} that satisfies the strict inequality (8), and that furthermore is a corner vertex of the polytope defined by (9)–(12). As we are working in $|\mathcal{P}_{st}|$-dimensional space, the corner vertex x^{**} satisfies $|\mathcal{P}_{st}|$ of the constraints (9)–(12) with equality; this implies that at most $|A| + 1$ of the coordinates in x^{**} are non-zero.

Now our coNP-certificate for LINEARIZABLE-QSPP simply lists all the paths $P \in \mathcal{P}_{st}$ with $x^{**}_P \neq 0$. As the certificate consists of at most $|A| + 1$ paths that each contain at most $|A|$ arcs, the size of the certificate is polynomially bounded in the instance size. The certificate is verified in polynomial time as follows: We consider the restriction of the system (6)–(7) to the paths P in the certificate. As this restricted system has a polynomial number of variables and a polynomial number of constraints, it can be solved (and found to be infeasible) in polynomial time by linear programming. Hence LINEARIZABLE-QSPP is indeed contained in coNP. □

The hardness reductions are done from TWO-VERTEX-DISJOINT-PATHS, and will be presented in the full version of the paper.

Theorem 4. *The problems VALID-LINEARIZATION and LINEARIZABLE-QSPP are coNP-complete.* □

Acknowledgement. This research has been supported by the Austrian Science Fund (FWF): W1230.

References

1. Bookhold, I.: A contribution to quadratic assignment problems. Optimization **21**, 933–943 (1990)
2. Çela, E., Deineko, V.G., Woeginger, G.J.: Linearizable special cases of the QAP. J. Comb. Optim. **31**(3), 1269–1279 (2014). https://doi.org/10.1007/s10878-014-9821-2
3. Ćustić, A., Punnen, A.P.: A characterization of linearizable instances of the quadratic minimum spanning tree problem. J. Comb. Optim. **35**(2), 436–453 (2017). https://doi.org/10.1007/s10878-017-0184-3
4. de Meijer, F., Sotirov, R.: The quadratic cycle cover problem: special cases and efficient bounds. J. Comb. Optim. **39**, 1096–1128 (2020). https://doi.org/10.1007/s10878-020-00547-7
5. Erdoğan, G.: Quadratic assignment problem: linearizations and polynomial time solvable cases. Ph.D. thesis, Bilkent University (2006)
6. Erdoğan, G., Tansel, B.C.: A branch-and-cut algorithm for quadratic assignment problems based on linearizations. Comput. Oper. Res. **34**, 1085–1106 (2007)
7. Erdoğan, G., Tansel, B.C.: Two classes of quadratic assignment problems that are solvable as linear assignment problems. Discrete Optim. **8**, 446–451 (2011)
8. Farkas, J.: Theorie der einfachen Ungleichungen. J. für die Reine und Angewandte Mathematik **124**, 1–27 (1902)
9. Fortune, S., Hopcroft, J., Wyllie, J.: The directed subgraph homeomorphism problem. Theor. Comput. Sci. **10**, 111–121 (1980)
10. Helly, E.: Über Mengen konvexer Körper mit gemeinschaftlichen Punkten. Jahresber. Deutsch. Math.-Verein. **32**, 175–176 (1923)
11. Hu, H., Sotirov, R.: Special cases of the quadratic shortest path problem. J. Comb. Optim. **35**(3), 754–777 (2017). https://doi.org/10.1007/s10878-017-0219-9
12. Hu, H., Sotirov, R.: The linearization problem of a binary quadratic problem and its applications. Working paper (2018). arXiv:1802.02426 [math.OC]
13. Hu, H., Sotirov, R.: On solving the quadratic shortest path problem. INFORMS J. Comput. **32**, 219–233 (2020)
14. Kabadi, S.N., Punnen, A.P.: An $O(n^4)$ algorithm for the QAP linearization problem. Math. Oper. Res. **36**, 754–761 (2011)
15. Punnen, A.P., Kabadi, S.N.: A linear time algorithm for the Koopmans-Beckmann QAP linearization and related problems. Discrete Optim. **10**, 200–209 (2013)
16. Punnen, A.P., Walter, M., Woods, B.: A characterization of linearizable instances of the quadratic travelling salesman problem. Working paper (2017). arXiv:1708.07217 [cs.DM]
17. Read, R.C., Tarjan, R.E.: Bounds on backtrack algorithms for listing cycles, paths, and spanning trees. Networks **5**, 237–252 (1975)
18. Rostami, B., et al.: The quadratic shortest path problem: complexity, approximability, and solution methods. Eur. J. Oper. Res. **268**, 473–485 (2018)

A Linear-Time Parameterized Algorithm for Computing the Width of a DAG

Manuel Cáceres[1(✉)], Massimo Cairo[1], Brendan Mumey[2], Romeo Rizzi[3], and Alexandru I. Tomescu[1]

[1] Department of Computer Science, University of Helsinki, Helsinki, Finland
{manuel.caceresreyes,alexandru.tomescu}@helsinki.fi

[2] School of Computer Science, Montana State University, Bozeman, USA
brendan.mumey@montana.edu

[3] Department of Computer Science, University of Verona, Verona, Italy
romeo.rizzi@univr.it

Abstract. The width k of a directed acyclic graph (DAG) $G = (V, E)$ equals the largest number of pairwise non-reachable vertices. Computing the width dates back to Dilworth's and Fulkerson's results in the 1950s, and is doable in quadratic time in the worst case. Since k can be small in practical applications, research has also studied algorithms whose complexity is parameterized on k. Despite these efforts, it is still open whether there exists a *linear-time* $O(f(k)(|V|+|E|))$ parameterized algorithm computing the width. We answer this question affirmatively by presenting an $O(k^2 4^k |V| + k 2^k |E|)$ time algorithm, based on a new notion of *frontier antichains*. As we process the vertices in a topological order, all frontier antichains can be maintained with the help of several combinatorial properties, paying only $f(k)$ along the way. The fact that the width can be computed by a single $f(k)$-sweep of the DAG is a new surprising insight into this classical problem. Our algorithm also allows deciding whether the DAG has width at most w in time $O(f(\min(w, k))(|V| + |E|))$.

Keywords: Directed acyclic graph · Maximum antichain · DAG width · Posets · Parameterized algorithms · Reachability queries

1 Introduction

An *antichain* in a directed acyclic graph (DAG) $G = (V, E)$ is a set of vertices that are pairwise non-reachable. The size k of a maximum-size antichain is also called the *width* of G. By Dilworth's theorem [9], the width of G also equals the minimum number of paths needed to cover all the vertices of G. As such, it can be computed with minimum path cover algorithms e.g., in time $O(\sqrt{|V|}|E^*|)$ by

This work was partially funded by the European Research Council (ERC) under the European Union's Horizon 2020 research and innovation programme (grant agreement No. 851093, SAFEBIO) and by the Academy of Finland (grants No. 322595, 328877).

L. Kowalik et al. (Eds.): WG 2021, LNCS 12911, pp. 257–269, 2021.
https://doi.org/10.1007/978-3-030-86838-3_20

a reduction to maximum matching [13,17] (where E^* is the set of edges in the transitive closure of G, and we assume that it is already computed), or in time $O(|V||E|)$ by another reduction to minimum flows [2,23].

Computing the width of a given DAG has applications in various fields. For example, in distributed computing, it is important to analyze if a distributed program can run so that no more than w processes have mutual access to some resource; this relies on testing whether a particular DAG inferred from of the program trace has width $k \leq w$ [18,25]; in bioinformatics, the problems of Perfect Phylogeny Haplotype [3], and of Perfect Path Phylogeny Haplotyping [16] are solved by recognizing special DAGs of width at most two; in evolutionary computation, the so-called dimension of a game between co-evolving agents [19] equals the width of a DAG defined from a minimum coordinate system of the game. For several practical applications, the width of the DAG may be small, for example, in [22] the DAG comes from a so-called *pan-genome* encoding genetic variation in a population: this has hundreds of millions of vertices, but yet it has a small width. Furthermore, there exist *fixed-parameter tractable (FPT)* algorithms for several problems on DAGs, which are parameterized by the width of the DAG (see examples in scheduling [8,26] and computational logic [5,14]), therefore, efficiently recognizing graphs of small width becomes vital for their application. It is thus natural to ask whether there exists a faster algorithm computing the width k of a DAG, when k is small. This question is also related to the line of research "FPT inside P" [15] of finding natural parameterizations for problems already in P (see also e.g., [1,12,21]).

Along this line, Felsner et al. [11] present the first algorithm parameterized on k, working for the special case of *transitive DAGs*, and running in time $O(k|V|^2)$. They also show how to recognize transitive DAGs of width 2 and 3 in time $O(|V|)$, and of width 4 in time $O(|V| \log |V|)$. The next parameterized algorithms for general DAGs are due to Chen and Chen: the first runs in time $O(|V|^2 + k\sqrt{k}|V|)$ [6], and the second one in time $O(\sqrt{|V|}|E| + k\sqrt{k}|V|)$ [7]. Recently, Mäkinen et al. [22] obtained a faster one for sparse graphs, running in time $O(k|E| \log |V|)$.

Despite these efforts, the time complexity of computing the width of a DAG parameterized on k is not fully settled, since all existing algorithms have either a superlinear dependence on $|E|$, or a quadratic dependence on $|V|$, in the worst case. We present here the first algorithm running in time $O(f(k)(|V| + |E|))$, where $f(k)$ is a function depending only on k. Thus, for constant k, this is the first algorithm to run in linear time. Moreover, if an integer w is also given in input, we can decide whether $k \leq w$ in time $O(f(\min(w,k))(|V| + |E|))$. Specifically, our main result is the following theorem:

Theorem 1. *Given a DAG $G = (V, E)$ of width k, we can compute a maximum antichain of it in time $O(k^2 4^k |V| + k 2^k |E|)$.*

Note that k corresponds to a property of the input graph that is unknown for the algorithm.

Approach. The main idea behind Theorem 1 is to traverse the graph in a topological order and have an antichain structure sweeping the vertices of the graph, while performing only $f(k)$ work per step. As such, it can also be viewed as an online algorithm receiving in every step a sink vertex and its incoming edges[1].

As a first attempt to obtain such a "sweeping" algorithm, one can think of maintaining only the (unique) *right-most* maximum antichain (recall that all maximum antichains form a lattice [10])[2]. However, it is difficult to update this antichain in time $f(k)$ since inherently we need to perform graph traversals. As a second attempt, one could maintain more structure at every step (in addition to the right-most maximum antichain), while still staying within the $f(k)$ budget. Along this line, for transitive DAGs Felsner et al. [11] propose to maintain a *tower* of right-most maximum antichains of decreasing size. That is, take the right-most maximum antichain of G, then consider the subgraph strictly reached by this antichain. Then take the right-most maximum antichain of this subgraph, and repeat. One thus obtains a tower of at most k antichains. Felsner et al. manage to maintain this structure based on an exhaustive combinatorial approach for $k = 2, 3, 4$, with the former two cases leading to $O(|V|)$ time algorithms, and the latter leading to an $O(|V| \log |V|)$ time algorithm. They also state that "the case $k = 5$ already seems to require an unpleasantly involved case analysis" [11, p. 359]. Moreover, the transitivity of the DAG is crucial in this approach, since reachability between two vertices is equivalent to the existence of an edge between them.

In order to break both of these barriers, we need a different and richer structure to maintain. As such, in Sect. 2 we introduce the notion of *frontier antichain*. A frontier antichain is one such that there is no other antichain *of the same size* and "to the right" of it (i.e., no one that *dominates* it, see Definition 2). Thus, the largest frontier antichain is also the (unique) right-most maximum antichain, and gives the width of G. Furthermore, since any antichain can take at most one vertex from any path in a path cover, there are at most $O(2^k)$ frontier antichains (Lemma 3).

In Sect. 3 we prove several combinatorial properties for maintaining all frontier antichains when a new vertex v in the topological order is added. We show that a frontier antichain of the new graph is either of the form $A \cup \{v\}$, where A is a frontier antichain of the old graph (Lemmas 4 and 6), or it is an old frontier antichain that is not dominated by a new frontier antichain (Lemmas 2 and 5). Thus, it suffices to check domination only between all old and new frontier antichains. However, since domination involves checking reachability (and the DAG is not assumed to be transitive), this might require $O(|V|+|E|)$ time, which we want to avoid. As such, in Sect. 4 we prove another key ingredient, namely

---

[1] Note that this notion of online algorithm is different from the "on-line chain partition" problem [4], where irrevocable decisions opt to be competitive against an optimal solution.

[2] Formally, we call right-most maximum antichain to the top element in the lattice of maximum antichains. If the graph is drawn with edges from left to right this element visually corresponds to the right-most maximum antichain.

that it is sufficient to know which vertices in the current frontier antichains reach v (Theorem 3). If we maintain this information for every added vertex ($O(k2^k)$ per vertex and edge[3], Theorem 2) we can answer the queries required to test domination. Finally, in Sect. 5, we combine these pieces into the main result if this paper, Algorithm 4.

Notation and Preliminaries. We say that a graph $S = (V_S, E_S)$ is a *subgraph* of G if $V_S \subseteq V$ and $E_S \subseteq E$. If $V' \subseteq V$, then $G[V']$ is the subgraph of G *induced* by V', defined as $G[V'] = (V', E_{V'})$, where $E_{V'} = \{(u, v) \in E \ : \ u, v \in V'\}$. A *path* P is a sequence of different vertices $v_1, \ldots, v_\ell$ of G such that $(v_i, v_{i+1}) \in E$, for all $i \in \{1, \ldots, \ell - 1\}$. We say that a path P is *proper* if $\ell \geq 2$. A *path cover* $\mathcal{P}$ is a set of paths such that every vertex belongs to some path of $\mathcal{P}$. A *cycle* is a proper path allowed to start and end at the same vertex. A *directed acyclic graph (DAG)* is a graph that does not contain cycles. For a DAG $G = (V, E)$ we can find in $O(|V| + |E|)$ time [20,24] an order of its vertices $v_1, \ldots, v_{|V|}$ such that for every edge (v_i, v_j), $i < j$, we call such an order a *topological order*. We say that v is reachable from u, or equivalently, that u *reaches* v, if there exists a path starting at u and ending at v. The problem of efficiently answering whether u reaches v is known as *reachability queries*, and if the queries are answered in constant time, *constant-time reachability queries*. An *antichain* A is a set of vertices such that for each $u, v \in A$ $u \neq v$ u, does not reach v. We say that A *reaches* a vertex v if there exists $u \in A$ such that u reaches v. Dilworth's theorem [9] states that the maximum size of an antichain equals the minimum size of a path cover in a DAG, this size is known as the *width* of the DAG and denoted by k. A *partially ordered set (poset)* is a set P and a *partial order* (reflexive, transitive and antisymmetric binary relation) over P. If P is finite, then there exists at least one maximal (minimal) element, and every element in the poset is comparable to some maximal (minimal) element [27]. A *maximal (minimal)* element of a poset is an element that is not smaller (greater) than any other element.

2 Frontier Antichains

We start by introducing the concept of *frontier antichains* and show a bound on the number of frontier antichains present in a DAG.

Definition 1 (Antichain domination). *Let A and B be antichains of the same size. We say that B* dominates *A if for all $b \in B$, A reaches b.*

Note that antichains can only dominate other antichains of the same size, since antichains of different size are, by definition, incomparable. Algorithm 1 shows a function determining whether an antichain dominates another. The following lemma shows that the set of antichains of G with the domination relation form a partial order.

[3] As a purely combinatorial inquiry, we leave open the question of whether the union of all frontier antichains of a given DAG has size $O(\text{poly}(k))$ (instead of $O(k2^k)$).

Algorithm 1: Function *dominates*$(B, A, \mathcal{S})$ checks if an antichain B dominates an antichain A (Definition 1), assuming that the structure $\mathcal{S}$ can compute reachability from vertices of A to vertices of B. If *reaches*$(A, v, \mathcal{S})$ takes $O(|A|)$ time (see Algorithm 2), then this function takes $O(|A||B|) = O(k^2)$ time.

```
Function dominates(B, A, S):
    isDominated ← |A| = |B|                // true if A is dominated by B
    for v ∈ B do
        if not reaches(A, v, S) then       // see Algorithm 2
            isDominated ← false
    return isDominated
```

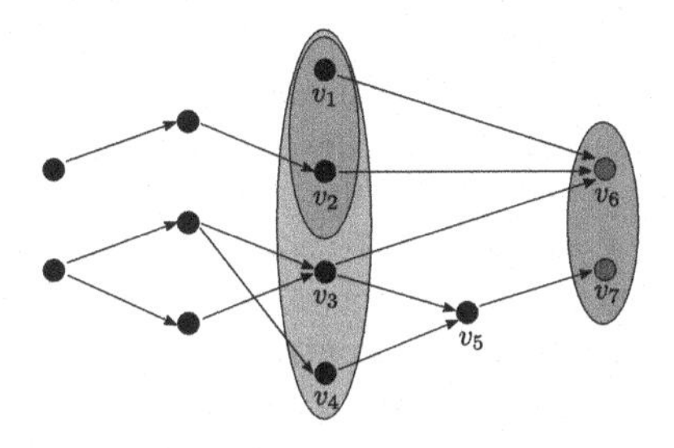

Fig. 1. A DAG and all its frontier antichains of size 1, 2 and 4, as colored sets. The sub-indices represent a topological order. The unique maximum-size frontier antichain is $\{v_1, v_2, v_3, v_4\}$, and is also the right-most maximum antichain. There are 2 frontier antichains of size 2, $\{v_1, v_2\}$ and $\{v_6, v_7\}$. The frontier antichains of size 1 are $\{v_6\}$ and $\{v_7\}$, and those of size 3 (not highlighted) are $\{v_1, v_2, v_7\}$, $\{v_1, v_3, v_4\}$ and $\{v_2, v_3, v_4\}$.

Lemma 1. *The antichains of G related with domination (Definition 1) form a partial order.*

Proof. Clearly, domination is reflexive and transitive (inherited by the transitivity of reachability between vertices). We argue that it is also antisymmetric: suppose A and B are antichains such that A dominates B and B dominates A. Suppose by contradiction that there exists $b \in B \setminus A$. Since B dominates A, there exists $a \in A$ such that a reaches b (note $a \neq b$). Since A dominates B, there exists $b' \in B$ such that b' reaches a. Thus, there is a proper path from b' to a to b in G. If $b' = b$, this implies a cycle exists in a DAG, a contradiction. If $b' \neq b$, this implies B is not an antichain, a contradiction. Thus, $B \setminus A = \emptyset$ and $A = B$, since $|A| = |B|$. Thus, domination is also antisymmetric. $\square$

Definition 2 (Frontier antichains). *Frontier antichains are the maximal elements of the domination partial order i.e., those antichains that are not dominated by any other antichain.*

Figure 1 shows frontier antichains of an example graph. Since frontier antichains are maximal elements of the domination relation we have the following observation.

Lemma 2. *Let A be a non-frontier antichain of G. Then, there exists a frontier antichain dominating A.*

Now we show that the number of such antichains only grows with k, thus there is no problem for our complexity bound to maintain them all. The following lemma shows that there are at most 2^k frontier antichains. The main idea is that there cannot be more than one frontier antichain whose vertices belong to the same set of paths in a minimum path cover of G.

Lemma 3. *If $G = (V, E)$ is a DAG of width k, then G has at most 2^k frontier antichains.*

Proof. By Dilworth's theorem [9], there exists a path cover of G of size k, $\mathcal{P} = \{P_1, \ldots, P_k\}$. Since any antichain can take at most one vertex from each of those paths, we show that for every size-ℓ subset of paths of $\mathcal{P}$, there is at most one frontier antichain of size ℓ whose vertices come from those paths, and thus there are at most 2^k frontier antichains. Without loss of generality consider the subset of paths $P_1, \ldots, P_\ell$, and suppose by contradiction that there are two frontier antichains A and B, $A \neq B$, $|A| = |B| = \ell$, whose vertices come from $P_1, \ldots, P_\ell$. Let us label the vertices in these antichains by the path they belong to. Namely, $A = \{a_1, \ldots, a_\ell\}$, $B = \{b_1, \ldots, b_\ell\}$, with a_i and b_i in P_i for all $i \in \{1, \ldots, \ell\}$. We define the following set of vertices:

$$M := \{m_i := (b_i, \text{ if } a_i \text{ reaches } b_i, \text{ and } a_i \text{ otherwise}) \mid i \in \{1, \ldots, \ell\}\}.$$

First, note that if $m_i = a_i$, then b_i reaches a_i, because a_i and b_i appear on the same path P_i. Next, note that M is an antichain of size ℓ. Otherwise, if there exists m_i that reaches m_j ($i \neq j$), then without loss of generality, suppose that $m_i = a_i$ and $m_j = b_j$. Since $m_i = a_i$, we have that b_i reaches a_i, and thus it reaches b_j, which contradicts B being an antichain. Second, note that $M \neq A$, since otherwise A would dominate B. Finally, M dominates A, since for all $m_i \in M$ there exists $a_i \in A$ such that a_i reaches m_i. □

3 Maintaining Frontier Antichains

Our algorithm will process the vertices in topological order $v_1, \ldots, v_{|V|}$, and maintain all *frontier* antichains (Definition 2) of the current subgraph $G_i := G[\{v_1, \ldots, v_i\}]$ (we say that $G_0 = (\emptyset, \emptyset)$). The following property allows us to upper bound the width of each of these induced subgraphs by the width k of the original graph.

Property 1. Let $G = (V, E)$ be a DAG of width k, and $v_1, \ldots, v_{|V|}$ a topological order of its vertices. Then, for all $i, j \in \{1, \ldots, |V|\}, i \leq j$, the width of $G_{i,j} := G[\{v_i, \ldots, v_j\}]$ is at most k.

Proof. We first show that the intersection of any path of G with the vertices of $G_{i,j}$ is a path in $G_{i,j}$. Consider a path P, and remove from it all the vertices from $V\backslash\{v_i, \ldots, v_j\}$. Thus, we obtain a (possibly empty) sequence $P_{i,j}$ of vertices from $\{v_i, \ldots, v_j\}$. We say that $P_{i,j}$ is the *intersection* of P with $G_{i,j}$. Since G is a DAG, $P_{i,j}$ is a sequence of consecutive vertices in P (otherwise, if it is not empty, we would have a vertex of smaller (bigger) topological index that is reached by v_i (reaches v_j)), and therefore a path in G. Since $P_{i,j}$ only contains vertices from $\{v_i, \ldots, v_j\}$ and $G_{i,j}$ is an induced subgraph, $P_{i,j}$ is a path also in $G_{i,j}$.

By Dilworth's theorem [9], there exists a path cover of G of size k, $\mathcal{P} = \{P_1, \ldots, P_k\}$. The intersection of each of those paths with $G_{i,j}$ forms a path cover of $G_{i,j}$, whose size is at least the width of $G_{i,j}$. □

We say that an antichain is *G_i-frontier* if it is a frontier antichain in the graph G_i. The following lemmas will show us how these frontier antichains evolve when processing the vertices of the graph i.e., when passing from G_{i-1} to G_i.

Lemma 4 (Type 1). *For every $i \in \{1, \ldots, |V|\}$, let A be a G_i-frontier antichain with $v_i \in A$. Then $A\backslash\{v_i\}$ is a G_{i-1}-frontier antichain.*

Proof. Otherwise, there would exist another antichain B dominating $A\backslash\{v_i\}$ in G_{i-1}. Consider $B \cup \{v_i\}$, which is an antichain (otherwise B would reach v_i, a contradiction, since B dominates $A\backslash\{v_i\}$, and A is an antichain). Finally, note that $B \cup \{v_i\}$ dominates A in G_i, which is a contradiction since A is G_i-frontier antichain. □

Lemma 5 (Type 2). *For every $i \in \{1, \ldots, |V|\}$, let A be a G_i-frontier antichain with $v_i \notin A$. Then A is a G_{i-1}-frontier antichain.*

Proof. Otherwise, there would exist another antichain B dominating A in G_{i-1}, and also in G_i, which is a contradiction. □

Looking at these two lemmas, we establish two types of G_i-frontier antichains: the ones containing v_i, called of *type 1*, and the ones that are also G_{i-1}-frontier antichains, called of *type 2*. We handle these two cases separately. First, we find all type-1 frontier antichains, then all of type 2.

Type-1 G_i-frontier antichains are made up of one G_{i-1}-frontier antichain and vertex v_i. A first requirement for a G_{i-1}-frontier antichain, A, to be a subset of a type-1 G_i-frontier antichain is that A does not reach v_i. We now show that this is enough to ensure that $A \cup \{v_i\}$ is a G_i-frontier antichain.

Lemma 6. *For every $i \in \{1, \ldots, |V|\}$, let A be a G_{i-1}-frontier antichain not reaching v_i. Then $A \cup \{v_i\}$ is a G_i-frontier antichain.*

Proof. If $A = \emptyset$, then $A \cup \{v_i\} = \{v_i\}$ is frontier antichain, because v_i is a sink of G_i. Otherwise $A \neq \emptyset$ and, by contradiction, take another antichain B dominating $A \cup \{v_i\}$ in G_i. Suppose that $v_i \in B$, then for all $b \in B$ there exists $a \in A \cup \{v_i\}$ such that a reaches b, but since v_i is a sink of G_i, for all $b \in B\backslash\{v_i\}$ there exists $a \in A$ such that a reaches b i.e., $B\backslash\{v_i\}$ dominates A in G_{i-1}, which is a contradiction. If $v_i \notin B$, then every vertex of B is reached by a vertex of A (it cannot be reached by v_i since it is a sink in G_i), and therefore take any subset of B of size $|A|$ different from A, which would dominate A, a contradiction. □

We use this lemma to find all type-1 G_i-frontier antichains by testing reachability from G_{i-1}-frontier antichains to v_i, with $O(k2^k)$ total reachability queries.

Type-2 G_i-frontier antichains are G_{i-1}-frontier antichains that are not dominated by any antichain in G_i containing v_i (this is sufficient since they are frontier in G_{i-1}). Moreover, by Lemma 2, if a G_{i-1}-frontier antichain is dominated in G_i, then it is dominated by a G_i-frontier antichain. Therefore, type-2 G_i-frontier antichains are G_{i-1}-frontier antichains that are not dominated by any type-1 G_i-frontier antichain. For every G_{i-1}-frontier antichain A we check if there exists a type-1 G_i-frontier antichain dominating A. We can do this in total $O(k^2 4^k)$ reachability queries from vertices in G_{i-1}-frontier antichains to vertices in G_{i-1}-frontier antichains and v_i.

Both type-1 and type-2 G_i-frontier antichains need answering reachability queries efficiently among vertices in G_{i-1}-frontier antichains and v_i. Next, we show how to maintain constant-time reachability queries among these vertices in $O(k2^k)$ time per vertex and edge.

4 Reachability Between Frontier Antichains

To complete our algorithm, we aim to maintain reachability queries among all vertices in G_{i-1}-frontier antichains and v_i. For this we rely on properties of the support of the frontier antichains, as detailed next.

Definition 3 (Support). *For every $i \in \{0, \ldots, |V|\}$, we define the* support S_i *of G_i as the set of all vertices belonging to some G_i-frontier antichain, that is,*

$$S_i := \bigcup_{A\,:\,G_i\text{-frontier antichain}} A.$$

Note that since $G_0 = (\emptyset, \emptyset)$, then $S_0 = \emptyset$, and Lemma 3 implies $|S_i| = O(k2^k)$. Also, $v_i \in S_i$, since $\{v_i\}$ is a G_i-frontier antichain. In Fig. 1, the vertex v_5 belongs to the support of S_5, but it does not belong to S_7, because there is no frontier antichain containing it. Another interesting fact is that if a vertex exits the support in some step, then it cannot re-enter. This is formalized as follows.

Lemma 7. *Let $v \in \{v_1, \ldots, v_i\}$. If $v \notin S_i$, then $v \notin S_j$ for all $j \in \{i, \ldots, |V|\}$.*

Proof. By induction on j. The base case $j = i$ is the hypothesis itself. Now, suppose that $v \notin S_j$ for some $j \in \{i, \ldots, |V| - 1\}$, and suppose by contradiction that $v \in S_{j+1}$. Then $v \in A$, for some G_{j+1}-frontier antichain A. If $v_{j+1} \notin A$, then by Lemma 5, A is a G_j-frontier antichain, and $v \in S_j$, which is a contradiction. But if $v_{j+1} \in A$, then by Lemma 4, $A \backslash \{v_{j+1}\}$ is a G_j-frontier antichain, and $v \in A \backslash \{v_{j+1}\} \subseteq S_j$, a contradiction. □

Algorithm 2: Function *reaches*$(A, v_t, \mathcal{S})$, with $\mathcal{S} = (S_0, \ldots, S_{i-1})$ for some $i \geq t$, and $A \cup \{v_t\} \subseteq S_{i-1} \cup \{v_i\}$. It checks if A reaches v_t. It assumes that for all the vertices $u \in S_{t-1}$, $S_{t-1}.u.reaches$ indicates if u reaches v_t. Reachability in $S_{i-1} \cup \{v_i\}$ is reduced to reachability from S_{j-1} to v_j for all $j \in \{1, \ldots, i\}$, according to Theorem 2. This function takes $O(|A|) = O(k)$ time.

```
Function reaches(A, v_t, S = (S_0, ..., S_{i-1})):
    isReached ← false   // true if v_t is reached from some vertex in A
    for v_s ∈ A do
        if v_s = v_t or (s < t and S_{t-1}.v_s.reaches) then
            isReached ← true
    return isReached
```

Lemma 8. *Let* $v_i \in \{v_1, \ldots, v_j\}$. *If* $v_i \in S_j$, *then* $v_i \in S_t$ *for all* $t \in \{i, \ldots, j\}$.

Proof. If this is not true, we have that there exists some $t \in \{i+1, \ldots, j-1\}$ such that $v_i \notin S_t$, which is a contradiction with $v_i \in S_j$ and Lemma 7. □

We now state that it is sufficient to support reachability queries from every S_{j-1} to v_j to answer queries among vertices in S_{i-1} and v_i. Then, we show how to maintain these reachability relations in $O(k2^k)$ time per vertex and edge.

Theorem 2. *If we know reachability from* S_{j-1} *to* v_j *for all* $j \in [1 \ldots i]$, *then we can answer reachability queries among vertices in* $S_{i-1} \cup \{v_i\}$.

Proof. Let $v_s, v_t \in S_{i-1} \cup \{v_i\}$. We can answer whether v_s reaches v_t by doing the following. If $s \geq t$ it is not possible that v_s reaches v_t unless they are the same vertex. In the other case, $s < t$, since $v_s \in S_{i-1}$, by Lemma 8, $v_s \in S_{t-1}$, and then we can use reachability from S_{t-1} to v_t to answer this query. □

Algorithm 2 shows a function deciding whether an antichain reaches a vertex, using the technique explained in Theorem 2. This function is used to implement Algorithm 1, and our final solution in Algorithm 4.

We will compute reachability from S_{j-1} to v_j for all $j \in \{1, \ldots, i\}$ incrementally when processing the vertices in topological order. That is, we assume that we have computed reachability from S_{j-1} to v_j for all $j \in \{1, \ldots, i-1\}$ and we want to compute reachability from S_{i-1} to v_i.

For this we do the following. Initially, we set reachability from u to v_i to `false` for all $u \in S_{i-1}$. Then, for every edge (v_j, v_i), if $v_j \in S_{i-1}$ we set reachability from v_j to v_i to `true`, and for each $u \in S_{i-1} \cap S_{j-1}$ such that u reaches v_j (known since $u \in S_{j-1}$) we set reachability from u to v_i to `true`. Note that we can compute the intersection $S_{i-1} \cap S_{j-1}$ in $O(|S_{i-1}|) = O(k2^k)$ time. For each $v_p \in S_{i-1}$ we decide whether $v_p \in S_{j-1}$ by testing if $p \leq j-1$, which is correct by Lemma 8.

Algorithm 3: Function *updateReachability* computes reachability from vertices in S_{i-1} to v_i. It assumes that $S_{i-1} \in \mathcal{S}$, for all $j \in \{1,\ldots,i-1\}$, $S_{j-1} \in \mathcal{S}$, and for all the vertices $u \in S_{j-1}$, $S_{j-1}.u.reaches$ indicates if u reaches v_j. This function takes $O(k2^k(|N^-(v_i)|+1))$ time.

```
Function updateReachability(v_i, S = (S_0, ..., S_{i-1})):
    for u ∈ S_{i-1} do
        S_{i-1}.u.reaches ← false                    // true if u reaches v_i
    for v_j ∈ N^-(v_i) do
        if v_j ∈ S_{i-1} then                        // Direct (by one edge) reachability
            S_{i-1}.v_j.reaches ← true
        for u ∈ S_{i-1} ∩ S_{j-1} do                 // More than one edge reachability
            if S_{j-1}.u.reaches then
                S_{i-1}.u.reaches ← true
```

Algorithm 3 shows a function that computes the reachability from S_{i-1} to v_i, according to what was explained in this section. The correctness of this procedure is guaranteed by the following theorem.

Theorem 3. *Algorithm 3 computes reachability from* S_{i-1} *to* v_i.

Proof. Clearly, what the algorithm sets to `true` is correct. Suppose by contradiction that there exists $y \in S_{i-1}$ reaching v_i such that reachability from y to v_i was not set to `true`. Since y reaches v_i, the in-neighborhood of v_i is not empty. Since y was not set to `true`, in particular, $y \notin N^-(v_i)$, thus it reaches v_i through a path whose last vertex previous to v_i is $v_j \in N^-(v_i)$. Again, since y was not set to `true`, $y \notin S_{i-1} \cap S_{j-1}$, thus $y \notin S_{j-1}$. But then, by Lemma 7 we have $y \notin S_{i-1}$, a contradiction, unless $y \notin G_{j-1}$ i.e., y is after v_j in topological order, which is a contradiction since y reaches v_j. □

5 A Linear-Time Parameterized Algorithm

We now have all the ingredients to prove the main theorem.

Theorem 1. *Given a DAG* $G = (V, E)$ *of width* k, *we can compute a maximum antichain of it in time* $O(k^2 4^k|V| + k2^k|E|)$.

Proof. We process the vertices in topological order. After processing v_i, we will have computed all G_i-frontier antichains (including the right-most maximum antichain of G_i), and constant-time reachability queries from S_{j-1} to v_j, for all $j \in \{1,\ldots,i\}$. Suppose we have this for $i-1$.[4] First, we obtain constant-time reachability queries from S_{i-1} to v_i, using the procedure from Theorem 3

[4] Since $G_0 = (\emptyset, \emptyset)$ and $S_0 = \emptyset$, there are no frontier antichains for the base case of the algorithm.

Algorithm 4: Our parameterized algorithm computing the right-most maximum antichain of a DAG $G = (V, E)$ in time $O(k^2 4^k |V| + k2^k |E|)$.

```
R ← ∅, F_0 ← {∅}, S ← (S_0 = ∅)
for v_i ∈ v_1, ..., v_|V| in topological order do
    updateReachability(v_i, S)
    F_i ← {∅}                                  // F_i stores G_i-frontiers
    for antichain A ∈ F_{i-1} do               // Compute type-1 G_i-frontiers
        if not reaches(A, v_i, S) then
            F_i.add(A ∪ {v_i})                 // A ∪ {v_i} is a type-1 G_i-frontier
            if |A ∪ {v_i}| > |R| then R ← A ∪ {v_i}
    T_1 ← F_i                                  // Contains type-1 G_i-frontiers
    for antichain A ∈ F_{i-1} do               // Compute type-2 G_i-frontiers
        isType2 ← true                         // true if A is a type-2 G_i-frontier
        for antichain B ∈ T_1 do
            if dominates(B, A, S) then isType2 ← false
        if isType2 then F_i.add(A)
    S.add(S_i ← ⋃_{A∈F_i} A)
return R
```

(Algorithm 3), spending $O(k2^k)$ time, and $O(k2^k)$ time per edge incoming to v_i. For the entire algorithm, this adds up to $O(k2^k(|V| + |E|))$.

By Lemma 6, we obtain all type-1 G_i-frontier antichains by taking every G_{i-1}-frontier antichain A, and testing if A reaches v_i using the reduction from Theorem 2 (Algorithm 2). This takes $O(k2^k)$ time, $O(k2^k|V|)$ in total.

We compute type-2 G_i-frontier antichains by taking every G_{i-1}-frontier antichain A and searching if there exists a type-1 G_i-frontier antichain B dominating A in time $O(k^2 4^k)$ ($O(k^2)$ constant-time reachability queries to test domination between a pair of antichains, $O(4^k)$ such pairs), $O(k^2 4^k |V|)$ in total. The total complexity of the algorithm is $O(k^2 4^k |V| + k2^k |E|)$. □

Algorithm 4 shows the pseudocode of the final solution explained in Theorem 1. It maintains reachability from the corresponding support to the newly added vertex using Algorithm 3. Type-1 frontier antichains are found by using Algorithm 2, and type-2 frontier antichains are confirmed using Algorithm 1.

Finally, if we are interested in recognizing whether G has width at most an additional input integer w we can adapt our algorithm to run in time $O(f(\min(w, k))\ (|V| + |E|))$ instead.

Remark 1. Given an additional input integer w we can determine whether $k \leq w$ in time $O(w'^2 4^{w'} |V| + w' 2^{w'} |E|)$ ($w' = \min(w, k)$) by stopping the computation of Algorithm 4 as soon as we find an antichain of size $w + 1$. If the algorithm does not stop by this reason, it means that $k \leq w$, and the opposite otherwise. In both cases maximum size of an observed antichain is not greater than $w' + 1$, obtaining the desired running time.

References

1. Abboud, A., Williams, V.V., Wang, J.: Approximation and fixed parameter subquadratic algorithms for radius and diameter in sparse graphs. In: Proceedings of the Twenty-Seventh Annual ACM-SIAM Symposium on Discrete Algorithms, pp. 377–391. SIAM (2016)
2. Bang-Jensen, J., Gutin, G.: Digraphs Theory, Algorithms and Applications, 1st edn. Springer, Berlin (2000)
3. Bonizzoni, P.: A linear-time algorithm for the perfect phylogeny haplotype problem. Algorithmica **48**(3), 267–285 (2007). https://doi.org/10.1007/s00453-007-0094-3
4. Bosek, B., Felsner, S., Kloch, K., Krawczyk, T., Matecki, G., Micek, P.: On-line chain partitions of orders: a survey. Order **29**(1), 49–73 (2012). https://doi.org/10.1007/s11083-011-9197-1
5. Bova, S., Ganian, R., Szeider, S.: Model checking existential logic on partially ordered sets. ACM Trans. Comput. Log. (TOCL) **17**(2), 1–35 (2015)
6. Chen, Y., Chen, Y.: An efficient algorithm for answering graph reachability queries. In: 2008 IEEE 24th International Conference on Data Engineering, pp. 893–902. IEEE (2008)
7. Chen, Y., Chen, Y.: On the graph decomposition. In: 2014 IEEE Fourth International Conference on Big Data and Cloud Computing, pp. 777–784. IEEE (2014)
8. Colbourn, C.J., Pulleyblank, W.R.: Minimizing setups in ordered sets of fixed width. Order **1**(3), 225–229 (1985). https://doi.org/10.1007/BF00383598
9. Dilworth, R.P.: A decomposition theorem for partially ordered sets. Ann. Math. **51**(1), 161–166 (1950). http://www.jstor.org/stable/1969503
10. Dilworth, R.P.: Some combinatorial problems on partially ordered sets. In: Bogart, K.P., Freese, R., Kung, J.P.S. (eds.) The Dilworth Theorems. CM, pp. 13–18. Springer, Boston (1990). https://doi.org/10.1007/978-1-4899-3558-8_2
11. Felsner, S., Raghavan, V., Spinrad, J.: Recognition algorithms for orders of small width and graphs of small Dilworth number. Order **20**(4), 351–364 (2003). https://doi.org/10.1023/B:ORDE.0000034609.99940.fb
12. Fomin, F.V., Lokshtanov, D., Pilipczuk, M., Saurabh, S., Wrochna, M.: Fully polynomial-time parameterized computations for graphs and matrices of low treewidth. In: Klein, P.N. (ed.) Proceedings of the Twenty-Eighth Annual ACM-SIAM Symposium on Discrete Algorithms, SODA 2017, Barcelona, Spain, Hotel Porta Fira, 16–19 January 2017, pp. 1419–1432. SIAM (2017). https://doi.org/10.1137/1.9781611974782.92
13. Fulkerson, D.R.: Note on Dilworth's decomposition theorem for partially ordered sets. Proc. Am. Math. Soc. **7**, 701–702 (1956)
14. Gajarský, J., et al.: FO model checking on posets of bounded width. In: 2015 IEEE 56th Annual Symposium on Foundations of Computer Science, pp. 963–974. IEEE (2015)
15. Giannopoulou, A.C., Mertzios, G.B., Niedermeier, R.: Polynomial fixed-parameter algorithms: a case study for longest path on interval graphs. Theor. Comput. Sci. **689**, 67–95 (2017)
16. Gramm, J., Nierhoff, T., Sharan, R., Tantau, T.: Haplotyping with missing data via perfect path phylogenies. Discrete Appl. Math. **155**(6–7), 788–805 (2007)
17. Hopcroft, J.E., Karp, R.M.: An $n^{5/2}$ algorithm for maximum matchings in bipartite graphs. SIAM J. Comput. **2**(4), 225–231 (1973)

18. Ikiz, S., Garg, V.K.: Efficient incremental optimal chain partition of distributed program traces. In: 26th IEEE International Conference on Distributed Computing Systems (ICDCS 2006), p. 18. IEEE (2006)
19. Jaśkowski, W., Krawiec, K.: Formal analysis, hardness, and algorithms for extracting internal structure of test-based problems. Evol. Comput. **19**(4), 639–671 (2011)
20. Kahn, A.B.: Topological sorting of large networks. Commun. ACM **5**(11), 558–562 (1962)
21. Koana, T., Korenwein, V., Nichterlein, A., Niedermeier, R., Zschoche, P.: Data reduction for maximum matching on real-world graphs: theory and experiments. J. Exp. Algorithmics (JEA) **26**, 1–30 (2021)
22. Mäkinen, V., Tomescu, A.I., Kuosmanen, A., Paavilainen, T., Gagie, T., Chikhi, R.: Sparse dynamic programming on DAGs with small width. ACM Trans. Algorithms (TALG) **15**(2), 1–21 (2019)
23. Orlin, J.B.: Max flows in $O(nm)$ time, or better. In: Proceedings of the Forty-Fifth Annual ACM Symposium on Theory of Computing, pp. 765–774 (2013)
24. Tarjan, R.E.: Edge-disjoint spanning trees and depth-first search. Acta Informatica **6**(2), 171–185 (1976). https://doi.org/10.1007/BF00268499
25. Tomlinson, A.I., Garg, V.K.: Monitoring functions on global states of distributed programs. J. Parallel Distrib. Comput. **41**(2), 173–189 (1997)
26. van Bevern, R., Bredereck, R., Bulteau, L., Komusiewicz, C., Talmon, N., Woeginger, G.J.: Precedence-constrained scheduling problems parameterized by partial order width. In: Kochetov, Y., Khachay, M., Beresnev, V., Nurminski, E., Pardalos, P. (eds.) DOOR 2016. LNCS, vol. 9869, pp. 105–120. Springer, Cham (2016). https://doi.org/10.1007/978-3-319-44914-2_9
27. Wallis, W.D., George, J.C.: Introduction to Combinatorics. CRC Press, Boca Raton (2016)

On Morphing 1-Planar Drawings

Patrizio Angelini[1], Michael A. Bekos[2(✉)], Fabrizio Montecchiani[3], and Maximilian Pfister[2]

[1] John Cabot University, Rome, Italy
pangelini@johncabot.edu
[2] Institut für Informatik, Universität Tübingen, Tübingen, Germany
bekos@informatik.uni-tuebingen.de, maximilian.pfister@uni-tuebingen.de
[3] Dipartimento di Ingegneria, Università degli Studi di Perugia, Perugia, Italy
fabrizio.montecchiani@unipg.it

Abstract. Computing a morph between two drawings of a graph is a classical problem in computational geometry and graph drawing. While this problem has been widely studied in the context of planar graphs, very little is known about the existence of topology-preserving morphs for pairs of non-planar graph drawings. We make a step towards this problem by showing that a topology-preserving morph always exists for drawings of a meaningful family of 1-planar graphs. While our proof is constructive, the vertices may follow trajectories of unbounded complexity.

1 Introduction

Computing a morph between two drawings of the same graph is a classical problem that attracted considerable attention over the years, also in view of its numerous applications in computer graphics and animations (refer to [1] for a short overview). At high level, given two drawings $\Gamma_a(G)$ and $\Gamma_b(G)$ of the same graph G, a *morph* between $\Gamma_a(G)$ and $\Gamma_b(G)$ is a continuously changing family of drawings such that the initial one coincides with $\Gamma_a(G)$ and the final one with $\Gamma_b(G)$. A standard assumption is that the two input drawings - as well as all intermediate ones - are *topologically equivalent*, i.e., they define the same set of cells (see Sect. 2 for formal definitions). The main challenge is to design morphing algorithms that maintain some additional geometric properties of the input drawings throughout the transformation, such as planarity with straight-line edges (see, e.g., [1,16,25]), convexity [6,34], orthogonality [12,26], and upwardness [20]. We point the reader to [4,8,10,11] for additional related work.

In this context, the most prominent research direction focuses on morphs of straight-line planar drawings: The topological equivalence condition implies that all drawings in the morph have the same planar embedding; in addition, it is also required that edges remain straight-line segments. Back in 1944, Cairns [16] proved that such morphs always exist if the input graphs are plane triangulations.

This work was partially supported by the Department of Engineering, University of Perugia, grants RICBA19FM and RICBA20ED and by DFG grant KA812/18-1.

L. Kowalik et al. (Eds.): WG 2021, LNCS 12911, pp. 270–282, 2021.
https://doi.org/10.1007/978-3-030-86838-3_21

This implies that, for a fixed plane triangulation, the space of its straight-line planar drawings is connected. The main drawback of Cairns result is in the underlying construction, which involves exponentially-many morphing steps. The extension of Cairns' result to all plane graphs was initially done by Thomassen [34], while later Floater and Gotsman [25], and Gotsman and Surazhsky [27,32] proposed different approaches using trajectories of unbounded complexity. More recently, Alamdari et al. [2] focused on the complexity of the morph. They described the first morphing algorithm for planar straight-line drawings that makes use of a polynomial number of steps, where in each step vertices move at uniform speed along linear trajectories. In a subsequent paper [1], a linear bound on the number of steps is shown, which is worst-case optimal.

Morphing non-planar drawings of graphs appears to be a more elusive problem. In particular, Angelini et al. [5] asked whether a morphing algorithm exists for pairs of non-planar straight-line drawings such that the topology of the crossings in the drawing is maintained throughout the morph. They stressed that a solution to this problem is not known even if the vertex trajectories are allowed to have arbitrary complexity. Note that the obvious idea of morphing the "planarizations" of the drawings (i.e., the planar drawings obtained by treating crossings as dummy vertices) does not trivially work. Namely, in order to guarantee that edges remain straight-line segments throughout the morph, one has to ensure that opposite edges incident to dummy vertices maintain the same slope. To the best of our knowledge, such requirement cannot be easily incorporated into any of the already known morphing algorithms for planar graphs.

One way of simplifying the problem is to consider graphs that are non-planar but still admit embeddings on surfaces of bounded genus. Chambers et al. [17] proved the existence of morphs for pairs of crossing-free drawings on the Euclidean flat torus (edges are still geodesics). Their technique is complex and the authors concluded that an extension to higher genus surfaces is fairly non-trivial.

We make a step towards settling the open problem in [5] by studying non-planar drawings of graphs with forbidden edge-crossing patterns. Our focus is on the family of 1-*planar* graphs, which naturally extends the notion of planarity by allowing each edge to be crossed at most once (see [29] for a survey). Note that 1-planar graphs form a well studied family of non-planar graphs with early results dating back to the 60's [9,31], while more recently they have gained considerable attention in the rapidly growing literature about beyond planarity [23,28].

Our Contribution. We provide a set of sufficient conditions under which any pair of 1-planar straight-line drawings admits a morph. At high-level, we require that if two edges cross, then they can be enclosed in a quadrilateral region whose boundary is uncrossed; although this region may contain further vertices in its interior, we require that any edge connecting an end-vertex of the crossing edges to a vertex inside the region is also uncrossed; refer to Fig. 1 for an illustration. A drawing that satisfies these requirements is called *kite-planar* 1-*planar* (see Definition 1). Our main result is summarized by the following theorem.

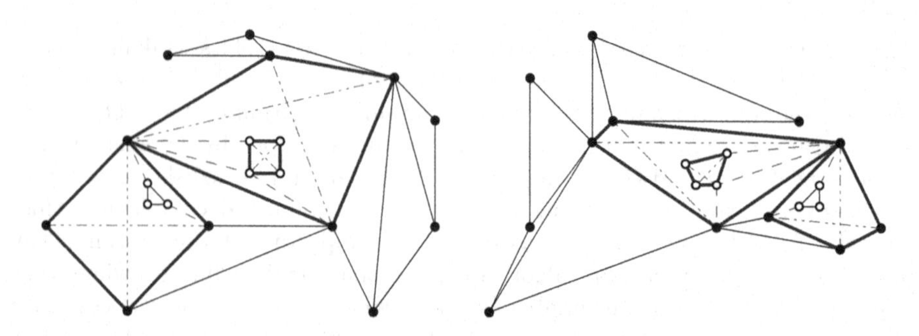

Fig. 1. Two topologically-equivalent kite-planar 1-planar drawings of the same graph.

Theorem 1. *There exists a morph between any pair of topologically-equivalent kite-planar* 1*-planar drawings.*

Theorem 1 implies that, for a fixed graph, the space of its topologically-equivalent kite-planar 1-planar drawings is connected. The proof is constructive, although the vertices may use trajectories of unbounded complexity. Concerning the definition of kite-planar 1-planar drawings, it may be worth observing that, due to a simple edge density argument, the graphs admitting such a drawing cannot be embedded on any surface of bounded genus. Indeed, as shown in Sect. 6, some well-known families of 1-planar graphs admit drawings that are kite-planar 1-planar and require arbitrary large genus to be embedded.

Paper Structure. Section 2 contains basic definitions and notation. Section 3 gives an overview of the proof technique, which exploits a recursive construction. The base case of the recursion is described in Sect. 4, while the recursive step is in Sect. 5. Implications of our result in terms of classes of 1-planar drawings that admit a morph are discussed in Sect. 6. Open problems are given in Sect. 7. For space reasons, the proofs of the statements marked with (◇) are omitted.

2 Preliminaries

Drawings. A *straight-line drawing* (or simply a *drawing*, for short) $\Gamma(G)$ of a graph G maps each vertex v of G to a distinct point p_v of the plane and each edge (u, v) of G to a straight-line segment connecting p_u and p_v without passing through any other point representing a vertex of G. When this creates no ambiguities, we will not distinguish between a vertex and the point representing it in $\Gamma(G)$, as well as between an edge and its segment. Note that, by definition, two edges of a drawing share at most one point, which is either a common endpoint or an interior point where the two edges properly cross. Drawing $\Gamma(G)$ partitions the plane into connected regions called *cells*. The boundary of a cell consists of vertices, crossing points, and (parts of) edges. The *external cell* of $\Gamma(G)$ is its (only) unbounded cell. Two drawings $\Gamma_a(G)$ and $\Gamma_b(G)$ of the same graph G are *topologically equivalent* if they define the same set of cells up to an orientation-preserving homeomorphism of the plane. An *embedding* of G is an equivalence class of drawings that are pairwise topologically equivalent.

A drawing $\Gamma(G)$ is *planar* if no two edges cross. In this case, the cells of $\Gamma(G)$ are called *faces* and their boundaries consist of just vertices and edges. A graph is *planar* if it admits a planar drawing. A planar graph together with an embedding defined by a planar drawing is a *plane graph.* A planar drawing is *strictly convex* if all its faces are strictly convex polygons.

A graph is 1-*planar* if it admits a (not necessarily straight-line) 1-*planar drawing* in which every edge crosses at most one other edge. A 1-planar graph together with an embedding defined by a 1-planar drawing is a 1-*plane graph.* A *kite* K in a 1-planar drawing $\Gamma(G)$ is a 1-planar drawing of K_4 in $\Gamma(G)$ whose external cell is a quadrilateral. The four edges on the boundary of the external cell of K are called *kite edges.* The other two edges are the *crossing edges* of K and are drawn inside the quadrilateral bounding K. Figure 1 shows three kites; the kite (crossing) edges are fat blue (dashed-dotted red, resp.).

Given a vertex v of G and a kite K, the following exclusive cases can occur: (i) v *belongs* to K, if it is a vertex of the K_4 defining K, or (ii) v is *inside* K (or K *contains* v) if v lies in the interior of the quadrilateral bounding K, or (iii) v is outside K, otherwise. A kite is *empty* if it contains no vertex; otherwise, it is *non-empty.* An edge (u, v) is a *binding edge* (dashed green in Fig. 1) if u belongs to a non-empty kite K and v is inside K. We can now introduce kite-planar 1-planar drawings.

Definition 1. *A straight-line drawing is* kite-planar 1-planar, *or* 1-kite-planar *for short, if:* **(P.1)** *every edge is crossed at most once,* **(P.2)** *the four kite edges of every kite are present and uncrossed, and* **(P.3)** *every binding edge is uncrossed.*

Let $\Gamma(G)$ be a 1-kite-planar drawing of G. We say that a vertex of G is of *level* 0 if no kite contains it, while it is of *level* $i > 0$ if the maximum level of the vertices belonging to a kite containing it is $i - 1$. In Fig. 1, the black (white) vertices are of level 0 (level 1, resp.). The next property follows from P.3 of Definition 1.

Property 1 ($\diamond$). *If two vertices belong to the same kite of a* 1*-kite-planar drawing* $\Gamma(G)$*, then they are of the same level.*

Morphs. Let $\Gamma_a(G)$ and $\Gamma_b(G)$ be two topologically-equivalent drawings of the same graph G. A *morph* between them is a continuously changing family of pairwise topologically-equivalent drawings of G indexed by time $t \in [0, 1]$, such that the drawing at time $t = 0$ is $\Gamma_a(G)$ and the drawing at time $t = 1$ is $\Gamma_b(G)$. Since edges are drawn as straight-line segments, a morph is uniquely specified by the vertex trajectories. Also, during the course of the morph, a vertex may coincide with neither another vertex nor an internal point of an edge.

3 Outline of the Proof of Theorem 1

In this section, we give an outline of the proof of Theorem 1, namely, that there exists a morph between any two topologically-equivalent 1-kite-planar drawings

$\Gamma_a(G)$ and $\Gamma_b(G)$ of a graph G. Recall that $\Gamma_a(G)$ and $\Gamma_b(G)$ define the same embedding of G. Hence, G is necessarily a 1-plane graph.

Our proof is by means of a recursive construction. The underlying idea is to compute a morph by keeping each kite boundary drawn as a strictly-convex polygon, so that, in the course of the morph, the drawing of the corresponding crossing edges will stay inside their boundary. The main challenge, however, stems from the fact that a kite may not be empty. Therefore, our approach is to remove the interior of each kite, recursively compute a morph that keeps the convexity of the kite boundaries, and suitably reinsert (and morph) the removed subdrawings. In the proof, we will use two key ingredients. The first one is a result by Aronov et al. [7], which guarantees that one can *compatibly triangulate* two topologically-equivalent planar drawings of a planar graph.

Theorem 2 (Aronov et al. [7]). *Given two topologically-equivalent planar drawings $\Gamma_a(P)$ and $\Gamma_b(P)$ of the same n-vertex planar graph P, it is possible to augment $\Gamma_a(P)$ and $\Gamma_b(P)$ to two topologically-equivalent planar drawings $\Gamma_a(P')$ and $\Gamma_b(P')$ of the same maximal planar graph P' such that $\Gamma_a(P) \subseteq \Gamma_a(P')$, $\Gamma_b(P) \subseteq \Gamma_b(P')$, and the order of $P' \setminus P$ is $O(n^2)$.*

The second ingredient is a result by Angelini et al. [6], which allows us to morph a pair of convex drawings by preserving the convexity of the faces. The main properties of this result are summarized in the next theorem.

Theorem 3 (Angelini et al. [6]). *Let $\langle \Gamma_a(P), \Gamma_b(P) \rangle$ be a pair of topologically-equivalent strictly-convex planar drawings of a graph P. There is a morph between $\Gamma_a(P)$ and $\Gamma_b(P)$ in which every intermediate drawing is strictly convex. If the outer face of G has only three vertices and each of them has the same position in $\Gamma_a(P)$ and $\Gamma_b(P)$, then these three vertices do not move during this morph.*

We apply recursion on the maximum level ℓ of a vertex of G. The base case ($\ell = 0$) is described in Sect. 4, while the recursive case ($\ell > 0$) in Sect. 5.

4 Base Case

In the base case of the recursion, all the vertices of G are of level 0, which implies that all the kites of G, if any, are empty. Let P be the graph obtained by removing both crossing edges from each kite of G. Let $\langle \Gamma_a(P), \Gamma_b(P) \rangle$ be the restrictions of $\langle \Gamma_a(G), \Gamma_b(G) \rangle$ to P, respectively; see Fig. 2. By construction, $\langle \Gamma_a(P), \Gamma_b(P) \rangle$ is a pair of planar and topologically-equivalent drawings, and P is a plane subgraph of G. The kite edges of each kite K of G are uncrossed (by P.2 of Definition 1) and bound a quadrangular face f_K in P, which we call *marked*.

Let P' and $\langle \Gamma_a(P'), \Gamma_b(P') \rangle$ be the graph and the corresponding pair of planar drawings obtained by applying Theorem 2 to $\langle \Gamma_a(P), \Gamma_b(P) \rangle$, except for the marked faces; see Fig. 2 for an illustration. This operation guarantees that every face in both drawings is a triangle, if not marked, or a quadrangle, if marked. We call a plane graph with such faces *almost triangulated*, and we next prove that it is triconnected.

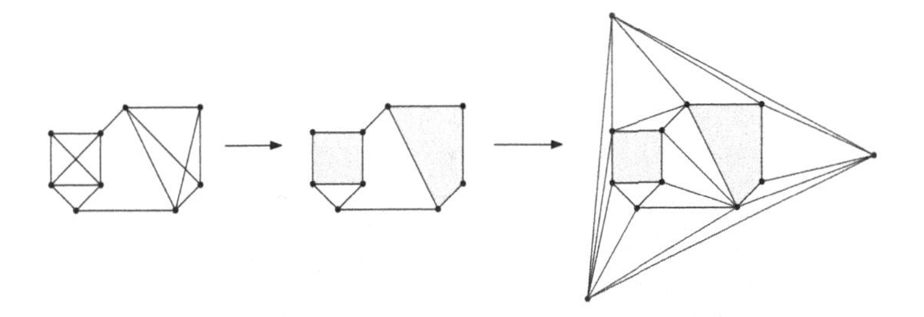

Fig. 2. Illustration of the transitions $\Gamma_a(G) \to \Gamma_a(P) \to \Gamma_a(P')$; marked faces are gray.

Lemma 1. *Every almost triangulated plane graph is triconnected.*

Proof. Let P' be an almost triangulated plane graph derived from a 1-kite-planar drawing of a 1-planar graph G. Suppose that P' contains a separation pair $\{u, v\}$. Then there exist at least two faces f_1 and f_2 that are incident to both u and v such that at least one, say f_2, is not triangular, by simplicity, and hence is marked with u and v not adjacent. Hence, the edge (u, v) exists in G and not in P'. Consequently, f_1 cannot be a triangle, as otherwise it would contain edge (u, v) on its boundary. On the other hand, if f_1 is marked, then G contains another copy of (u, v) drawn inside the kite that yielded f_1, which is impossible since G is simple. Hence, P' contains no separation pair. The absence of cutvertices stems from the fact that each face is either a triangle or a quadrangle (if marked). □

Since each kite contains two crossing edges in G, its boundary is drawn strictly convex in both $\Gamma_a(G)$ and $\Gamma_b(G)$. Hence, $\Gamma_a(P')$ and $\Gamma_b(P')$ are two strictly convex planar drawings of P'. This property allows to apply Theorem 3 to compute a morph of $\langle \Gamma_a(P'), \Gamma_b(P') \rangle$ that maintains the strict convexity of the drawing at any time instant. Since each marked face f_K remains strictly convex, adding back the two crossing edges of the corresponding kite K in P' yields a morph of a supergraph of G (and thus of G) in which these crossing edges remain inside the boundary of K at any time instant. This concludes the base case.

5 Recursive Case

In this section, we focus on the recursive step of the proof of Theorem 1, in which the maximum level of a vertex in G is $\ell > 0$. Let Q be the graph obtained by removing all the vertices of level ℓ from G, and let $\langle \Gamma_a(Q), \Gamma_b(Q) \rangle$ be the restriction of $\langle \Gamma_a(G), \Gamma_b(G) \rangle$ to Q. Clearly, the two drawings of Q are topologically equivalent and the maximum level of a vertex is $\ell - 1$. Thus, we can recursively compute a morph of $\langle \Gamma_a(Q), \Gamma_b(Q) \rangle$. In what follows, we describe how to incorporate the trajectories of the level-ℓ vertices into the morph of $\langle \Gamma_a(Q), \Gamma_b(Q) \rangle$, so to obtain the desired morph of $\langle \Gamma_a(G), \Gamma_b(G) \rangle$.

Setting Up the Morph. We begin by observing that, by Property 1, each vertex of level ℓ is contained in a kite whose vertices are all of level $\ell - 1$, which

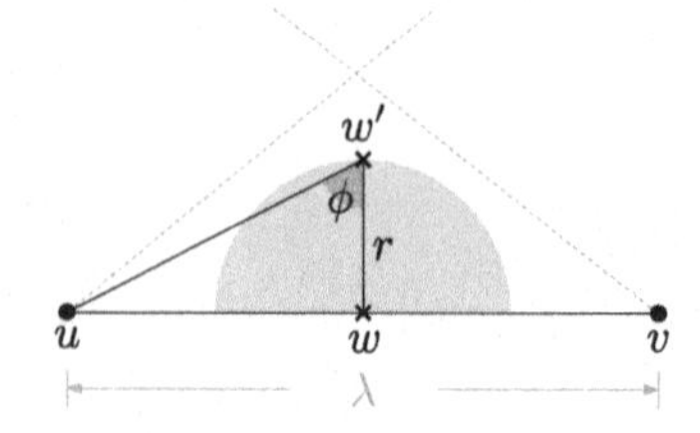

Fig. 3. Illustration of the half-disk D of $\triangle$ and their geometric properties.

implies that this kite is empty in Q (but not in G). Consider such a kite K, and note that its two crossing edges define four triangular regions that remain non-degenerate during the morph of $\langle \Gamma_a(Q), \Gamma_b(Q) \rangle$. We refer to each of these four triangular regions as a *piece of a kite.* The natural idea of applying recursion to every piece of a kite does not work, since the algorithm in [6] does not allow prescribing the trajectories of the vertices of the outer face, which would be required in the base case of this approach. Thus, we describe a more elaborated approach.

Consider a piece of kite K and denote it by $\triangle$. The unique edge (u, v) of $\triangle$ that belongs to the boundary of K is called the *base edge* of $\triangle$. Since $\triangle$ remains non-degenerate during the morph, there exists a half-disk D that, throughout the whole morph, has the following properties (see also Fig. 3 for an illustration):

- half-disk D lies in $\triangle$ and is centered at the midpoint w of (u, v), and
- the length of its radius is positive and it does not change.

Let λ be the smallest length of the base edge (u, v) during the morph, let r be the radius of D perpendicular to (u, v), and let w' be the endpoint of r different from w. Also, denote by t^* any time instant of the morph when the length of (u, v) equals λ, and let ϕ be the internal angle at w' of the triangle formed by u, w and w' at time t^*. In particular, ϕ satisfies $\tan(\phi) = \frac{\lambda}{2} \cdot \frac{1}{|r|}$.

Consider the graph $\mathcal{H} = G \setminus Q$ induced by the level-ℓ vertices of G, and let $H_\triangle$ be the subgraph of $\mathcal{H}$ that lies inside $\triangle$. We proceed to compute a drawing of $H_\triangle$ that, intuitively, will be "small" enough to fit inside D and "skinny" enough to avoid crossings with the binding edges that connect u or v to $H_\triangle$. To ease the notation, from now on we will refer to $H_\triangle$ as H.

To compute this drawing, we first augment H as well as its drawings in $\langle \Gamma_a(G), \Gamma_b(G) \rangle$, as follows. We add a dummy vertex d connected to u and to v, which is drawn sufficiently close to the crossing point of the two diagonals of K in both $\Gamma_a(G)$ and $\Gamma_b(G)$, so that the triangle formed by u, v, and d contains H.

As in the transition from P to P' in Sect. 4, we remove the crossing edges of every (empty) kite of $H \cup \{u, v, d\}$ and we mark the resulting quadrangular face. Then we apply Theorem 2 to the resulting planar subgraph of $H \cup \{u, v, d\}$ and to its drawings in $\langle \Gamma_a(G), \Gamma_b(G) \rangle$, except for its marked faces. This results in an almost triangulated plane graph H' and in a pair of topologically-equivalent strictly convex drawings $\langle \Gamma_a(H'), \Gamma_b(H') \rangle$ of H'. The following observation is directly implied by Property P.3 of Definition 1.

Observation 1. *Every face incident to u or to v in H' is triangular.*

Consider the plane graph obtained from H' by removing u and v, and let $\mathcal{C}$ be the graph formed by the vertices and the edges of its outer face. In the following lemma, we investigate some properties of $\mathcal{C}$. The *BC-tree* $\mathcal{T}$ of a connected graph G represents the decomposition of G into its biconnected components, called *blocks*. Namely, $\mathcal{T}$ has a *B-node* for each block of G and a *C-node* for each cutvertex of G, such that each B-node is connected to the C-nodes that are part of its block.

Lemma 2 (◇). *The following properties of $\mathcal{C}$ hold: (i) $\mathcal{C}$ is outerplane and connected. (ii) Each block of $\mathcal{C}$ is a cycle, possibly degenerated to a single edge. (iii) Every cutvertex of $\mathcal{C}$ is connected to both u and v in H'. (iv) The BC-tree of $\mathcal{C}$ is a path. (v) Every non cutvertex of $\mathcal{C}$ is connected to exactly one of u and v in H', with the exception of exactly two vertices (one of them is d) which belong to the blocks of $\mathcal{C}$ corresponding to degree-1 B-nodes in the BC-tree of $\mathcal{C}$.*

In view of Properties (ii) and (iv) of Lemma 2, we refer to $\mathcal{C}$ as a *chain of cycles* and to its blocks as *cycles*, even when degenerated to single edges. Moreover, we denote by d' the non cutvertex of $\mathcal{C}$ different from d that is incident to both u and v, as specified in Property (v) of Lemma 2.

Making Each Chain of Cycles Skinny. In order to incorporate the level-ℓ vertices that lie inside $\triangle$ into the morph of $\langle \Gamma_a(Q), \Gamma_b(Q) \rangle$, we perform a preliminary morph of $\Gamma_a(H')$ to a strictly convex drawing $\Gamma_a^s(H')$ of H' that is *skinny*, in the sense that it satisfies the following requirements with respect to the disk D and the angle ϕ associated with the base edge (u, v) derived from the morph of $\langle \Gamma_a(Q), \Gamma_b(Q) \rangle$ (see also Fig. 4 for an illustration).

- R.1 Every cycle of $\mathcal{C}$ is drawn inside the disk D.
- R.2 Every cycle of $\mathcal{C}$ is drawn strictly convex.
- R.3 The cutvertices of $\mathcal{C}$, as well as d and d', lie on the radius r of D.
- R.4 For every cycle of $\mathcal{C}$ and for every segment on its boundary, the smaller of the two angles formed at the intersection of the line through r and the line through the segment is smaller than ϕ.

The existence of such a drawing is proven in the following lemma by means of a construction that exploits the properties of $\mathcal{C}$ given in Lemma 2.

Lemma 3. *There exists a drawing $\Gamma_a^s(H')$ of H' that is strictly convex, skinny, and topologically equivalent to $\Gamma_a(H')$.*

Proof. We prove the statement by construction. Initially, we place u and v in the same positions as they are in $\Gamma_a(H')$. Further, we place the cutvertices of the chain of cycles $\mathcal{C}$ as well as d and d' on the radius r in the order they appear in

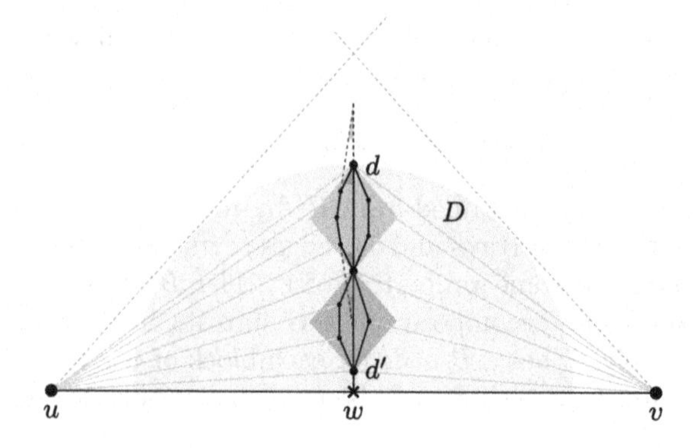

Fig. 4. Illustration of the requirements R.1–R.4 of a skinny drawing.

the chain to satisfy R.3. For each cycle c of $\mathcal{C}$, we proceed as follows. Let x and y be the two vertices of c that have already been placed on r. Let T_c^u and T_c^v be two isosceles triangles sharing the same base $\overline{xy}$, such that the third vertex of each of them lies inside D and on opposite sides of r and such that the internal angles at x and at y are smaller than ϕ; refer to the colored triangles in Fig. 4. We place the vertices of c that are incident only to u (only to v) equidistant along a circular arc connecting x and y that lies completely inside T_c^u (inside T_c^v, respectively). By the definition of T_c^u and T_c^v, and also by the fact that the two circular arcs are drawn completely inside T_c^u and T_c^v, it follows that R.1, R.2, and R.4 are satisfied for the drawing of c.

To complete the drawing of $\Gamma_a^s(H')$, we describe how to draw the subgraph H'_c of H' that is contained inside or on the boundary of c such that every internal face of H'_c is strictly convex. Since H'_c is drawn convex in $\Gamma_a(H')$, it admits a strictly convex drawing for any given strictly convex drawing of its outer face [18]. Thus, we can apply the algorithm in [18] to construct a strictly convex drawing of H'_c, whose outerface is the drawing of c satisfying R.1-R.4. Finally, we add the edges incident to u and v that are contained inside $\triangle$ to the resulting drawing, which does not introduce crossings due to R.4. This completes the construction of $\Gamma_a^s(H')$. Since every cycle in $\mathcal{C}$ satisfies R.1–R.4 and since by Observation 1 all faces incident to u and v in H' are triangular, the drawing $\Gamma_a^s(H')$ is strictly convex and skinny as desired. Since our construction and the algorithm in [18] maintain the cyclic order of the edges around each vertex, we have that $\Gamma_a^s(H')$ is topologically equivalent to $\Gamma_a(H')$. This concludes the proof. □

To describe the morph between $\Gamma_a(H')$ and $\Gamma_a^s(H')$, we need some more work. Since both drawings are strictly convex and topologically equivalent, the preconditions of Theorem 3 are met. However, to ensure that this morph can be done independently for each piece of a kite, we further need to guarantee that vertices u and v do not move and that all vertices of H' remain inside $\triangle$ throughout the morph. As stated in Theorem 3, this can be achieved if the (triangular) outer face is drawn the same in the two input drawings, which is not necessarily the case for $\Gamma_a(H')$ and $\Gamma_a^s(H')$ because of the position of d (recall

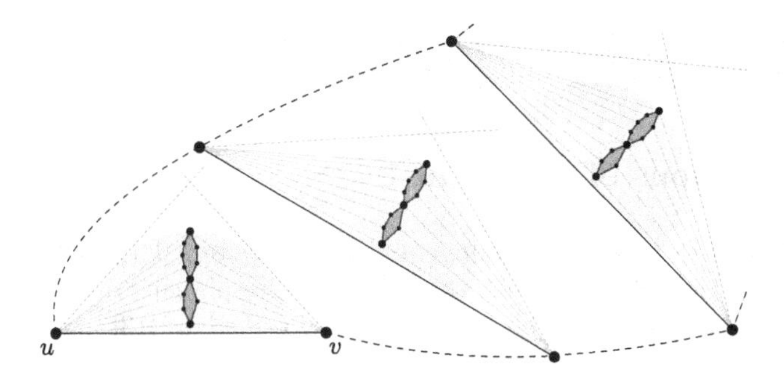

Fig. 5. Computing the trajectories for the vertices of $H' \setminus \{u, v\}$ based on the, already computed, trajectories of u and v.

that u and v have the same position in $\Gamma_a(H')$ and $\Gamma_a^s(H')$). To this end, we augment $\Gamma_a(H')$ and $\Gamma_a^s(H')$ by adding a new vertex d^* in the outer face of H' and connect it to u, v, and d. Moreover, we place d^* at the same position inside $\triangle$ in both $\Gamma_a(H')$ and $\Gamma_a^s(H')$ so that the triangle formed by u, v, and d^* contains all the other vertices of H' (in particular, d). The edge (d, d^*) can always be drawn without crossings, as u, v, and d were the vertices on the outer face of H' before. After this augmentation, we apply Theorem 3 to compute the desired morph of $\langle \Gamma_a(H'), \Gamma_a^s(H') \rangle$, and then we remove d^* from the drawings.

Performing the Global Morph. Applying the above procedure for each piece of a kite yields a drawing $\Gamma_a(G')$ of the supergraph G' of G that is the union of Q and all the graphs $H'_\triangle$ corresponding to every piece of a kite $\triangle$. Observe that $\Gamma_a(G')$ is composed of $\Gamma_a(Q)$ and the skinny drawing $\Gamma_a^s(H'_\triangle)$ of every graph $H'_\triangle$. To perform the global morph, recall that the vertices of the subgraph Q of G' follow the same trajectories as in the morph between $\Gamma_a(Q)$ and $\Gamma_b(Q)$ (which has been recursively computed). The level-ℓ vertices of each subgraph $H'_\triangle$ are moved inside $\triangle$, which again ensures that this can be done independently for each piece of a kite. In the following we describe the trajectories of one such subgraph. We denote this subgraph as H' and adopt the same notation as before.

Since the trajectories of u and v are specified by the morph between $\Gamma_a(Q)$ and $\Gamma_b(Q)$, we only describe the trajectories of $H' \setminus \{u, v\}$, i.e., the vertices of level ℓ; see Fig. 5 for an example. The drawing of $H' \setminus \{u, v\}$ is a copy of $\Gamma_a^s(H' \setminus \{u, v\})$ rotated and translated so that the cutvertices of $\mathcal{C}$ as well as d and d' lie on the radius of D perpendicular to (u, v), and the distance between w and d' is the same as in $\Gamma_a^s(H')$. This ensures that the drawing of H' remains skinny, planar (by R.4), and strictly convex at every time instant.

Let $\Gamma_b(G')$ be the drawing of G' obtained so far. The next step of the morph is to transform, for each subgraph $H'_\triangle$, the current skinny drawing $\Gamma_b^s(H'_\triangle)$ in $\Gamma_b(G')$ to $\Gamma_b(H'_\triangle)$. By construction, $\Gamma_b^s(H'_\triangle)$ and $\Gamma_b(H'_\triangle)$ are topologically equivalent and strictly convex. Similarly as for $\Gamma_a(H'_\triangle)$, we insert vertex d^* so that the outer face of $H'_\triangle$ is drawn the same in both $\Gamma_b^s(H'_\triangle)$ and $\Gamma_b(H'_\triangle)$, which allows to apply Theorem 3 independently for each $H'_\triangle$. The target drawing $\Gamma_b(G)$

is obtained by removing the vertices and edges in $G' \setminus G$ and by reinserting the crossed edges in the marked faces. This concludes the proof of Theorem 1.

6 Implications of Theorem 1

In this section, we discuss the applicability of Theorem 1 by presenting meaningful families of 1-planar graphs that admit 1-kite-planar drawings.

An n-vertex 1-planar graph has at most $4n - 8$ edges [13], and if it achieves exactly this density, then it is called *optimal*. Moreover, any 1-planar drawing of an optimal 1-planar graph G is such that the uncrossed edges induce a plane triconnected quadrangulation P, while each pair of crossing edges of G is drawn inside a corresponding face of P [33]. When restricting to straight-line drawable 1-planar graphs, this bound is reduced to $4n - 9$ [21]. Similarly to the general case, an optimal 1-planar straight-line drawing is one in which the uncrossed edges induce a plane triconnected graph whose every inner face is a quadrangle, while the outer face is a triangle [21]. As a consequence, we obtain that each kite is empty and its kite edges are present and uncrossed. Therefore, any optimal 1-planar straight-line drawing is 1-kite-planar.

Another family of 1-planar graphs that recently attracted considerable attention is the one of *IC-planar* graphs [3,15,19,30], which admit 1-planar drawings where the crossed edges induce a matching. Note that both the binding edges and the kite edges that are part of an IC-planar drawing are uncrossed. It follows that, if an IC-planar drawing is *kite-augmented* [14], i.e., it contains all kite edges, then it is 1-kite-planar. Observe that kite-augmented graphs are also known as *locally maximal* [24]. Overall, the following result is a corollary of Theorem 1.

Corollary 1. *There exists a morph between any pair of topologically-equivalent optimal* 1*-planar or kite-augmented IC-planar straight-line drawings.*

We conclude this section with a remark. As already mentioned, Chambers et al. [17] studied morphs of toroidal graphs and asked to generalize their result to surfaces of higher genus. We note that, since an n-vertex graph embeddable on a surface of genus g has at most $3n + 6(g - 1)$ edges, while n-vertex optimal 1-planar straight-line drawable graphs have $4n - 9$ edges, it follows that the latter do not admit an embedding (without edge crossings) on any surface of bounded genus. Thus, a solution to the open problem by Chambers et al. would not provide morphs of 1-kite-planar drawings.

7 Open Problems

We made a first step towards the problem of morphing pairs of non-planar drawings. Besides the general open problem of morphing any two such drawings [5], the main questions that stem from our research are as follows: (i) Is it possible to compute morphs of 1-kite-planar drawings where the vertex trajectories have bounded complexity? (ii) Regardless of the complexity, can we drop P.2 or P.3

of Definition 1? Observe that dropping both would extend Theorem 1 to all 1-planar drawings. (iii) On the other hand, as a relaxation of P.1, further families of beyond-planar graphs [23] could be considered, for instance, does every pair of RAC drawings [22] admit a morph?

References

1. Alamdari, S., et al.: How to morph planar graph drawings. SIAM J. Comput. **46**(2), 824–852 (2017)
2. Alamdari, S., et al.: Morphing planar graph drawings with a polynomial number of steps. In: Khanna, S. (ed.) ACM-SIAM Symposium on Discrete Algorithms, (SODA 2013). pp. 1656–1667. SIAM (2013)
3. Albertson, M.O.: Chromatic number, independence ratio, and crossing number. Ars Math. Contemp. **1**(1) (2008)
4. Angelini, P., Chaplick, S., Cornelsen, S., Lozzo, G.D., Roselli, V.: Morphing contact representations of graphs. In: Symposium on Computational Geometry (SoCG 2019). LIPIcs, vol. 129, pp. 10:1–10:16. Schloss Dagstuhl - Leibniz-Zentrum für Informatik (2019)
5. Angelini, P., Da Lozzo, G., Di Battista, G., Frati, F., Patrignani, M., Roselli, V.: Morphing planar graph drawings optimally. In: Esparza, J., Fraigniaud, P., Husfeldt, T., Koutsoupias, E. (eds.) International Colloquium on Automata, Languages, and Programming (ICALP 2014). LNCS, vol. 8572, pp. 126–137. Springer (2014)
6. Angelini, P., Da Lozzo, G., Frati, F., Lubiw, A., Patrignani, M., Roselli, V.: Optimal morphs of convex drawings. In: Arge, L., Pach, J. (eds.) Symposium on Computational Geometry (SoCG 2015). LIPIcs, vol. 34, pp. 126–140. Schloss Dagstuhl - Leibniz-Zentrum für Informatik (2015)
7. Aronov, B., Seidel, R., Souvaine, D.: On compatible triangulations of simple polygons. Comput. Geom. Theory Appl. **3**(1), 27–35 (1993)
8. Arseneva, E., Bose, P., Cano, P., D'Angelo, A., Dujmovic, V., Frati, F., Langerman, S., Tappini, A.: Pole dancing: 3D morphs for tree drawings. J. Graph Algorithms Appl. **23**(3), 579–602 (2019)
9. Avital, S., Hanani, H.: Graphs. Gilyonot Lematematika **3**, 2–8 (1966)
10. Barrera-Cruz, F., et al.: How to morph a tree on a small grid. In: Friggstad, Z., Sack, J.-R., Salavatipour, M.R. (eds.) WADS 2019. LNCS, vol. 11646, pp. 57–70. Springer, Cham (2019). https://doi.org/10.1007/978-3-030-24766-9_5
11. Barrera-Cruz, F., Haxell, P., Lubiw, A.: Morphing Schnyder drawings of planar triangulations. Discrete Comput. Geom. **61**, 1–24 (2018)
12. Biedl, T., Lubiw, A., Petrick, M., Spriggs, M.: Morphing orthogonal planar graph drawings. ACM Trans. Algorith. **9**(4), 29:1–29:24 (2013)
13. Bodendiek, R., Schumacher, H., Wagner, K.: Bemerkungen zu einem Sechsfarbenproblem von G. Ringel. Abhandlungen aus dem Mathematischen Seminar der Universitaet Hamburg **53**(1), 41–52 (1983)
14. Brandenburg, F.J.: Characterizing and recognizing 4-map graphs. Algorithmica **81**(5), 1818–1843 (2019)
15. Brandenburg, F.J., Didimo, W., Evans, W.S., Kindermann, P., Liotta, G., Montecchiani, F.: Recognizing and drawing IC-planar graphs. Theor. Comput. Sci. **636**, 1–16 (2016)

16. Cairns, S.S.: Deformations of plane rectilinear complexes. Am. Math. Monthly **51**(5), 247–252 (1944)
17. Chambers, E.W., Erickson, J., Lin, P., Parsa, S.: How to morph graphs on the torus. In: ACM-SIAM Symposium on Discrete Algorithms (SODA 2021) (To appear 2021)
18. Chiba, N., Yamanouchi, T., Nishizeki, T.: Linear algorithms for convex drawings of planar graphs. Prog. Graph Theory **173**, 153–173 (1984)
19. Czap, J., Šugerek, P.: Drawing graph joins in the plane with restrictions on crossings. Filomat **31**(2), 363–370 (2017)
20. Da Lozzo, G., Di Battista, G., Frati, F., Patrignani, M., Roselli, V.: Upward planar morphs. Algorithmica **82**(10), 2985–3017 (2020)
21. Didimo, W.: Density of straight-line 1-planar graph drawings. Inf. Process. Lett. **113**(7), 236–240 (2013)
22. Didimo, W.: Right angle crossing drawings of graphs. In: Beyond Planar Graphs, pp. 149–169. Springer (2020). https://doi.org/10.1007/978-981-15-6533-5
23. Didimo, W., Liotta, G., Montecchiani, F.: A survey on graph drawing beyond planarity. ACM Comput. Surv. **52**(1), 4:1–4:37 (2019)
24. Fabrici, I., Harant, J., Madaras, T., Mohr, S., Soták, R., Zamfirescu, C.T.: Long cycles and spanning subgraphs of locally maximal 1-planar graphs. J. Graph Theory **95**(1), 125–137 (2020)
25. Floater, M., Gotsman, C.: How to morph tilings injectively. J. Comput. Appl. Math. **101**, 117–129 (1999)
26. van Goethem, A., Speckmann, B., Verbeek, K.: Optimal morphs of planar orthogonal drawings II. In: Archambault, D., Tóth, C.D. (eds.) GD 2019. LNCS, vol. 11904, pp. 33–45. Springer, Cham (2019). https://doi.org/10.1007/978-3-030-35802-0_3
27. Gotsman, C., Surazhsky, V.: Guaranteed intersection-free polygon morphing. Comput Graph. **25**(1), 67–75 (2001)
28. Hong, S., Tokuyama, T. (eds.): Beyond Planar Graphs, Communications of NII Shonan Meetings. Springer (2020). https://doi.org/10.1007/978-981-15-6533-5
29. Kobourov, S.G., Liotta, G., Montecchiani, F.: An annotated bibliography on 1-planarity. Comput. Sci. Rev. **25**, 49–67 (2017)
30. Liotta, G., Montecchiani, F.: L-visibility drawings of IC-planar graphs. Inf. Process. Lett. **116**(3), 217–222 (2016)
31. Ringel, G.: Ein Sechsfarbenproblem auf der Kugel. Abh. Math. Sem. Univ. Hamb. **29**, 107–117 (1965)
32. Surazhsky, V., Gotsman, C.: Controllable morphing of compatible planar triangulations. ACM Trans. Graph. **20**(4), 203–231 (2001)
33. Suzuki, Y.: Optimal 1-planar graphs which triangulate other surfaces. Discrete Math. **310**(1), 6–11 (2010)
34. Thomassen, C.: Deformations of plane graphs. J. Comb. Theory Ser. B **34**(3), 244–257 (1983)

Bears with Hats and Independence Polynomials

Václav Blažej[1], Pavel Dvořák[2], and Michal Opler[2](✉)

[1] Faculty of Information Technology, Czech Technical University in Prague, Prague, Czech Republic

[2] Faculty of Mathematics and Physics, Charles University, Prague, Czech Republic
{koblich,opler}@iuuk.mff.cuni.cz

Abstract. Consider the following hat guessing game. A bear sits on each vertex of a graph G, and a demon puts on each bear a hat colored by one of h colors. Each bear sees only the hat colors of his neighbors. Based on this information only, each bear has to guess g colors and he guesses correctly if his hat color is included in his guesses. The bears win if at least one bear guesses correctly for any hat arrangement.

We introduce a new parameter—fractional hat chromatic number $\hat{\mu}$, arising from the hat guessing game. The parameter $\hat{\mu}$ is related to the hat chromatic number which has been studied before. We present a surprising connection between the hat guessing game and the independence polynomial of graphs. This connection allows us to compute the fractional hat chromatic number of chordal graphs in polynomial time, to bound fractional hat chromatic number by a function of maximum degree of G, and to compute the exact value of $\hat{\mu}$ of cliques, paths, and cycles.

Keywords: Hat guessing game · Independence polynomial · Chordal graphs

1 Introduction

In this paper, we study a variant of a hat guessing game. In these types of games, there are some entities—players, pirates, sages, or, as in our case, bears. A bear sits on each vertex of graph G. There is some adversary (a demon in our case) that puts a colored hat on the head of each bear. A bear on a vertex v sees only the hats of bears on the neighboring vertices of v but he does not know the color of his own hat. Now to defeat the demon, the bears should guess correctly

V. Blažej—Acknowledges the support of the OP VVV MEYS funded project CZ.02.1.01/0.0/0.0/16_019/0000765 "Research Center for Informatics". This work was supported by the Grant Agency of the Czech Technical University in Prague, grant No. SGS20/208/OHK3/3T/18.

P. Dvořák—Supported by Czech Science Foundation GAČR grant #19-27871X.

M. Opler—The work was supported by the grant SVV–2020–260578.

The original version of this chapter was revised: this chapter was previously published non-open access. The correction to this chapter is available at
https://doi.org/10.1007/978-3-030-86838-3_31

Ł. Kowalik et al. (Eds.): WG 2021, LNCS 12911, pp. 283–295, 2021.
https://doi.org/10.1007/978-3-030-86838-3_22

the color of their hats. However, the bears can only discuss their strategy before they are given the hats. After they get them, no communication is allowed, each bear can only guess his hat color. The variants of the game differ in the bears' winning condition.

The first variant was introduced by Ebert [8]. In this version, each bear gets a red or blue hat (chosen uniformly and independently) and they can either guess a color or pass. The bears see each other, i.e. they stay on vertices of a clique. They win if at least one bear guesses his color correctly and no bear guesses a wrong color. The question is what is the highest probability that the bears win achievable by some strategy. Soon, the game became quite popular and it was even mentioned in NY Times [26].

Winkler [29] studied a variant where the bears cannot pass and the objective is how many of them guess correctly their hat color. A generalization of this variant for more than two colors was studied by Feige [11] and Aggarwal [1]. Butler et al. [6] studied a variant where the bears are sitting on vertices of a general graph, not only a clique. For a survey of various hat guessing games, we refer to theses of Farnik [10] or Krzywkowski [23].

In this paper, we study a variant of the game introduced by Farnik [10], where each bear has to guess and they win if at least one bear guesses correctly. He introduced a hat guessing number HG of a graph G (also named as hat chromatic number and denoted μ in later works) which is defined as the maximum h such that bears win the game with h hat colors. We study a variant where each bear can guess multiple times and we consider that a bear guesses correctly if the color of his hat is included in his guesses. We introduce a parameter *fractional hat chromatic number* $\hat{\mu}$ of a graph G, which we define as the supremum of $\frac{h}{g}$ such that each bear has g guesses and they win the game with h hat colors.

Albeit the hat guessing game looks like a recreational puzzle, connections to more "serious" areas of mathematics and computer science were shown—like coding theory [9,19], network coding [14,25], auctions [1], finite dynamical systems [12], and circuits [30]. In this paper, we exhibit a connection between the hat guessing game and the independence polynomial of graphs, which is our main result. This connection allows us to compute the optimal strategy of bears (and thus the value of $\hat{\mu}$) of an arbitrary chordal graph in polynomial time. We also prove that the fractional hat chromatic number $\hat{\mu}$ is asymptotically equal, up to a logarithmic factor, to the maximum degree of a graph. Finally, we compute the exact value of $\hat{\mu}$ of graphs from some classes, like paths, cycles, and cliques.

We would like to point out that the existence of the algorithm computing $\hat{\mu}$ of a chordal graph is far from obvious. Butler et al. [6] asked how hard is to compute $\mu(G)$ and the optimal strategy for the bears. Note that a trivial non-deterministic algorithm for computing the optimal strategy (or just the value of $\mu(G)$ or $\hat{\mu}(G)$) needs exponential time because a strategy of a bear on v is a function of hat colors of bears on neighbors of v (we formally define the strategy in Sect. 2). It is not clear if the existence of a strategy for bears would imply a strategy for bears where each bear computes his guesses by some efficiently computable function (like linear, computable by a polynomial circuit, etc.). This would allow us to put the problem of computing μ into some level of the polynomial hierarchy, as noted by Butler et al. [6]. However, we are not aware of any hardness results for the hat guessing games. The maximum degree bound for $\hat{\mu}$ does not imply an exact efficient algorithm computing $\hat{\mu}(G)$ as well. This phenomenon can be illustrated by the edge chromatic number χ' of graphs. By Vizing's theorem [7, Chapter 5], it holds for any graph G that $\Delta(G) \leq \chi'(G) \leq \Delta(G) + 1$. However, it is NP-hard to distinguish between these two cases [18].

Organization of the Paper. We finish this section with a summary of results about the variant of the hat guessing game we are studying. In the next section, we present notions used in this paper and we define formally the hat guessing game. In Sect. 3, we formally define the fractional hat chromatic number $\hat{\mu}$ and compare it to μ. In Sect. 4, we generalize some previous results to the multi-guess setting. We use these tools to prove our main result in Sect. 5 including the poly-time algorithm that computes $\hat{\mu}$ for chordal graphs. The maximum degree bound for $\hat{\mu}$ and computation of exact values of paths and cycles are provided in Sect. 6.

1.1 Related Works

As mentioned above, Farnik [10] introduced a hat chromatic number $\mu(G)$ of a graph G as the maximum number of colors h such that the bears win the hat guessing game with h colors and played on G. He proved that $\mu(G) \leq O(\Delta(G))$ where $\Delta(G)$ is the maximum degree of G.

Since then, the parameter $\mu(G)$ was extensively studied. The parameter μ for multipartite graphs was studied by Gadouleau and Georgiu [13] and by Alon et al. [2]. Szczechla [28] proved that μ of cycles is equal to 3 if and only if the length of the cycle is 4 or it is divisible by 3 (otherwise it is 2). Bosek et al. [5] gave bounds of μ for some graphs, like trees and cliques. They also provided some connections between $\mu(G)$ and other parameters like chromatic number and degeneracy. They conjectured that $\mu(G)$ is bounded by some function of the degeneracy $d(G)$ of the graph G. They showed that such function has to be at least exponential as they presented a graph G of $\mu(G) \geq 2^{d(G)}$. This result was improved by He and Li [16] who showed there is a graph G such that $\mu(G) \geq 2^{2^{d(G)-1}}$. Since $\hat{\mu}(G)$ is upper-bounded $O(\Delta(G))$ [10] it holds that $\hat{\mu}$ can not be bounded by any function of degeneracy as there are graph classes of unbounded maximum degree and bounded degeneracy (e.g. trees or planar graphs). Recently, Kokhas et al. [21,22] studied a non-uniform version of the game, i.e., for each bear, there could be a different number of colors of the hat. They considered cliques and almost cliques. They also provided a technique to build a strategy for a graph G whenever G is made up by combining G_1 and G_2 with known strategies. We generalize some of their results and use them as "basic blocks" for our main result.

2 Preliminaries

We use standard notions of the graph theory. For an introduction to this topic, we refer to the book by Diestel [7]. We denote a clique as K_n, a cycle as C_n, and a path as P_n, each on n vertices. The maximum degree of a graph G is denoted by $\Delta(G)$, where we shorten it to Δ if the graph G is clear from the context. The neighbors of a vertex v are denoted by $N(v)$. We use $N^+(v)$ to denote the closed neighborhood of v, i.e. $N^+(v) = N(v) \cup \{v\}$. For a set U of vertices of a graph G, we denote $G \setminus U$ a graph induced by vertices $V(G) \setminus U$, i.e., a graph arising from G by removing the vertices in U.

A *hat guessing game* is a triple $\mathcal{H} = (G, h, g)$ where

- $G = (V, E)$ is an undirected graph, called the *visibility graph*,
- $h \in \mathbb{N}$ is a *hatness* that determines the number of different possible hat colors for each bear, and
- $g \in \mathbb{N}$ is a *guessing number* that determines the number of guesses each bear is allowed to make.

The rules of the game are defined as follows. On each vertex of G sits a bear. The demon puts a hat on the head of each bear. Each hat has one of h colors. We would like to point out, that it is allowed that bears on adjacent vertices get a hat of the same color. The only information the bear on a vertex v knows are the colors of hats put on bears sitting on neighbors of v. Based on this information only, the bear has to guess a set of g colors according to a deterministic strategy agreed to in advance. We say bear *guesses correctly* if he included the color of his hat in his guesses. The bears win if at least one bear guesses correctly.

Formally, we associate the colors with natural numbers and say that each bear can receive a hat colored by a color from the set $S = [h] = \{0, \ldots, h-1\}$. A *hats arrangement* is a function $\varphi : V \to S$. A strategy of a bear on v is a function $\Gamma_v : S^{|N(v)|} \to \binom{S}{g}$, and a *strategy for* $\mathcal{H}$ is a collection of strategies for all vertices, i.e. $(\Gamma_v)_{v \in V}$. We say that a strategy is *winning* if for any possible hats arrangement $\varphi : V \to S$ there exists at least one vertex v such that $\varphi(v)$ is contained in the image of Γ_v on φ, i.e., $\varphi(v) \in \Gamma_v\big((\varphi(u))_{u \in N(v)}\big)$. Finally, the game $\mathcal{H}$ is *winning* if there exists a winning strategy of the bears.

As a classical example, we describe a winning strategy for the hat guessing game $(K_3, 3, 1)$. Let us denote the vertices of K_3 by v_0, v_1 and v_2 and fix a hats arrangement φ. For every $i \in [3]$, the bear on the vertex v_i assumes that the sum $\sum_{j \in [3]} \varphi(v_j)$ is equal to i modulo 3 and computes its guess accordingly. It follows that for any hat arrangement φ there is always exactly one bear that guesses correctly, namely the bear on the vertex v_i for $i = \sum_j \varphi(v_j) \pmod 3$.

Some of our results are stated for a non-uniform variant of the hat guessing game. A non-uniform game is a triple $\big(G = (V, E), \mathbf{h}, \mathbf{g}\big)$ where $\mathbf{h} = (h_v)_{v \in V}$ and $\mathbf{g} = (g_v)_{v \in V}$ are vectors of natural numbers indexed by the vertices of G and a bear on v gets a hat of one of h_v colors and is allowed to guess exactly g_v colors. Other rules are the same as in the standard hat guessing game. To distinguish between the uniform and non-uniform games, we always use plain letters h and g for the hatness and the guessing number, respectively, and bold letters (e.g. $\mathbf{h}, \mathbf{g}$) for vectors indexed by the vertices of G.

3 Fractional Hat Chromatic Number

From the hat guessing games, we can derive parameters of the underlying visibility graph G. Namely, the *hat chromatic number* $\mu(G)$ is the maximum integer h for which the hat guessing game $(G, h, 1)$ is winning, i.e., each bear gets a hat colored by one of h colors and each bear has only one guess—we call such game a single-guessing game. In this paper, we study a parameter *fractional hat chromatic number* $\hat{\mu}(G)$ arising from the hat multi-guessing game and defined as

$$\hat{\mu}(G) = \sup \left\{ \frac{h}{g} \;\middle|\; (G, h, g) \text{ is a winning game} \right\}$$

Observe that $\mu(G) \leq \hat{\mu}(G)$. Farnik [10] and Bosek et al. [5] also study multi-guessing games. They considered a parameter $\mu_g(G)$ that is the maximum number of colors h such that the bears win the game (G, h, g). The difference between μ_g and $\hat{\mu}$ is the following. If $\mu_g(G) \geq k$, then the bears win the game (G, k, g) and $\hat{\mu} \geq \frac{k}{g}$. If $\hat{\mu}(G) \geq \frac{p}{q}$, then there are $h, g \in \mathbb{N}$ such that $\frac{p}{q} = \frac{h}{g}$ and the bears win the game (G, h, g). However, it does not imply that the bears would win the game (G, p, q). It is easy to prove that if the bears win the game (G, h, g) then they win the game (G, kh, kg) for any constant $k \in \mathbb{N}$ (see the full version [4] for the details). The opposite implication does not hold– we discuss a counterexample at the end of this section. Unfortunately, this property prevents us from using our algorithm, which computes $\hat{\mu}$, to compute also μ of chordal graphs.

Moreover, by definition, the parameter $\hat{\mu}$ does not even have to be a rational number. In such a case, for each $p, q \in \mathbb{N}$, it holds that

- If $\frac{p}{q} < \hat{\mu}(G)$ then there are $h, g \in \mathbb{N}$ such that $\frac{p}{q} = \frac{h}{g}$ and the bears win the game (G, h, g).
- If $\frac{p}{q} > \hat{\mu}(G)$ then the demon wins the game (G, p, q).

For example, the fractional hat chromatic number $\hat{\mu}(P_3)$ of the path P_3 is irrational. We discuss path P_3 the full version [4]. In the case of an irrational $\hat{\mu}(G)$, our algorithm computing the value of $\hat{\mu}$ of chordal graphs outputs an estimate of $\hat{\mu}(G)$ with arbitrary precision. The next lemma state that the multi-guessing game is in some sense monotone. The proof is in the full version [4].

Lemma 1. *Let $\big(G = (V, E), h, g\big)$ be a winning hat guessing game. Let r' be a rational number such that $r' \leq h/g$. Then, there exist numbers $h', g' \in \mathbb{N}$ such that $h'/g' = r'$ and the hat guessing game (G, h', g') is winning.*

It is straight-forward to prove a generalization of Lemma 1 for non-uniform games. However, for simplicity, we state it only for the uniform games. By the proof of the previous lemma, we know that we can use a strategy for (G, h, g) to create a strategy for a game $(G, k \cdot h, k \cdot g + \ell)$ for arbitrary $k, \ell \in \mathbb{N}$. A question is if we can do it in general: Can we derive a winning strategy if we decrease the fraction h/g, but the hatness h and the guessing number g are changed arbitrarily? It is true for cliques. We show in Sect. 4 that the bears win the game (K_n, h, g) if and only if $h/g \leq n$. However, it is not true in general. For example, for n large enough it holds that $\hat{\mu}(P_n) \geq 3$, as we show in Sect. 6 that $\hat{\mu}(P_n)$ converges to 4 when n goes to infinity. However, Butler et al. [6] proved that $\mu(T) = 2$ for any tree T. Thus, the bears lose the game $(P_n, 3, 1)$.

4 Basic Blocks

In this section, we generalize some results of Kokhas et al. [21,22] about cliques and strategies for graph products, which we use for proving our main result. The single-guessing version of the next theorem (without the algorithmic consequences) was proved by Kokhas et al. [21,22]. The proof of the following theorem is stated in the full version [4].

Theorem 1. *Bears win a game* $\big(K_n = (V,E), \mathbf{h}, \mathbf{g}\big)$ *if and only if*

$$\sum_{v \in V} \frac{g_v}{h_v} \geq 1.$$

Moreover, if there is a winning strategy, then there is a winning strategy $(\Gamma_v)_{v \in V}$ *such that each* Γ_v *can be described by two linear inequalities whose coefficients can be computed in linear time.*

By Theorem 1, we can conclude the following corollary.

Corollary 1. *For each* $n \in \mathbb{N}$, *it holds that* $\hat{\mu}(K_n) = n$.

Further, we generalize a result of Kokhas and Latyshev [21]. In particular, we provide a new way to combine two hat guessing games on graphs G_1 and G_2 into a hat guessing game on graph obtained by gluing G_1 and G_2 together in a specific way.

Let $G_1 = (V_1, E_1)$ and $G_2 = (V_2, E_2)$ be graphs, let $S \subseteq V_1$ be a set of vertices inducing a clique in G_1, and let $v \in V_2$ be an arbitrary vertex of G_2. The *clique join of graphs* G_1 *and* G_2 *with respect to* S *and* v is the graph $G = (V, E)$ such that $V = V_1 \cup V_2 \setminus \{v\}$; and E contains all the edges of E_1, all the edges of E_2 that do not contain v, and an edge between every $w \in S$ and every neighbor of v in G_2. See Fig. 1 for an example of a clique join and the application of the following lemma.

Lemma 2. *Let* $\mathcal{H}_1 = \big(G_1 = (V_1, E_1), \mathbf{h}^1, \mathbf{g}^1\big)$ *and* $\mathcal{H}_2 = \big(G_2 = (V_2, E_2), \mathbf{h}^2, \mathbf{g}^2\big)$ *be two hat guessing games and let* $S \subseteq V_1$ *be a set inducing a clique in* G_1 *and* $v \in V_2$. *Set* G *to be the clique join of graphs* G_1 *and* G_2 *with respect to* S *and* v. *If the bears win the games* $\mathcal{H}_1$ *and* $\mathcal{H}_2$, *then they also win the game* $\mathcal{H} = (G, \mathbf{h}, \mathbf{g})$ *where*

$$h_u = \begin{cases} h^1_u & u \in V_1 \setminus S \\ h^2_u & u \in V_2 \setminus \{v\} \\ h^1_u \cdot h^2_v & u \in S, \text{ and} \end{cases} \qquad g_u = \begin{cases} g^1_u & u \in V_1 \setminus S \\ g^2_u & u \in V_2 \setminus \{v\} \\ g^1_u \cdot g^2_v & u \in S. \end{cases}$$

Proof Idea. For every bear $u \in S$, we interpret his color as a tuple (c_u^1, c_u^2) where $c_u^1 \in [h_u^1]$ and $c_u^2 \in [h_v^2]$. The bears in $G_1 \setminus S$ or $G_2 \setminus \{v\}$ use the strategies for $\mathcal{H}_1$ or $\mathcal{H}_2$, respectively. The bears in S combine the winning strategies for $\mathcal{H}_1$ and $\mathcal{H}_2$. The full proof is in the full version [4]. □

We remark that Lemma 2 generalizes Theorem 3.1 and Theorem 3.5 of [21] not only by introducing multiple guesses but also by allowing for more general ways to glue two graphs together. Thus, it provides new constructions of winning games even for single-guessing games.

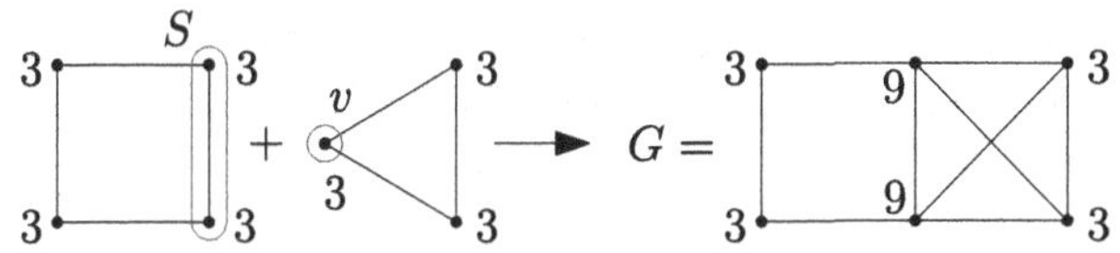

Fig. 1. Applying Lemma 2 on winning hat guessing games $(C_4, 3, 1)$ (see [28]) and $(K_3, 3, 1)$, we obtain a winning hat guessing game $(G, \mathbf{h}, 1)$ where G is the result of identifying an edge in C_4 and K_4, and $\mathbf{h}$ is given in the picture.

5 Independence Polynomial

The multivariate *independence polynomial* of a graph $G = (V, E)$ on variables $\mathbf{x} = (x_v)_{v \in V}$ is

$$P_G(\mathbf{x}) = \sum_{\substack{I \subseteq V \\ I \text{ independent set}}} \prod_{v \in I} x_v.$$

First, we describe informally the connection between the multi-guessing game and the independence polynomial. Consider the game (G, h, g) and fix a strategy of bears. Suppose that the demon put on the head of each bear a hat of random color (chosen uniformly and independently). Let A_v be an event that the bear on the vertex v guesses correctly. Then, the probability of A_v is exactly g/h. Moreover, for any independent set I holds that A_v is independent on all events A_w for $w \in I, w \neq v$. Thus, we can use the inclusion-exclusion principle to compute the probability that A_v occurs for at least one $v \in I$, i.e., at least one bear sitting on some vertex of I guesses correctly.

Assume that no two bears on adjacent vertices guess correctly their hat colors at once; it turns out that if we plug $-g/h$ into all variables of the non-constant terms of $-P_G$, then we get exactly the fraction of all hat arrangements on which the bears win. The non-constant terms of P_G correspond (up to sign) to the terms of the formula from the inclusion-exclusion principle. Because of that, we have to plug $-g/h$ into the polynomial P_G.

To avoid confusion with the negative fraction $-g/h$, we define *signed independence polynomial* as $Z_G(\mathbf{x}) = P_G(-\mathbf{x})$, i.e.,

$$Z_G(\mathbf{x}) = \sum_{\substack{I \subseteq V \\ I \text{ independent set}}} (-1)^{|I|} \prod_{v \in I} x_v.$$

We also introduce the monovariate signed independence polynomial $U_G(x)$ obtained by plugging x for each variable x_v of Z_G.

Note that the constant term of any independence polynomial $P_G(\mathbf{x})$ equals to 1, arising from taking $I = \emptyset$ in the sum from the definition of P_G. When $U_G(g/h) = 0$ and no two adjacent bears guess correctly at the same time, then the bears win the game (G, h, g) because the fraction of all hat arrangements, on which at least one bear guesses correctly, is exactly 1, however, the proof is far from trivial.

Slightly abusing the notation, we use $Z_{G'}(\mathbf{x})$ to denote the independence polynomial of an induced subgraph G' with variables $\mathbf{x}$ restricted to the vertices of G'. The independence polynomial P_G can be expanded according to a vertex $v \in V$ in the following way.

$$P_G(\mathbf{x}) = P_{G \setminus \{v\}}(\mathbf{x}) + x_v P_{G \setminus N^+(v)}(\mathbf{x})$$

The analogous expansions hold for the polynomials Z_G and U_G as well. This expansion follows from the fact that for any independent set I of G, it holds that either v is not in I (the first term of the expansion), or v is in I but in that case, no neighbor of v is in I (the second term). The formal proof of this expansion of P_G was provided by Hoede and Li [17].

For a graph G, we let $\mathcal{R}(G)$ denote the set of all vectors $\mathbf{r} \in [0, \infty)^V$ such that $Z_G(\mathbf{w}) > 0$ for all $0 \leq \mathbf{w} \leq \mathbf{r}$, where the comparison is done entry-wise. For the monovariate independence polynomial U_G, an analogous set to $\mathcal{R}(G)$ would be exactly the real interval $[0, r)$ where r is the smallest positive root of U_G. (Note that $Z_G(\mathbf{0}) = 1$ and $U_G(0) = 1$.)

Our first connection of the independence polynomial to the hat guessing game comes in the shape of a sufficient condition for bears to lose. Consider the following beautiful connection between Lovász Local Lemma and independence polynomial due to Scott and Sokal [27].

Theorem 2 ([27] Theorem 4.1). *Let $G = (V, E)$ be a graph and let $(A_v)_{v \in V}$ be a family of events on some probability space such that for every v, the event A_v is independent of $\{A_w \mid w \notin N^+(v)\}$. Suppose that $\mathbf{p} \in [0,1]^V$ is a vector of real numbers such that for each v we have $P(A_v) \leq p_v$ and $\mathbf{p} \in \mathcal{R}(G)$. Then*

$$P\Big(\bigcap_{v \in V} \bar{A}_v\Big) \geq Z_G(\mathbf{p}) > 0.$$

The full proofs omitted in this section are stated in the full version [4].

Proposition 1. *A hat guessing game $\mathcal{H} = (G = (V, E), \mathbf{h}, \mathbf{g})$ is losing whenever $\mathbf{r} \in \mathcal{R}(G)$ where $\mathbf{r} = (g_v/h_v)_{v \in V}$.*

Proof Idea. We let the demon assign a hat to each bear uniformly at random and independently from the other bears. Let A_v be the event that the bear on the vertex v guesses correctly. Applying Theorem 2 to G and the events A_v, we conclude that the bears lose (no event A_v occurs) with a non-zero probability. □

A strategy for a hat guessing game $\mathcal{H}$ is *perfect* if it is winning and in every hat arrangement, no two bears that guess correctly are on adjacent vertices. We remark that perfect strategies exist, for example the strategy for a single-guessing game on a clique K_n and exactly n colors [20], or for a multi-guessing game on a clique K_n and $h/g = n$ (Corollary 1). The following proposition shows that a perfect strategy can occur only when $\mathbf{r} = (g_v/h_v)_{v \in V}$ lies in some sense just outside of $\mathcal{R}(G)$.

Proposition 2. *If there is a perfect strategy for the hat guessing game* $(G, \mathbf{h}, \mathbf{g})$ *then for* $\mathbf{r} = (g_v/h_v)_{v \in V}$ *we have that* $Z_G(\mathbf{r}) = 0$ *and* $Z_G(\mathbf{w}) \geq 0$ *for every* $0 \leq \mathbf{w} \leq \mathbf{r}$.

Proof Idea. We fix a perfect strategy and show that if we plug the vector $\mathbf{r}$ into Z_G then the non-constant terms of Z_G compute exactly the negative fraction of hat arrangements for which at least one bear guesses his hat color correctly. We point out that the assumption of the perfect strategy is crucial and this step would not be true without this assumption. Since the constant term of Z_G is always equal to 1, it follows that $Z_G(\mathbf{r}) = 0$.

Scott and Sokal [27, Corollary 2.20] proved that $Z_G(\mathbf{w}) \geq 0$ for every $0 \leq \mathbf{w} \leq \mathbf{r}$ if and only if $\mathbf{r}$ lies in the closure of $\mathcal{R}(G)$. Therefore, Proposition 2 further implies that if a perfect strategy for game $(G, \mathbf{h}, \mathbf{g})$ exists, then $\mathbf{r} = (g_v/h_v)_{v \in V}$ lies in the closure of $\mathcal{R}(G)$. And since $\mathbf{r}$ cannot lie inside $\mathcal{R}(G)$ due to Proposition 1, it must belong to the boundary of the set $\mathcal{R}(G)$.

The natural question is what happens outside of the closure of $\mathcal{R}(G)$. We proceed to answer this question for chordal graphs.

A graph G is *chordal* if every cycle of length at least 4 has a chord. For our purposes, it is more convenient to work with a different equivalent definition of chordal graphs. For a graph $G = (V, E)$, a *clique tree of* G is a tree T whose vertex set is precisely the subsets of V that induce maximal cliques in G and for each $v \in V$ the vertices of T containing v induces a connected subtree. Gavril [15] showed that G is chordal if and only if there exists a clique tree of G.

Theorem 3. *Let* $G = (V, E)$ *be a chordal graph and let* $\mathbf{r} = (r_v)_{v \in V}$ *be a vector of rational numbers from the interval* $[0, 1]$. *If* $\mathbf{r} \notin \mathcal{R}(G)$ *then there are vectors* $\mathbf{g}, \mathbf{h} \in \mathbb{N}^V$ *such that* $g_v/h_v \leq r_v$ *for every* $v \in V$ *and the hat guessing game* $(G, \mathbf{h}, \mathbf{g})$ *is winning.*

Proof Idea. The proof is done by induction over the vertices of a clique tree T of G. We take a leaf of T, which represents a clique C of G. If the vector $\mathbf{r}$ is such that the bears win on C by Theorem 1, then we are done. Otherwise, let G' be a graph arising from G by removing vertices that are only in C and no other maximal clique. We define new vectors $\mathbf{g}_1, \mathbf{g}_2, \mathbf{g}_1$, and $\mathbf{h}_2$ arising from $\mathbf{g}$

and $\mathbf{h}$ in such a way that the bears would win the game $\mathcal{H}_1 = (C, \mathbf{h}_1, \mathbf{g}_1)$ by Theorem 1 and the game $\mathcal{H}_2 = (G', \mathbf{h}_2, \mathbf{h}_1)$ by induction hypothesis. We use the winning strategies for $\mathcal{H}_1$ and $\mathcal{H}_2$ and combine them into a winning strategy for the game $(G, \mathbf{h}, \mathbf{g})$ using Lemma 2. See Fig. 2 for an illustration of the proof. □

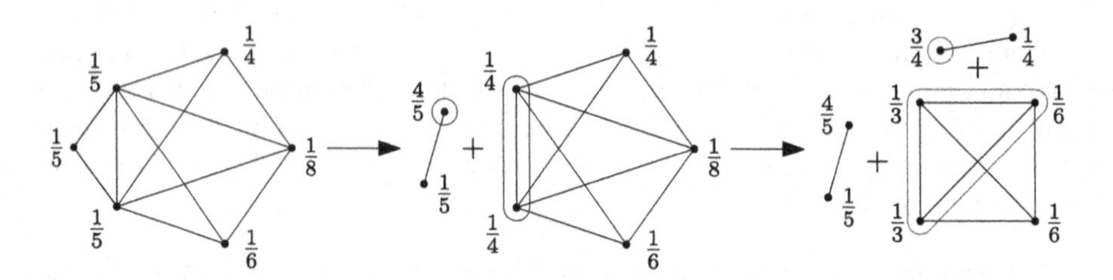

Fig. 2. Application of Theorem 1 on a chordal graph G with vector $\mathbf{r} \in \mathcal{R}(G)$. In each step, we highlight the clique S and vertex w that are used for Lemma 2 to inductively build a strategy for G from strategies on cliques given by Theorem 1.

Theorem 3 applied for the uniform polynomial U_G immediately gives us the following corollary.

Corollary 2. *For any chordal graph G, the fractional hat chromatic number $\hat{\mu}(G)$ is equal to $1/r$ where r is the smallest positive root of $U_G(x)$.*

Proof. Theorem 3 implies that $\hat{\mu}(G) \geq 1/r$. For the other direction, let $(w_i)_{i \in \mathbb{N}}$ be a sequence of rational numbers such that $w_i < r$ for every i and $\lim_{i \to \infty} w_i = r$. Set $\mathbf{w}_i = (w_i)_{v \in V}$. Scott and Sokal [27, Thereom 2.10] prove that $\mathbf{r} \in \mathcal{R}(G)$ if and only if there is a path in $[0, \infty)^V$ connecting $\mathbf{0}$ and $\mathbf{r}$ such that $Z_G(\mathbf{p}) > 0$ for any $\mathbf{p}$ on the path. Taking the path $\{\lambda \mathbf{w}_i \mid \lambda \in [0, 1]\}$, we see that $Z_G(\lambda \mathbf{w}_i) = U_G(\lambda \cdot w_i) > 0$ and thus $\mathbf{w}_i \in \mathcal{R}(G)$ for every i. Therefore by Proposition 1, the hat guessing game (G, h, g) is losing for any h, g such that $g/h = w_i$ and $\hat{\mu}(G) \leq 1/w_i$ for every i. It follows that $\hat{\mu}(G) \leq 1/r$. □

We would like to remark that the proof of Theorem 3 (and also Theorem 1) is constructive in the sense that given a graph G and a vector $\mathbf{r}$ it either greedily finds vectors $\mathbf{g}, \mathbf{h} \in \mathbb{N}^V$ such that $g_v/h_v \leq r_v$ together with a succinct representation of a winning strategy on $(G, \mathbf{h}, \mathbf{g})$ or it reaches a contradiction if $\mathbf{r} \in \mathcal{R}(G)$. Moreover, it is easy to see that it can be implemented to run in polynomial time if the clique tree of G is provided. Combining it with the well-known fact that a clique tree of a chordal graph can be obtained in polynomial time (see Blair and Peyton [3]) we get the following corollary.

Corollary 3. *There is a polynomial-time algorithm that for a chordal graph $G = (V, E)$ and vector $\mathbf{r}$ decides whether $\mathbf{r} \in \mathcal{R}(G)$. Moreover, if $\mathbf{r} \notin \mathcal{R}(G)$ it outputs vectors $\mathbf{h}, \mathbf{g} \in \mathbb{N}^V$ such that $g_v/h_v \leq r_v$ for every $v \in V$, together with a polynomial-size representation of a winning strategy for the hat guessing game $(G, \mathbf{h}, \mathbf{g})$.*

This result is consistent with the fact that chordal graphs are in general well-behaved with respect to Lovász Local Lemma—Pegden [24] showed that for a chordal graph G, we can decide in polynomial time whether a given vector $\mathbf{r}$ belongs to $\mathcal{R}(G)$. We finish this section by presenting an algorithm that computes hat chromatic number of chordal graphs.

Theorem 4. *There is an algorithm* $\mathcal{A}$ *such that given a chordal graph* G *as an input, it approximates* $\hat{\mu}(G)$ *up to an additive error* $1/2^k$*. The running time of* $\mathcal{A}$ *is* $2k \cdot poly(n)$*, where* n *is the number of vertices of* G*. Moreover, if* $\hat{\mu}(G)$ *is rational, then the algorithm* $\mathcal{A}$ *outputs the exact value of* $\hat{\mu}(G)$*.*

Proof Idea. We start with an interval $I_0 = [0, 1]$. We repeatedly use the algorithm given by Corollary 3 to produce intervals I_j such that $1/\hat{\mu}(G)$ is in I_j. We gradually decrease the length of the intervals I_j until it is small enough to determine $\hat{\mu}(G)$ with the sought precision $1/2^k$. □

6 Applications

In this section, we present applications of the relation between the hat guessing game and independence polynomials which was presented in the previous section.

First, we prove that $\hat{\mu}(G)$ is asymptotically equal to $\Delta(G)$ up to a logarithmic factor. Since the bears can use a strategy for trees on a star with a central vertex of degree $\Delta(G)$ (which is always a subgraph of any graph G), we deduce a lower bound stated as Proposition 3. The formal proof is in the full version [4].

Proposition 3. *The fractional hat chromatic number of any graph* $G = (V, E)$ *is at least* $\Omega(\Delta/\log\Delta)$*.*

Farnik [10] proved that $\mu_g(G) \in O\big(g \cdot \Delta(G)\big)$, from which we can deduce that $\hat{\mu}(G) \in O\big(\Delta(G)\big)$. It gives with Proposition 3 the following corollary that $\hat{\mu}(G)$ is almost linear in $\Delta(G)$.

Corollary 4. *For any graph* G*, it holds that* $\hat{\mu}(G) \in \Omega(\Delta/\log\Delta)$ *and* $\hat{\mu}(G) \in O(\Delta)$*.*

It follows from Corollary 4, that $\hat{\mu}(P_n)$ and $\hat{\mu}(C_n)$ are some constants. In the full version [4] we prove the following proposition that the fractional hat chromatic number of paths and cycles goes to 4 with their increasing length.

Proposition 4. $\lim_{n\to\infty} \hat{\mu}(P_n) = \lim_{n\to\infty} \hat{\mu}(C_n) = 4$

We remark that Proposition 4 follows also from the results of Scott and Sokal [27] as they proved that the small positive roots of U_{P_n} and U_{C_n} go to $1/4$ when n goes to infinity. However, their proof is purely algebraic whereas we provide a combinatorial proof.

Acknowledgments. We would like to thanks to Miloš Chromý, Michał Dębski, Sophie Rehberg, and Pavel Valtr for fruitful discussions at early stage of this project during workshop KAMAK 2019.

References

1. Aggarwal, G., Fiat, A., Goldberg, A.V., Hartline, J.D., Immorlica, N., Sudan, M.: Derandomization of auctions. In: Proceedings of the Thirty-Seventh Annual ACM Symposium on Theory of Computing, pp. 619–625. STOC 2005. Association for Computing Machinery, New York, NY, USA, May 2005. https://doi.org/10.1145/1060590.1060682
2. Alon, N., Ben-Eliezer, O., Shangguan, C., Tamo, I.: The hat guessing number of graphs. J. Comb. Theory Ser. B **144**, 119–149 (2020). https://doi.org/10.1016/j.jctb.2020.01.003
3. Blair, J.R.S., Peyton, B.: An introduction to chordal graphs and clique trees. In: George, A., Gilbert, J.R., Liu, J.W.H. (eds.) Graph Theory and Sparse Matrix Computation, pp. 1–29. Springer, New York (1993). https://doi.org/10.1007/978-1-4613-8369-7_1
4. Blažej, V., Dvořák, P., Opler, M.: Bears with hats and independence polynomials (2021). https://arxiv.org/abs/2103.07401
5. Bosek, B., Dudek, A., Farnik, M., Grytczuk, J., Mazur, P.: Hat chromatic number of graphs, May 2019. https://arxiv.org/abs/1905.04108
6. Butler, S., Hajiaghayi, M.T., Kleinberg, R.D., Leighton, T.: Hat guessing games. SIAM J. Discrete Math. **22**(2), 592–605 (2008). https://doi.org/10.1137/060652774
7. Diestel, R.: Graph Theory. Graduate Texts in Mathematics, Springer, Heidelberg (2017). https://doi.org/10.1007/978-3-662-53622-3
8. Ebert, T.: Applications of Recursive Operators to Randomness and Complexity. Ph.D. thesis, University of California, Santa Barbara (1998)
9. Ebert, T., Merkle, W., Vollmer, H.: On the autoreducibility of random sequences. SIAM J. Comput. **32**(6), 1542–1569 (2003). https://doi.org/10.1137/S0097539702415317
10. Farnik, M.: A hat guessing game. Ph.D. thesis, Jagiellonian University (2015)
11. Feige, U.: You can leave your hat on (if you guess its color). Technical Report MCS04-03, The Weizmann Institute, Rehovot, Israel (2004)
12. Gadouleau, M.: Finite dynamical systems, hat games, and coding theory. SIAM J. Discrete Math. **32**(3), 1922–1945 (2018). https://doi.org/10.1137/15M1044758
13. Gadouleau, M., Georgiou, N.: New constructions and bounds for Winkler's hat game. SIAM J. Discrete Math. **29**(2), 823–834 (2015). https://doi.org/10.1137/130944680
14. Gadouleau, M., Riis, S.: Graph-theoretical constructions for graph entropy and network coding based communications. IEEE Trans. Inf. Theory **57**(10), 6703–6717 (2011). https://doi.org/10.1109/TIT.2011.2155618
15. Gavril, F.: The intersection graphs of subtrees in trees are exactly the chordal graphs. J. Comb. Theory Ser. B **16**(1), 47–56 (1974). https://doi.org/10.1016/0095-8956(74)90094-X
16. He, X., Li, R.: Hat guessing numbers of degenerate graphs. Electron. J. Comb. **27**(3), P3.58 (2020). https://doi.org/10.37236/9449
17. Hoede, C., Li, X.: Clique polynomials and independent set polynomials of graphs. Discrete Math. **125**(1), 219–228 (1994). https://doi.org/10.1016/0012-365X(94)90163-5
18. Holyer, I.: The NP-completeness of edge-coloring. SIAM J. Comput. **10**(4), 718–720 (1981). https://doi.org/10.1137/0210055

19. Jin, K., Jin, C., Gu, Z.: Cooperation via codes in restricted hat guessing games. In: Proceedings of the 18th International Conference on Autonomous Agents and MultiAgent Systems, pp. 547–555. AAMAS 2019. International Foundation for Autonomous Agents and Multiagent Systems, Richland, SC (2019)
20. Kokhas, K., Latyshev, A., Retinsky, V.: Cliques and constructors in "Hats" game (2020). https://arxiv.org/abs/2004.09605
21. Kokhas, K., Latyshev, A.: Cliques and Constructors in "Hats" Game. I. J. Math. Sci. **255**(1), 39–57 (2021). https://doi.org/10.1007/s10958-021-05348-9
22. Kokhas, K.P., Latyshev, A.S., Retinskiy, V.I.: Cliques and Constructors in "Hats" Game. II. J. Math. Sci. **255**(1), 58–70 (2021). https://doi.org/10.1007/s10958-021-05349-8
23. Krzywkowski, M.P.: Hat problem on a graph. Ph.D. thesis, University of Exeter (2012)
24. Pegden, W.: The lefthanded local lemma characterizes chordal dependency graphs. Random Struct. Algorithms **41**(4), 546–556 (2012)
25. Riis, S.: Information flows, graphs and their guessing numbers. Electron. J. Comb. **14**(1), 962 (2007). https://doi.org/10.37236/962
26. Robinson, S.: Why mathematicians now care about their hat color. New York Times, 10 April 2001. https://www.nytimes.com/2001/04/10/science/why-mathematicians-now-care-about-their-hat-color.html
27. Scott, A.D., Sokal, A.D.: On dependency graphs and the lattice gas. Comb. Probab. Comput. **15**(1–2), 253–279 (2006). https://doi.org/10.1017/S0963548305007182
28. Szczechla, W.: The three colour hat guessing game on cycle graphs. Electron. J. Comb. **24**, 135 (2017). https://doi.org/10.37236/5135
29. Winkler, P.: Games people don't play. In: Wolfe, D., Rodgers, T. (eds.) Puzzlers' Tribute: A Feast for the Mind, pp. 301–313 (2002). https://doi.org/10.1201/9781439864104-50
30. Wu, T., Cameron, P., Riis, S.: On the guessing number of shift graphs. J. Discrete Algorithms **7**(2), 220–226 (2009). https://doi.org/10.1016/j.jda.2008.09.009

The Largest Connected Subgraph Game

Julien Bensmail[1], Foivos Fioravantes[1(✉)], Fionn Mc Inerney[2], and Nicolas Nisse[1]

[1] Université Côte d'Azur, CNRS, Inria, I3S, Valbonne, France
foivos.fioravantes@inria.fr
[2] CISPA Helmholtz Center for Information Security, Saarbrücken, Germany

Abstract. We introduce the *largest connected subgraph game* played on an undirected graph G. Each round, Alice colours an uncoloured vertex of G red, and then, Bob colours one blue. Once every vertex is coloured, Alice (Bob, resp.) wins if there is a red (blue, resp.) connected subgraph whose order is greater than that of any blue (red, resp.) connected subgraph. If neither player wins, it is a draw. We prove that Alice can ensure Bob never wins, and define a class of graphs (*reflection graphs*) in which the game is a draw. We show that the game is PSPACE-complete in bipartite graphs of diameter 5, and that recognising reflection graphs is GI-hard. We prove that the game is a draw in paths if and only if the path has even order or at least 11 vertices, and that Alice wins in cycles if and only if the cycle is of odd order. We also give an algorithm computing the outcome of the game in cographs in linear time.

Keywords: Games on graphs · Scoring games · Connection games · PSPACE-complete

1 Introduction

Games where players strive to make connected structures are *connection games.* Several of these games are well-known, like the game of *Hex*, introduced by Hein in 1942, and independently by Nash in 1948 [9]. Hex is played by two players on a hexagon-tiled board with two of its opposing sides coloured red and the other two blue. Each round, the first player colours an uncoloured tile red, and then, the second player colours one blue. The player that connects the two sides with his colour wins. Another famous connection game is the *Shannon switching game*, invented by Shannon in the 1950s [10]. In this game, the first player wants to connect two marked vertices in a graph, and the second player wants to prevent this. The players take turns selecting edges of the graph, and the first player wins if there is a path consisting of only his edges between the two marked vertices.

This work has been supported by the European Research Council (ERC) consolidator grant No. 725978 SYSTEMATICGRAPH, the STIC-AmSud project GALOP, the PHC Xu Guangqi project DESPROGES, and the UCA[JEDI] Investments in the Future project managed by the National Research Agency (ANR-15-IDEX-01). See [2] for the full version of the paper.

Ł. Kowalik et al. (Eds.): WG 2021, LNCS 12911, pp. 296–307, 2021.
https://doi.org/10.1007/978-3-030-86838-3_23

A variant where the players select vertices (and obtain their incident edges) also exists. However, not all connection games involve connecting sides of a board or two vertices in a graph. For example, in *Havannah*, a board game invented by Freeling in 1981, the players may also win by forming closed loops, with the board and the rules similar to Hex. Connection games tend to be difficult complexity-wise, which is a main reason they are played and studied. For example, Reisch proved that generalised Hex is PSPACE-complete [14], Even and Tarjan proved that the Shannon switching game on vertices (players select vertices, not edges) is PSPACE-complete [8], and Bonnet et al. proved that (generalised) Havannah is PSPACE-complete [3]. That being said, the Shannon switching game on edges is polynomial-time solvable [5]. For more on connection games, see [3,4].

Games in which the player with the largest *score* wins are *scoring games*. The score in these games is measured in a unit called *points*. Players may gain points in a myriad of ways depending on the rules of the game, *i.e.*, in the orthogonal colouring game [1], each player gets a point for each coloured vertex in their copy of the graph, and a player's final score is their total number of points. Recently, the papers [11,12] started to build a general theory around scoring games, and there have been many papers on different scoring games, such as [7,13,16]. In this paper, we introduce the following 2-player game linking connection and scoring games on graphs. For any graph G, the *largest connected subgraph game* is played between *Alice* and *Bob*. Initially, no vertices are coloured. Each round, Alice first colours an uncoloured vertex of G red, and then, Bob colours an uncoloured vertex blue. Each vertex can only be coloured once and its colour cannot be modified. The game ends when every vertex in G is coloured. If there is a connected red (blue, resp.) subgraph whose order (number of vertices) is strictly greater than the order of any connected blue (red, resp.) subgraph, then Alice (Bob, resp.) wins. If the order of the largest connected red subgraph equals the order of the largest connected blue subgraph, then the game is a draw.

Notations and first results for the game are given in Sect. 2, *i.e.*, we show that Alice can ensure that Bob never wins, that the game is a draw in a class of graphs we call *reflection graphs*, and that recognising these graphs is GI-hard. In Sect. 3, we prove that the game is PSPACE-complete in bipartite graphs of diameter 5. We then study the game in some graph classes, with the resolution of the game for paths and cycles in Sect. 4, and a linear-time algorithm for solving the game in cographs in Sect. 5. These graph classes interestingly illustrate different types of strategies for Alice and Bob. Lastly, we finish with open questions in Sect. 6.

2 Notations and First Results

In this section, we define notations and give preliminary results for the game. For any graph G, if Alice (Bob, resp.) has a winning strategy in the largest connected subgraph game, then G is *A-win* (*B-win*, resp.). If neither Alice nor Bob has a winning strategy in the largest connected subgraph game, *i.e.*, it is a draw if both players use optimal strategies, then G is *AB-draw*. Since it is never a disadvantage to play an extra turn, by the classic strategy stealing argument:

Theorem 1 [2]. *There does not exist a graph G that is B-win.*

Since there are no B-win graphs, the next natural question to ask is whether there are A-win (AB-draw, resp.) graphs. There are an infinite number of A-win graphs as any star is A-win (Alice first colours the universal vertex). This also shows that there are an infinite number of graphs for which the order of the largest connected red subgraph is arbitrarily bigger than that of the blue one. There are also an infinite number of AB-draw graphs, since any graph of even order with two universal vertices is AB-draw. In Sect. 4, we show that any path of order at least 10 is AB-draw, and hence, that there are an infinite number of graphs of odd order that are AB-draw. We can actually define a much richer class of AB-draw graphs. A *reflection graph* is any graph G, whose vertices can be partitioned into two sets $U = \{u_1, \dots, u_n\}$ and $V = \{v_1, \dots, v_n\}$ such that:

1. there is an isomorphism between the subgraph induced by U and the subgraph induced by V, where v_i is the image of u_i, for all $1 \leq i \leq n$;
2. for any two vertices $u_i \in U$ and $v_j \in V$, if $u_i v_j \in E(G)$, then $u_j v_i \in E(G)$.

Theorem 2 [2]. *Any reflection graph G is AB-draw.*

Indeed, a drawing strategy for Bob is to colour v_i (u_i, resp.) when Alice colours u_i (v_i, resp.). Any even-order graph that is a path, cycle, or Cartesian product of two graphs, is a reflection graph. We prove that recognising reflection graphs is not in P unless the Graph Isomorphism problem is:

Theorem 3 [2]. *Given a graph G, deciding if G is a reflection graph is GI-hard.*

3 Complexity

In this section, we show that the largest connected subgraph game is PSPACE-complete, even in bipartite graphs of small diameter. Our reduction is via POS CNF, which was shown to be PSPACE-complete in [15], and is as follows:

POS CNF: *2-player game whose input is a set of variables $X = \{x_1, \dots, x_n\}$ and a conjunctive normal form (CNF) formula ϕ made up of clauses $C_1, \dots, C_m$ comprised of variables from X in their positive form. In each round, the first player, Alice, sets a variable (that is not yet set) to true, and then, the second player, Bob, sets a variable (that is not yet set) to false. Once each variable has been assigned a truth value, Alice wins if ϕ is true, and Bob wins if ϕ is false.*

Theorem 4. *Given a graph G, deciding if G is A-win is PSPACE-complete, even if G is bipartite and has a diameter of* 5.

Proof. As there are $\lceil |V(G)|/2 \rceil$ rounds and at most $|V(G)|$ possible moves for a player in a round, the problem is in PSPACE. To prove it is PSPACE-hard, we give a reduction from POS CNF. By adding a dummy variable, POS CNF remains PSPACE-hard if the number of variables n is odd. From an instance ϕ of POS CNF where n is odd, we construct, in polynomial time, an instance G

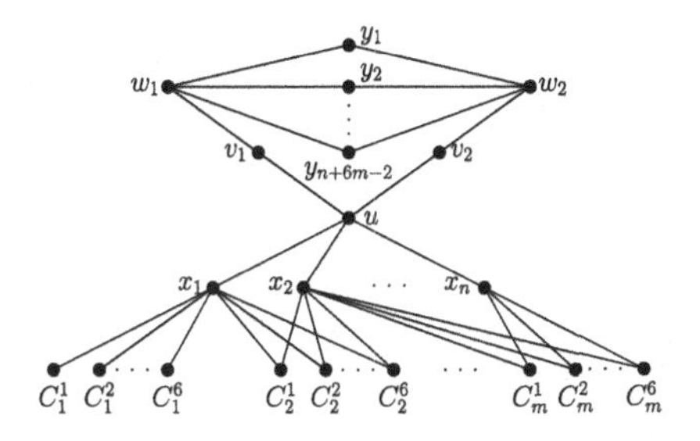

Fig. 1. An example of the construction of the graph G in the proof of Theorem 4, where, among other variables, the clause C_1 contains the variable x_1, the clause C_2 contains the variables x_1 and x_2, and the clause C_m contains the variables x_2 and x_n.

of the largest connected subgraph game such that Alice wins in ϕ if and only if G is A-win. Let $x_1, \ldots, x_n$ be the variables and $C_1, \ldots, C_m$ be the clauses in ϕ. The construction of G is as follows (see Fig. 1): for each variable x_i ($1 \leq i \leq n$), there is a vertex x_i, and for each clause C_j ($1 \leq j \leq m$), there are 6 vertices $C_j^1, \ldots, C_j^6$. For all $1 \leq i \leq n$ and $1 \leq j \leq m$, if the variable x_i appears in the clause C_j, then there is the edge $x_iC_j^q$ for all $1 \leq q \leq 6$. There are the vertices u, v_1, v_2, w_1, w_2, and $y_1, \ldots, y_{n+6m-2}$, and the edges w_1v_1, v_1u, uv_2, and v_2w_2, and, for all $1 \leq i \leq n$, there is the edge ux_i, and, for all $1 \leq \ell \leq n+6m-2$, there are the edges w_1y_ℓ and w_2y_ℓ. To simplify the proof, let P be the subgraph of G induced by the vertices x_i ($1 \leq i \leq n$) and C_j^q ($1 \leq q \leq 6$ and $1 \leq j \leq m$), and let Q be the subgraph of G induced by the vertices $V(G) \setminus (V(P) \cup \{u\})$.

First, we prove that, if Alice wins in ϕ, then G is A-win. We give a winning strategy for Alice. In what follows, whenever Alice cannot follow her strategy, she colours an arbitrary vertex and resumes her strategy for the subsequent rounds. Alice first colours u. Now, Bob can only construct connected blue subgraphs in P or Q since u separates them. For all $1 \leq j \leq m$, whenever Bob colours a vertex in $\{C_j^1, \ldots, C_j^6\}$, then Alice also colours a vertex in $\{C_j^1, \ldots, C_j^6\}$, so in what follows, we assume that Bob does not colour such a vertex. There are two cases depending on Bob's next move.

Case 1: Bob colours a vertex in Q. Then, Alice colours the vertex x_i corresponding to the variable x_i she would set to true in her winning strategy in ϕ. Now, whenever Bob colours a vertex x_p ($1 \leq p \leq n$ and $p \neq i$), Alice assumes Bob set the variable x_p to false in ϕ and colours the vertex in $\{x_1, \ldots, x_n\}$ corresponding to her winning strategy in ϕ. Otherwise, whenever Bob colours a vertex in Q, Alice colours a vertex in Q. By this strategy, Alice ensures a connected red subgraph of order at least $\lceil n/2 \rceil + 3m + 1$ as she colours half the variable vertices (rounded up), half the clause vertices, and u, and since she followed a winning strategy in ϕ, this subgraph is connected. Moreover, she ensures that any connected blue subgraph in P is of order at most $\lfloor n/2 \rfloor + 3m$, and hence, Bob must build his largest connected blue subgraph in Q to manage a draw. Also note that, if Alice colours v_1 or v_2 she wins, since then she ensures a connected red subgraph of order at least $\lceil n/2 \rceil + 3m + 2$, while she ensures that any connected blue subgraph in Q is of order at most $\lfloor(n+6m-2+3-2)/2\rfloor+2 = \lfloor(n-1)/2\rfloor+3m+2$. Thus, Bob must have coloured v_1 and v_2 in the first two rounds. Now, Alice colours w_2, and she wins since she ensures that any connected blue subgraph in Q is of order at most $\lfloor(n+6m-2+2-2)/2\rfloor+2 = \lfloor n/2\rfloor+3m+1$.

Case 2: Bob colours a vertex in $\{x_1,\dots,x_n\}$. Then, Alice colours w_2. This forces Bob to colour v_2, as otherwise, Alice will colour v_2 in the next round and win with the following strategy: whenever Bob colours a vertex

- in $\{w_1, v_1\}$, then Alice colours the other vertex in $\{w_1, v_1\}$;
- y_ℓ, then Alice colours a vertex y_k ($\ell \neq k$);
- x_i ($1 \leq i \leq n$), then Alice colours a vertex x_p ($1 \leq p \leq n$ and $i \neq p$).

In this way, Alice guarantees a connected red subgraph of order at least $\lceil(n+6m-2+n-3)/2\rceil+3 = n+3m+1$ without counting any of the vertices C_j^q ($1 \leq q \leq 6$, $1 \leq j \leq m$). Regarding Bob, any connected blue subgraph in P has at most $\lfloor(n-3)/2\rfloor+3+3m = \lfloor(n-1)/2\rfloor+3m+2$ vertices, and any connected blue subgraph in Q has at most $\lfloor(n+6m-2+2-3)/2\rfloor+3 = \lfloor(n-1)/2\rfloor+3m+2$ vertices. Hence, Alice wins in this case, and thus, we can assume Bob colours v_2. Now, Alice colours w_1 and Bob is forced to colour v_1 for the same reasons as above. Alice now colours y_1 and then she follows the strategy just previously described above (as in the case where Bob did not colour v_2). In this way, Alice ensures a connected red subgraph of order at least $\lceil(n+6m-2+2-2)/2\rceil+2 = \lceil n/2\rceil+3m+1$ in Q. Regarding Bob, any connected blue subgraph in P has at most $\lfloor(n-2)/2\rfloor+2+3m = \lfloor n/2\rfloor+3m+1$ vertices, and any connected blue subgraph in Q has at most one vertex. Hence, Alice wins in this case as well (recall that n is odd), and this concludes the proof of the first direction.

Now, we prove that if Bob wins in ϕ, then G is AB-draw. We give a drawing strategy for Bob. In what follows, whenever Bob cannot follow his strategy, he colours an arbitrary vertex and resumes his strategy for the subsequent moves of Alice. Part of Bob's strategy is as follows: whenever Alice colours a vertex

- in $\{C_j^1,\dots,C_j^6\}$ for $1 \leq j \leq m$, then Bob colours a vertex in $\{C_j^1,\dots,C_j^6\}$;
- x_i for $1 \leq i \leq n$, then Bob assumes Alice set x_i to true in ϕ and colours the vertex in $\{x_1,\dots,x_n\}$ corresponding to his winning strategy in ϕ.

So, we just need to give a strategy for Bob in Q', the subgraph of G induced by $V(Q)\cup\{u\}$. W.l.o.g., we may assume that the first vertex Alice colours in Q' is not v_2 nor w_2. Bob colours w_2. If the first two vertices Alice colours in Q' are:

- w_1 and v_1, then Bob colours u. Now, Alice must colour v_2, as otherwise, Bob wins as in the proof of the first direction where Alice wins if she manages to colour w_2, v_2, and u. Then, Bob colours y_k for some $1 \leq k \leq n+6m-2$. Now, whenever Alice colours a vertex y_ℓ ($1 \leq \ell \leq n+6m-2$), then Bob colours a vertex y_k ($1 \leq k \leq n+6m-2$ and $\ell \neq k$);
- w_1 and v_2, then Bob colours y_ℓ for some $1 \leq \ell \leq n+6m-2$. Now, whenever Alice colours a vertex in $\{v_1, u\}$, then Bob colours the other vertex in $\{v_1, u\}$. Otherwise, whenever Alice colours a vertex y_ℓ ($1 \leq \ell \leq n+6m-2$), then Bob colours a vertex y_k ($1 \leq k \leq n+6m-2$ and $\ell \neq k$);

- w_1 and u, then Bob colours v_1. Now, whenever Alice colours a vertex in $\{y_1, \ldots, y_{n+6m-2}, v_2\}$, Bob colours another vertex in $\{y_1, \ldots, y_{n+6m-2}, v_2\}$;
- w_1 and y_k for some $1 \leq k \leq n + 6m - 2$, then Bob colours v_2. Now, Alice must colour u, as otherwise, Bob wins as in the proof of the first direction where Alice wins if she manages to colour w_2, v_2, and u. Then, Bob colours v_1. Now, whenever Alice colours a vertex y_ℓ $(1 \leq \ell \leq n + 6m - 2)$, then Bob colours a vertex y_p $(1 \leq p \leq n + 6m - 2$ and $\ell \neq p)$;
- any other combination, then Bob colours w_1. Now, whenever Alice colours a vertex in $\{y_1, \ldots, y_{n+6m-2}, v_1, v_2, u\}$, then Bob colours a different vertex in $\{y_1, \ldots, y_{n+6m-2}, v_1, v_2\}$ (note that u is not included here).

In the first two cases above, there is a connected blue subgraph in Q of order at least $\lfloor (n+6m-2+1-2)/2 \rfloor + 2 = \lfloor (n-1)/2 \rfloor + 3m + 1$. In the third case above, there is a connected blue subgraph in Q of order at least $\lfloor (n+6m-2+1)/2 \rfloor + 1 = \lfloor (n-1)/2 \rfloor + 3m + 1$. In the fourth case above, there is a connected blue subgraph in Q of order at least $\lfloor (n + 6m - 2 + 2 - 3)/2 \rfloor + 2 = \lfloor (n - 1)/2 \rfloor + 3m + 1$. In the last case above, there is a connected blue subgraph in Q of order at least $\lfloor (n + 6m - 2 + 4 - 4)/2 \rfloor + 2 = \lfloor n/2 \rfloor + 3m + 1 = \lfloor (n - 1)/2 \rfloor + 3m + 1$ (since n is odd). To summarise, in each of the cases, Bob has ensured that there is a connected blue subgraph in Q of order at least $\lfloor (n - 1)/2 \rfloor + 3m + 1$.

Regarding Alice, in the first two cases above, any connected red subgraph in Q is of order at most $\lceil (n + 6m - 2 + 2 - 3)/2 \rceil + 2 = \lceil (n - 1)/2 \rceil + 3m + 1$. In the third case above, any connected red subgraph in Q is of order at most $\lceil (n + 6m - 2 + 1 - 1)/2 \rceil + 1 = \lceil n/2 \rceil + 3m = \lceil (n - 1)/2 \rceil + 3m + 1$ (since n is odd). In the fourth case above, any connected red subgraph in Q is of order at most $\lceil (n + 6m - 2 + 1 - 2)/2 \rceil + 2 = \lceil (n - 1)/2 \rceil + 3m + 1$. In the last case above, any connected red subgraph in Q is of order at most 1. Thus, in each of the cases, Bob ensured that any connected red subgraph in Q is of order at most $\lceil (n - 1)/2 \rceil + 3m + 1 = \lfloor (n - 1)/2 \rfloor + 3m + 1$ (since n is odd). Hence, for Alice to win, she must have a connected red subgraph of order at least $\lfloor (n-1)/2 \rfloor + 3m + 2$ in P', the subgraph of G induced by $V(P) \cup \{u, v_1, v_2\}$ (since, by Bob's strategy, it can never be that u, v_1, and w_1 (u, v_2, and w_2, resp.) are all red). Since Bob follows a winning strategy in ϕ whenever Alice colours a vertex in $\{x_1, \ldots, x_n\}$, there is a j for which no vertex in $C_j^1, \ldots, C_j^6$ is adjacent to a red vertex. Thus, any connected red subgraph in P' has order at most $\lceil (n+6m-6)/2 \rceil + 3 = \lceil n/2 \rceil + 3m = \lfloor (n-1)/2 \rfloor + 3m + 1$ (since n is odd). Thus, in G, there is a connected blue subgraph of order at least $\lfloor (n - 1)/2 \rfloor + 3m + 1$ and any connected red subgraph has order at most $\lfloor (n-1)/2 \rfloor + 3m + 1$, so Alice does not win in any of the cases. This ends the proof of the second direction. □

4 Paths and Cycles

In this section, we deal with the case of n-vertex paths $P_n = (v_1, \ldots, v_n)$ and cycles $C_n = (v_1, \ldots, v_n)$. We begin with two lemmas for specific cases in paths, which we use in the proofs for paths and cycles of odd order. In the following proofs in this section, we often divide the main path P_n into two subpaths Q and Q', and say that Alice "follows" Bob, that is, when Bob plays in Q (in Q', resp.), Alice then plays in Q (in Q', resp.). The way Alice answers to Bob's moves in Q (in Q', resp.) is given in the proofs and depends on the different cases. Note that, when following this strategy, Alice may be unable to colour a desired vertex (because Q, resp., Q', has no uncoloured vertex anymore, or because the desired vertex is already red). In this case, Alice colours an arbitrary uncoloured vertex of P_n. The same applies for when we say that Bob "follows" Alice.

Lemma 1 [2]. *For all $n \geq 1$, for the path P_n, Bob has a strategy that ensures that the largest connected red subgraph is of order at most 2, even if one of the path's vertices of degree 1 is initially coloured red and it is Alice's turn.*

Lemma 2 [2]. *Let $x \geq 1$ and $n \geq x$. For any path P_n with x vertices initially coloured blue, let y be the maximum order of an initial connected blue subgraph.*

- *if $y = x$ and, either the blue subgraph contains no ends of P_n or $x = 1$, then, if Alice starts, she has a strategy ensuring that Bob cannot create a connected blue subgraph of order more than $x + 1$;*
- *otherwise, if Alice starts, she has a strategy ensuring that Bob cannot create a connected blue subgraph of order more than x.*

Theorem 5. *For all $n \geq 1$, the path P_n is A-win if and only if $n \in \{1, 3, 5, 7, 9\}$.*

Proof. By Theorem 1, we must prove that P_n is A-win if $n \in \{1, 3, 5, 7, 9\}$, and P_n is AB-draw otherwise. By Theorem 2, P_n is AB-draw if n is even. If $n \leq 9$ is odd, by a case analysis, Alice has a winning strategy that first colours the center of P_n. So, let us assume that $n \geq 11$ is odd. We orient the path from left to right (from v_1 to v_n), so we can use the notions of left and right. We now give a drawing strategy for Bob when $n \geq 11$. Let v_j $(1 \leq j \leq n)$ be the first vertex coloured by Alice. Since $n \geq 11$, there are at least 5 vertices to the left or right of v_j, say to the left of v_j, *i.e.*, $5 \leq j \leq n$. Bob colours v_{j-1}. Let $Q = (v_1, \ldots, v_{j-1})$ and $Q' = (v_j, \ldots, v_n)$. Now, Bob "follows" Alice, that is, when Alice plays in Q (Q', resp.), Bob then plays in Q (Q', resp.), and both games are considered independently (since v_{j-1} is blue and v_j is red). Considering Q' as a path with one of its ends initially coloured red, and applying Lemma 1 to it, Bob can ensure that Alice cannot create a connected red subgraph of order more than 2 in Q'. Let v_ℓ be the first vertex that Alice colours in Q. We distinguish two cases:

Case 1: $\ell \neq j - 2$. Bob colours v_{j-2}. Now, whenever Alice plays in Q, while it is possible, Bob colours a neighbour of the connected blue subgraph containing v_{j-1} and v_{j-2}. If it is not possible anymore, either the connected blue subgraph is of order $\lceil (j-1)/2 \rceil \geq 2$ (in which case the largest connected red subgraph in Q is of order $\lfloor (j-1)/2 \rfloor$ and so, the game is a draw) or it is of order $2 \leq x < (j-1)/2$ and it is Bob's turn. In the latter case, the connected blue subgraph in Q consists of the vertices $v_{j-x}, \ldots, v_{j-1}$, and v_{j-x-1} is red since Bob cannot colour a neighbour of the connected blue subgraph. Let $R = (v_1, \ldots, v_{j-x-1})$ and note that there are exactly x red vertices in R including v_{j-x-1} (one of its ends). Then, applying Lemma 2 to R (but with Bob as the first player), Bob has a strategy ensuring that Alice cannot create a connected red subgraph of order more than x in R. Hence, the game in P_n ends in a draw in this case.

Case 2: $\ell = j-2$. Bob colours v_{j-4}. If Alice colours v_{j-3}, then Bob colours v_{j-5}, and *vice versa*, and this ensures a connected blue subgraph of order at least 2. Otherwise, if Alice colours a vertex v_t with $1 \leq t \leq j-6$, then Bob colours v_{t+1}, unless v_{t+1} is already coloured, in which case, Bob colours v_{t-1}. In the latter case, Bob can ensure a draw since he can ensure that Alice cannot create a connected red subgraph of order more than 2 in $R^* = (v_1, \ldots, v_{t-1})$ by Lemma 1. So, assume we are in the former case. Let $R = (v_1, \ldots, v_t)$ and $R' = (v_{t+1}, \ldots, v_{j-5})$. From now on, Bob "follows" Alice (unless Alice colours v_{j-5}, in which case, Bob colours v_{j-3}), that is, when Alice plays in R (in R', resp.), Bob then plays in R (in R', resp.), and both games are considered independently (since v_t is coloured red and v_{t+1} is coloured blue). Considering R as a path with one of its ends initially coloured red, and applying Lemma 1 to it, Bob has a strategy ensuring that Alice cannot create a connected red subgraph of order more than 2 in R. Bob plays in R' assuming that v_{j-5} is already coloured red, and applying Lemma 1 to it, Bob has a strategy ensuring that Alice cannot create a connected red subgraph of order more than 2 in R'. It is easy to see that, in this case, the largest connected blue (red, resp.) subgraph is of order 2 (at most 2, resp.). □

Now, we address the largest connected subgraph game in cycles. We start with a lemma for a specific case in paths, which we use in the proof for cycles.

Lemma 3 [2]. *Let $x \geq 3$, $n \geq x+1$, and $n-x$ be odd. For any path P_n with x vertices, including both ends, initially blue, if Alice starts, then she can ensure that no connected blue subgraph of order more than $x-1$ is created in P_n.*

Theorem 6. *For all $n \geq 3$, the cycle C_n is A-win if and only if n is odd.*

Proof. If n is even, then C_n is a reflection graph, and thus, is AB-draw by Theorem 2. So let n be odd. We describe a winning strategy for Alice. If $n \leq 5$, the result is obvious, so let us assume that $n > 5$. First, let us assume (independently of how this configuration appears) that after $x \geq 3$ rounds, the vertices $v_1, \ldots, v_x$ are red, the vertices v_n and v_{x+1} are blue, and any $x-2$ other vertices in $\{v_{x+2}, \ldots, v_{n-1}\}$ are blue. Note that it is Alice's turn. By Lemma 3, Alice may ensure that Bob cannot create a connected blue subgraph of order at least x in the subgraph induced by $(v_{x+1}, \ldots, v_n)$. Thus, in this case, Alice wins.

Now, let Alice first colour the vertex v_1. If Bob does not colour a neighbour of v_1 (say Bob colours v_j with $3 < j < n$, since $n \geq 5$ and odd), then, on her second turn, Alice colours v_2. Then, while it is possible, Alice colours a neighbour of the connected red subgraph. When it is not possible anymore, either the connected red subgraph is of order $\lceil n/2 \rceil$ or it is of order at least 3 and we are in the situation of the above paragraph. In both cases, Alice wins.

Therefore, after Alice colours her first vertex (call it v_2), Bob must colour some neighbour of it (say v_1). By induction on the number $t \geq 1$ of rounds, let us assume that the game reaches, after t rounds, a configuration where, for every $1 \leq i \leq t$, vertices v_{2i-1} are coloured blue and vertices v_{2i} are coloured red. If $t = \lfloor n/2 \rfloor$, then Alice finally colours v_n (recall that n is odd) and wins. Otherwise, let Alice colour v_{2t+2}. If Bob then colours v_{2t+1}, then we are back to the previous situation for $t' = t+1$. Then, eventually, Alice wins by induction on $n-2t$. If Bob does not colour v_{2t+1}, then Alice colours v_{2t+1} and then continues to grow the connected red subgraph containing v_{2t+1} while possible. When it is not possible anymore, note that removing (or contracting) the vertices v_2 to v_{2t}, we are back to the situation of the first paragraph of this proof (with a connected red subgraph of order at least 3) and, therefore, Alice wins. □

5 Cographs

For paths and cycles, optimal play depended on positional play with respect to the previously coloured vertices since the graphs are sparse, making it easy for the players to stop the expansion of the opponent's largest connected subgraph. As a consequence, in such cases, players must stop growing their largest connected subgraph, and start growing a new one. Such a strategy is likely to be less viable in denser graphs, in which the game tends to turn into a different one, where the players grow a single connected subgraph each, that they have to keep "alive" for as long as possible. We illustrate this with the case of *cographs*, which leads us to introduce a few more notations (see $\mathcal{A}^*$ below) to describe a linear-time algorithm deciding the outcome of the game in such instances.

A graph G is a cograph if it is P_4-free, *i.e.*, it does not contain P_4 as an induced subgraph. The class of cographs can be defined recursively as follows. The single-vertex graph K_1 is a cograph. Let G_1 and G_2 be two cographs. Then, the disjoint union $G_1 + G_2$ is a cograph. Moreover, the join $G_1 \oplus G_2$, obtained from $G_1 + G_2$ by adding all the possible edges between the vertices of G_1 and G_2, is a cograph. Recall that a decomposition, *i.e.*, a sequence of disjoint unions and joins from single vertices, of a cograph can be computed in linear time [6].

To simplify notation in Theorem 7 and its proof, let $\mathcal{A}^*$ be the set of graphs such that there exists a strategy for Alice that ensures a connected red subgraph of order $\lceil |V(G)|/2 \rceil$, regardless of Bob's strategy. *I.e.*, $\mathcal{A}^*$ is the set of graphs in which Alice has a strategy to ensure a single connected red subgraph.

Theorem 7. *Let G be a cograph. There exists a linear-time algorithm that decides whether G is A-win or AB-draw, and whether $G \in \mathcal{A}^*$ or not.*

Proof. The proof is by induction on $n = |V(G)|$. More precisely, we describe a recursive algorithm. If $n = 1$, then G is clearly A-win and $G \in \mathcal{A}^*$.

Let us assume that $n > 1$. There are two cases to be considered. Either $G = G_1 \oplus G_2$ for some cographs G_1 and G_2, or $G = G_1 + G_2 + \dots, G_m$, where, for every $1 \leq i \leq m$ ($m \geq 2$), G_i is either a single vertex or is a cograph obtained from the join of two other cographs. For every $1 \leq i \leq m$, let us assume by induction that it can be computed in time linear in $|V(G_i)|$, whether G_i is A-win or AB-draw and whether $G_i \in \mathcal{A}^*$ or not. Let us show how to decide if G is A-win or AB-draw, and whether $G \in \mathcal{A}^*$ or not, in constant time.

1. Let us first assume that $\boldsymbol{G = G_1 \oplus G_2}$. We prove that (see [2]):
   (a) If n is odd, then G is A-win and $G \in \mathcal{A}^*$.
   (b) If $|V(G_1)|, |V(G_2)| \geq 2$ and n is even, then G is AB-draw and $G \in \mathcal{A}^*$.
   (c) If $|V(G_1)| = 1$ and n is even, there are two cases to consider:
      i. If $G_2 \notin \mathcal{A}^*$, then G is A-win and $G \in \mathcal{A}^*$.
      ii. If $G_2 \in \mathcal{A}^*$, then G is AB-draw and $G \in \mathcal{A}^*$.
2. Now, let us assume that $\boldsymbol{G = G_1 + \ldots + G_m}$ where, for all $1 \leq i \leq m$ ($m \geq 2$), G_i is either a single vertex or the join $G_i' \oplus G_i''$ of two cographs G_i' and G_i'' with $|V(G_i')| \geq |V(G_i'')|$. Let $n_i = |V(G_i)|$ for all $1 \leq i \leq m$, and let $n_1 \geq \cdots \geq n_m$. To simplify the proof to follow, first note that, if $n_1 = 1$, then G is AB-draw (since $n_2 = 1$ as $m \geq 2$) and $G \in \mathcal{A}^*$ if and only if $G = G_1 + G_2$. Second, if $n_2 = 1$, then the result of the game in G is the same as in G_1, which is known, by Case 1, since G_1 is a join. Moreover, in this case, $G \in \mathcal{A}^*$ if and only if n_1 is odd and $G = G_1 + G_2$. Hence, we may assume that $n_1 > 1$ and $n_2 > 1$. Lastly, in what follows, for any of the winning strategies described for Alice, whenever Bob colours a vertex in G_j for $3 \leq j \leq m$, Alice also colours a vertex in G_j on her next turn. The same holds for any of the drawing strategies for Bob (with Bob and Alice reversed), except for Case 2(e)ii, for which the same only holds for $4 \leq j \leq m$. This guarantees that a player never has a connected subgraph of order more than $\lceil n_j/2 \rceil$ in G_j for $3 \leq j \leq m$ ($4 \leq j \leq m$ for Case 2(e)ii). Alice (Bob, resp.) always has a connected red (blue, resp.) subgraph of order at least $\lceil n_1/2 \rceil$ in all of her winning (his drawing, resp.) strategies below. Hence, for all the cases except Case 2(e)ii, we can assume that $G = G_1 + G_2$, and for Case 2(e)ii, we can assume that $G = G_1 + G_2 + G_3$. In what follows, if a player cannot follow their strategy in a round, unless otherwise stated, they colour an arbitrary vertex and then resume their strategy for the subsequent rounds.
   There are 5 cases to consider. Since we assume that $n_1 > 1$ and $n_2 > 1$, both G_1'' and G_2'' exist. In Case 2(e)iii, the statement involves n_3, so if $m = 2$, then $n_3 = 0$. Also, since $n_2 > 1$ and Bob has a strategy where, for all $1 \leq i \leq m$, he colours at least $\lfloor n_i/2 \rfloor$ vertices of G_i, then $G \notin \mathcal{A}^*$ in each case. Thus, we just need to show the outcome of the game on G for each case.
   (a) If $n_1 = n_2$, then G is AB-draw.
   Assume, w.l.o.g., that Alice first colours a vertex in G_1. Bob colours a vertex in G_2''. Now, whenever Alice colours a vertex in G_1 (G_2, resp.), Bob colours a vertex in G_1 (G_2, resp.). In particular, if Bob is to colour a vertex in G_2, then he colours one in G_2' if he can, and if not, then he colours one in G_2'', and, if that is not possible, he colours one in G_1. Similarly, if Bob is to colour a vertex in G_1 but cannot, then he colours one in G_2' first if possible, and if not, then he colours one in G_2''. If n_1 is odd, then Bob's strategy ensures a connected blue subgraph of order $\frac{n_2-1}{2} + 1 = \frac{n_1-1}{2} + 1$ in G_2 and that the largest connected red subgraph in G is of order at most $\frac{n_1-1}{2} + 1$. If n_1 is even and Alice colours the last vertex in G_1, then Bob's strategy ensures a connected blue subgraph of order $\lceil \frac{n_2-1}{2} \rceil + 1 = \lceil \frac{n_1-1}{2} \rceil + 1$ in G_2 and that the largest connected

red subgraph in G is of order at most $\lceil \frac{n_1-1}{2} \rceil + 1$. If Alice coloured the last vertex in G_2 instead, then Bob ensures a connected blue subgraph of order $\lceil \frac{n_2-2}{2} \rceil + 1 = \frac{n_1}{2}$ in G_2 and that the largest connected red subgraph in G is of order at most $\frac{n_1}{2}$. Hence, G is AB-draw.

(b) If $n_1 > n_2$ and n_1 is odd, then G is A-win.
Alice first colours a vertex in G_1. Then, whenever Bob colours a vertex in G_1 (G_2, resp.), Alice colours a vertex in G_1 (G_2, resp.). By Case 1(a), Alice has a winning strategy in G_1 ensuring a connected red subgraph of order at least $\lceil \frac{n_1}{2} \rceil$. By Case 1, Alice ensures that any connected blue subgraph in G_2 is of order at most $\lceil \frac{n_2}{2} \rceil < \lceil \frac{n_1}{2} \rceil$. Hence, G is A-win.

(c) If $n_1 > n_2$, n_1 is even, and $|V(G_1'')| \geq 2$, then G is AB-draw.
Whenever Alice colours a vertex in G_1 (G_2, resp.), Bob also colours a vertex in G_1 (G_2, resp.). By Case 1(b), Bob has a drawing strategy in G_1 ensuring a connected blue subgraph of order at least $\frac{n_1}{2}$. By Case 1, Bob ensures that any connected red subgraph in G_2 is of order at most $\lceil \frac{n_2}{2} \rceil \leq \frac{n_1}{2}$. Hence, G is AB-draw.

(d) If $n_1 > n_2$, n_1 is even, $|V(G_1'')| = 1$, and $G_1' \in \mathcal{A}^*$, then G is AB-draw.
Whenever Alice colours a vertex in G_1 (G_2, resp.), Bob also colours a vertex in G_1 (G_2, resp.). By Case 1(c)ii, Bob has a drawing strategy in G_1 ensuring a connected blue subgraph of order at least $\frac{n_1}{2}$. By Case 1, Bob ensures that any connected red subgraph in G_2 is of order at most $\lceil \frac{n_2}{2} \rceil \leq \frac{n_1}{2}$. Hence, G is AB-draw.

(e) If $n_1 > n_2$, n_1 is even, $|V(G_1'')| = 1$, and $G_1' \notin \mathcal{A}^*$, then:

i. If $n_1 > n_2 + 1$, then G is A-win.
Alice first colours a vertex in G_1. Then, whenever Bob colours a vertex in G_1 (G_2, resp.), Alice colours a vertex in G_1 (G_2, resp.). By Case 1(c)i, Alice ensures a connected red subgraph of order at least $\frac{n_1}{2}$ in G_1, and that any connected blue subgraph in G_1 is of order less than $\frac{n_1}{2}$. By Case 1, Alice ensures that any connected blue subgraph in G_2 is of order at most $\lceil \frac{n_2}{2} \rceil < \frac{n_1}{2}$. Hence, G is A-win.

ii. If $n_1 = n_2 + 1 = n_3 + 1$, then G is AB-draw.
Whenever Alice colours a vertex in G_1, Bob colours a vertex in G_1. By Case 1, this ensures that $\frac{n_1}{2}$ of the vertices in G_1 are red and $\frac{n_1}{2}$ are blue. The first time that Alice colours a vertex $v \in V(G_2) \cup V(G_3)$, assume, w.l.o.g., that $v \in V(G_2)$. Bob then colours a vertex in G_3''. Then, whenever Alice colours a vertex in G_2 (G_3, resp.), Bob colours a vertex in G_2 (G_3, resp.). In particular, if Bob is to colour a vertex in G_3, then he colours one in G_3' first if possible, if not, then he colours a vertex in G_3'', and, if that is not possible, he colours a vertex in G_2. As in Case 2(a), Bob ensures a connected blue subgraph of order $\lceil \frac{n_3}{2} \rceil = \frac{n_1}{2}$ in G_3 and that any connected red subgraph in G_2 is of order at most $\lceil \frac{n_2}{2} \rceil = \frac{n_1}{2}$. Hence, G is AB-draw.

iii. If $n_1 = n_2 + 1$ and $n_2 > n_3$, then G is A-win.
Alice first colours the vertex in G_1''. Then, Alice colours vertices in G_1 as long as she can. By Case 1(c)i, she ensures that any connected blue subgraph in G_1 is of order less than $\frac{n_1}{2}$. If it is Alice's turn, there is

a connected red subgraph of order $n_1 - k$ in G_1 for some $0 \leq k \leq \frac{n_1}{2}$, and it is the first round in which she can no longer colour vertices in G_1, then Bob coloured k vertices in G_1 and $n_1 - 2k$ vertices in G_2. Then, any connected blue subgraph in G_2 is of order at most $\lceil \frac{n_2 - n_1 + 2k - 1}{2} \rceil + n_1 - 2k = n_1 - k - 1 < n_1 - k$. Hence, G is A-win.

We get the result as a decomposition of a cograph is computed in linear time. □

6 Further Work

It would be interesting to study the game in other graph classes such as trees and interval graphs. Also, since grids of even order are AB-draw by Theorem 2, it would be intriguing to look at grids of odd order. Just as reflection graphs are a large class of graphs that are AB-draw, another direction would be to find a diverse class of graphs that are A-win. Any graph $G \in \mathcal{A}^*$ of odd order is A-win, and so, perhaps a class of dense graphs of odd order would be a prime candidate.

References

1. Andres, S.D., Huggan, M., Mc Inerney, F., Nowakowski, R.J.: The orthogonal colouring game. Theoret. Comput. Sci. **795**, 312–325 (2019)
2. Bensmail, J., Fioravantes, F., Mc Inerney, F., Nisse, N.: The largest connected subgraph game. Research report (2021). https://hal.inria.fr/hal-03137305
3. Bonnet, E., Jamain, F., Saffidine, A.: On the complexity of connection games. Theoret. Comput. Sci. **644**, 2–28 (2016)
4. Browne, C.: Connection Games: Variations on a Theme. AK Peters (2005)
5. Bruno, J., Weinberg, L.: A constructive graph-theoretic solution of the Shannon switching game. IEEE Trans. Circuit Theory **17**(1), 74–81 (1970)
6. Corneil, D.G., Perl, Y., Stewart, L.K.: A linear recognition algorithm for cographs. SIAM J. Comput. **14**(4), 926–934 (1985)
7. Duchêne, É., Gonzalez, S., Parreau, A., Rémila, E., Solal, P.: INFLUENCE: a partizan scoring game on graphs. CoRR abs/2005.12818 (2020)
8. Even, S., Tarjan, R.E.: A combinatorial problem which is complete in polynomial space. J. ACM **23**(4), 710–719 (1976)
9. Gardner, M.: The Scientific American Book of Mathematical Puzzles & Diversions. Simon and Schuster, New York (1959)
10. Gardner, M.: The Second Scientific American Book of Mathematical Puzzles and Diversions. Simon and Schuster, New York (1961)
11. Larsson, U., Nowakowski, R.J., Neto, J.P., Santos, C.P.: Guaranteed scoring games. Electron. J. Comb. **23**(3) (2016)
12. Larsson, U., Nowakowski, R.J., Santos, C.P.: Games with guaranteed scores and waiting moves. Int. J. Game Theory **47**(2), 653–671 (2017). https://doi.org/10.1007/s00182-017-0590-x
13. Micek, P., Walczak, B.: A graph-grabbing game. Comb. Probab. Comput. **20**(4), 623–629 (2011)
14. Reisch, S.: Hex ist PSPACE-vollständig. Acta Informatica **15**(2), 167–191 (1981)
15. Schaefer, T.J.: On the complexity of some two-person perfect-information games. J. Comput. Syst. Sci. **16**(2), 185–225 (1978)
16. Shapovalov, A.: Occupation games on graphs in which the second player takes almost all vertices. Discrete Appl. Math. **159**(15), 1526–1527 (2011)

Can Romeo and Juliet Meet? or Rendezvous Games with Adversaries on Graphs

Fedor V. Fomin[1], Petr A. Golovach[1](✉), and Dimitrios M. Thilikos[2]

[1] Department of Informatics, University of Bergen, Bergen, Norway
{Fedor.Fomin,Petr.Golovach}@uib.no
[2] LIRMM, Univ Montpellier, CNRS, Montpellier, France
sedthilk@thilikos.info

For never was a story of more woe than
this of Juliet and her Romeo.

— *William Shakespeare, Romeo and Juliet*

Abstract. We introduce the rendezvous game with adversaries. In this game, two players, *Facilitator* and *Divider*, play against each other on a graph. Facilitator has two agents, and Divider has a team of k agents located in some vertices of the graph. They take turns in moving their agents to adjacent vertices (or staying put). Facilitator wins if his agents meet in some vertex of the graph. The goal of Divider is to prevent the rendezvous of Facilitator's agents. Our interest is to decide whether Facilitator can win. It appears that, in general, the problem is PSPACE-hard and, when parameterized by k, co-W[2]-hard. Moreover, even the game's variant where we ask whether Facilitator can ensure the meeting of his agents within τ steps is co-NP-complete already for $\tau = 2$. On the other hand, for chordal and P_5-free graphs, we prove that the problem is solvable in polynomial time. These algorithms exploit an interesting relation of the game and minimum vertex cuts in certain graph classes. Finally, we show that the problem is fixed-parameter tractable parameterized by both the graph's neighborhood diversity and τ.

Keywords: Rendezvous games · Dynamic separators · Complexity

1 Introduction

We introduce the *Rendezvous Game with Adversaries* on graphs. In our game, a team of dividers tries to prevent two passionate lovers, say Romeo and Juliet,

This research was supported by the French Ministry of Europe and Foreign Affairs, via the Franco-Norwegian project PHC AURORA. The two first authors have been supported by the Research Council of Norway via the project "MULTIVAL" (grant no. 263317). The last author was supported by the ANR projects DEMOGRAPH (ANR-16-CE40-0028), ESIGMA (ANR-17-CE23-0010), and the French-German Collaboration ANR/DFG Project UTMA (ANR-20-CE92-0027).

Ł. Kowalik et al. (Eds.): WG 2021, LNCS 12911, pp. 308–320, 2021.
https://doi.org/10.1007/978-3-030-86838-3_24

from meeting each other. We are interested in the minimum size of the team of dividers sufficient to obstruct their romantic encounter. In the static setting, when dividers do not move, this is just the problem of computing the minimum vertex cut between the pair of vertices occupied by Romeo and Juliet. But in the dynamic variant, when dividers are allowed to change their position, the team's size can be significantly smaller than the size of the minimum cut. In fact, this gives rise to a new interactive form of connectivity that is much more challenging both from the combinatorial and the algorithmic point of view.

Our rendezvous game rules are very similar to the rules of the classical COPS-AND-ROBBER game of Nowakowski-Winkler and Quillioit [23,24], see also the book of Bonato and Nowakowski [4]. The difference is that in the COPS-AND-ROBBER game, a team of k cops tries to capture a robber in a graph, while in our game, the group of k dividers tries to keep the two lovers separated.

A bit more formally. The game is played on a finite undirected connected graph G by two players: *Facilitator* and *Divider*. Facilitator has two agents R and J that are initially placed in designated vertices s and t of G. Divider has a team of $k \geq 1$ agents $D_1, \ldots, D_k$ that are initially placed in some vertices of $V(G) \setminus \{s, t\}$ selected by Divider. Several divider agents can occupy the same vertex. Then the players make their moves by turn, starting with Facilitator. At every move, each player moves some of his agents to adjacent vertices or keeps them in their old positions. No agent can be moved to a vertex that is currently occupied by adversary agents. Both players have complete information about G and the positions of all the agents. Facilitator aims to ensure that R and J meet; that is, they are in the same vertex. The task of Divider is to prevent the rendezvous of R and J by maintaining $D_1, \ldots, D_k$ in positions that block the possibility to meet. Facilitator wins if R and J meet, and Divider wins if he succeed in preventing the meeting of R and J forever.

We define the following problem:

RENDEZVOUS

Input:	A graph G with two given vertices s and t, and a positive integer k.
Task:	Decide whether Facilitator can win on G starting from s and t against Divider with k agents.

Another variant of the game is when the number of moves of the players is at most a given integer parameter τ. Then Facilitator wins if R and J meet within the first τ moves, and Divider wins otherwise. We call this problem RENDEZVOUS IN TIME. We also consider the version of the problem where τ is a fixed constant. This generates a family of problems, one for each different value of τ, and we refer to each of them as the τ-RENDEZVOUS IN TIME problem.

Our Results. We start with combinatorial results. If $s = t$ or if s and t are adjacent, then Facilitator wins by a trivial strategy. However, if s and t are distinct nonadjacent vertices, then Divider wins provided that he has sufficiently many agents. For example, the agents can be placed in the vertices of an

(s, t)-separator and stay there. Then R and J never meet. We call the minimum number k of the agents of Divider that is sufficient for his winning, the (s, t)-*dynamic separation* number and denote it by $d_G(s, t)$. We put $d_G(s, t) = +\infty$ for $s = t$ or when s and t are adjacent. The dynamic separation number can be seen as a dynamic analog of the minimum size $\lambda_G(s, t)$ of a vertex (s, t)-separator in G. (The minimum number of vertices whose removal leaves s and t in different connected components.) Then RENDEZVOUS can be restated as the problem of deciding whether $d_G(s, t) > k$.

The first natural question is: What is the relation between $d_G(s, t)$ and $\lambda_G(s, t)$? Clearly, $d_G(s, t) \leq \lambda_G(s, t)$. We show that $d_G(s, t) = 1$ if and only if $\lambda_G(s, t) = 1$. If $d_G(s, t) \geq 2$, then we construct examples demonstrating that the difference $\lambda_G(s, t) - d_G(s, t)$ can be arbitrary even for sparse graphs. Interestingly, there are graph classes where both parameters are equal. In particular, we show that $\lambda_G(s, t) = d_G(s, t)$ holds for P_5-free graphs and chordal graphs. This also yields a polynomial time algorithm computing $d_G(s, t)$ on these classes of graphs.

Then we turn to the computational complexity of RENDEZVOUS and RENDEZVOUS IN TIME on general graphs. Both problems can be solved it $n^{\mathcal{O}(k)}$ time by using a backtracking technique. We show that this running time is asymptotically tight by proving that they are both co-W[2]-hard when parameterized by k (we prove that it is W[2]-hard to decide whether $d_G(s, t) \leq k$) and cannot be solved in $f(k) \cdot n^{o(k)}$ time for any function f of k, unless ETH fails. Moreover, τ-RENDEZVOUS IN TIME is W[2]-hard, for every $\tau \geq 2$. If τ is a constant, then τ-RENDEZVOUS IN TIME is in co-NP and our co-W[2]-hardness proof implies that τ-RENDEZVOUS IN TIME is co-NP-complete for every $\tau \geq 2$. For the general case, the problems are even harder as we prove that RENDEZVOUS and RENDEZVOUS IN TIME are both PSPACE-hard.

Finally, we initiate the study of the complexity of the problems under structural parameterization of the input graphs. We show that RENDEZVOUS IN TIME is FPT when parameterized by the neighborhood diversity of the input graph and τ.

Related Work. The classical rendezvous game introduced by Alpern [2] is played by two agents that are placed in some unfamiliar area and whose task is to develop strategies that maximize the probability that they meet. We refer to the book of Alpern and Gal [3] for detailed study of the subject. Deterministic rendezvous problem was studied by Ta-Shma and Zwick [26].

RENDEZVOUS is closely related to the COPS-AND-ROBBER game. The game was defined (for one cop) by Winkler and Nowakowski [23] and Quilliot [24] who also characterized graphs for which one cop can catch the robber. Aigner and Fromme [1] initiated the study of the problem with several cops. The minimum number of cops that are required to capture the robber is called the cop number of a graph. This problem was studied intensively and we refer to the book of Bonato and Nowakowski [4] for further references. Kinnersley [19] established that the problem is EXPTIME-complete. The COPS-AND-ROBBER game can be seen as a special case of search games played on graphs, surveys [5,12] provide further references on search and pursuit-evasion games on graphs. A related variant of COPS-AND-ROBBER game is the guarding game studied in [9,10,22,25]. Here the set of cops is trying to prevent the robber from entering a specified subgraph in a graph.

Due to space constraints the majority of the proofs are omitted in this extended abstract. The details are available in [11].

2 Preliminaries

Graphs. All graphs considered in this paper are finite undirected graphs without loops or multiple edges, unless it is said explicitly that we consider directed graphs. We follow the standard graph theoretic notation and terminology (see, e.g., [7]). For each of the graph problems considered in this paper, we let $n = |V(G)|$ and $m = |E(G)|$ denote the number of vertices and edges, respectively, of the input graph G if it does not create confusion. For a graph G and a subset $X \subseteq V(G)$ of vertices, we write $G[X]$ to denote the subgraph of G induced by X. For a set of vertices S, $G - S$ denotes the graph obtained by deleting the vertices of S, that is, $G - S = G[V(G) \setminus S]$; for a vertex v, we write $G - v$ instead of $G - \{v\}$. For a vertex v, we denote by $N_G(v)$ the *(open) neighborhood* of v, i.e., the set of vertices that are adjacent to v in G. For two nonadjacent vertices s and t, a set of vertices $S \subseteq V(G) \setminus \{s, t\}$ is an *(s,t)-separator* if s and t are in distinct connected components of $G - S$. We use $\lambda_G(s,t)$ to denote the minimum size of an (s,t)-separator of G; $\lambda_G(s,t) = +\infty$ if $s = t$ or s and t are adjacent. A *path* is a connected graph with at leat one and most two vertices (called *end-vertices*) of degree at most one whose remaining vertices (called *internal*) have degrees two. We say that a path with end-vertices u and v is an *(u,v)-path.* The *length* of a path P, denoted by $\ell(P)$, is the number of its edges. The *distance* $\mathsf{dist}_G(u,v)$ between two vertices u and v of G in the length of a shortest (u,v)-path. We use $v_1 \cdots v_k$ to denote the path with the vertices $v_1, \ldots, v_k$ and the edges $v_{i-1}v_i$ for $i \in \{2, \ldots, k\}$. A *cycle* is a connected graph with all the vertices of degree two. The *length* $\ell(C)$ of a cycle C is the number of edges of C.

Parameterized Complexity. We obtain a number of results about the parameterized complexity of Rendezvous and Rendezvous in Time. We refer to the recent book of Cygan et al. [6] for the introduction to the area. Here we just remind that an instanse of the parameterized version Π_p of a decision problem Π is a pair (I, k), where I is an instance of Π and k is an integer *parameter* associated with I. It is said that Π_p is *fixed-parameter tractable* (FPT) if it can be solved in time $f(k)|I|^{\mathcal{O}(1)}$ for a computable function $f(k)$ of the parameter k. The Parameterized Complexity theory also provides tools that allow to show that a parameterized problem cannot be solved in FPT time (up to some reasonable complexity assumptions). For this, Downey and Fellows (see [8]) introduced a hierarchy of parameterized complexity classes, namely $\mathsf{FPT} \subseteq \mathsf{W}[1] \subseteq \mathsf{W}[2] \subseteq \cdots \subseteq \mathsf{XP}$, and the basic conjecture is that all inclusions in the hierarchy are proper. The usual way to show that it is unlikely that a parameterized problem admits an FPT algorithm is to show that it is W[1] or W[2]-hard using a *parameterized reduction* from a known hard problem in the corresponding class. The most common tool for establishing fine-grained complexity lower bound for parameterized problems is the *Exponential Time Hypothesis* (ETH) proposed by Impagliazzo, Paturi, and Zane [16,17]. This is the conjecture stating that there is $\varepsilon > 0$ such that 3-SATISFIABILITY cannot be solved in $\mathcal{O}^*(2^{\varepsilon n})$ time on formulas with n variables.

We conclude this section by giving some easy observations about complexity of the considered games. As it is common for various games on graphs (see, e.g., the book of Bonato and Nowakowski [4] about COPS-AND-ROBBER games), our Rendezvous Game with Adversaries can be resolved by backtracking.

Theorem 1 **(∗).**[1] RENDEZVOUS *and* RENDEZVOUS IN TIME *can be solved in* $n^{\mathcal{O}(k)}$ *time.*

If the number of steps τ is bounded by a constant, then the number of possible moves of Facilitator in the first τ steps is polynomial. This implies that a winning strategy of Divider can be given by a certificate of polynomial size encoding the responses of Divider on each move of Facilitator. Thus, we can observe the following.

Observation 1 **(∗).** *For every fixed constant* τ*, the problem* τ-RENDEZVOUS IN TIME *is in* co-NP.

3 Dynamic Separation vs. Separators

In this section we investigate relations between $d_G(s,t)$ and $\lambda_G(s,t)$. Given a connected graph G and two vertices s and t, it is straightforward to see that $d_G(s,t) \leq \lambda_G(s,t)$. Indeed, if $S \subseteq V(G) \setminus \{s,t\}$ is an (s,t)-separator of size $k = \lambda_G(s,t)$, then Divider with k agents can put then in the vertices of S in the beginning of the game. Then he can use the trivial strategy that keeps the agents $D_1, \ldots, D_k$ in their positions. However, $d_G(s,t)$ and $\lambda_G(s,t)$ can be far apart. Still, $d_G(s,t) = 1$ if and only if $\lambda_G(s,t) = 1$, and this is the first result of the section.

Theorem 2. *Let G be a connected graph and let $s,t \in V(G)$. Then $d_G(s,t) = 1$ if and only if $\lambda_G(s,t) = 1$.*

Proof. As we already observed, $d_G(s,t) \leq \lambda_G(s,t)$. Hence, if $\lambda_G(s,t) = 1$, then $d_G(s,t) = 1$. This means that it is sufficient to show that if $d_G(s,t) = 1$, then $\lambda_G(s,t) = 1$. We prove this by contradiction. Assume that $\lambda_G(s,t) \geq 2$. We show that Facilitator has a winning strategy when starting from s and t on G against Divider with one agent.

Let C be a cycle in G. For every two distinct vertices u and v of C, C has two internally vertex disjoint (u,v)-paths P_1 and P_2 in C. We say that C has a (u,v)-*shortcut* if there is a (u,v)-path P in $G - (V(C) \setminus \{u,v\})$ that is shorter than P_1 and P_2. That is, $\ell(P) < \ell(P_1)$ and $\ell(P) < \ell(P_2)$. We say that C has a *shortcut* if there are distinct $u,v \in V(C)$ that have a (u,v)-shortcut.

We claim the following.

[1] The proofs of the statements labeled by (∗) are omitted.

Claim 1. *If R and J occupy vertices of a cycle C of G that has a shortcut, then Facilitator has a strategy such that in at most $\ell(C)$ steps R and J are moved into vertices of a cycle C' with $\ell(C') < \ell(C)$.*

Proof (of Claim 1). Suppose that R and J occupy vertices x and y of C, respectively. Assume that a path P is a (u, v)-shortcut for some distinct $u, v \in V(G)$. Denote by P_1 and P_2, respectively, the internally vertex disjoint (u, v)-paths in C. Let C_1 be the cycle of G composed by P_1 and P, and let C_2 be the cycle composed by P_2 and P. Because P is a shortcut for C, we have that $\ell(C_1) < \ell(C)$ and $\ell(C_2) < \ell(C)$. If $x, y \in V(P_1)$, then $x, y \in V(C_1)$ and the claim holds trivially, since R and J are already on cycle C_1 with $\ell(C_1) < \ell(C)$. Symmetrically, if $x, y \in V(P_2)$, then the claim holds. Assume that this is not the case. Then x and y are internal vertices of P_1 and P_2 belonging to distinct paths. We assume without loss of generality that x is an internal vertex of P_1 and y is an internal vertex of P_2.

Facilitator uses the following strategy. In each step, R is moved along P_1 toward u, unless the next vertex is occupied by D_1. In the last case, R stays in the current position. Similarly, J moves toward v in P_2 whenever this is possible and stays in the current position if the way is blocked. Notice that, since the unique agent D_1 of Divider occupies a unique vertex in each step, at least one of the agents R or J moves to an adjacent vertex. Therefore, either R reaches u or J reaches v in at most $\ell(C)$ steps. If R is in u, then R and J are in the vertices of C_2 and $\ell(C_2) < \ell(C)$. Symmetrically, if J reaches v, then R and J reach C_1 with $\ell(C_1) < \ell(C)$. ◀

Next, we show that Facilitator can win if R and J are in a cycle without shortcuts and D_1 is in the same cycle.

Claim 2. *If R and J occupy vertices of a cycle C of G without a shortcut, and the unique agent D_1 of Divider is in a vertex of C as well, then Facilitator has a winning strategy with at most $\ell(C)/2$ steps.*

Proof (of Claim 2). Suppose that R and J occupy vertices x and y of C, respectively, and that D_1 occupies $z \in V(C)$. Denote by P the unique (x, y)-path in $C - z$. Facilitator uses the following strategy. In every step, R and J move towards each other along P except if they appear to occupy adjacent vertices. In the last case, R stays and J moves to the vertex occupied by R. We show that this strategy is a feasible winning strategy.

The proof is by induction on the length of P. The claim is trivial when $\ell(P) \leq 2$. Assume that $\ell(P) \geq 3$ and the claim holds for all positions x', y' and z' of R, J and D_1, respectively, if the length of the (x', y')-path in $C - z'$ is at most $\ell(P) - 1$.

In the first step, R and J move to the neighbors x' and y' of x and y, respectively, in P. If D_1 moves to a vertex $z' \in V(C)$, then we apply the inductive assumption and, since the length of the (x', y')-subpath P' is $\ell(P) - 2$ and $z' \notin V(P')$, obtain that the strategy of Facilitator is winning. Assume that by the first move Divider removes D_1 from C. If D_1 does not return to a vertex of C in $\ell(P)/2$ steps, Facilitator wins. Hence for some $h \leq \ell(P)/2$, at the h-th move, D_1 steps back on a vertex $z' \in V(C)$.

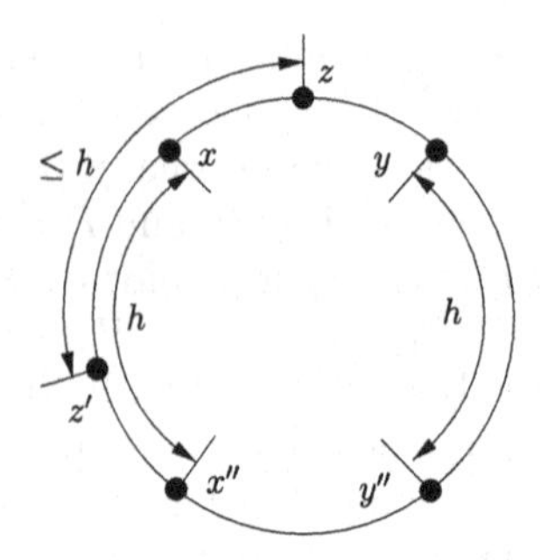

Fig. 1. The position placement after h steps (up to symmetry).

By the assumption, cycle C has no shortcuts. In particular, there is no (z, z')-shortcut. This implies, that the length of one of the two (z, z')-paths in C is at most h. Observe that in h steps, R and J reach vertices x'' and y'' that are at the distance h in P from x and y, respectively. Therefore (see Fig. 1), the (x'', y'')-subpath P'' of P does not contain z'. Since $\ell(P'') < \ell(P)$, we can apply the inductive assumption. This proves that the Facilitator's strategy is a feasible winning strategy and the claim holds.

Notice that the total number of steps is $\lceil \ell(P)/2 \rceil \leq \ell(C)/2$. This completes the proof. ◀

Now we are ready to complete the proof of the theorem. If $s = t$ or s and t are adjacent, then Facilitator has a straightforward winning strategy. Assume that s and t are distinct and nonadjacent. Since $\lambda_G(s, t) \geq 2$, by Menger's theorem (see, e.g., [7]), there are two internally vertex disjoint (s, t)-paths P_1 and P_2. The union of these two paths forms cycle C. If the agent D_1 of Divider occupies a vertex of C', then Facilitator has a winning strategy by Claim 2. If D_1 is outside C', then Facilitator moves R and J along C' towards each other. Then either R and J meet or D_1 steps on C' at some moment. In this case, Facilitator switches to the strategy from Claim 2 that guarantees him to win. □

We observed that $d_G(s, t) \leq \lambda_G(s, t)$ and, by Theorem 2, $d_G(s, t) = 1$ if and only if $\lambda_G(s, t) = 1$. However, if $d_G(s, t) \geq 2$, then the difference betweem $\lambda_G(s, t)$ and $d_G(s, t)$ may be arbitrary. To see this, consider the following example for $p \geq 2$.

- Construct a set $U = \{u_1, \ldots, u_p\}$ of pairwise adjacent vertices.
- Add a vertex s and join s with each vertex $u_i \in U$ by a path sx_iu_i.
- Add a vertex t and join t with each vertex $u_i \in U$ by a path ty_iu_i.

Denote the obtained graph by G (see the left part of Fig. 2). Observe that $\lambda_G(s,t) = p$. We show that $d_G(s,t) = 2$ by demonstrating a winning strategy for Divider with two agents D_1 and D_2. Initially, D_1 and D_2 are placed in arbitrary vertices of the clique U. Then D_1 "shadows" R and D_2 "shadows" J in U in the following sense. If R moves to x_i for some $i \in \{1,\ldots,p\}$, Divider responds by moving D_1 to u_i. Symmetrically, if R moves to y_j for some $j \in \{1,\ldots,p\}$, then D_2 is moved to u_j. It is easy to verify that if Divider follows this strategy, then neither R no J can enter U. Therefore, Divider wins. Since p can be arbitrary, we have that $\lambda_G(s,t) - d_G(s,t) = p - 2$ can be arbitrary large.

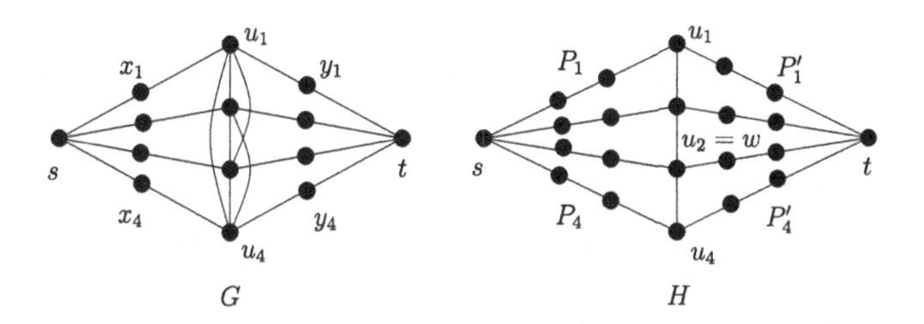

Fig. 2. The construction of G and H for $p = 4$.

The family of graphs G for $p \geq 2$ in the above example is a family of dense graph, because G contains a clique with p vertices. However, exploiting the same idea as for G, we can show that there are sparse graphs with the same property. For this, we considered the following more complicated example.

- Construct a path $P = u_1 \cdots u_p$ on p vertices.
- Add a vertex s and join s with each vertex $u_i \in V(P)$ by an (s,u_i)-path P_i of length $h = \lfloor p/2 \rfloor + 1$.
- Add a vertex t and join t with each vertex $u_i \in V(P)$ by an (t,u_i)-path P_j' of length $h = \lfloor p/2 \rfloor + 1$.

Denote the obtained graph by H (see the right part of Fig. 2). Clearly, $\lambda_H(s,t) = p$. We claim that $d_H(s,t) = p$. The idea behind the winning strategy for Divider with two agents D_1 and D_2 is similar to the one from the first example: D_1 "shadows" R and D_2 "shadows" J on P. Let $w = u_{\lfloor p/2 \rfloor}$. Initially, D_1 and D_2 are placed in w. Then D_1 is moved as follows. If R moves to/stays in s, then D_1 moves to/stays in w. If R is moved into an internal vertex x of P_i for some $i \in \{1,\ldots,p\}$, then Divider responds my moving D_1 toward u_i or keeping D_1 in the current position maintaining the following condition: D_1 is in a vertex u_j at minimum distance from w such that the distance between x and u_i in P_i is more than the distance between u_j and u_i in P. The construction of the strategy for D_2 is symmetric. It is easy to see that the described strategy for Divider is feasible, and the strategy allows neither R no J to enter a vertex of P. Therefore, $d_H(s,t) = 2$.

Notice that the graph H for each $p \geq 2$ is planar and it can be seen that the treewidth of H is at most 3 (we refer to [6,7] for the formal treewidth definition), that is, graphs H are, indeed, sparse.

Our examples indicate that $\lambda_G(s,t)$ may differ from $d_G(s,t)$ if G has sufficiently long induced paths and cycles. We observe that $\lambda_G(s,t) = d_G(s,t)$ if G belongs to graph classes that have no graphs of this type. A graph G is *P_5-free* if G has no induced subgraph isomorphic to the path with 5 vertices. A graph G is *chordal* if G does not contain induced cycles on at least 4 vertices, that is, if C is a cycle in G of length at least 4, then there is a *chord*, i.e., an edge of G with end-vertices in two nonconsecutive vertices of C.

Proposition 1 (*). *If G is a connected P_5-free (chordal, respectively) graph, then for every $s, t \in V(G)$, $d_G(s,t) = \lambda_G(s,t)$.*

The *chordality* of a graph is biggest smallest size of a induced cycle in it. Clearly, chordal graphs are the graphs of chordality three. It is natural to ask whether for graphs of bigger chordality the difference betweem $\lambda_G(s,t)$ and $d_G(s,t)$ may be arbitrary. In the example we gave after the Proof of Claim 2, we have seen that, for the graph G (depicted in the left part of Fig. 2), it holds that $\lambda_G(s,t) - d_G(s,t) = p - 2$ and is easy to see that any such G has chordality five. Notice that this graph G can be further enhanced so to obtain chordality four: just add a clique between the vertices in $\{x_1, \ldots, x_p\}$ and a clique between the vertices in $\{y_1, \ldots, y_p\}$. This indicates a sharp transition of d_G away from λ_G when graphs are not chordal any more.

Since $\lambda_G(s,t)$ can be computed in polynomial time by the standard maximum flow algorithms (see, e.g., the recent textbook [27]), we obtain the following corollary.

Corollary 1. Rendezvous *can be solved in polynomial time on the classes of P_5-free and chordal graphs.*

4 Hardness of Rendezvous Game with Adversaries

In this section, we discuss algorithmic lower bounds for Rendezvous and Rendezvous in Time.

We proved in Theorem 1 that Rendezvous and Rendezvous in Time can be solved in $n^{\mathcal{O}(k)}$. We show that it is unlikely that the dependance on k can be improved. For this, we show that both problems are co-W[2]-hard (i.e., it is W[2]-hard to decide whether the input is a no-instance; in fact, we show that it is W[2]-hard to decide whether $d_G(s,t) \leq k$) and, therefore, cannot be solved in time $f(k) \cdot n^{\mathcal{O}(1)}$ for any computable function $f(k)$, unless FPT = W[2]; the result for Rendezvous in Time holds also for τ-Rendezvous in Time when $\tau \geq 2$. Our proof also implies that neither Rendezvous nor τ-Rendezvous in Time, for $\tau \geq 2$, cannot be solved in time $f(k) \cdot n^{o(k)}$ unless ETH fails.

Observe that Rendezvous in Time can be solved in polynomial time if $\tau = 1$, because of the following straightforward observation.

Observation 2. *Facilitator can win in Rendezvous Game with Adversaries game in one step on G starting from s and t against Divider with k agents if and only if one of the following holds: (i) $s = t$, (ii) s and t are adjacent, or (iii) $|N_G(s) \cap N_G(t)| > k$.*

However, if $\tau \geq 2$, τ-Rendezvous in Time becomes hard.

Theorem 3 (*). Rendezvous *and* τ-Rendezvous in Time *for every constant* $\tau \geq 2$ *are* co-W[2]-*hard when parameterized by* k. *Moreover, these problems cannot be solved in time* $f(k) \cdot n^{o(k)}$ *unless* ETH *fails.*

We prove Theorem 3 by a polynomial reduction from Set Cover. Since Set Cover is NP-complete (see [15]), we obtain the following corollary using Observation 1.

Corollary 2. τ-Rendezvous in Time *is* co-NP-*complete for every fixed constant* $\tau \geq 2$.

Using the reduction from the proof of Theorem 3, we can conclude that Rendezvous and τ-Rendezvous in Time, for $\tau \geq 2$, are co-NP-hard. However, the general problems are harder. By a reduction from the Quantified Boolean Formula in Conjunctive Normal Form problem with alternating quantifiers that is well-known to be PSPACE-complete (see, e.g., [15]), we show that our games are PSPACE-hard.

Theorem 4 (*). Rendezvous *and* Rendezvous in Time *are* PSPACE-*hard.*

5 Rendezvous in Time for Graphs of Bounded Neighborhood Diversity

In this section, we show that Rendezvous in Time is FPT when parameterized by τ and the neighborhood diversity of the input graph.

The notion of neighborhood diversity was introduced by Lampis in [20]. It is convenient for us to define this notion in terms of modules. Let G be a graph. A set of vertices $U \subseteq V(G)$ is a *module* if for every $v \in V(G) \setminus U$, either $N_G(v) \cap U = \emptyset$ or $U \subseteq N_G(v)$. It is said that is a module U is a *clique module* if U is a clique, and U is an *independent module* if U is an independent set. We say that a partition $\{U_1, \ldots, U_\ell\}$ of $V(G)$ into clique and independent modules is a *neighborhood decomposition*. The *neighborhood diversity* of a graph G is the minimum ℓ such that G has a neighborhood decomposition with ℓ modules; we use $\mathsf{nd}(G)$ to denote the neighborhood diversity of G. The value of $\mathsf{nd}(G)$ and the corresponding partition of $V(G)$ into clique and independent modules can be computed in polynomial (linear) time [20].

Theorem 5. Rendezvous in Time *can be solved in* $2^{\ell^{\mathcal{O}(\tau)}} \cdot n^{\mathcal{O}(1)}$ *time on graphs of neighborhood diversity* ℓ.

The theorem is proved by giving an FPT Turing reduction of Rendezvous in Time to Integer Linear Programming Feasibility such that each constructed instance has $\ell^{\mathcal{O}(\tau)}$ variables and the bit-size $\ell^{\mathcal{O}(\tau)} \cdot \log n$. Then we apply the celebrated results of Lenstra [21] (see also [13,18] for further improvements) about Integer Linear Programming Feasibility parameterized by the number of variables to solve the problem.

6 Conclusion

We initiated the study of the Rendezvous Game with Adversaries on graphs. We proved that in several cases, the dynamic separation number $d_G(s,t)$, the minimum number of agents needed for Divider to win against Facilitator, could be equal to the minimum size $\lambda_G(s,t)$ of an (s,t)-separator in G. In particular, this equality holds on for P_5-free and chordal graphs. In general, the difference $\lambda_G(s,t) - d_G(s,t)$ could be arbitrary large. Are there other natural graph classes with this property? Is it is possible to characterize hereditary graph classes for which the equality holds?

Further, we investigated the computational complexity of Rendezvous and Rendezvous in Time. Both problems can be solved it $n^{\mathcal{O}(k)}$ time. However, they are co-W[2]-hard when parameterized by k and cannot be solved in $n^{o(k)}$ time unless FPT = W[1]. In fact, τ-Rendezvous in Time is co-W[2]-hard for every $\tau \geq 2$. We also proved that Rendezvous and Rendezvous in Time are PSPACE-hard. We conjecture that these twoproblems are EXPTIME-complete.

Finally, we initiated the study of the complexity of Rendezvous and Rendezvous in Time under structural parameterization of the input graphs. We proved that Rendezvous in Time is FPT when parameterized by the neighborhood diversity of the input graph and τ. Can this result be generalized for the parameterization by *modular width* (see, e.g., [14] for the definition and the discussion of this parameterization) and τ? Is Rendezvous in Time FPT when parameterized by the neighborhood diversity only? The same question is open for Rendezvous. We believe that this problem is interesting even for the more restrictive parameterization by the vertex cover number. Another question is about Rendezvous and Rendezvous in Time parameterized by the treewidth of a graph. Are the problems FPT or XP for this parameterization? Notice that if the initial positions s and t are not is the same bag of a tree decomposition of width w, then the upper bound for the dynamic separation number by $\lambda_G(s,t)$ together with Theorem 1 imply that the problems can be solved it time $n^{\mathcal{O}(w)}$. Can the problems be solved in this time if s and t are in the same bag?

References

1. Aigner, M., Fromme, M.: A game of cops and robbers. Discrete Appl. Math. **8**(1), 1–11 (1984)
2. Alpern, S.: The rendezvous search problem. SIAM J. Control Optim. **33**(3), 673–683 (1995). https://doi.org/10.1137/S0363012993249195

3. Alpern, S., Gal, S.: The Theory of Search Games and Rendezvous. International Series in Operations Research & Management Science, vol. 55, Kluwer Academic Publishers, Boston (2003)
4. Bonato, A., Nowakowski, R.J.: The Game of Cops and Robbers on Graphs, Student Mathematical Library, vol. 61. American Mathematical Society, Providence (2011). https://doi.org/10.1090/stml/061
5. Bonato, A., Yang, B.: Graph searching and related problems. In: Handbook of Combinatorial Optimization, pp. 1511–1558 (2013)
6. Cygan, M., et al.: Parameterized Algorithms. Springer, Cham (2015). https://doi.org/10.1007/978-3-319-21275-3
7. Diestel, R.: Graph Theory, 4th edn, Graduate texts in mathematics, vol. 173. Springer (2012)
8. Downey, R.G., Fellows, M.R.: Fundamentals of Parameterized Complexity. TCS, Springer, London (2013). https://doi.org/10.1007/978-1-4471-5559-1
9. Fomin, F.V., Golovach, P.A., Hall, A., Mihalák, M., Vicari, E., Widmayer, P.: How to guard a graph? Algorithmica **61**(4), 839–856 (2010). https://doi.org/10.1007/s00453-009-9382-4
10. Fomin, F.V., Golovach, P.A., Lokshtanov, D.: Guard games on graphs: Keep the intruder out! Theor. Comput. Sci. **412**(46), 6484–6497 (2011). https://doi.org/10.1016/j.tcs.2011.08.024
11. Fomin, F.V., Golovach, P.A., Thilikos, D.M.: Can Romeo and Juliet meet? or rendezvous games with adversaries on graphs. CoRR abs/2102.13409 (2021), http://arxiv.org/abs/2102.13409
12. Fomin, F.V., Thilikos, D.M.: An annotated bibliography on guaranteed graph searching. Theor. Comput. Sci. **399**(3), 236–245 (2008). https://doi.org/10.1016/j.tcs.2008.02.040
13. Frank, A., Tardos, É.: An application of simultaneous diophantine approximation in combinatorial optimization. Combinatorica **7**(1), 49–65 (1987). https://doi.org/10.1007/BF02579200
14. Gajarský, J., Lampis, M., Ordyniak, S.: Parameterized algorithms for modular-width. In: Gutin, G., Szeider, S. (eds.) IPEC 2013. LNCS, vol. 8246, pp. 163–176. Springer, Cham (2013). https://doi.org/10.1007/978-3-319-03898-8_15
15. Garey, M.R., Johnson, D.S.: Computers and Intractability: A Guide to the Theory of NP-Completeness. W.H. Freeman, New York (1979)
16. Impagliazzo, R., Paturi, R.: On the complexity of k-sat. J. Comput. Syst. Sci. **62**(2), 367–375 (2001). https://doi.org/10.1006/jcss.2000.1727
17. Impagliazzo, R., Paturi, R., Zane, F.: Which problems have strongly exponential complexity? J. Comput. Syst. Sci. **63**(4), 512–530 (2001). https://doi.org/10.1006/jcss.2001.1774
18. Kannan, R.: Minkowski's convex body theorem and integer programming. Math. Oper. Res. **12**(3), 415–440 (1987). https://doi.org/10.1287/moor.12.3.415
19. Kinnersley, W.B.: Cops and robbers is exptime-complete. J. Comb. Theory, Ser. B **111**, 201–220 (2015). https://doi.org/10.1016/j.jctb.2014.11.002
20. Lampis, M.: Algorithmic meta-theorems for restrictions of treewidth. Algorithmica **64**(1), 19–37 (2012). https://doi.org/10.1007/s00453-011-9554-x
21. Lenstra, H.W.: Integer programming with a fixed number of variables. Math. Oper. Res. **8**(4), 538–548 (1983). https://doi.org/10.1287/moor.8.4.538
22. Nagamochi, H.: Cop-robber guarding game with cycle robber-region. Theor. Compu. Sci. **412**(4–5), 383–390 (2011)
23. Nowakowski, R., Winkler, P.: Vertex-to-vertex pursuit in a graph. Discrete Math. **43**(2–3), 235–239 (1983)

24. Quilliot, A.: A short note about pursuit games played on a graph with a given genus. J. Comb. Theory, Ser. B **38**(1), 89–92 (1985)
25. Sámal, R., Valla, T.: The guarding game is E-complete. Theor. Comput. Sci. **521**, 92–106 (2014). https://doi.org/10.1016/j.tcs.2013.11.034
26. Ta-Shma, A., Zwick, U.: Deterministic rendezvous, treasure hunts, and strongly universal exploration sequences. ACM Trans. Algorith. **10**(3), 12:1–12:15 (2014). https://doi.org/10.1145/2601068
27. Williamson, D.: Network Flow Algorithms. Cambridge University Press, Cambridge (2019)

Beyond Helly Graphs: The Diameter Problem on Absolute Retracts

Guillaume Ducoffe(✉)

University of Bucharest and National Institute for Research and Development in Informatics, Bucureşti, Romania
guillaume.ducoffe@ici.ro

Abstract. A subgraph H of a graph G is called a retract of G if it is the image of some idempotent endomorphism of G. We say that H is an absolute retract of some graph class $\mathcal{C}$ if it is a retract of any $G \in \mathcal{C}$ of which it is an isochromatic and isometric subgraph. In this paper, we study the complexity of computing the diameter within the absolute retracts of various hereditary graph classes. First, we show how to compute the diameter within absolute retracts of bipartite graphs in randomized $\tilde{\mathcal{O}}(m\sqrt{n})$ time. Even on the proper subclass of cube-free modular graphs it is, to our best knowledge, the first subquadratic-time algorithm for diameter computation. For the special case of chordal bipartite graphs, it can be improved to linear time, and the algorithm even computes all the eccentricities. Then, we generalize these results to the absolute retracts of k-chromatic graphs, for every $k \geq 3$. Finally, we study the diameter problem within the absolute retracts of planar graphs and split graphs.

Keywords: Absolute retract · Chordal bipartite graphs · Split graphs · Planar graphs · Diameter computation

1 Introduction

One of the most basic graph properties is the diameter of a graph (maximum number of edges on a shortest path). It is a rough estimate of the maximum delay in order to send a message in a communication network [32], but it also got used in the literature for various other purposes [2,74]. The complexity of computing the diameter has received tremendous attention in the Graph Theory community [1,14,18,20,24–26,29–31,34,41,43–45,47,49,65]. Indeed, while this can be done in $\mathcal{O}(nm)$ time for any n-vertex m-edge graph, via a simple reduction to breadth-first search, breaking this quadratic barrier (in the size $n+m$ of the input) happens to be a challenging task. In fact, under plausible complexity assumptions such as the Strong Exponential-Time Hypothesis (SETH), the

This work was supported by project PN-19-37-04-01 "New solutions for complex problems in current ICT research fields based on modelling and optimization", funded by the Romanian Core Program of the Ministry of Research and Innovation (MCI) 2019–2022.

Ł. Kowalik et al. (Eds.): WG 2021, LNCS 12911, pp. 321–335, 2021.
https://doi.org/10.1007/978-3-030-86838-3_25

optimal running time for computing the diameter is essentially in $\mathcal{O}(nm)$—up to sub-polynomial factors [71]. This negative result holds even if we restrict ourselves to bipartite graphs or split graphs [1,13]. However, on the positive side, several recent works have identified important graph classes for which we can achieve for the diameter problem $\mathcal{O}(m^{2-\epsilon})$ time, or even better $\mathcal{O}(mn^{1-\epsilon})$ time, for some $\epsilon > 0$. Next, we focus on a few such classes that are most relevant to our work. Specifically, we call $G = (V, E)$ a Helly graph if every family of pairwise intersecting balls of G (of arbitrary radius and center) have a nonempty common intersection. The Helly graphs are a broad generalization of many better-known graph classes in Structural Graph Theory, such as: trees, interval graphs, strongly chordal graphs and dually chordal graphs [4]. Furthermore, a celebrated theorem in Metric Graph Theory is that every graph is an isometric (distance-preserving) subgraph of some Helly graph [40,57]. Other properties of Helly graphs were also thoroughly investigated in prior works [7,8,10,23,35–38,63,69,70]. In particular, as far as we are concerned here, there is a randomized $\tilde{\mathcal{O}}(m\sqrt{n})$-time algorithm in order to compute the diameter within n-vertex m-edge Helly graphs [43].

Recall that an endomorphism of a graph G is an edge-preserving mapping of G to itself. A retraction is an idempotent endomorphism. If H is the image of G by some retraction (in particular, H is a subgraph of G) then, we call H a retract of G. The notion of retract has applications in some discrete facility location problems [56], and it is useful in characterizing some important graph classes. For instance, the median graphs are exactly the retracts of hypercubes [3]. We here focus on the relation between retracts and Helly graphs, that is as follows (for other classes related to the Helly graphs and considered recently, see [15,21,28,39,42,45]). For some class $\mathcal{C}$ of reflexive graphs (*i.e.*, with a loop at every vertex), let us define the absolute retracts of $\mathcal{C}$ as those H such that, whenever H is an isometric subgraph of some $G \in \mathcal{C}$, H is a retract of G. Absolute retracts find their root in Geometry, where they got studied for various metric spaces [60]. In the special case of the class of all reflexive graphs, the absolute retracts are exactly the Helly (reflexive) graphs [55]. Motivated by this characterization of Helly graphs, and the results obtained in [43] for the diameter problem on this graph class, we here consider the following notion of absolute retracts, for irreflexive graphs. – Unless stated otherwise, all graphs considered in this paper are irreflexive. – Namely, let us first recall that a subgraph H of a graph G is isochromatic if it has the same chromatic number as G. Then, given a class of (irreflexive) graphs $\mathcal{C}$, the absolute retracts of $\mathcal{C}$ are those H such that, whenever H is an isometric and isochromatic subgraph of some $G \in \mathcal{C}$, H is a retract of G. We refer the reader to [5,6,9,54,56,59,61,64,66–68], where this notion got studied for various graph classes.

Our Results. In this paper, we prove new structural and algorithmic properties of the absolute retracts of various hereditary graph classes, such as: bipartite graphs, k-chromatic graphs (for any $k \geq 3$), split graphs and planar graphs. Our focus is about the diameter problem on these graph classes but, on our way, we uncover several nice properties of the shortest-path distribution of their absolute retracts, that may be of independent interest.

First, in Sect. 2, we consider the absolute retracts of bipartite graphs and some important subclasses of the latter. Recall that the diameter of a bipartite graph can unlikely be computed in subquadratic time. We prove that the diameter of absolute bipartite retracts can be computed in $\tilde{\mathcal{O}}(m\sqrt{n})$ time (Theorem 2). For that, we observe that in the square of such graph G, its two partite sets induce Helly graphs. This result complements the known relations between Helly graphs and absolute retracts of bipartite graphs [6]. Then, roughly, we show how to compute the diameter of G from the diameter of both Helly graphs (actually, from the knowledge of the peripheral vertices in these graphs, *i.e.*, their vertices with maximal eccentricity). Absolute bipartite retracts properly contain all cube-free modular graphs, and so, the cube-free median graphs and chordal bipartite graphs [5]. Therefore, as a byproduct of our Theorem 2, we get the first truly subquadratic-time algorithm for computing the diameter within the cube-free modular graphs. However, the structure of absolute bipartite retracts is far more complex than cube-free modular graphs: in fact, every bipartite graph is an isometric subgraph of some absolute bipartite retract [66].

Recently [39], we announced an $\mathcal{O}(m\sqrt{n})$-time algorithm in order to compute all the eccentricities in a Helly graph. However, extending this result to the absolute retracts of bipartite graphs appears to be a more challenging task. We manage to do so for the subclass of chordal bipartite graphs, for which we achieve a linear-time algorithm in order to compute all the eccentricities. For that, we use the stronger result that in the square of such graph, its two partite sets induce strongly chordal graphs.

In Sect. 3, we generalize our above framework to the absolute retracts of k-chromatic graphs, for any $k \geq 3$. Notice that we are not aware of any prior work showing the usefulness of (efficiently computable) proper colorings for faster diameter computation. Our positive results in Sects. 2 and 3 rely on some Helly-type properties of the graph classes considered. We complement those with a hardness result in Sect. 4, that hints that the weaker property of being an absolute retract of some well-structured graph class is not sufficient on its own for faster diameter computation. Specifically, we prove that under SETH, there is no $\mathcal{O}(mn^{1-\epsilon})$-time algorithm for the diameter problem, for any $\epsilon > 0$, on the class of absolute retracts of split graphs. This negative result follows from an elegant characterization of this subclass of split graphs in [59].

Finally, in Sect. 5, we consider the absolute planar retracts. While there now exist several truly subquadratic-time algorithms for the diameter problem on all planar graphs [20,45,49] the existence of a quasi linear-time algorithm for this problem has remained so far elusive, and it is sometimes conjectured that no such algorithm exists [20]. We give evidence that finding such algorithm for the absolute retracts of planar graphs is already a hard problem on its own. Specifically, we prove that every planar graph is an isometric subgraph of some absolute retract of planar graphs. This result mirrors the aforementioned property that every graph isometrically embeds in a Helly graph [40,57].

Let us mention that all graph classes considered here are polynomial-time recognizable. For all that, we do *not* need to execute these recognition algorithms before we can compute the diameter of these graphs.

Notations. We mostly follow the graph terminology from [12,33]. All graphs considered are finite, simple, unweighted and connected. For a graph $G = (V, E)$, let the (open) neighbourhood of a vertex v be defined as $N_G(v) = \{u \in V \mid uv \in E\}$ and its closed neighbourhood as $N_G[v] = N_G(v) \cup \{v\}$. Similarly, for a vertex-subset $S \subseteq V$, let $N_G(S) = \bigcup_{v \in S} N_G(v) \setminus S$, and let $N_G[S] = N_G(S) \cup S$. The distance between two vertices $u, v \in V$ equals the minimum number of edges on a uv-path, and it is denoted $d_G(u, v)$. Let $I_G(u, v) = \{w \in V \mid d_G(u, v) = d_G(u, w) + d_G(w, v)\}$. The ball of center v and radius r is defined as $N_G^r[v] = \{u \in V \mid d_G(u, v) \leq r\}$. Furthermore, let the eccentricity of a vertex v be defined as $e_G(v) = \max_{u \in V} d_G(u, v)$. The diameter and the radius of a graph G are defined as $diam(G) = \max_{v \in V} e_G(v)$ and $rad(G) = \min_{v \in V} e_G(v)$, respectively. A vertex $v \in V$ is called central if $e_G(v) = rad(G)$, and peripheral if $e_G(v) = diam(G)$. We introduce additional terminology where it is needed throughout the paper.

2 Bipartite Graphs

The study of the absolute retracts of bipartite graphs dates back from Hell [53], and since then many characterizations of this graph class were proposed [5]. This section is devoted to the diameter problem on this graph class. In Sect. 2.1, we propose a randomized $\tilde{\mathcal{O}}(m\sqrt{n})$-time algorithm for this problem. Then, we consider the chordal bipartite graphs in Sect. 2.2, that have been proved in [5] to be a subclass of the absolute retracts of bipartite graphs. For the chordal bipartite graphs, we present a deterministic linear-time algorithm in order to compute all the eccentricities. Before going further, let us introduce a few additional terminology. For a connected bipartite graph G, we denote its two partite sets by V_0 and V_1. A half-ball is the intersection of a ball with one of the two partite sets of G. Finally, for $i \in \{0, 1\}$, let H_i be the graph with vertex-set V_i and an edge between every two vertices with a common neighbour in G.

2.1 Faster Diameter Computation

We start with the following characterization of the absolute bipartite retracts:

Theorem 1 ([5]). *$G = (V, E)$ is an absolute retract of bipartite graphs if and only if the collection of half-balls of G satisfies the Helly property.*

This above Theorem 1 leads us to the following simple observation about the internal structure of the absolute retracts of bipartite graphs:

Lemma 1. *If $G = (V_0 \cup V_1, E)$ is an absolute retract of bipartite graphs then both H_0 and H_1 are Helly graphs.*

Next, we prove that in order to compute $diam(G)$, with G an absolute retract of bipartite graphs, it is sufficient to compute the peripheral vertices of the Helly graphs H_0 and H_1.

Lemma 2. *If $G = (V_0 \cup V_1, E)$ is an absolute bipartite retract such that $diam(H_0) \leq diam(H_1)$ then, $diam(G) \in \{2diam(H_1), 2diam(H_1)+1\}$. Moreover, if $diam(G) \geq 3$ then we have $diam(G) = 2diam(H_1)+1$ if and only if:*

- *$diam(H_1) = 1$;*
- *or $diam(H_0) = diam(H_1)$ and, for some $i \in \{0,1\}$, there exists a peripheral vertex of H_i whose all neighbours in G are peripheral vertices of H_{1-i}.*

The remaining of Sect. 2.1 is devoted to the computation of all the peripheral vertices in both Helly graphs H_0 and H_1. While there exists a truly subquadratic-time algorithm for computing the diameter of a Helly graph [43], we observe that in general, we cannot compute H_0 and H_1 in truly subquadratic time from G. Next, we adapt [43, Theorem 2], for the Helly graphs, to our needs.

Lemma 3. *If $G = (V_0 \cup V_1, E)$ is an absolute bipartite retract then, for any k, we can compute in $\mathcal{O}(km)$ time the set of vertices of eccentricity at most k in H_0 (resp., in H_1).*

Proof (Sketch). By symmetry, we only need to prove the result for H_0. Let $U = \{v \in V_0 \mid e_{H_0}(v) \leq k\}$ be the set to be computed. We consider the more general problem of computing, for any t, a partition $\mathcal{P}_t = (A_1^t, A_2^t, \ldots, A_{p_t}^t)$ of V_0, in an arbitrary number p_t of subsets, subject to the following constraints:

- For every $1 \leq i \leq p_t$, let $C_i^t := \bigcap_{v \in A_i^t} N_G^t[v]$. Let $B_i^t := C_i^t \cap V_0$ if t is even and let $B_i^t := C_i^t \cap V_1$ if t is odd (for short, $B_i^t = C_i^t \cap V_{t \pmod 2}$). *We impose the sets B_i^t to be <u>nonempty</u> and <u>pairwise disjoint</u>.*

Indeed, under these two conditions above, we have $U \neq \emptyset$ if and only if, for any partition $\mathcal{P}_{2k}$ as described above, $p_{2k} = 1$. Furthermore if it is the case then $U = B_1^{2k}$. To construct the desired partition, we proceed by induction over t. If $t = 0$ then, let $V_0 = \{v_1, v_2, \ldots, v_{p_0}\}$. We just set $\mathcal{P}_0 = (\{v_0\}, \{v_1\}, \ldots, \{v_{p_0}\})$ (each set is a singleton), and for every $1 \leq i \leq p_0$ let $B_i^0 = A_i^0 = \{v_i\}$. Else, we construct $\mathcal{P}_t$ from $\mathcal{P}_{t-1}$. Specifically, for every $1 \leq i \leq p_{t-1}$, we let $W_i^t := N_G(B_i^{t-1})$. Then, starting from $j := 0$ and $\mathcal{F} := \mathcal{P}_{t-1}$, we proceed as follows until we have $\mathcal{F} = \emptyset$. We pick a vertex u s.t. $\#\{i \mid A_i^{t-1} \in \mathcal{F},\ u \in W_i^t\}$ is maximized (the maximality of u ensures that all sets B_i^t will be pairwise disjoint). Then, we set $A_j^t := \bigcup\{A_i^{t-1} \mid A_i^{t-1} \in \mathcal{F},\ u \in W_i^t\}$ and $B_j^t := \bigcap\{W_i^t \mid A_i^{t-1} \in \mathcal{F},\ u \in W_i^t\}$. We add the new subset A_j^t to $\mathcal{P}_t$, we remove all the subsets $A_i^{t-1}, u \in W_i^t$ from $\mathcal{F}$, then we set $j := j+1$. Overall, by using standard lists and pointer structures, each inductive step takes $\mathcal{O}(n+m)$ time.

The base case of our above induction is trivially correct. In order to prove correctness of our inductive step, we use Theorem 1 in order to prove that for each $1 \leq i \leq p_t$ we get $W_i^t = V_{t \pmod 2} \cap \left(\bigcap_{v \in A_i^{t-1}} N_G^t[v]\right)$. Doing so, for each subset A_j^t created at step t, we have $B_j^t = V_{t \pmod 2} \cap \left(\bigcap_{v \in A_j^t} N_G^t[v] \right)$, as desired. Finally, observe that all the subsets B_j^t are nonempty since they at least contain the vertex $u \in V_{t \pmod 2}$ that is selected in order to create A_j^t. □

We use Lemma 3 when the diameters of H_0 and H_1 are in $\mathcal{O}(\sqrt{n})$. For larger values of diameters, we use a randomized procedure.

Lemma 4 (Theorem 3 in [43]**).** *For a Helly graph H s.t. $diam(H) > 3k = \omega(\log |V(H)|)$, one can compute with high probability its diameter and all the peripheral vertices in $\tilde{\mathcal{O}}(|E(H)| \cdot |V(H)|/k)$ time.*

It is important to note that, in the algorithmic procedure of Lemma 4, we just need to perform a BFS from randomly selected vertices. As any BFS in H_0 or H_1 can be simulated with a BFS in G, we can implement this procedure in order to compute $diam(H_i)$, for $i \in \{0, 1\}$, in $\tilde{\mathcal{O}}(mn/diam(H_i))$ time with high probability. Combined with Lemma 3, we get:

Theorem 2. *If $G = (V_0 \cup V_1, E)$ is an absolute retract of bipartite graphs then, with high probability, we can compute $diam(G)$ in $\tilde{\mathcal{O}}(m\sqrt{n})$ time.*

We suspect that Theorem 2 can be derandomized by using a recent technique from [39, Theorem 3]. This is left for future work.

2.2 Chordal Bipartite Graphs

We improve Theorem 2 for the special case of chordal bipartite graphs. Recall (amongst many characterizations) that a bipartite graph is chordal bipartite if and only if every induced cycle has length four [51]. It was proved in [5] that every chordal bipartite graph is an absolute retract of bipartite graphs.

Theorem 3. *If $G = (V, E)$ is chordal bipartite then we can compute all the eccentricities (and so, the diameter) in linear time.*

We subdivide our proof of Theorem 3 into four main steps.

The Chordal Structure of the Partite Sets. A graph is chordal if it has no induced cycle of length more than three. It is strongly chordal if it is chordal and it does not contain any n-sun ($n \geq 3$) as an induced subgraph [46]. We use the following characterization of the half-sets of chordal bipartite graphs:

Lemma 5 ([62])**.** *If $G = (V_0 \cup V_1, E)$ is chordal bipartite, then H_0 and H_1 are strongly chordal graphs.*

Computation of a Clique-Tree. The same as in Sect. 2.1, in general we cannot compute H_0 and H_1 from G in subquadratic time. In order to overcome this issue, we use a more compact representation of the latter. Specifically, for a graph $H = (V, E)$, a clique-tree is a tree T whose nodes are the maximal cliques of H and such that, for every $v \in V$, the maximal cliques of H containing v induce a connected subtree T_v of T. It is well-known that H is chordal if and only if it has a clique-tree [19,48,73]. By using standard results on dual hypertrees [11,72], we obtain that:

Lemma 6. *If $G = (V_0 \cup V_1, E)$ is chordal bipartite then, we can compute a clique-tree for H_0 and H_1 in linear time.*

Computation of all the Eccentricities in the Partite Sets. Next, we propose a new algorithm in order to compute all the eccentricities of a strongly chordal graph H, being given a clique-tree. There already exist linear-time algorithms for computing all the eccentricities of a strongly chordal graph, being given by its adjacency list [17,39,43]. However, in general these algorithms do not run in time linear in the size of a clique-tree. We often use in our proof the *clique-vertex incidence graph* of H, *i.e.*, the bipartite graph I_H whose partite sets are the vertices and the maximal cliques of H, and such that there is an edge between every vertex of H and every maximal clique of H containing it.

Let us first recall the following result about Helly graphs:

Lemma 7 ([35]). *If H is Helly then, for every vertex v we have $e_H(v) = d_H(v, C(H)) + rad(H)$, where $C(H)$ denotes the set of central vertices of H.*

Hence, by Lemma 7, we are left computing $C(H)$. It starts with computing one central vertex. Define, for every vertex v and vertex-subset C, $d_H(v, C) = \min_{c \in C} d_H(v, c)$. Following [27], we call a set C gated if, for every $v \notin C$, there exists a vertex $v^* \in N_H^{d_H(v,C)-1}[v] \cap (\bigcap\{N_H(c) \mid c \in C,\ d_H(v,c) = d_H(v,C)\})$ (such vertex v^* is called a gate of v).

Lemma 8 ([22]). *Every clique in a chordal graph is a gated set.*

Lemma 9 ([43]). *If T is a clique-tree of a chordal graph H then, for every clique C of H, for every $v \notin C$ we can compute $d_H(v, C)$ and a corresponding gate v^* in total $\mathcal{O}(w(T))$ time, where $w(T)$ denotes the sum of cardinalities of all the maximal cliques of H.*

For every $u, v \in V$ and $k \leq d_H(u, v)$, the set $L_H(u, k, v) = \{x \in I_H(u, v) \mid d_H(u, x) = k\}$ is called a slice. We also need the following result:

Lemma 10 ([22]). *Every slice in a chordal graph is a clique.*

Now, consider the procedure described in Algorithm 1.

Lemma 11 (special case of Theorem 5 in [43]**).** *Algorithm 1 outputs a central vertex of H.*

By using dynamic programming on a clique-tree in order to compute, for each candidate vertex $c \in C$, its number of neighbours in S, we get:

Lemma 12. *If T is a clique-tree of a strongly chordal graph H then, we can implement Algorithm 1 in order to run in $\mathcal{O}(w(T))$ time, where $w(T)$ denotes the sum of cardinalities of all the maximal cliques of H.*

We need one more result about the center of strongly chordal graphs:

Lemma 13 ([35,36]). *If H is strongly chordal then, its center $C(H)$ induces a strongly chordal graph of radius ≤ 1.*

Algorithm 1. Computation of a central vertex.

Require: A strongly chordal graph H.
1: $v \leftarrow$ an arbitrary vertex of H
2: $u \leftarrow$ a furthest vertex from v, *i.e.*, $d_H(u,v) = e_H(v)$
3: $w \leftarrow$ a furthest vertex from u, *i.e.*, $d_H(u,w) = e_H(u)$
4: **for all** $r \in \{\lceil e_H(u)/2 \rceil, \lceil (e_H(u)+1)/2 \rceil, 1 + \lceil e_H(u)/2 \rceil\}$ **do**
5: Set $C := L(w,r,u)$ //*C is a clique by Lemma 10*
6: **for all** $v \notin C$ **do**
7: Compute $d_H(v,C)$ and a corresponding gate v^* //*whose existence follows from Lemma 8*
8: Set $S := \{v^* \mid d_H(v,C) = r\}$ //*gates of vertices at max. distance from C*
9: **for all** $c \in C$ **do**
10: **if** $S \subseteq N_H(c)$ **then**
11: **return** c

By Lemma 13, given a central vertex c of H, we can compute $C(H)$ by local search in the neighbourhood at distance two around c. For doing that efficiently, we also need the following nice characterization of strongly chordal graphs. Recall that the *clique-vertex incidence graph* of H is a bipartite graph whose partite sets are the vertices and the maximal cliques of H, respectively; there is an edge between every vertex and every maximal clique in which this vertex is contained.

Lemma 14 ([16,46]). *H is strongly chordal if and only if its clique-vertex incidence graph I_H is chordal bipartite.*

By Lemma 14, we can apply the techniques of Sect. 2.1 to the clique-vertex incidence graph of any strongly chordal H. In particular, by combining Lemma 3 with the dynamic programming technique of Lemma 12, we obtain:

Proposition 1. *If T is a clique-tree of a strongly chordal graph $H = (V,E)$ then, we can compute its center $C(H)$ in $\mathcal{O}(w(T))$ time.*

Computation of all the eccentricities in G. Before proving Theorem 3, we need a final ingredient. Let us first generalize Lemma 2 as follows.

Lemma 15. *If $G = (V_0 \cup V_1, E)$ is an absolute retract of bipartite graphs then, the following holds for every $i \in \{0,1\}$ and $v \in V_i$:*

- *If $e_{H_i}(v) \leq rad(H_{1-i}) - 1$ then, $e_G(v) = 2e_{H_i}(v) + 1 = 2rad(H_{1-i}) - 1$.*
- *If $e_{H_i}(v) = rad(H_{1-i})$ then, $e_G(v) = 2rad(H_{1-i})$ if and only if $N_G(v) \subseteq C(H_{1-i})$ and, for every $u \in V_{1-i}$, we have $d_{H_{1-i}}(u, N_G(v)) \leq rad(H_{1-i}) - 1$ (otherwise, $e_G(v) = 2rad(H_{1-i}) + 1$).*
- *If $e_{H_i}(v) \geq rad(H_{1-i}) + 1$ then, $e_G(v) = 2e_{H_i}(v)$ if and only if we have $e_{H_{1-i}}(u) < e_{H_i}(v)$ for some $u \in N_G(v)$ (otherwise, $e_G(v) = 2e_{H_i}(v) + 1$).*

Of the three cases in the above Lemma 15, the real algorithmic challenge is the case $e_{H_i}(v) = rad(H_{1-i})$, for some $i \in \{0,1\}$. We solve this case by using similar techniques as for Proposition 1, which concludes the proof of Theorem 3.

3 k-Chromatic Graphs

Recall that a proper k-coloring of $G = (V, E)$ is any mapping $c : V \rightarrow \{1, 2, \ldots, k\}$ such that $c(u) \neq c(v)$ for every edge $uv \in E$. The chromatic number of G is the least k such that it has a proper k-coloring, and a k-chromatic graph is a graph whose chromatic number is equal to k. We study the diameter problem within the absolute retracts of k-chromatic graphs, for every $k \geq 3$.

Our approach requires such graphs to be equipped with a proper k-coloring. While this is a classic NP-hard problem for every $k \geq 3$ [58], it can be done in polynomial time for absolute retracts of k-chromatic graphs [9]. By using a standard greedy coloring approach, we first improve this result as follows:

Proposition 2. *There is a linear-time algorithm such that, for every $k \geq 3$, if the input G is an absolute retract of k-chromatic graphs, then it computes a proper k-coloring of G.*

In the remainder of the section, we always assume the input graph G to be given with a proper k-coloring. We sometimes use the fact that, for an absolute retract, such proper k-coloring is unique up to permuting the colour classes [68]. Now, let us recall the following characterization of absolute retracts:

Theorem 4 ([68]). *Let $k \geq 3$. The graph $G = (V, E)$ is an absolute retract of k-chromatic graphs if and only if for any proper k-coloring c, every peripheral vertex v is adjacent to all vertices u with $c(u) \neq c(v)$, or it is covered*[1] *and $G \setminus v$ is an absolute retract of k-chromatic graphs.*

A special case of Theorem 4 leads to a linear-time algorithm in order to decide whether an absolute k-chromatic retract has diameter at most two. For those graphs with diameter at least three, we propose a generalization of Lemma 2. Specifically, for each colour i, let $V_i := \{v \in V \mid c(v) = i\}$ be called a colour class. For every $v \in V_i$, $e_i(v) := \max\{d_G(u, v) \mid u \in V_i\}$. A vertex $v \in V_i$ is i-peripheral if it maximizes $e_i(v)$. Finally, let $d_i := \max\{e_i(v) \mid v \in V_i\}$.

Lemma 16. *Let $G = (V, E)$ be an absolute retract of k-chromatic graphs for some $k \geq 3$, and let c be a corresponding proper k-coloring. Then, $\max_{1 \leq i \leq k} d_i \leq diam(G) \leq 1 + \max_{1 \leq i \leq k} d_i$. Moreover, if $diam(G) \geq 3$, then we have $diam(G) = 1 + \max_{1 \leq i \leq k} d_i$ if and only if:*

- *either $\max_{1 \leq i \leq k} d_i = 2$;*
- *or, for some $i \neq j$ s.t. $d_i = d_j$ is maximized, there is some i-peripheral vertex whose all neighbours coloured j are j-peripheral.*

We end up sketching the computation, for each colour i, of the value d_i and of the i-peripheral vertices. Our strategy is as follows. First, we prove that we can reduce our study to the case $k = 3$. This is done by using another, more algorithmic, characterization of absolute retracts [9].

[1] A vertex v is covered by another vertex w if $N_G(v) \subseteq N_G(w)$ (a covered vertex is called embeddable in [68]).

Lemma 17. *Let $G = (V, E)$ be an absolute retract of k-chromatic graphs for some $k \geq 3$, and let c be a corresponding proper k-coloring. For every distinct colours i_1, i_2, i_3, the subgraph $H := G[V_{i_1} \cup V_{i_2} \cup V_{i_3}]$ is isometric. Moreover, H is an absolute retract of 3-chromatic graphs.*

Next, we deal with the case when d_i is sufficiently small. For that, we extend the techniques of Lemma 3 to the absolute 3-chromatic retracts. Correctness of our approach follows from the following property of these graphs: if $v_1, v_2, \ldots, v_t$ are vertices coloured i then, for any $r \geq 2$ and *any* colour j, the balls $N_G^r[v_1], N_G^r[v_2], \ldots, N_G^r[v_t]$ intersect in colour j if and only if they also intersect in colour i.

Lemma 18. *Let $G = (V, E)$ be an absolute retract of 3-chromatic graphs, and let c be a corresponding proper 3-coloring. For each colour i and $D \geq 2$, we can compute in $\mathcal{O}(Dm)$ time the set $U_i := \{v \in V_i \mid e_i(u) \leq D\}$.*

Finally, we address the case when d_i is large. A function is called unimodal if every local minimum is also a global minimum. It is known that the eccentricity function of a Helly graph is unimodal [35], and this property got used in [39] in order to compute all the eccentricities in this graph class in subquadratic time. We prove that a similar, but weaker property holds for each colour class of absolute retracts:

Lemma 19. *Let $G = (V, E)$ be an absolute retract of k-chromatic graphs for some $k \geq 3$, and let c be a corresponding proper k-coloring. For each colour i and any $u \in V_i$ s.t. $e_i(u) \geq (d_i+5)/2 \geq 7$, there exists a $u' \in V_i$ s.t. $d_G(u, u') = 2$ and $e_i(u') = e_i(u) - 2$.*

We apply this almost-unimodality property to the computation of the d_i's:

Lemma 20. *Let $G = (V, E)$ be an absolute retract of k-chromatic graphs for some $k \geq 3$, let c be a corresponding proper k-coloring, and let i be such that $d_i \geq 8D + 5 = \omega(\log n)$. Then, with high probability, we can compute in total $\tilde{\mathcal{O}}(mn/D)$ time the value d_i and the i-peripheral vertices.*

By combining Lemmas 17–20, we get:

Theorem 5. *If $G = (V, E)$ is an absolute k-chromatic retract, for some $k \geq 3$, then we can compute its diameter with high probability in $\tilde{\mathcal{O}}(m\sqrt{n})$ time.*

4 Split Graphs

Recall that $G = (V, E)$ is a split graph if its vertex-set V can be partitioned into a clique K and a stable set S. Such partition, that may not be unique, can be computed in linear time [50]. In contrast to Sects. 2 and 3, we prove that:

Theorem 6. *For any $\epsilon > 0$, there exists a $c(\epsilon)$ s.t., under SETH, we cannot compute the diameter in $\mathcal{O}(n^{2-\epsilon})$ time on the absolute retracts of split graphs of order n and clique-number at most $c(\epsilon) \log n$.*

Proof (Sketch). The result holds for general split graphs [13]. Let $G = (K+S, E)$ be any split graph. In order to decide whether $diam(G) \leq 2$ or $diam(G) = 3$, we may remove first all vertices v s.t. $N_G(v) = K \setminus v$ (*i.e.*, because $e_G(v) \leq 2$ and v is simplicial). By applying this above pruning rule until it can no more be done, we get a split graph G' with a unique partition $K' + S'$ [50]. All such graphs are absolute split retracts [59]. □

5 Planar Graphs

Our last (non-algorithmic) section is about the absolute retracts of planar graphs

Theorem 7 ([54]). *A planar graph G is an absolute retract of planar graphs if and only if it is maximal planar and, in an embedding of G in the plane, any triangle bounding a face of G belongs to a subgraph of G isomorphic to K_4.*

To our best knowledge, there has been no relation uncovered between the absolute retracts of planar graphs and other important planar graph subclasses. We make a first step in this direction. Specifically, we prove the following two results.

Proposition 3. *Every planar 3-tree is an absolute retract of planar graphs.*

Theorem 8. *Every connected planar graph is an isometric subgraph of some absolute planar retract. In particular, there are absolute retracts of planar graphs with arbitrarily large treewidth.*

We stress that the proof of Theorem 8 is constructive, and that it leads to a polynomial-time algorithm in order to construct an absolute planar retract in which the input planar graph G isometrically embeds. In contrast to our result, the smallest Helly graph in which a graph G isometrically embeds may be exponential in its size [52].

The existence of an almost linear-time algorithm for computing the diameter of planar graphs is an important open problem. We see our Theorem 8 as evidence that answering to this problem for the absolute planar retracts would be already an important intermediate step toward a full resolution.

References

1. Abboud, A., Vassilevska Williams, V., Wang, J.R.: Approximation and fixed parameter subquadratic algorithms for radius and diameter in sparse graphs. In: SIAM, pp. 377–391 (2016). https://doi.org/10.1137/1.9781611974331.ch28
2. Albert, R., Jeong, H., Barabási, A.L.: Diameter of the world-wide web. Nature **401**(6749), 130–131 (1999)
3. Bandelt, H.J.: Retracts of hypercubes. Journal of graph theory **8**(4), 501–510 (1984)

4. Bandelt, H.J., Chepoi, V.: Metric graph theory and geometry: a survey. Contem. Math. **453**, 49–86 (2008)
5. Bandelt, H.J., Dählmann, A., Schütte, H.: Absolute retracts of bipartite graphs. Discre. Appl. Math. **16**(3), 191–215 (1987)
6. Bandelt, H.J., Farber, M., Hell, P.: Absolute reflexive retracts and absolute bipartite retracts. Discre. Appl. Math. **44**(1–3), 9–20 (1993)
7. Bandelt, H.J., Pesch, E.: Dismantling absolute retracts of reflexive graphs. Eur. J. Combin. **10**(3), 211–220 (1989)
8. Bandelt, H.-J., Pesch, E.: A Radon theorem for Helly graphs. Archiv der Mathematik **52**(1), 95–98 (1989). https://doi.org/10.1007/BF01197978
9. Bandelt, H.J., Pesch, E.: Efficient characterizations of n-chromatic absolute retracts. J. Combin. Theor. Ser. B **53**(1), 5–31 (1991)
10. Bandelt, H.J., Prisner, E.: Clique graphs and Helly graphs. J Combin. Theor. Ser. B **51**(1), 34–45 (1991)
11. Beeri, C., Fagin, R., Maier, D., Yannakakis, M.: On the desirability of acyclic database schemes. J. ACM (JACM) **30**(3), 479–513 (1983)
12. Bondy, J.A., Murty, U.S.R.: Graph Theory, Graduate Texts in Mathematics, vol. 244. Springer-Verlag, London (2008)
13. Borassi, M., Crescenzi, P., Habib, M.: Into the square: On the complexity of some quadratic-time solvable problems. Electr. Notes Theor. Comput. Sci. **322**, 51–67 (2016)
14. Borassi, M., Crescenzi, P., Trevisan, L.: An axiomatic and an average-case analysis of algorithms and heuristics for metric properties of graphs. In: Proceedings of the Twenty-Eighth Annual ACM-SIAM Symposium on Discrete Algorithms (SODA), pp. 920–939. SIAM (2017)
15. Bousquet, N., Thomassé, S.: VC-dimension and Erdős-Pósa property. Discre. Math. **338**(12), 2302–2317 (2015)
16. Brandstädt, A.: Classes of bipartite graphs related to chordal graphs. Discre. Appl. Math. **32**(1), 51–60 (1991)
17. Brandstädt, A., Dragan, F., Chepoi, V., Voloshin, V.: Dually chordal graphs. SIAM J. Discre. Math. **11**(3), 437–455 (1998)
18. Bringmann, K., Husfeldt, T., Magnusson, M.: Multivariate analysis of orthogonal range searching and graph distances. Algorithmica **82**(8), 2292–2315 (2020). https://doi.org/10.1007/s00453-020-00680-z
19. Buneman, P.: A characterisation of rigid circuit graphs. Discre. Math. **9**(3), 205–212 (1974)
20. Cabello, S.: Subquadratic algorithms for the diameter and the sum of pairwise distances in planar graphs. ACM Tran. Algorith. (TALG) **15**(2), 1–38 (2018)
21. Chalopin, J., Chepoi, V., Genevois, A., Hirai, H., Osajda, D.: Helly groups. Tech. Rep. arXiv:2002.06895, arXiv (2020)
22. Chang, G., Nemhauser, G.: The k-domination and k-stability problems on sun-free chordal graphs. SIAM J. Algebr. Discre. Methods **5**(3), 332–345 (1984)
23. Chastand, M., Laviolette, F., Polat, N.: On constructible graphs, infinite bridged graphs and weakly cop-win graphs. Discre. Math. **224**(1–3), 61–78 (2000)
24. Chepoi, V.: On distances in benzenoid systems. J. Chem. Inf. Comput. Sci. **36**(6), 1169–1172 (1996)
25. Chepoi, V., Dragan, F., Estellon, B., Habib, M., Vaxès, Y.: Diameters, centers, and approximating trees of δ-hyperbolic geodesic spaces and graphs. In: Symposium on Computational Geometry (SocG), pp. 59–68. ACM (2008)

26. Chepoi, V., Dragan, F., Vaxès, Y.: Center and diameter problems in plane triangulations and quadrangulations. In: Symposium on Discrete Algorithms (SODA 2002), pp. 346–355 (2002)
27. Chepoi, V., Dragan, F.: A linear-time algorithm for finding a central vertex of a chordal graph. In: van Leeuwen, J. (ed.) ESA 1994. LNCS, vol. 855, pp. 159–170. Springer, Heidelberg (1994). https://doi.org/10.1007/BFb0049406
28. Chepoi, V., Estellon, B., Vaxès, Y.: Covering planar graphs with a fixed number of balls. Discre. Comput. Geom. **37**(2), 237–244 (2007)
29. Corneil, D., Dragan, F., Habib, M., Paul, C.: Diameter determination on restricted graph families. Discre. Appl. Math. **113**(2–3), 143–166 (2001)
30. Coudert, D., Ducoffe, G., Popa, A.: Fully polynomial FPT algorithms for some classes of bounded clique-width graphs. ACM Trans. Algorith. (TALG) **15**(3), 1–57 (2019)
31. Damaschke, P.: Computing giant graph diameters. In: Mäkinen, V., Puglisi, S.J., Salmela, L. (eds.) IWOCA 2016. LNCS, vol. 9843, pp. 373–384. Springer, Cham (2016). https://doi.org/10.1007/978-3-319-44543-4_29
32. De Rumeur, J.: Communications dans les réseaux de processeurs. Masson, Paris (1994)
33. Diestel, R.: Graph Theory. GTM, vol. 173. Springer, Heidelberg (2017). https://doi.org/10.1007/978-3-662-53622-3
34. Dragan, F.: Almost diameter of a house-hole-free graph in linear time via LexBFS. Discr. Appl. Mathe **95**(1–3), 223–239 (1999)
35. Dragan, F.: Centers of graphs and the Helly property. Ph.D. thesis, Moldova State University (1989)
36. Dragan, F.: Domination in quadrangle-free Helly graphs. Cybern. Syst. Anal. **29**(6), 822–829 (1993)
37. Dragan, F., Brandstädt, A.: r-dominating cliques in graphs with hypertree structure. Discre. Math **162**(1–3), 93–108 (1996)
38. Dragan, F., Guarnera, H.: Obstructions to a small hyperbolicity in Helly graphs. Discre. Math **342**(2), 326–338 (2019)
39. Dragan, F., Guarnera, H.: Helly-gap of a graph and vertex eccentricities. Tech. Rep. arXiv:2005.01921, arXiv (2020)
40. Dress, A.: Trees, tight extensions of metric spaces, and the cohomological dimension of certain groups: a note on combinatorial properties of metric spaces. Adv. Math **53**(3), 321–402 (1984)
41. Ducoffe, G.: A new application of orthogonal range searching for computing giant graph diameters. In: Symposium on Simplicity in Algorithms (SOSA) (2019)
42. Ducoffe, G.: Distance problems within Helly graphs and k-Helly graphs. Tech. Rep. arXiv:2011.00001, arXiv preprint (2020)
43. Ducoffe, G., Dragan, F.: A story of diameter, radius and Helly property. Networks **77**(3), 435–453 (2021)
44. Ducoffe, G., Habib, M., Viennot, L.: Fast diameter computation within split graphs. In: Li, Y., Cardei, M., Huang, Y. (eds.) COCOA 2019. LNCS, vol. 11949, pp. 155–167. Springer, Cham (2019). https://doi.org/10.1007/978-3-030-36412-0_13
45. Ducoffe, G., Habib, M., Viennot, L.: Diameter computation on H-minor free graphs and graphs of bounded (distance) VC-dimension. In: Proceedings of the Fourteenth Annual ACM-SIAM Symposium on Discrete Algorithms (SODA), pp. 1905–1922. SIAM (2020)
46. Farber, M.: Characterizations of strongly chordal graphs. Discre. Math. **43**(2–3), 173–189 (1983)

47. Farley, A., Proskurowski, A.: Computation of the center and diameter of outerplanar graphs. Discre. Appl. Math. **2**(3), 185–191 (1980)
48. Gavril, F.: The intersection graphs of subtrees in trees are exactly the chordal graphs. J. Comb. Theory Ser. B **16**(1), 47–56 (1974)
49. Gawrychowski, P., Kaplan, H., Mozes, S., Sharir, M., Weimann, O.: Voronoi diagrams on planar graphs, and computing the diameter in deterministic $\tilde{O}(n^{5/3})$ time. In: Symposium on Discrete Algorithms (SODA), pp. 495–514. SIAM (2018)
50. Golumbic, M.: Algorithmic Graph Theory and Perfect Graphs, vol. 57. Elsevier, Amsterdam (2004)
51. Golumbic, M., Goss, C.: Perfect elimination and chordal bipartite graphs. J. Graph Theor. **2**(2), 155–163 (1978)
52. Guarnera, H., Dragan, F., Leitert, A.: Injective hulls of various graph classes. Tech. Rep. arXiv:2007.14377, arXiv (2020)
53. Hell, P.: Rétractions de graphes. Ph.D. thesis, Thèse (Ph. D.: Mathématiques)-Université de Montréal. 1972. (1972)
54. Hell, P.: Absolute planar retracts and the four color conjecture. J. Combin. Theor, Ser. B **17**(1), 5–10 (1974)
55. Hell, P., Rival, I.: Absolute retracts and varieties of reflexive graphs. Canadian journal of mathematics **39**(3), 544–567 (1987)
56. Hell, P.: Absolute retracts in graphs. In: In: Bari R.A., Harary, F. (eds.) Graphs and Combinatorics. Lecture Notes in Mathematics, vol. 406, pp. 291–301 (1974). Springer, Berlin (1974). https://doi.org/10.1007/BFb0066450
57. Isbell, J.: Six theorems about injective metric spaces. Commentarii Mathematici Helvetici **39**(1), 65–76 (1964)
58. Johnson, D.S., Garey, M.R.: Computers and Intractability: A Guide to the Theory of NP-Completeness. WH Freeman, San Francisco (1979)
59. Klavžar, S.: Absolute retracts of split graphs. Discre. Math **134**(1–3), 75–84 (1994)
60. Klisowski, J.: A survey of various modifications of the notions of absolute retracts and absolute neighborhood retracts. In: Colloquium Mathematicum, vol. 46, pp. 23–35. Institute of Mathematics Polish Academy of Sciences (1982)
61. Kloks, T., Wang, Y.-L.: On retracts, absolute retracts, and foldings in cographs. Optim. Lett. **12**(3), 535–549 (2017). https://doi.org/10.1007/s11590-017-1126-9
62. Le, H.O., Le, V.: Hardness and structural results for half-squares of restricted tree convex bipartite graphs. Algorithmica **81**(11), 4258–4274 (2019)
63. Lin, M., Szwarcfiter, J.: Faster recognition of clique-Helly and hereditary clique-Helly graphs. Inf. Process. Lett. **103**(1), 40–43 (2007)
64. Loten, C.: Absolute retracts and varieties generated by chordal graphs. Discre Math. **310**(10–11), 1507–1519 (2010)
65. Olariu, S.: A simple linear-time algorithm for computing the center of an interval graph. Int. J. Comput. Math. **34**(3–4), 121–128 (1990)
66. Pesch, E.: Minimal extensions of graphs to absolute retracts. J. Graph Theor. **11**(4), 585–598 (1987)
67. Pesch, E.: Products of absolute retracts. Discre. Math. **69**(2), 179–188 (1988)
68. Pesch, E., Poguntke, W.: A characterization of absolute retracts of n-chromatic graphs. Discre. Math. **57**(1–2), 99–104 (1985)
69. Polat, N.: Convexity and fixed-point properties in Helly graphs. Discre. Math. **229**(1–3), 197–211 (2001)
70. Polat, N.: On constructible graphs, locally Helly graphs, and convexity. J. Graph Theor. **43**(4), 280–298 (2003)

71. Roditty, L., Vassilevska Williams, V.: Fast approximation algorithms for the diameter and radius of sparse graphs. In: Proceedings of the Forty-fifth Annual ACM Symposium on Theory of Computing (STOC), pp. 515–524 (2013)
72. Tarjan, R., Yannakakis, M.: Simple linear-time algorithms to test chordality of graphs, test acyclicity of hypergraphs, and selectively reduce acyclic hypergraphs. SIAM J. Computi. **13**(3), 566–579 (1984)
73. Walter, J.R.: Representations of rigid cycle graphs. Ph.D. thesis, Wayne State University, Department of Mathematics (1972)
74. Watts, D., Strogatz, S.: Collective dynamics of 'small-world' networks. Nature **393**(6684), 440–442 (1998)

Acyclic, Star, and Injective Colouring: Bounding the Diameter

Christoph Brause[1], Petr Golovach[2], Barnaby Martin[3], Daniël Paulusma[3(✉)], and Siani Smith[3]

[1] TU Bergakademie Freiberg, Freiberg, Germany
brause@math.tu-freiberg.de
[2] University of Bergen, Bergen, Norway
petr.golovach@ii.uib.no
[3] Durham University, Durham, UK
{barnaby.d.martin,daniel.paulusma,siani.smith}@durham.ac.uk

Abstract. We examine the effect of bounding the diameter for well-studied variants of the COLOURING problem. A colouring is acyclic, star, or injective if any two colour classes induce a forest, star forest or disjoint union of vertices and edges, respectively. The corresponding decision problems are ACYCLIC COLOURING, STAR COLOURING and INJECTIVE COLOURING. The last problem is also known as $L(1,1)$-LABELLING and we also consider the framework of $L(a,b)$-LABELLING. We prove a number of (almost-)complete complexity classifications, in particular, for ACYCLIC 3-COLOURING, STAR 3-COLOURING and $L(1,2)$-LABELLING.

1 Introduction

A natural way of increasing our understanding of NP-complete graph problems is to restrict the input. The *diameter* of a graph G is the maximum distance between any two vertices of G. We look at graph classes of *bounded* diameter, that is, with diameter at most d for some constant d. Such a graph class is closed under vertex deletion (hereditary) only if $d = 1$. Many graph problems stay NP-complete even if $d = 2$. The reason usually is that from a general instance we can obtain an instance of diameter 2 by adding a dominating vertex. For example, in this way, CLIQUE, INDEPENDENT SET and COLOURING all stay NP-complete for graphs of diameter 2. The latter problem is to decide if for a graph G and integer k, there is a mapping $c : V(G) \rightarrow \{1, \ldots, k\}$ with $c(u) \neq c(v)$ for each $uv \in E(G)$. If k is *fixed*, i.e., not part of the input, we write k-COLOURING.

Let $d \geq 2$ and $k \geq 3$. It is readily seen that k-COLOURING for graphs of diameter at most d is NP-complete for every $(d, k) \notin \{(2, 3), (3, 3)\}$. Mertzios and Spirakis [18] gave a highly non-trivial NP-hardness proof for the case $(3, 3)$. The case $(2, 3)$ is a notorious open problem, see, for example, [2,8,16–19].
The ith *colour class* in a graph $G = (V, E)$ with a colouring c is the set $V_i = \{u \in V \mid c(u) = i\}$. For $i \neq j$, let $G_{i,j}$ be the (bipartite) subgraph of

Daniël Paulusma—Supported by the Leverhulme Trust (RPG-2016-258).

Ł. Kowalik et al. (Eds.): WG 2021, LNCS 12911, pp. 336–348, 2021.
https://doi.org/10.1007/978-3-030-86838-3_26

G induced by $V_i \cup V_j$. If every $G_{i,j}$ is a forest, then c is an *acyclic colouring*. If every $G_{i,j}$ is P_4-free, i.e., a disjoint union of stars, then c is a *star colouring*. If every $G_{i,j}$ is P_3-free, i.e., a disjoint union of vertices and edges, then c is an *injective colouring*. The three decision problems are Acyclic Colouring, Star Colouring and Injective Colouring, respectively; for the last problem it is sometimes allowed for adjacent vertices to be coloured alike (see, e.g., [12–14]) but we do *not* permit this: as can be observed from the aforementioned definitions, all colourings considered in this paper are proper. If k is fixed we write Acyclic k-Colouring, Star k-Colouring and Injective k-Colouring.

Injective colourings are also known as *distance-2 colourings* and as $L(1,1)$-labelings. Namely, a colouring of a graph G is injective if the neighbours of every vertex of G are coloured differently, i.e., also vertices of distance 2 from each other must be coloured differently. The distance constrained labelling problem $L(a_1,\ldots,a_p)$-Labelling is to decide if a graph G has an $L(a_1,\ldots,a_p)$-(k-)labelling, i.e., a mapping $c : V(G) \to \{1,\ldots,k\}$ for some $k \geq 1$, such that for every two vertices u and v and every integer $1 \leq i \leq p$: if G contains a path of length i between u and v, then $|c(u) - c(v)| \geq a_i$; see also [9] (if $a_1 \geq a_2 \geq \ldots \geq a_p$, the condition is equivalent to "if u and v are of distance i").

The above problems are NP-complete, even for very restricted graph classes, see the survey [9] and very recent papers, such as [4,5,15,20]. We consider graph classes of bounded diameter. In contrast to many other problems, bounding the diameter does help for colouring variants. For instance, the problem Near Bipartiteness is to determine if a graph has a 3-colouring such that (only) two colour classes induce a forest. This problem, on graphs of diameter at most d, is polynomial-time solvable if $d \leq 2$ [21] and NP-complete if $d \geq 3$ [6]. Or consider the $L(a_1,\ldots,a_p)$-Labelling problem. The degree of every vertex of a graph G with an $L(a_1,\ldots,a_p)$-k-labelling is at most k. Hence, $|V(G)| \leq 1 + k + \ldots + k^d$, where d is the diameter of G, and we can make the following observation.

Proposition 1. *Let $a_1,\ldots,a_p,d \geq 1$. Then, for every $k \geq 1$, $L(a_1,\ldots,a_p)$-k-Labelling is constant-time solvable for graphs of diameter at most d.*

This led us to the question: *How much does bounding the diameter help for obtaining polynomial-time algorithms for well-known graph colouring variants?*

Our Results. By using a very recent NP-completeness result on Acyclic 3-Colouring for graphs of diameter at most 4 [7] we obtain the following two almost-complete dichotomies; note that the case where $k \leq 2$ is trivial.

Theorem 1. *Let $d \geq 1$ and $k \geq 3$. Then Acyclic k-Colouring on graphs of diameter at most d is*

- *polynomial-time solvable if $d \leq 2$, $k = 3$ and NP-complete if $d \geq 4$, $k = 3$.*
- *polynomial-time solvable if $d = 1$, $k \geq 4$ and NP-complete if $d \geq 2$, $k \geq 4$.*

Theorem 2. *Let $d \geq 1$ and $k \geq 3$. Then Star k-Colouring on graphs of diameter at most d is*

- *polynomial-time solvable if $d \leq 3$, $k = 3$ and NP-complete if $d \geq 8$, $k = 3$.*
- *polynomial-time solvable if $d = 1$, $k \geq 4$ and NP-complete if $d \geq 2$, $k \geq 4$.*

Finally, we consider $L(a,b)$-LABELLING for the most studied values of (a,b), namely when $1 \leq a \leq b \leq 2$. We now assume that k is part of the input, due to Proposition 1. Every two non-adjacent vertices in a graph G of diameter 2 have a common neighbour. Hence, an $(1,1)$-labelling of G colours each vertex uniquely, and $L(1,1)$-LABELLING, on graph of diameter $d \leq 2$, is trivial. The problem is NP-complete if $d = 3$, as it is NP-complete for the subclass of split graphs [3]. Griggs and Yeh [11] proved that $L(2,1)$-LABELLING is NP-complete for graphs of diameter 2 via a relation with HAMILTONIAN PATH. We also connect the remaining case $(a,b) = (1,2)$ to HAMILTONIAN PATH in order to prove NP-completeness in Sect. 4. To summarize, we obtained the following dichotomy:

Theorem 3. *Let $a,b \in \{1,2\}$ and $d \geq 1$. Then $L(a,b)$-LABELLING on graphs of diameter at most d is*

- *polynomial-time solvable if $a = b$ and $d \leq 2$, or $d = 1$.*
- *NP-complete if either $a = b$ and $d \geq 3$, or $a \neq b$ and $d \geq 2$.*

Future Work. It would be interesting to close the gaps in Theorems 1 and 2, but this seems challenging. The NP-hardness construction of Mertzios and Spirakis [18] for 3-COLOURING of graphs of diameter 3 does lead to NP-hardness for NEAR BIPARTITENESS for graphs of diameter 3 [6]. However, it cannot be used for ACYCLIC 3-COLOURING and STAR 3-COLOURING.

2 The Proof of Theorem 1

We show the following result (proof omitted) and also recall a very recent result.

Lemma 1. ACYCLIC 3-COLOURING *is polynomial-time solvable for graphs of diameter at most* 2.

Lemma 2 [7]**.** ACYCLIC 3-COLOURING *is* NP*-complete on triangle-free* 2*-degenerate graphs of diameter at most* 4.

The Proof of Theorem 1. The first statement follows from Lemmas 1 and 2. For the second statement, the case $d = 1$ is trivial, and for the case $d \geq 2$, $k \geq 4$ we reduce from ACYCLIC 3-COLOURING: to an instance G of ACYCLIC k-COLOURING, we add a clique of $k - 3$ vertices, which we make adjacent to every vertex of G.

3 The Proof of Theorem 2

A *list assignment* of a graph G is a function L that gives each vertex $u \in V(G)$ a *list of admissible colours* $L(u) \subseteq \{1, 2, \ldots\}$. A colouring c *respects* L if $c(u) \in L(u)$ for every $u \in V$. If $|L(u)| \leq 2$ for each $u \in V$, then L is a 2*-list assignment.* The 2-LIST COLOURING problem is the corresponding decision problem.

Theorem 4 [10]. *The* 2-LIST COLOURING *problem is solvable in time* $O(n+m)$ *on graphs with* n *vertices and* m *edges.*

We will use Theorem 4 in the proof of Lemma 6, which is the main result of the section. In order to do this, we must first be able to modify an instance of STAR 3-COLOURING into an equivalent instance of 3-COLOURING. We can do this as follows. Let $G = (V, E)$ be a graph. We construct a supergraph G_s of G as follows. For each edge $e = uv$ of G we add a vertex z_{uv} that we make adjacent to both u and v. We also add an edge between two vertices z_{uv} and $z_{u'v'}$ if and only if u, v, u', v' are four distinct vertices such that G has at least one edge with one end-vertex in $\{u, v\}$ and the other one in $\{u', v'\}$. We say that G_s is the *edge-extension* of G. Observe that we constructed G_s in $O(m^2)$ time. It is readily seen that G has a star 3-colouring if and only if G_s has a 3-colouring.

Now suppose G has a 2-list assignment L. We extend L to a list assignment L_s of G_s. We first set $L_s(u) = L(u)$ for every $u \in V(G)$. Initially, we set $L_s(z_e) = \{1, 2, 3\}$ for each edge $e \in E(G)$. We now adjust a list $L_s(z_e)$ as follows. Let $e = uv$. If $L(u) = L(v)$ or $L(u)$ has size 1, then we set $L_s(z_{uv}) = \{1, 2, 3\} \setminus L(u)$. If $L(v)$ has size 1, then we set $L_s(z_{uv}) = \{1, 2, 3\} \setminus L(v)$. If $z_{u'v'}$ is adjacent to a vertex z_{uv} with $|L'(z_{uv})| = 1$, then we set $L_s(z_{u'v'}) = \{1, 2, 3\} \setminus L'(z_{uv})$. We apply the rules exhaustively. We call the resulting list assignment L_s of G_s the *edge-extension* of L. We say that an edge uv of G is *unsuitable* if $|L(u)| = |L(v)| = 2$ but $L(u) \neq L(v)$, whereas uv is *list-reducing* if $|L(u)| = |L(v)| = 1$ and $L(u) \neq L(v)$. Note that in G_s, we may have $|L_s(z_e)| = 3$ if e is unsuitable, whereas $|L_s(z_e)| = 1$ if e is list-reducing. We say that an end-vertex u of an unsuitable edge e is a *fixer* for e if u is adjacent to an end-vertex of a list-reducing edge $u'v'$ (note that $\{u, v\} \cap \{u', v'\} = \emptyset$). We make the following observation.

Lemma 3. *Let* G *be a graph on* m *edges with a* 2*-list assignment* L*. Then we can construct in* $O(m^2)$ *time the edge-extension* G_s *of* G *and the edge-extension* L_s *of* L*. Moreover,* G *has a star* 3*-colouring that respects* L *if and only if* G_s *has a* 3*-colouring that respects* L_s*. Furthermore,* L_s *is a* 2*-list assignment of* G_s *if every unsuitable edge* uv *of* G *has a fixer.*

Let $d_G(u)$ be the degree of a vertex u in G. We omit the proofs of two lemmas.

Lemma 4. *Let* G *be a graph of diameter at most* 3*. If* G *has a star* 3*-colouring, then*

1. *for every* 4*-cycle* $v_0v_1v_2v_3v_0$ *of* G*,* $d_G(v_0) = d_G(v_2) = 2$ *or* $d_G(v_1) = d_G(v_3) = 2$*, and*
2. *there is no* 5*-cycle in* G*.*

Lemma 5. *Let* G *be a graph of diameter at most* 3 *that has two vertices* u *and* v *with at least three common neighbours. Let* $w \in N(u) \cap N(v)$*. Then* G *has a star* 3*-colouring if and only if* $G - w$ *has a star* 3*-colouring. Moreover,* $G - w$ *has diameter at most* 3 *as well.*

Two non-adjacent vertices in a graph G that have the same neighbourhood are *false twins* of G. We are now ready to give our algorithm.

Lemma 6. Star 3-Colouring *is polynomial-time solvable for graphs of diameter at most* 3.

Proof. Let G be a graph of diameter 3. We may assume without loss of generality that G is connected. We first determine in $O(nm^2)$ time all 4-cycles and all 5-cycles in G. If G has a 4-cycle with two adjacent vertices of degree at least 3 in G or if G has a 5-cycle, then G is not star 3-colourable by Lemma 4. We continue by assuming that G satisfies the two properties of Lemma 4. We reduce G by applying Lemma 5 exhaustively. Let G' be the resulting graph, which has diameter at most 3 (by Lemma 5). We can determine in $O(n)$ time all vertices of degree 2 in G. For each vertex of degree 2 we can compute in $O(n)$ time all its false twins. Hence, we found G' in $O(n^2)$ time. As we only removed vertices, G' also satisfies the two properties of Lemma 4.

If G' has maximum degree at most 4, then $|V(G')| \leq 53$, as G' has diameter at most 3. We check in constant time if $|V(G')| \leq 53$ and if so, whether G' has a star 3-colouring. Otherwise, we found a vertex v of degree at least 5 in G'.

Let N_i be the set of vertices of distance i from v. Then, $N_1 = N(v)$ and as G' has diameter at most 3, $V(G') = \{v\} \cup N_1 \cup N_2 \cup N_3$. We assume without loss of generality that if G' has a star 3-colouring c, then $c(v) = 1$. We will examine the following situations: c gives each vertex in N_1 colour 3; or c gives at least one vertex of N_1 colour 2 and at least three vertices of N_1 colour 3. As v has degree at least 5, at least one of colours 2, 3 must occur three times on $N(v)$, and we may assume without loss of generality that this colour is 3. Hence, G' has a star 3-colouring if and only if one of these two cases holds.

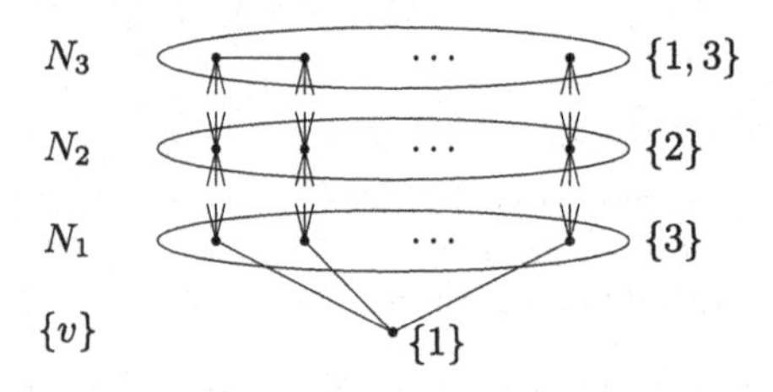

Fig. 1. The pair (G', L') in Case 1.

Case 1. Check if G' has a star 3-colouring that gives every vertex of N_1 colour 3. As $|N_1| \geq 5$, such a star 3-colouring c must assign each vertex of N_2 colour 2. This means that every vertex of N_3 gets colour 1 or 3. Hence, we obtained, in $O(n)$ time, a 2-list assignment L' of G'. We construct the pair (G'_s, L'_s). By Lemma 3 this take $O(m^2)$ time. As every list either has size 1 or is equal to $\{1, 3\}$, we find that the edge-extension L'_s of L' is a 2-list assignment of G'_s. By Lemma 3, it remains to solve 2-List-Colouring on (G'_s, L'_s). We can do this in $O(m^2)$ time using Theorem 4 as the size of G'_s is $O(m^2)$. Hence, the total running time for dealing with Case 1 is $O(m^2)$. See also Fig. 1.

Case 2. Check if G' has a star 3-colouring that gives at least one vertex of N_1 colour 2 and at least three vertices of N_1 colour 3.

We set $L'(v) = \{1\}$. This gives us the property: **P0.** $N_0 = \{v\}$ and $L'(v) = \{1\}$. We now select four arbitrary vertices of $N(v)$. We consider all possible colourings of these four vertices with colours 2 and 3, where we assume without loss of generality that colour 3 is used on these four vertices at least as many times as colour 2. For the case where colour 2 is not used we consider each of the $O(n)$ options of colouring another vertex from $N(v)$ with colour 2. For the cases where colour 3 is used exactly twice, we consider each of the $O(n)$ options of colouring another vertex from $N(v)$ with colour 3. Hence, the total number of options is $O(n)$, and in each option we have a neighbour x of v with colour 2 and a set $W = \{w_1, w_2, w_3\}$ of three distinct neighbours of v with colour 3. That is, we set $L'(x) = \{2\}$ and $L'(w_i) = \{3\}$ for $1 \leq i \leq 3$.

For each set $\{x\} \cup W$ we do as follows. We first check if W is independent; otherwise we discard the option. If W is independent, then initially we set $L'(u) = \{1, 2, 3\}$ for each $u \notin \{x, v\} \cup W$. We now show that we can reduce the list of every such vertex u by at least 1. As an *implicit step*, we will discard the instance (G', L') if one of the lists has become empty. In doing this we will use the following *Propagation Rule*:

Whenever a vertex has only one colour in its list, we remove that colour from the list of each of its neighbours.

By the Propagation Rule, we obtain the following property, in which we updated the set W:

P1. N_1 can be partitioned into sets W, X, Y with $|W| \geq 3$, $|X| \geq 1$ and $|Y| \geq 0$, such that no vertex of Y is adjacent to any vertex of $X \cup W$, and moreover, X is an independent set with $x \in X$ and W is an independent set with $\{w_1, w_2, w_3\} \subseteq W$, such that
- every vertex $w \in W$ has list $L'(w) = \{3\}$,
- every vertex $x \in X$ has list $L'(x) = \{2\}$, and
- every vertex $y \in Y$ has list $L'(y) = \{2, 3\}$.

Note that by the Propagation Rule, we removed colour 3 from the list of every neighbour of a vertex of W in N_2. We now also remove colour 1 from the list of every neighbour of a vertex of W in N_2; the reason for this is that if a neighbour y of, say, w_1 is coloured 1, then the vertices y, w_1, v, w_2 form a bichromatic P_4. Hence, any neighbour of every vertex in W in N_2 has list $\{2\}$.

Now consider a vertex $z \in N_2$ that still has a list of size 3. Then z is not adjacent to any vertex in N_1 with a singleton list (as otherwise we applied the Propagation Rule), but by definition z still has a neighbour z' in N_1. This means that $z' \in Y$ and thus z' has list $\{2, 3\}$. Hence, z cannot be coloured 1: if z' gets colour 2, the vertices x, v, z', z will form a bichromatic P_4, and if z' gets colour 3, the vertices w_1, v, z', z will form a bichromatic P_4. Hence, we may remove colour 1 from $L'(z)$, so $L'(z)$ will have size at most 2.

We make some more observations. First, we recall that every neighbour of a vertex in W in N_2 has list $\{2\}$, and every vertex in X has list $\{2\}$ as well. Hence, no vertex in N_2 has both a neighbour in W and a neighbour in X; otherwise this vertex would have an empty list by the Propagation Rule and we would have discarded this option.

Due to the above, we can partition N_2 into sets W^*, X^*, and Y^* such that the vertices of W^* are the neighbours of W and the vertices of X^* are the neighbours of X, whereas $Y^* = N_2 \setminus (X^* \cup W^*)$. Consequently, the neighbours in N_1 of every vertex of Y^* belong to Y.

Recall that G' has no 5-cycles. Hence, there is no edge between vertices from two different sets of $\{W^*, X^*, Y^*\}$. Furthermore, every vertex $w^* \in W^*$ has list $L'(w^*) = \{2\}$, every vertex $x^* \in X^*$ has list $L'(x^*) = \{1,3\}$, and every vertex $y^* \in Y^*$ has list $L'(y^*) = \{2,3\}$. If a vertex $y \in Y$ has a neighbour $w^* \in W^*$, then vww^*yv is a 4-cycle where $w \in W$ is a neighbour of w^*. Recall that G' satisfies the properties of Lemma 4. As v has degree at least 5 in G', this means that y has degree 2 in G'. Hence, v and w^* are the only neighbours of y. In particular, we find that every vertex in Y with a neighbour in W^* has no neighbour in $X^* \cup Y^*$.

We now apply the Propagation Rule again. As a consequence, we update the lists of the vertices in $Y \cup N_3$, the sets Y and W in **P1**. The latter is because some vertices might have moved from Y to W; in particular it now holds that no vertex in W^* is adjacent to any vertex in Y.

We summarize the above in the following property:

P2. N_2 can be partitioned into sets W^*, X^* and Y^*, such that
- every vertex $w^* \in W^*$ has list $L'(w^*) = \{2\}$ and all its neighbours in N_1 belong to W,
- every vertex $x^* \in X^*$ has list $L'(x^*) \subseteq \{1,3\}$ and at least one of its neighbours in N_1 belong to X and none of them belong to W,
- every vertex $y^* \in Y^*$ has list $L'(y^*) \subseteq \{2,3\}$ and all its neighbours in N_1 belong to Y, and
- there is no edge between vertices from two different sets of $\{W^*, X^*, Y^*\}$.

We now consider N_3. We let T_1 be the set consisting of all vertices in N_3 that have at least two neighbours in W^*. We let T_2 be the set consisting of all vertices in N_3 that have exactly one neighbour in W^*. Moreover, we let S_1 be the set of vertices of $N_3 \setminus (T_1 \cup T_2)$ that have at least one neighbour in T_1. We let S_2 be the set of vertices of $N_3 \setminus (T_1 \cup T_2)$ that have no neighbours in T_1 but at least two neighbours in T_2. If for a vertex $s \in N_3$, there is a vertex $w \in W$ and a 4-path from s to w whose internal vertices are in X and X^*, then we let $s \in R$.

We note that the sets S_1, S_2, T_1 and T_2 are pairwise disjoint by definition, whereas the set R may intersect with $S_1 \cup S_2 \cup T_1 \cup T_2$. We now show that $N_3 = R \cup S_1 \cup S_2 \cup T_1 \cup T_2$. For contradiction, assume that s is a vertex of N_3 that does not belong to any of the five sets R, S_1, S_2, T_1, T_2. As $s \notin T_1 \cup T_2$, we find that the distance from s to every vertex of W is at least 3. Then, as G' has diameter 3, there exists a 4-path P_i from s to each $w_i \in W$ (by **P1** we can write

$W^* = \{w_1, \dots, w_a\}$ for some $a \geq 3$). Every P_i must be of one of the following forms: $s - N_2 - N_1 - w_i$ or $s - N_2 - N_2 - w_i$ or $s - N_3 - N_2 - w_i$.

First assume there is some P_i that is of the form $s - N_2 - N_1 - w_i$, that is, $P_i = szz'w_i$ for some $z \in N_2$ and $z' \in N_1$. As z' is a neighbour of both w_i and v, we find that $z' \in X$ and $z' \in X^\star$, and consequently, $s \in R$, a contradiction.

Now assume that there exists some P_i that is of the form $s - N_2 - N_2 - w_i$, that is, $P_i = szz'w_i$ for some z and z' in N_2. By definition, z must have a neighbour in N_1. As G' has no 5-cycle, this is only possible if z is adjacent to w_i. However, now s is no longer of distance 3 from w_i in G', a contradiction.

Finally, assume that no path from s to any w_i is of one of the two forms above. Hence, every P_i is of the form $s - N_3 - N_2 - w_i$. We write $P_i = st_iw_i^*w_i$ where $t_i \in T_1 \cup T_2$ and $w_i^* \in W^*$. We consider the paths P_1, P_2, P_3, which exist as $|W| \geq 3$. As $s \notin S_1$, we find that $t_i \notin T_1$. Moreover, as $s \notin S_2$, we find that $t_1 = t_2 = t_3$, and so $w_1^* = w_2^* = w_3^*$. In particular, the latter implies that w_1^* is adjacent to w_1, w_2 and w_3 and thus has degree at least 3. Recall that G' satisfies Property 1 of Lemma 4. As w_1^* and v each have degree at least 3 in G', this means that each w_i must only be adjacent to v and w_1^*. However, then w_1, w_2 and w_3 are three false twins of degree 2 in G', and by construction of G' we would have removed one of them, a contradiction. We conclude that $N_3 = R \cup S_1 \cup S_2 \cup T_1 \cup T_2$.

We now reduce the lists of the vertices in N_3. Let $s \in N_3$. If $s \in T_1 \cup T_2$ (that is, s is adjacent to a vertex $w^* \in W^*$) then, as $L'(w^*) = \{2\}$, we find that $L'(s) \subseteq \{1, 3\}$. If $s \in T_1$, then we can reduce the list of s as follows. By the definition of T_1, s is adjacent to a second vertex $w' \neq w^*$ in W^*. By **P2**, we find that w' has a neighbour $w \in W$. We find that $L'(w^*) = L(w') = \{2\}$ and $L(w) = \{3\}$. Then s cannot be assigned colour 3, as otherwise w^*, s, w', w would form a bichromatic P_4. Hence, we can reduce the list of s from $\{1, 3\}$ to $\{1\}$.

Now suppose that $s \in S_1$. Then, by the definitions of the sets S_1 and T_1 and **P2**, there exists a path $P = stw^*w$ where $t \in T_1$, $w^* \in W^*$ and $w \in W$. We deduced above that t has list $L'(t) = \{1\}$. Consequently, we can delete colour 1 from the list of s by the Propagation Rule, so $L'(s) \subseteq \{2, 3\}$. Now suppose that $s \in S_2$. Then, by the definition of S_2 and **P2**, there exist two paths $P_1 = st_1w_1^*w_1$ and $P_2 = st_2w_2^*w_2$ where $t_1, t_2 \in T_2$, $w_1^*, w_2^* \in W^*$, $w_1, w_2 \in W$, and $t_1 \neq t_2$. We claim that s cannot be assigned colour 2. For contradiction, suppose that s has colour 2. Then t_1, which has list $\{1, 3\}$, must receive colour 1, as otherwise t_1 will have colour 3 and s, t_1, w_1^*, w_1 is a bichromatic P_4 (recall that w_1^* and w_1 can only be coloured with colours 2 and 3, respectively). For the same reason, t_2 must get colour 1 as well. However, now w_1^*, t_1, s, t_2 is a bichromatic P_4, a contradiction. Hence, we can remove colour 2 from $L'(s)$. Afterwards, $L'(s) \subseteq \{1, 3\}$.

Finally, suppose that $s \in R$. By the definition of R, there is some path $P_i = sx^*x'w$ where $x^* \in X^*$, $x' \in X$, and $w \in W$. By **P1** and **P2**, respectively, it holds that $L'(x') = \{2\}$ and $L'(x^*) \subseteq \{1, 3\}$. Hence, s cannot be coloured 2: if x^* gets colour 1, the vertices v, x', x^*, s will form a bichromatic P_4, and if x^* gets colour 3, the vertices w_1, x', x^*, s will form a bichromatic P_4. In other words, we may remove colour 2 from $L'(s)$, so $L'(s) \subseteq \{1, 3\}$.

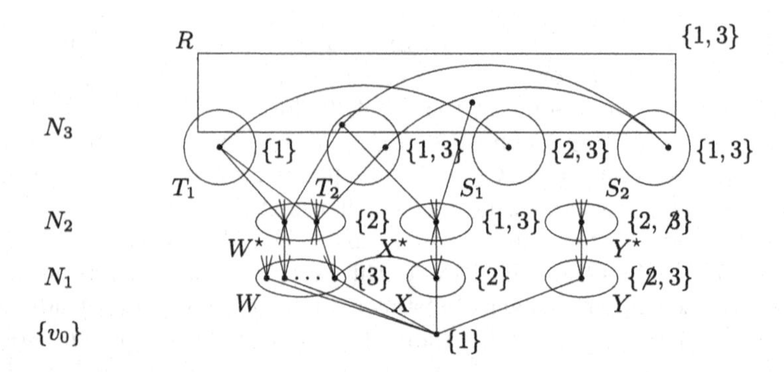

Fig. 2. An example of a pair (G', L') in Case 2a. The colours crossed out show the difference between the general situation in Case 2 and what we show holds in Case 2a.

As $N_3 = R \cup S_1 \cup S_2 \cup T_1 \cup T_2$, we obtained the following property:

P3. N_3 only consists of vertices whose lists are a subset of $\{1, 3\}$ or $\{2, 3\}$, and N_3 can be split into sets R, S_1, S_2, T_1, T_2, such that S_1, S_2, T_1 and T_2 are pairwise disjoint, and

- every vertex $r \in R$ has list $L'(r) \subseteq \{1, 3\}$ and there is a 4-path from r to a vertex in W that has its two internal vertices in X^* and X, respectively,
- every vertex $t \in T_1$ has list $L'(t) = \{1\}$ and has at least two neighbours in W^*,
- every vertex $t \in T_2$ has list $L'(t) \subseteq \{1, 3\}$ and has exactly one neighbour in W^*,
- every vertex $s \in S_1$ has list $L'(s) \subseteq \{2, 3\}$, has no neighbours in W^* but is adjacent to at least one vertex in T_1, and
- every vertex $s \in S_2$ has list $L'(s) \subseteq \{1, 3\}$ and has no neighbours in $T_1 \cup W^*$ but at least two neighbours in T_2.

Hence, we constructed a set $\mathcal{L}'$ of 2-list assignments of G', such that $\mathcal{L}'$ is of size $O(n)$ and G' has a star 3-colouring if and only if G' has a star 3-colouring that respects L' for some $L' \in \mathcal{L}'$. Moreover, we can find each $L' \in \mathcal{L}$ in $O(m + n)$ time by a bread-first search for detecting the 4-paths. For each $L' \in \mathcal{L}$, we do as follows. We still need to construct the edge-extension G'_s of G'. However, the edge-extension L'_s of L' might not be a 2-list assignment. The reason is that G' may have an edge ss' for some vertex $s \in N_2$ with $L'(s) = \{2, 3\}$ and some vertex $s' \in N_3$ with $L'(s') = \{1, 3\}$ such that $L'_s(z_{ss'}) = \{1, 2, 3\}$. We distinguish between two cases; see also Fig. 2 and Fig. 3.

Case 2a. Check if G' has a star 3-colouring that gives x colour 2 and every other vertex of N_1 colour 3.

We only consider this case if $|X| = 1$. We give every vertex in Y list $\{3\}$. Then, by the Propagation Rule, we can delete colour 3 from every list of a vertex in Y^*. We construct G'_s and L'_s in $O(m^2)$ time by Lemma 3. Then L'_s is a 2-list assignment of G'_s. This can be seen as follows. Let $e = ss'$ be an unsuitable

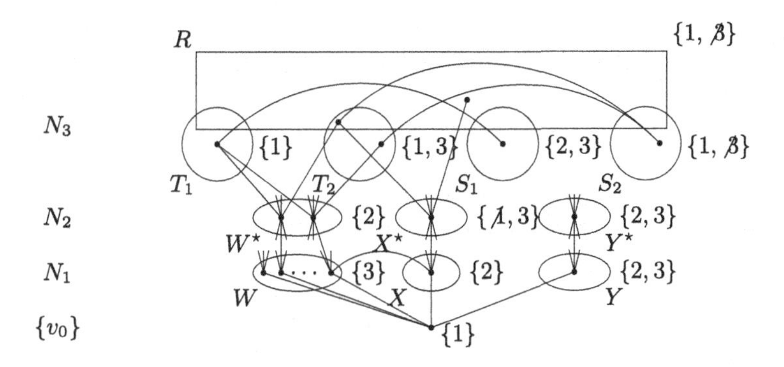

Fig. 3. An example of a pair (G', L') in Case 2b. The colours crossed out show the difference between the general situation in Case 2 and what we show holds in Case 2b.

edge of G'. As G' has no vertices with list $\{1,2\}$, we find that $L'(s) = \{2,3\}$ and $L'(s') = \{1,3\}$. Then s must be in S_1. By definition, it follows that there exist vertices $t \in T_1$ and $w^* \in W^*$ such that st and tw^* are edges of G'. As $L'(t) = \{1\}$ and $L'(w^*) = \{2\}$, the edge tw^* is list-reducing. Hence, s is a fixer for the edge ss'. The claim now follows from Lemma 3, and by the same lemma, it remains to check if G'_s has a 3-colouring that respects L'_s. We can do the latter in $O(m^2)$ time by Theorem 4.

Case 2b. Check if G' has a star 3-colouring that gives at least one other vertex of N_1, besides x, colour 2.

If $|X| \geq 2$, then we found a vertex of $N_1 \setminus \{x\}$ that gets colour 2. If $X = \{x\}$, we will not try to find this vertex; for our algorithm its existence will suffice. By **P2**, every $x^* \in X^*$ has list $L(x^*) \subseteq \{1,3\}$ and a neighbour $x' \in X$ with $L'(x') = \{2\}$. By the Case 2b assumption, there is at least one other vertex x'' in N_1 that gets colour 2. Then x^* cannot be coloured 1, as otherwise x'', v, x', x^* would form a bichromatic P_4. Hence, we remove colour 1 from the list of every vertex of X^* so that afterwards $L(x^*) = \{3\}$ for every $x^* \in X^*$. We remove colour 3 from the list of every neighbour of a vertex of X^*. As L' is a 2-list assignment that does not assign any vertex of G' the list $\{1,2\}$, afterwards every neighbour of every vertex of X^* in N_3 has list $\{1\}$ or $\{2\}$. Moreover, X^* is an independent set (as otherwise we discard (G', L')). No vertex of $W^* \cup Y^*$ is adjacent to any vertex in X^* (by **P2**). Hence, every vertex in X^* has no neighbours in N_2.

We now prove that no vertex in S_2 can receive colour 3. For contradiction, assume that c is a star 3-colouring of G that respects L' and that assigns a vertex $s \in S_2$ colour $c(s) = 3$. As G' has diameter 3, there is a path P from s to $x \in X$ of length at most 3. Then P is of the form $s - N_2 - x$ or $s - N_3 - N_2 - x$ or $s - N_2 - N_2 - x$ or $s - N_2 - N_1 - x$. If P is of the form $s - N_2 - x$, then s has a neighbour in X^*, which has list $\{3\}$. Hence, as s received colour 3, this is not possible. We show that the other three cases are not possible either.

First suppose that P is of the form $s - N_3 - N_2 - x$, say $P = szx^*x$ for some $z \in N_3$ and $x^* \in N_2$. As no vertex of $W^* \cup Y^*$ is adjacent to any vertex in X, we find that $x^* \in X^*$. This means that z must receive colour 1, as otherwise the vertices x, x^*, z, s would form a bichromatic P_4. As $s \in S_2$, we find that s has two neighbours t_1 and t_2 in T_2. Both t_1 and t_2 have list $\{1,3\}$, so they must receive colour 1. At least one of them, say t_1, is not equal to z. However, now x^*, z, s, t_1 form a bichromatic P_4, a contradiction. Hence, this case cannot happen.

Now suppose that P is of the form $s - N_2 - N_2 - x$, say $P = szx^*x$ for some $z, x^* \in N_2$. As no vertex of $W^* \cup Y^*$ is adjacent to any vertex in X, $x^* \in X^*$. However, no vertex in X^* has a neighbour in N_2. Hence, this case cannot happen.

Finally, suppose that P is of the form $s - N_2 - N_1 - x$, say $P = sw^*wx$ for some $w^* \in N_2$ and $w \in N_1$. As X is independent and no vertex of Y is adjacent to a vertex of X, we find that $w \in W$ and thus $w^* \in W^*$. However, this is not possible, as $s \in S_2$ is not adjacent to any vertex in W^* by definition. Hence, this case cannot happen either, so we have proven the claim. So, we can remove colour 3 from the list of every vertex $s \in S_2$. Hence, $L'(s) = \{1\}$ for every $s \in S_2$.

We construct G'_s and L'_s in $O(m^2)$ time by Lemma 3. We claim that L'_s is a 2-list assignment of G'_s. This can be seen as follows. Let $e = ab$ be an unsuitable edge of G'. As G' has no vertices with list $\{1,2\}$, we may assume that $L'(a) = \{1,3\}$ and $L'(b) = \{2,3\}$. As every vertex in R is adjacent to a vertex in X^* with list $\{3\}$, no vertex in R has list $\{1,3\}$. We just deduced that no vertex in S_2 has list $\{1,3\}$ either. Hence, the only vertices with list $\{1,3\}$ belong to T_2, so $a \in T_2$. Then, by definition, we find that a has a neighbour $w \in W^*$, which has a neighbour $w \in W$. As w^* has list $\{2\}$ and w has list $\{3\}$, the edge w^*w is list-reducing. Hence, a is a fixer for the edge ab. The claim now follows from Lemma 3, and by the same lemma, it remains to check if G'_s has a 3-colouring that respects L'_s. We can do the latter in $O(m^2)$ time by Theorem 4.

This concludes the description of our algorithm. The correctness of our algorithm follows from the correctness of the branching steps. Its running time is $O(nm^2)$, as there are $O(n)$ branches, and we deal with each branch in $O(m^2)$ time. □

We also need an observation on a known construction [1] (proof omitted).

Lemma 7. Star 3-Colouring *is* NP*-complete on graphs of diameter at most* 8.

The Proof of Theorem 2. The first statement follows from Lemmas 6 and 7. For the second statement, the case $d = 1$ is trivial, and for the case $d \geq 2$, $k \geq 4$ we reduce from Star 3-Colouring: to an instance G of Star k-Colouring, we add a clique of $k - 3$ vertices, which we make adjacent to every vertex of G.

4 L(1, 2)-Labelling for Graphs of Diameter 2

We show that an n-graph G of diameter 2 has an $L(1,2)$-n-labelling if and only if G has a Hamiltonian path, no edge of which is contained in a triangle, and that the latter problem is NP-complete (proofs omitted). This yields:

Theorem 5. $L(1,2)$-LABELLING *is* NP*-complete for graphs of diameter at most* 2.

References

1. Albertson, M.O., Chappell, G.G., Kierstead, H.A., Kündgen, A., Ramamurthi, R.: Coloring with no 2-colored P_4's. Electron. J. Combinat. **11**, R26 (2004)
2. Bodirsky, M., Kára, J., Martin, B.: The complexity of surjective homomorphism problems - a survey. Discrete Appl. Math. **160**, 1680–1690 (2012)
3. Bodlaender, H.L., Kloks, T., Tan, R.B., van Leeuwen, J.: Approximations for lambda-colorings of graphs. Comput. J. **47**, 193–204 (2004)
4. Bok, J., Jedlicková, N., Martin, B., Paulusma, D., Smith, S.: Acyclic, star and injective colouring: a complexity picture for H-free graphs. In: Proceedings ESA 2020, LIPIcs 173, 22:1–22:22 (2020)
5. Bok, J., Jedličková, N., Martin, B., Paulusma, D., Smith, S.: Injective colouring for H-free graphs. In: Santhanam, R., Musatov, D. (eds.) CSR 2021. LNCS, vol. 12730, pp. 18–30. Springer, Cham (2021). https://doi.org/10.1007/978-3-030-79416-3_2
6. Bonamy, M., Dabrowski, K.K., Feghali, C., Johnson, M., Paulusma, D.: Independent feedback vertex sets for graphs of bounded diameter. Inf. Process. Lett. **131**, 26–32 (2018)
7. Brause, C., Golovach, P.A., Martin, B., Ochem, P., Paulusma D., Smith S.: Acyclic, star and injective colouring: bounding the diameter, Manuscript. arXiv:2104.10593 (2021)
8. Broersma, H., Fomin, F.V., Golovach, P.A., Paulusma, D.: Three complexity results on coloring P_k-free graphs. Eur. J. Combinat. **34**(3), 609–619 (2013)
9. Calamoneri, T.: The $L(h,k)$-labelling problem: an updated survey and annotated bibliography. Comput. J. **54**, 1344–1371 (2011)
10. Edwards, K.: The complexity of colouring problems on dense graphs. TCS **43**, 337–343 (1986)
11. Griggs, J.R., Yeh, R.K.: Labelling graphs with a condition at distance 2. SIAM J. Discrete Math. **5**, 586–595 (1992)
12. Hahn, G., Kratochvíl, J., Širáň, J., Sotteau, D.: On the injective chromatic number of graphs. Discrete Math. **256**, 179–192 (2002)
13. Hell, P., Raspaud, A., Stacho, J.: On injective colourings of chordal graphs. In: Laber, E.S., Bornstein, C., Nogueira, L.T., Faria, L. (eds.) LATIN 2008. LNCS, vol. 4957, pp. 520–530. Springer, Heidelberg (2008). https://doi.org/10.1007/978-3-540-78773-0_45
14. Jin, J., Xu, B., Zhang, X.: On the complexity of injective colorings and its generalizations. Theoret. Comput. Sci. **491**, 119–126 (2013)
15. Karthick, T.: Star coloring of certain graph classes. Graphs Combinat. **34**, 109–128 (2018)
16. Martin, B., Paulusma, D., Smith, S.: Colouring graphs of bounded diameter in the absence of small cycles. In: Calamoneri, T., Corò, F. (eds.) CIAC 2021. LNCS, vol. 12701, pp. 367–380. Springer, Cham (2021). https://doi.org/10.1007/978-3-030-75242-2_26
17. Martin, B., Paulusma, D., Smith, S.: Colouring H-free graphs of bounded diameter. In: Proceedings MFCS 2019, LIPIcs 138, 14:1–14:14 (2019)
18. Mertzios, G.B., Spirakis, P.G.: Algorithms and almost tight results for 3-Colorability of small diameter graphs. Algorithmica **74**, 385–414 (2016)

19. Paulusma, D.: Open problems on graph coloring for special graph classes. In: Mayr, E.W. (ed.) WG 2015. LNCS, vol. 9224, pp. 16–30. Springer, Heidelberg (2016). https://doi.org/10.1007/978-3-662-53174-7_2
20. Shalu, M.A., Antony, C.: Complexity of restricted variant of star colouring. In: Changat, M., Das, S. (eds.) CALDAM 2020. LNCS, vol. 12016, pp. 3–14. Springer, Cham (2020). https://doi.org/10.1007/978-3-030-39219-2_1
21. Yang, A., Yuan, J.: Partition the vertices of a graph into one independent set and one acyclic set. Discrete Math. **306**, 1207–1216 (2006)

The Graphs of Stably Matchable Pairs

David Eppstein(✉)

Computer Science Department, University of California, Irvine, USA
eppstein@uci.edu

Abstract. We study the graphs formed from instances of the stable matching problem by connecting pairs of elements with an edge when there exists a stable matching in which they are matched. Our results include the NP-completeness of recognizing these graphs, a recognition algorithm that is singly exponential in the number of edges of the given graph, and an algorithm whose time is linear in the number of vertices of the graph but exponential in a polynomial of its carving width.

1 Introduction

A stable matching instance consists of two sets of elements to be matched (e.g., medical students and medical residencies), with each element in one set having linearly-ordered preferences for elements of the other set. Somewhat more generally, an element may have a partial preference list, preferring to remain unmatched over being matched to the omitted elements. A matching is *stable* when no matched element prefers being unmatched to its assigned match, and no pair of elements both prefer being matched to each other over their assigned outcomes. A stable matching always exists and can be found in time linear in the input size by the Gale–Shapley algorithm [7].

A given instance of stable matching may have many stable matchings, forming a distributive lattice in which the matching found by the Gale–Shapley algorithm is the top element. Although this lattice may consist of exponentially many matchings [12], it has a concise description as a partially ordered set of alternating cycles, called *rotations*, which can be constructed in polynomial time [11]. A given pair of elements participates in at least one stable matching if and only if it is either part of the top stable matching found by the Gale–Shapley algorithm, or part of one of these rotations. Therefore, from a given instance of stable matching, we may construct in polynomial time a bipartite graph, the graph of pairs that can be matched to each other in at least one stable matching.

This naturally raises many questions. Can we reverse this process, determine whether a given graph comes from an instance of stable matching in this way, and construct a stable matching instance that has the given graph as its stably matchable pairs? What structural properties do these graphs have? For instance, by the rural hospitals theorem [16,19] we know that every stable matching has the same set of matched elements; in graph-theoretic terms, this means that removing the isolated vertices from the graph of stably matchable pairs (the elements that cannot be stably matched) leaves a balanced bipartite graph (one with

L. Kowalik et al. (Eds.): WG 2021, LNCS 12911, pp. 349–360, 2021.
https://doi.org/10.1007/978-3-030-86838-3_27

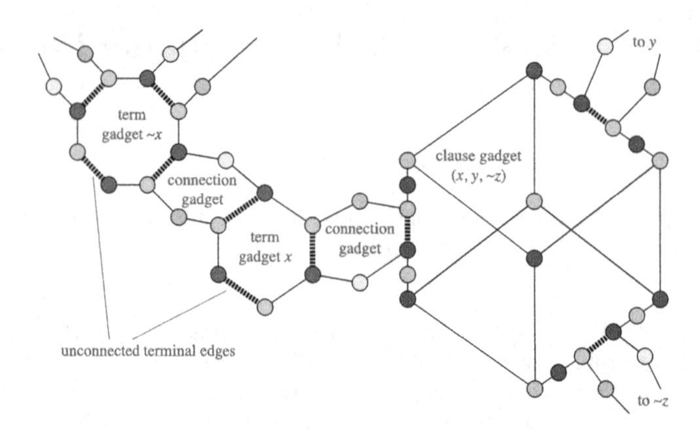

Fig. 1. Gadgets and their connections for a reduction from NAE3SAT to recognizing the graphs of stably matchable pairs.

equal numbers of vertices on each side of the bipartition), and this graph is a *matching-covered graph* meaning that each edge participates in a perfect matching. Matching-covered balanced bipartite graphs can be recognized efficiently, in the same asymptotic time as finding a single perfect matching [18,20]. Are these necessary conditions sufficient? Is every matching-covered balanced bipartite graph the graph of stably matchable pairs of some stable matching instance?

New Results. In this paper we provide algorithms for recognizing the graphs of stably matchable pairs, more efficiently than a brute-force search through all $(n/2)!^n$ preference systems on the given graph. Our main results are algorithms that can test whether an arbitrary graph is a graph of stably matchable pairs, either in time singly exponential in the number of edges, or in time that is fixed-parameter tractable in the carving width of the given graph. For space reasons we have omitted many additional results on these graphs, which can be found in the full preprint version of this paper [6]. These results include:

- Figure 1 sketches an NP-completeness reduction from NAE3SAT to recognizing graphs of stably-matchable pairs. The full version provides a stronger reduction proving this problem NP-complete for subcubic planar graphs.
- We investigate the graphs of stably matchable pairs that belong to special classes of graphs, showing in particular that all regular bipartite graphs are such graphs. Every bipartite graph is an induced subgraph of a regular bipartite graph, implying that graphs of stably matchable pairs have no forbidden induced subgraphs. We prove that, for subcubic graphs to be graphs of stably matchable pairs, it is necessary that the induced subgraphs of their degree-one and degree-three vertices have perfect matchings, and sufficient that in addition their induced subgraph of degree-two vertices have a perfect matching. We prove that outerplanar bipartite graphs are graphs of stably matchable pairs if and only if they have no articulation vertex, and that realizability is preserved by Cartesian products. We characterize the rectangular grid graphs that are graphs of stably matchable pairs.

- Extending prior work of Gusfield, Irving, Leather, and Saks [9], we study the distributive lattices of stable matchings associated with graphs of stably matchable pairs, and we examine how the structure of the graph affects the structure of the lattice. We prove that every finite distributive lattice is the lattice of stable matchings of an instance whose graph is subcubic. We prove that the lattices of stable matchings associated with subcubic outerplanar and subcubic series-parallel graphs are exactly the lattices of closures of oriented forests, and that the lattices of stable matchings associated with planar graphs are exactly the lattices of closures of oriented string graphs.

Related Work. The graphs of stably matchable pairs for random instances with uniformly-random preferences have been investigated by Knuth, Motwani, and Patel, who proved that with n elements on each side the number of edges in these graphs is $\Theta(n \log n)$ with high probability [14], and by Pittel, Shepp, and Veklerov, who bounded the numbers of vertices with degree d in these graphs, for constant d, with high probability [17]. In contrast to these results, random instances for which the elements on one side have preference lists of bounded size produce graphs in which almost all vertices have degree one [10].

2 Preliminaries

We summarize below known and standard results on the structure of the system of stable matchings of a given stable matching instance. For convenience we will follow the convention that the elements being matched are students s_i and residencies r_j. For surveys of the stable matching problem see [8,13,15].

Stable matchings for given preferences may be partially ordered, with $M \leq M'$ in this order when every student prefers M to M' (or has the same match in both and is indifferent), or equivalently when every residency prefers M' to M or is indifferent. This order forms a distributive lattice, in which the join (least upper bound) $M \vee M'$ of matchings M and M' gives every residency its more-preferred match from M and M' and gives every student their least-preferred match from M and M'. Symmetrically, the meet (greatest lower bound) $M \wedge M'$ gives every residency its least-preferred match and gives every student their most-preferred match. Joins and meets are associative and can be extended in the obvious way from pairs of stable matchings to arbitrary sets of stable matchings.

If two stable matchings M and M' are adjacent in the distributive lattice (they are comparable, say as $M \leq M'$, and no third matching M'' is between them with $M \leq M'' \leq M'$) then their symmetric difference is a *rotation*, a cycle of matched pairs that alternates between pairs of M and of M'. Within a rotation, pairs alternate between *upper edges* (pairs in M') and *lower edges* (pairs in M), with stable matchings that differ by the same rotation having the same upper and lower edges. Define an *extended rotation* to be either a rotation or one of the two stable matchings $\top$ and $\bot$ (the top and bottom elements of the distributed lattice of stable matchings), with every pair in $\top$ a lower edge, and with every pair in $\bot$ an upper edge. Then every pair that can be stably matched is a lower edge in one extended rotation and an upper edge in one extended rotation.

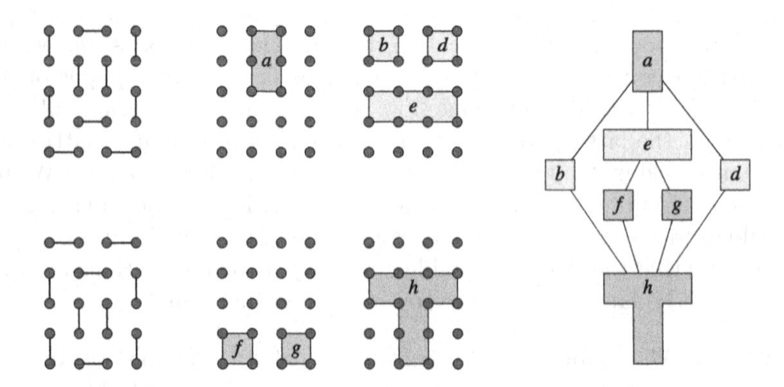

Fig. 2. A rotation system realizing a 4×5 grid as a graph of stably matchable pairs. The two matchings at left are the matchings $\top, \bot$ of this realization, its cycles are the colored regions in the center of the figure, and their partial order is at right.

We define a *rotation system* for given sets of students and residencies to be a partially ordered collection C of cycles that alternate between students and residencies, together with two matchings $\top$ and $\bot$ from students to residencies, such that each pair (s_i, r_j) belongs to either zero or two members of $C \cup \{\top, \bot\}$, and such that each cycle alternates between upper edges (belonging to $\top$ or to an element above the cycle in the partial order) and lower edges (belonging to $\bot$ or to an element below the cycle in the partial order). An example is depicted in Fig. 2. We call the elements of C *rotations* and the elements of $C \cup \{\top, \bot\}$ *extended rotations.* Every stable matching instance can be described in polynomial time by a rotation system, and the graph of stably matchable pairs equals the graph of pairs participating in extended rotations. Conversely:

Lemma 1. *Every rotation system describes the collection of stable matchings for a system of partial preferences. If in addition the rotation system has the property that there are either no unmatched students, or no unmatched residencies, then it describes the collection of stable matchings for a system of complete preferences.*

See the full version of the paper for the proof.

3 Singly-Exponential Algorithm

Our singly exponential time and space bounds will depend on the following analysis of matching-related structures in graphs. The following lemma concerns a general bound for structures consisting of a perfect matching and something else disjoint from the matching. The "something else" (another matching, a spanning tree, an edge-coloring, or some other structure) enters the notation of the lemma in the following way. If T is a type of structure associated with a graph, we define $\mathrm{count}_T(G)$ to be the number of different choices of a structure of type T in a graph G, and we define

$$\mathrm{base}_T = \limsup_{m \to \infty} \max_{G, |E(G)|=m} \mathrm{count}_T(G)^{1/m}.$$

Intuitively, this means that a graph with m edges has $O(\mathrm{base}_T^m)$ structures of type T. We consider only structures for which this definition gives a finite nonzero value to base_T. Finally, we define $\mathrm{match}_T(G)$ to be the number of ways that one can form a structure in G consisting of a perfect matching M together with a structure of type T in the disjoint subgraph $G \setminus M$.

Lemma 2. *With the notation above, for a graph G with m edges,* $\mathrm{match}_T(G) = O(c_M^m)$, *where c_M is at most $1 + \mathrm{base}_T$ and can be upper-bounded as*

$$c_M \leq \max\left\{\left(i\,\mathrm{base}_T^{2i-2}\right)^{1/(2i-1)} \mid 1 \leq i\right\}.$$

Proof. The $1 + \mathrm{base}_T$ bound follows from the binomial theorem, and the more precise bound comes from a recurrence-based analysis of a degree-based backtracking enumeration algorithm, where i is the worst-case degree encountered by the algorithm. We defer details to the full version of the paper. □

In cases where the structures of type T can be efficiently enumerated the proof of the lemma can be made into an algorithm whose running time is obtained by replacing the bound on the number of structures by the time to enumerate them.

Lemma 3. *A graph with m edges has at most c_0^m perfect matchings, where*

$$c_0 = 6^{1/9} \approx 1.22028.$$

Proof. This follows from a result of Alon and Friedman [1] bounding the number of perfect matchings in a graph with degree sequence d_i to be at most $\prod(d_i)^{1/2d_i}$. See the full versions for details. □

Lemma 4. *In a graph with m edges, the number of structures consisting of a perfect matching and an arbitrary subset of the remaining edges is $O(c_1^m)$, and the number of structures consisting of two perfect matchings (not necessarily disjoint) and an arbitrary subset of the remaining edges is $O(c_2^m)$, where*

$$c_1 = 1280^{1/9} \approx 2.21435$$

and

$$c_2 = (4c_0(1 + c_0)^6)^{1/7} \approx 2.48475.$$

Proof. The bound on c_1 follows from Lemma 2, counting matchings plus a structure T consisting of an arbitrary set of edges, for which $\mathrm{base}_T = 2$. Calculation shows that the maximum value for the upper bound in the lemma is achieved at $i = 5$, giving the value shown.

The bound on c_2 follows by using the minimum-degree algorithm from the proof of Lemma 2 to choose a matching and disjoint set of edges, and applying Lemma 3 to bound the number of matchings in the edges that were not chosen in the disjoint set of edges. See the full version for details. □

Theorem 1. *We can test whether a given graph G with m edges and n vertices is a graph of stably matchable pairs in space $O\big(\min\{c_1^m, (n/2)!2^{m-n/2}\}\big)$ and time $O\big(\min\{mc_2^m, (n/2)!^2 2^{m-n/2}\}\big)$, where c_1 and c_2 are the numbers from Lemma 4.*

Proof. We will search for a sequence of perfect matchings, differing from one to the next in the sequence by a single rotation, that together cover all edges of G and such that each edge of G is rotated into a matching at most once and rotated out of a matching at most once. If such a sequence exists, we can produce a rotation system that has the first matching in the sequence as its bottom matching, the last matching in the sequence as its top matching, and the rotations of the sequence as the rotations of the rotation system, partially ordered consistently with the sequence order.

We search for this sequence as a path in a graph Γ whose vertices are pairs (M, S) where M is a perfect matching and S is any subset of the unmatched edges. (This is essentially a dynamic programming algorithm but we find it more convenient to describe it as a graph search.) In our search, we will use M to represent the matching at the current endpoint of a partial sequence of matchings of G, and S to represent the edges that have already been covered by matchings earlier in the partial sequence. We make an edge in Γ from vertex (M, S) to vertex (M', S') when matchings M and M' differ by a single rotation, when $M' \cap S = \emptyset$, and when $S' = S \cup (M \setminus M')$. With this definition, paths in Γ automatically correspond to sequences of matchings that never re-use any edges: an edge, once used, gets added to S and its presence there and the requirement that S be disjoint from M' prevent later matchings in the sequence from re-using it. Therefore, the sequences we seek are exactly the paths from any vertex of the form $(M, \emptyset)$ to any other vertex of the form $(M', E(G) \setminus M')$. We can find such a path by breadth first or depth first search of Γ, starting from the vertices $(M, \emptyset)$ and terminating whenever a vertex of the form $(M', E(G) \setminus M')$ is reached.

By Lemma 4, Γ has $O(c_1^m)$ vertices and $O(c_2^m)$ edges. It remains to describe how to construct Γ and search it in a space-efficient way, bearing in mind that we do not want to use the amount of memory it would require to store all the edges of Γ. We have enough memory that (by the assumption that machine words are large enough to store a single memory address) we can store a single matching M or a single set of edges S in a single word, and perform set operations like unions and intersections of these sets in constant time per operation. We will index the vertices of Γ by consecutive integers in the range from 0 to $|V(\Gamma)| - 1$. Based on these indexes, we store the following information:

- An array A_1, addressed by vertex index, of the pairs (M_i, S_i) of structures associated with vertex i.
- An array A_2, addressed by vertex index, of the predecessor of vertex i in the first path to reach that vertex found by the search, or a special flag value to indicate that it has not yet been reached.
- An array A_3 of indexes of vertices on the current frontier of the search, arranged either as a stack for depth first search or a queue for breadth first search.

- An array A_4 of Boolean values indexed by sets of edges, true when that set of edges is a simple cycle and false otherwise.
- An array A_5 of lists indexed by sets S of edges, listing the matchings M disjoint from S in sorted order, together with the vertex index for (M, S) for each matching.

Array A_4 takes space 2^m to store, and the remaining arrays (including the total length of the lists in A_5 take space $O(c_1^m)$. Initializing A_1 can be accomplished in linear time using the structure enumeration algorithm of the proof of Lemma 4. Array A_2 is easily initialized to contain only flag values, and A_3 is initialized to list all vertices of the form $(M, \emptyset)$. Initializing A_4 can be done by a naive algorithm that tests each set for being a cycle, in time $O(m2^m)$, smaller than our stated time and space bounds. A_5 can be initialized by using A_1 to count how many matchings are associated with each disjoint set in order to allocate storage for the lists within A_5, using a second pass over A_1 to place matchings into lists, and then sorting each list.

Each step of the search of Γ is obtained by removing a vertex (M, S) of Γ from the search frontier A_3, terminating the algorithm if $M \cup S = E(G)$, otherwise looping through the neighbors of (M, S) in Γ, and for each previously-unreached neighbor setting its predecessor in A_2 and adding it to the frontier A_3. It remains to describe how to find the outgoing neighbors of a vertex (M, S) of Γ, since these are not explicitly represented by the data structures described above (and storing them explicitly would use too much space). We do this simply by using A_5 to find all of the matchings M' disjoint from S, performing set operations on M and M' to find their symmetric difference, and using A_4 to check that this symmetric difference is a simple cycle. When it is, we have found a neighbor (M', S') of (M, S) in Γ , with $S' = S \cup (M \setminus M')$. We can find the vertex index of this neighbor by a binary search in the list for S' of A_5. The bottleneck of the algorithm is performing these binary searches; each takes time $O(m)$ and the number of searches performed is $O(c_2^m)$.

The space and time bounds $O((n/2)!2^{m-n/2})$ and $O((n/2)!^2 2^{m-n/2})$ are better for dense graphs. They follow from the same algorithm, after a preprocessing stage that eliminates vertices of degree less than two, using a more naive analysis in which we count pairs of a matching and a disjoint set, or two matchings and a disjoint set, by bounding the number of matchings by $(n/2)!$ and the number of disjoint sets by two to the power of the number of unmatched edges. □

4 Parameterized by Carving Width

In this section we parameterize the recognition of graphs of stably matchable pairs by *carving width*. The precise definition of carving width involves hierarchical clusterings of the vertices of a graph so that each cluster boundary is crossed by a small number of edges, but we do not need the details for this section; instead, it suffices to know that for graphs of treewidth w and maximum degree d, the carving width is bounded below by $\Omega(\max(w, d))$ and bounded above $O(wd)$ [5]. Therefore, we can equivalently think of the graphs of small carving width as being graphs that simultaneously have small treewidth and small maximum degree.

Our parameterized algorithm for recognizing graphs of stably matchable pairs uses dynamic programming on a *tree decomposition*, a tree whose nodes are associated with sets of vertices called *bags*, with every graph edge having both endpoints in at least one bag and every vertex belonging to bags that form a connected subtree. The *width* of the decomposition is one less than the number of vertices in its largest bag, and the treewidth of the graph is the minimum width of a tree decomposition. It will be convenient to use a *nice tree decomposition*, in which the tree of bags is rooted and each bag has one of three types [4]:

- An *introduce node* is either a leaf node whose bag contains a single vertex, or a node with exactly one child whose bag is formed by adding one vertex to its child's bag.
- A *forget node* is a node with exactly one child whose bag is formed by removing one vertex from its child's bag.
- A *join node* has exactly two children, with the bags of it and its children all being equal.

Each vertex can be introduced in many introduce nodes, but forgotten only from one, which must be the parent of the common ancestor of the introduce nodes. It will be convenient to introduce one more type of node, an *edge node*, associated bijectively with a specific edge in the given graph whose endpoints belong to the bag of the edge node. In this way, each node of the tree decomposition can be associated with a unique subgraph of the given graph, the subgraph whose edges and vertices are associated with the given node and its descendants. By introducing additional forget nodes we may assume without loss of generality that the root of the tree has an empty bag; its associate is the whole graph. We describe a tree decomposition that meets these minor modifications to the usual definition of a nice tree decomposition as a *good tree decomposition*.

Minimum-width tree decompositions can be found in time linear in the number of vertices of an input graph but exponential in the cube of the width [2]; it is also possible to find an approximate tree decomposition whose width is within a constant factor of minimum, in time linear in the number of vertices of the input graph and single-exponential in the width [3]. An arbitrary tree decomposition can be transformed into a nice tree decomposition of the same width, blowing up the number of nodes by a factor polynomial in the width, in time linear in the number of vertices of the graph and polynomial in the width [4]. Adding edge nodes between the highest node whose bag contains both endpoints of each edge and the forget nodes above them introduces a number of nodes which is again linear in vertices and polynomial in width, within the same time bounds.

Our dynamic programming algorithm will consider two different kinds of partial information about a rotation system for the given graph, restricted to the subgraph associated with a node of a good tree decomposition, which we call a *rotation subsystem* and a *rotation state*. A rotation subsystem completely describes the part of a rotation system within the subgraph associated with a node, while a rotation state describes only the parts of the rotation system involving the vertices of the current bag. A rotation subsystem and a rotation state are *compatible* when the rotation state is obtained from the rotation subsystem by forgetting the information about vertices that are not in the bag. For a given node x of the tree decomposition, let G_x denote the subgraph associated with that node, B_x denote the set of vertices of the bag associated with that node, $F_x = V(G_x) \setminus B_x$ denote the set of forgotten vertices, and $G_x[B_x]$ denote the part of G_x having both endpoints in B_x. More specifically, a rotation subsystem for a node x consists of the following information:

- Top and bottom matchings $\top \cap G_x$ and $\bot \cap G_x$ in which all vertices of F_x are matched, but vertices of B_x may or may not be matched.
- A collection of rotations, an abstract partially ordered set whose elements will represent the subset of rotations of the rotation system that include at least one edge of G_x.
- A collection of *pieces* of rotations, intersections of rotations with G_x. Each piece is either a full cycle, or a path whose endpoints both belong to B_x. Each two pieces of the same rotation must be disjoint. A rotation can either have a full cycle as its piece or one or more paths as pieces, but not both.

Each edge in the subgraph associated with a node must be part of two elements (either pieces or $\top$ or $\bot$), and be upper in one and lower in the other (as determined by the partial order of the rotations). Each piece must alternate between upper and lower edges. We do not require that this structure can be extended to a full rotation system, in general; however, we have:

Observation 1. *A rotation subsystem at the root node of a good tree decomposition coincides with a rotation system for the given graph.*

We define a *rotation state* at a node of the tree decomposition to consist of the following information, corresponding to the information in a rotation subsystem but restricted to bag vertices. Then a rotation state for x consists of:

- Top and bottom matchings $\top$ and $\bot$ in $G_x[B_x]$, together with two bits of information for each remaining bag node specifying whether it has a match in $\top \cap G_x$ or $\bot \cap G_x$.
- A collection of rotations, an abstract partially ordered set whose elements will represent the subset of rotations of the rotation system that either include at least one edge of G_x or at least one piece that is a path with an endpoint in B_x. Because of the assumption that the input graph has low degree, the number of rotations in this collection will also be low.
- A collection of pieces of rotations. For each piece we store the rotation that it is part of, the edges of the piece that belong to $G_x[B_x]$, and the endpoints of the piece in B_x, but not the edges in the rest of G_x. At each endpoint in B_x, we store whether the edge of the piece that is incident to that endpoint is an upper or lower edge in its rotation.

We require that each edge of $G_x[B_x]$ belong to two elements (pieces, $\top$, or $\bot$), upper in one and lower in the other, and that two edges of $G_x[B_x]$ that are consecutive within a piece they alternate between upper and lower. We do not require that this structure can be extended to a rotation subsystem for the same node of the tree decomposition, but we call it *valid* when it can be so extended.

Our algorithm constructs a tree decomposition and works bottom up calculating each node's set of valid rotation states. It uses the following lemmas, which describe the valid rotation states for each type of node in our decomposition, and allow each set to be computed in time exponential in a polynomial of the carving width, independent of the overall graph size. See the full version for their proofs.

Lemma 5. *In a tree decomposition of width w for a graph of degree d, the number of possible rotation states is at most exponential in $O((wd)^2)$.*

Lemma 6. *At an introduce node x of the tree decomposition, with child y, a state S is valid if and only if the vertex v that is introduced at x does not take part in any pieces of rotations, v is marked as not matched in $\top \cap G_x$ and $\bot \cap G_x$, and the state S' formed by removing v from S is valid for y.*

Lemma 7. *At a forget node x of the tree decomposition, with child y, a state S is valid if and only if the vertex v that is forgotten at x does not take part in any pieces of rotations, and there exists a valid state S' formed from S by adding v to S, marking it as matched in both $\top \cap G_y$ and $\bot \cap G_y$, and possibly adding an edge incident to v in $G_y[B_y]$ to $\top$ or $\bot$.*

Lemma 8. *At an edge node x which introduces edge uv as the only edge of graph G_x, a state S is valid if and only if one of the following three conditions is met:*

- *Edge uv is included in $\top$, vertices u and v are marked as matched in $\top$ and unmatched in $\bot$, and there are no rotations or pieces of rotations.*
- *Edge uv is included in $\bot$, vertices u and v are marked as matched in $\top$ and unmatched in $\bot$, and there are no rotations or pieces of rotations.*
- *Edge uv is not included in $\top$ or $\bot$, vertices u and v are marked as unmatched in both $\top$ and $\bot$, and there is a single rotation having a single piece, consisting of this edge, marked as either upper or lower.*

Lemma 9. *At an edge node x with child y which introduces edge uv to a non-empty subgraph G_y, a state S is valid if and only if it would be valid for a join node x' with the same bag as x whose two children are y and an edge node introducing uv as the only edge in its subgraph.*

Lemma 10. *At a join node x with children y and z, a rotation state S is valid if and only if there exist valid states S' for y and S'' for z such that:*

- *The matchings $\bot$ and $\top$ in S are disjoint unions of the corresponding matchings in S' and S'', and the sets of matched vertices in $\bot$ and $\top$ in S are disjoint unions of the corresponding sets of matched vertices in S' and S''.*
- *The rotations in S are a union (not necessarily disjoint) of the rotations in S' and S'', with the partial order on these rotations in S' and S'' equalling the restriction to those subsets of the partial order on rotations in S.*

– *Each piece of a rotation in S is formed as a concatenation of one or more pieces of rotations in S' and S'', all pieces of rotations in S' and S'' are used to form pieces of rotations in this way, and at each point of concatenation the two concatenated pieces alternate between an upper edge and a lower edge.*

Theorem 2. *We can test whether a graph G with n vertices and carving width w is a graph of stably matchable pairs in time $O(n) \cdot \exp O(w^4)$.*

Proof. We can find a good tree-decomposition of width $O(w)$ with $O(wn)$ edge nodes and join nodes and n forget nodes in the stated time bound. Removing the forget nodes and join nodes partitions the decomposition into subtrees each having $O(w)$ introduce nodes, so there are $O(w^2 n)$ introduce nodes total.

Because G has carving width w, its treewidth and maximum degree are $O(w)$, so by Lemma 5 the number of states at each node of this decomposition is exponential in $O(w^4)$. By the lemmas above, we can compute the set of valid states at each node by examining each valid state at the child node (when there is one child) or each pair of valid states at the two children (when there are two children), in time polynomial in w per state or pair of states. The total time per node is bounded by this polynomial multiplied by the square of the number of states, which remains exponential in $O(w^4)$. □

5 Conclusions and Open Problems

We have investigated the graphs of stably matchable pairs of stable matching instances, with results including the characterization of special classes of these graphs, characterization of the lattices of stable matchings corresponding to these classes of graphs, the NP-completeness of recognizing graphs of stably matchable pairs in general, and exact algorithms for recognizing these graphs that run in singly exponential or fixed-parameter tractable time.

It is open whether our algorithm for carving width can be extended to treewidth for graphs of unbounded degree. As a first step, what is the complexity of recognizing series-parallel graphs of stably matchable pairs? Another problem concerns the analysis of numbers of combinations of matchings and subsets of edges in Lemma 4, which we used to analyze our exponential-time recognition algorithm. This analysis seems unlikely to be tight; can it be improved?

References

1. Alon, N., Friedland, S.: The maximum number of perfect matchings in graphs with a given degree sequence. Electron. J. Combinatorics **15**(1), N13 (2008). https://doi.org/10.37236/888
2. Bodlaender, H.L.: A linear-time algorithm for finding tree-decompositions of small treewidth. SIAM J. Comput. **25**(6), 1305–1317 (1996). https://doi.org/10.1137/S0097539793251219
3. Bodlaender, H.L., Drange, P.G., Dregi, M.S., Fomin, F.V., Lokshtanov, D., Pilipczuk, M.: A $c^k n$ 5-approximation algorithm for treewidth. SIAM J. Comput. **45**(2), 317–378 (2016). https://doi.org/10.1137/130947374

4. Dorn, F., Telle, J.A.: Semi-nice tree-decompositions: the best of branchwidth, treewidth and pathwidth with one algorithm. Discrete Appl. Math. **157**(12), 2737–2746 (2009). https://doi.org/10.1016/j.dam.2008.08.023
5. Eppstein, D.: The effect of planarization on width. J. Graph Algorithms Appl. **22**(3), 461–481 (2018). https://doi.org/10.7155/jgaa.00468
6. Eppstein, D.: The graphs of stably matchable pairs. Electronic preprint arxiv:2010.09230 (2020)
7. Gale, D., Shapley, L.S.: College admissions and the stability of marriage. Am. Math. Mon. **69**(1), 9–14 (1962). https://doi.org/10.2307/2312726
8. Gusfield, D., Irving, R.: The Stable Marriage Problem: Structure and Algorithms. MIT Press, Foundations of Computing Series (1989)
9. Gusfield, D., Irving, R., Leather, P., Saks, M.: Every finite distributive lattice is a set of stable matchings for a small stable marriage instance. J. Combinat. Theory Ser. A **44**(2), 304–309 (1987). https://doi.org/10.1016/0097-3165(87)90037-9
10. Immorlica, N., Mahdian, M.: Marriage, honesty, and stability. In: Proceedings of the 16th ACM-SIAM Symposium on Discrete Algorithms (SODA 2005), pp. 53–62 (2005)
11. Irving, R., Leather, P.: The complexity of counting stable marriages. SIAM J. Comput. **15**(3), 655–667 (1986). https://doi.org/10.1137/0215048
12. Karlin, A.R., Gharan, S.O., Weber, R.: A simply exponential upper bound on the maximum number of stable matchings. In: Diakonikolas, I., Kempe, D., Henzinger, M. (eds.) Proceedings of the 50th Symposium on Theory of Computing (STOC 2018), pp. 920–925. Association for Computing Machinery (2018). http://arxiv.org/abs/1711.01032,https://doi.org/10.1145/3188745.3188848
13. Knuth, D.E.: Stable marriage and its relation to other combinatorial problems, volume 10 of CRM Proceedings & Lecture Notes. American Mathematical Society (1997). https://doi.org/10.1090/crmp/010
14. Knuth, D.E., Motwani, R., Pittel, B.: Stable husbands. Random Struct. Algorithms **1**(1), 1–14 (1990). https://doi.org/10.1002/rsa.3240010102
15. Manlove, D.F.: Algorithmics of Matching under Preferences, volume 2 of Series on Theoretical Computer Science. World Scientific (2013). https://doi.org/10.1142/8591
16. McVitie, D.G., Wilson, L.B.: Stable marriage assignment for unequal sets. BIT Numerical Math. **10**(3), 295–309 (1970). https://doi.org/10.1007/BF01934199
17. Pittel, B., Shepp, L., Veklerov, E.: On the number of fixed pairs in a random instance of the stable marriage problem. SIAM J. Discrete Math. **21**(4), 947–958 (2007). https://doi.org/10.1137/070696155
18. Régin, J.-C.: A filtering algorithm for constraints of difference in CSPs. In: Hayes-Roth, B., Korf, R.E. (eds.) Proceedings of the 12th National Conference on Artificial Intelligence, Seattle, WA, USA, 31 July–4 August 1994, vol. 1, pp. 362–367. AAAI Press and MIT Press (1994). https://www.aaai.org/Library/AAAI/1994/aaai94-055.php
19. Roth, A.E.: On the allocation of residents to rural hospitals: a general property of two-sided matching markets. Econometrica **54**(2), 425–427 (1986). https://doi.org/10.2307/1913160
20. Tassa, T.: Finding all maximally-matchable edges in a bipartite graph. Theoretical Comput. Sci. **423**, 50–58 (2012). https://doi.org/10.1016/j.tcs.2011.12.071

On Additive Spanners in Weighted Graphs with Local Error

Reyan Ahmed[1(✉)], Greg Bodwin[2], Keaton Hamm[3], Stephen Kobourov[1], and Richard Spence[1]

[1] Department of Computer Science, University of Arizona, Tucson, USA
abureyanahmed@email.arizona.edu

[2] Department of Computer Science, University of Michigan, Ann Arbor, USA

[3] Department of Mathematics, University of Texas at Arlington, Arlington, USA

Abstract. An *additive* $+\beta$ *spanner* of a graph G is a subgraph which preserves distances up to an additive $+\beta$ error. Additive spanners are well-studied in unweighted graphs but have only recently received attention in weighted graphs [Elkin et al. 2019 and 2020, Ahmed et al. 2020]. This paper makes two new contributions to the theory of weighted additive spanners.

For weighted graphs, [Ahmed et al. 2020] provided constructions of sparse spanners with *global* error $\beta = cW$, where W is the maximum edge weight in G and c is constant. We improve these to *local* error by giving spanners with additive error $+cW(s,t)$ for each vertex pair (s,t), where $W(s,t)$ is the maximum edge weight along the shortest s–t path in G. These include pairwise $+(2+\varepsilon)W(\cdot,\cdot)$ and $+(6+\varepsilon)W(\cdot,\cdot)$ spanners over vertex pairs $\mathcal{P} \subseteq V \times V$ on $O_\varepsilon(n|\mathcal{P}|^{1/3})$ and $O_\varepsilon(n|\mathcal{P}|^{1/4})$ edges for all $\varepsilon > 0$, which extend previously known unweighted results up to ε dependence, as well as an all-pairs $+4W(\cdot,\cdot)$ spanner on $\widetilde{O}(n^{7/5})$ edges.

Besides sparsity, another natural way to measure the quality of a spanner in weighted graphs is by its *lightness*, defined as the total edge weight of the spanner divided by the weight of an MST of G. We provide a $+\varepsilon W(\cdot,\cdot)$ spanner with $O_\varepsilon(n)$ lightness, and a $+(4+\varepsilon)W(\cdot,\cdot)$ spanner with $O_\varepsilon(n^{2/3})$ lightness. These are the first known additive spanners with nontrivial lightness guarantees. All of the above spanners can be constructed in polynomial time.

1 Introduction

Given an undirected graph $G(V,E)$, a *spanner* is a subgraph H which approximately preserves distances in G up to some error. Spanners are an important primitive in the literature on network design and shortest path algorithms, with applications in motion planning in robotics [10,16,29,33], asynchronous protocol design [31], approximate shortest path algorithms [17], and much more; see survey [4]. One general goal in research on spanners is to minimize the size of the spanner (measured by the number of edges $|E(H)|$), given some error by which distances can be distorted. For weighted graphs, another desirable goal is

L. Kowalik et al. (Eds.): WG 2021, LNCS 12911, pp. 361–373, 2021.
https://doi.org/10.1007/978-3-030-86838-3_28

to minimize the *lightness*, defined as the total weight of the spanner divided by the weight of a minimum spanning tree (MST) of G.

Spanners were introduced in the 1980s by Peleg and Schäffer [30], who first considered *multiplicative* error. A subgraph H is a (multiplicative) α-spanner of $G(V, E)$ if $d_H(s,t) \leq \alpha \cdot d_G(s,t)$ for all vertices $s, t \in V$, where $d_G(s,t)$ is the distance between s and t in G. Since H is a subgraph, we also have $d_G(s,t) \leq d_H(s,t)$ by definition, so the distances in H approximate those of G within a multiplicative factor of k, sometimes called the stretch factor. Althöfer et al. [7] showed that all n-vertex graphs have multiplicative $(2k-1)$-spanners on $O(n^{1+1/k})$ edges with $O(n/k)$ lightness, and this edge bound is the best possible assuming the girth conjecture by Erdős [22] from extremal combinatorics. Meanwhile, this initial lightness bound has been repeatedly improved in follow-up work, and the optimal bound still remains open [12,14,20,23,28].

Multiplicative spanners are extremely well applied in computer science. However, they are typically applied to very large graphs where it may be undesirable to take on errors that scale with the (possibly very large) distances in the input graph. A more desirable error is *additive* error which does not depend on the original graph distances at all:

Definition 1 (Additive $+\beta$ spanner). *Given a graph $G(V, E)$ and $\beta \geq 0$, a subgraph H is a $+\beta$ spanner of G if*

$$d_G(s,t) \leq d_H(s,t) \leq d_G(s,t) + \beta \tag{1}$$

for all vertices $s, t \in V$.

A pairwise $+\beta$ spanner is a subgraph H for which (1) only needs to hold for specific vertex pairs $\mathcal{P} \subseteq V \times V$ given on input, and a subsetwise spanner is a pairwise spanner with $\mathcal{P} = \mathcal{S} \times \mathcal{S}$ for some $\mathcal{S} \subseteq V$ (if $\mathcal{P} = V \times V$, these are sometimes called all-pairs spanners, for clarity). It is known that all unweighted graphs $G(V, E)$ with $|V| = n$ have (all-pairs) $+2$ spanners on $O(n^{3/2})$ edges [6,27], $+4$ spanners on $\widetilde{O}(n^{7/5})$ edges [13], and $+6$ spanners on $O(n^{4/3})$ edges [8,27,34]. On the negative side, for all $\varepsilon > 0$ there exist graphs which have no $+\beta$ spanner on $O(n^{4/3-\varepsilon})$ edges even for arbitrarily large constant β [1]. This presents a barrier to using additive spanners in applications where a very sparse subgraph, say on $O(n^{1.001})$ edges, is needed. However, the lower bound construction in [1] is rather pathological; tradeoffs do continue for certain natural classes of graphs with good girth or expansion properties [8], and recent experimental work [5] showed that tradeoffs seem to continue for graphs constructed from common random graph models.

A more serious barrier preventing the applicability of additive spanners is that classic constructions only apply to *unweighted* graphs, while many naturally-occurring metrics are not expressible by a unit-weight graph. To obtain additive spanners of weighted graphs $G = (V, E, w)$ where $w : E \to \mathbb{R}^+$, the error term $+\beta$ needs to scale somehow with the edge weights of the input graph. Prior work [2] has considered *global* error of type $\beta = cW$, where $W = \max_{e \in E} w(e)$ is the maximum edge weight in the input graph and c is a constant.

However, a more desirable paradigm studied by Elkin, Gitlitz, and Neiman [18, 19] is to consider *local* error in terms of the maximum edge weight along a shortest s–t path:

Definition 2 (Local $+cW(\cdot,\cdot)$ spanner). *Given a graph $G(V, E)$, subgraph H is a (local) $+cW(\cdot,\cdot)$ spanner if $d_H(s,t) \leq d_G(s,t) + cW(s,t)$ for all $s, t \in V$, where $W(s,t)$ is the maximum edge weight along a shortest path $\pi(s,t)$ in G.*[1]

It is often the case that $W(s,t) \ll W$ for many vertex pairs (s,t) in which a $+cW(\cdot,\cdot)$ spanner has much less additive error for such vertex pairs. Additionally, a $+cW(\cdot,\cdot)$ spanner is also a multiplicative $(c+1)$-spanner, whereas a $+cW$ spanner can have unbounded multiplicative stretch. This relationship between additive and multiplicative stretch is thematic in the area [8,21].

Sparse Local Additive Spanners. Weighted additive spanners were first studied by Elkin et al. [18], who gave a local $+2W(\cdot,\cdot)$ spanner with $O(n^{3/2})$ edges, as well as a "mixed" spanner with a similar error type. Ahmed, Bodwin, Sahneh, Kobourov, and Spence [2] gave a comprehensive study of weighted additive spanners, including a *global* $+4W$ spanner with $\widetilde{O}(n^{7/5})$ edges and $+8W$ spanner with $O(n^{4/3})$ edges, analogous to the previously-known unweighted constructions. The $+6$ unweighted error vs. $+8W$ global weighted error left a gap to be closed; this was mostly closed in a recent follow-up work of Elkin et al. [19], who gave a *local* all-pairs $+(6+\varepsilon)W(\cdot,\cdot)$ spanner on $O_\varepsilon(n^{4/3})$ edges[2] by generalizing the $+6$ spanner by Knudsen [27], and a subsetwise $+(2+\varepsilon)W(\cdot,\cdot)$ spanner on $O_\varepsilon(n\sqrt{|S|})$ edges [19] (following a similar $+2$ subsetwise spanner in unweighted graphs by Elkin (unpublished), later published in [15,32]). Our first contribution is the improvement of several remaining known constructions of weighted additive spanners from global to local error.

Theorem 1. *Let $\varepsilon > 0$. Then every weighted graph G and set $\mathcal{P} \subseteq V \times V$ of vertex pairs has:*

1. *a deterministic pairwise $+(2+\varepsilon)W(\cdot,\cdot)$ spanner on $O_\varepsilon(n|\mathcal{P}|^{1/3})$ edges,*
2. *a deterministic pairwise $+(6+\varepsilon)W(\cdot,\cdot)$ spanner on $O_\varepsilon(n|\mathcal{P}|^{1/4})$ edges,*
3. *a pairwise $+2W(\cdot,\cdot)$ spanner on $O(n|\mathcal{P}|^{1/3})$ edges,*
4. *a pairwise $+4W(\cdot,\cdot)$ spanner on $O(n|\mathcal{P}|^{2/7})$ edges, and*
5. *an all-pairs $+4W(\cdot,\cdot)$ spanner on $\widetilde{O}(n^{7/5})$ edges.*

The first two pairwise constructions are deterministic unlike the randomized constructions from [4]. Unweighted versions of these results were proved in [24,25], and weighted versions with global error but without ε dependence were proved in [2]. This theorem is the first to provide versions with local error. Together with [18,19], the above results complete the task of converting unweighted additive spanners to weighted additive spanners with local error; see Tables 1 and 2.

[1] If there are multiple shortest s–t paths, then we break ties consistently so that subpaths of shortest paths are also shortest paths.

[2] We use $O_\varepsilon(f(n))$ as shorthand for $O(\text{poly}(\frac{1}{\varepsilon})f(n))$.

Table 1. All-pairs additive spanner constructions for unweighted and weighted graphs.

Unweighted			Weighted		
$+\beta$	Size	Ref.	$+\beta$	Size	Ref.
$+2$	$O(n^{3/2})$	[6,8,15,27]	$+2W(\cdot,\cdot)$	$O(n^{3/2})$	[18]
$+4$	$\widetilde{O}(n^{7/5})$	[13]	$+4W(\cdot,\cdot)$	$\widetilde{O}(n^{7/5})$	**[this paper]**
$+6$	$O(n^{4/3})$	[8,27]	$+(6+\varepsilon)W(\cdot,\cdot)$	$O_\varepsilon(n^{4/3})$	[19]
$+n^{o(1)}$	$\Omega(n^{4/3-\varepsilon})$	[1]			

Table 2. Pairwise and subsetwise (purely) additive spanner constructions for unweighted and weighted graphs.

Unweighted				Weighted		
Type	$+\beta$	Size	Ref.	$+\beta$	Size	Ref.
Subset	$+2$	$O(n\sqrt{\lvert S\rvert})$	[15]	$+(2+\varepsilon)W(\cdot,\cdot)$	$O_\varepsilon(n\sqrt{\lvert S\rvert})$	[19]
Pairwise	$+2$	$O(n\lvert\mathcal{P}\rvert^{1/3})$	[11,25]	$+2W(\cdot,\cdot)$	$O(n\lvert\mathcal{P}\rvert^{1/3})$	**[this paper]**
Pairwise	$+4$	$O(n\lvert\mathcal{P}\rvert^{2/7})$	[9,24]	$+4W(\cdot,\cdot)$	$O(n\lvert\mathcal{P}\rvert^{2/7})$	[3]
Pairwise	$+6$	$O(n\lvert\mathcal{P}\rvert^{1/4})$	[24]	$+(6+\varepsilon)W(\cdot,\cdot)$	$O_\varepsilon(n\lvert\mathcal{P}\rvert^{1/4})$	**[this paper]**

Lightweight Local Additive Spanners. All aforementioned results are in terms of the number of edges $|E(H)|$ of the spanner. If minimizing the total edge weight is more desirable than constructing a sparse spanner, a natural problem is to construct *lightweight* spanners. Given a connected graph $G = (V, E)$ with positive edge weights $w : E \to \mathbb{R}^+$, the *lightness* of a subgraph H is defined by

$$\text{lightness}(H) := \frac{w(H)}{w(\text{MST}(G))} \tag{2}$$

where $w(H)$ and $w(\text{MST}(G))$ are the sum of edge weights in H and an MST of G, respectively. Section 4 highlights why none of the aforementioned sparse spanners have good lightness guarantees. Our second contribution is the following:

Theorem 2. *Let $\varepsilon > 0$. Then every weighted graph G has:*

1. *a deterministic all-pairs $+\varepsilon W(\cdot,\cdot)$ spanner with $O_\varepsilon(n)$ lightness, and*
2. *a deterministic all-pairs $+(4+\varepsilon)W(\cdot,\cdot)$ spanner with $O_\varepsilon(n^{2/3})$ lightness.*

To the best of our knowledge, these are the first nontrivial lightness results known for additive spanners. For comparison on the first result, it is easy to show that every graph G has an all-pairs distance preserver H with lightness$(H) = O(n^2)$. It follows from the seminal work of Khuller, Raghavachari, and Young [26] on shallow-light trees that every graph G has a subgraph H that preserves distances up to a $(1+\varepsilon)$ multiplicative factor, with lightness $O_\varepsilon(n)$. Our first result implies that the same lightness bound (up to the specifics of the ε dependence) can be achieved with *additive* error. Theorem 2.1 strictly strengthens this consequence of [26], since local $+\varepsilon W(\cdot,\cdot)$ error implies multiplicative $(1+\varepsilon)$ stretch as previously mentioned. Additionally, Theorem 2.1 is tight, in the sense that an unweighted complete graph K_n has lightness $\Theta(n)$ and no nontrivial $+\varepsilon W(\cdot,\cdot)$ spanner. Table 3 summarizes these lightness results compared with those in unweighted graphs.

Table 3. Lightweight all-pairs additive spanners in unweighted and weighted graphs.

Unweighted		Weighted		
$+\beta$	Lightness	$+\beta$	Lightness	Ref.
0	$O(n)$	$+\varepsilon W(\cdot,\cdot)$	$O_\varepsilon(n)$	**[this paper]**
+2	$O(n^{1/2})$		?	
+4	$\widetilde{O}(n^{2/5})$	$+(4+\varepsilon)W(\cdot,\cdot)$	$O_\varepsilon(n^{2/3})$	**[this paper]**
+6	$O(n^{1/3})$		?	

The lightweight spanners are based on a new initialization technique which we call *d-lightweight initialization*, in which an initial set of lightweight edges is added to the spanner starting from the MST of the input graph. Nearly all of the above spanners have a common theme in the construction and analysis: add to H an initial set of edges oblivious to the distances in the graph, then add shortest paths for any vertex pairs which do not satisfy inequality (1). The size or lightness bounds are then analyzed by determining how many pairs of nearby vertices there are whose distances sufficiently improve upon adding a shortest path; this method was also used by Elkin et al. [19]. Note that such improvements in unweighted graphs must be by at least 1; in weighted graphs, distances may improve by arbitrarily small amount which leads to the ε dependence. We remark that the above spanner results hold if we assign each vertex pair (u,v) a shortest path $\pi(u,v)$ which minimizes the maximum edge weight. We leave as open questions the lightness bounds for $+2W$ and $+6W$ spanners (with or without ε dependence), whether the lightness for the $+(4+\varepsilon)W(\cdot,\cdot)$ spanner can be improved to $\widetilde{O}(n^{2/5})$, and whether we can construct additive spanners in weighted graphs which are simultaneously sparse *and* lightweight.

2 Preliminaries

Most additive spanner constructions begin with either a clustering [8,15,24] or initialization phase [2,19,27], where an initial set of edges is added to the spanner oblivious to distances or vertex pairs $\mathcal{P}$ in the graph. Experimental results [5] suggest that initialization is preferred over clustering in terms of runtime and spanner size, and all constructions in this paper are initialization-based. Given a weighted graph G and $d \geq 0$, a *d-light initialization* of G is a subgraph obtained by selecting the d lightest edges incident to every vertex, or all edges if the degree is less than d. We exploit the following lemma from [2]:

Lemma 1 ([2]). *Let H be a d-light initialization of a weighted graph G, and let $\pi(s,t)$ be a shortest path in G. If there are ℓ edges of $\pi(s,t)$ absent from H, then there is a set N of $\frac{d\ell}{6} = \Omega(d\ell)$ vertices, where each vertex in N is adjacent to a vertex on $\pi(s,t)$, connected via an edge of weight at most $W(s,t)$.*

We refer to vertices in N as the *d-light neighbors* of $\pi(s,t)$. The fact that d-light neighbors are connected to $\pi(s,t)$ via light edges of weight $\leq W(s,t)$ was not explicitly stated in [2] but follows directly from the proof, as N is constructed by taking the d lightest edges incident to vertices on $\pi(s,t)$ which are incident to a missing edge of weight at most $W(s,t)$. After d-light initialization, additional shortest paths are added to H in order to "satisfy" the remaining unsatisfied vertex pairs[3]. We will use the standard method of *$+\beta$ spanner completion*, where we iterate over each vertex pair (s,t) in nondecreasing order of maximum weight $W(s,t)$ (then by nondecreasing distance $d_G(s,t)$ in case of a tie) and add $\pi(s,t)$ to the spanner if (s,t) is unsatisfied.

3 Sparse Local Additive Spanners

Instead of $+cW$ additive error considered in [2], we consider $+cW(\cdot,\cdot)$ local error.

3.1 Pairwise $+(2+\varepsilon)W(\cdot,\cdot)$ and $+(6+\varepsilon)W(\cdot,\cdot)$ Spanners

For the pairwise $+(2+\varepsilon)W(\cdot,\cdot)$ and $+(6+\varepsilon)W(\cdot,\cdot)$ spanners (Theorem 1.1-2), we describe a deterministic construction. The analysis behind the edge bounds uses a set-off and improving strategy also used in [19]. The constructions involve one additional step of adding a certain number of edges along every vertex pair's shortest path before spanner completion.

Let $\ell, d \geq 0$ be parameters which are defined later. Let H be a d-light initialization. Then for each vertex pair $(s,t) \in \mathcal{P}$, consider the shortest path $\pi(s,t)$ in G and add the first ℓ missing edges and the last ℓ missing edges to H (if $\pi(s,t)$ is missing at most 2ℓ edges, all missing edges from $\pi(s,t)$ are added to H). We remark that if (s,t) is already satisfied, we can skip this step for the pair (s,t).

After this phase, we perform $+(2+\varepsilon)W(\cdot,\cdot)$ or $+(6+\varepsilon)W(\cdot,\cdot)$ spanner completion as described in Sect. 2. This construction clearly outputs a valid pairwise spanner, and $O(nd + \ell|\mathcal{P}|)$ edges are added in the "distance-oblivious" phase before spanner completion. It remains to determine the number of edges added in spanner completion.

Let $(s,t) \in \mathcal{P}$ be a vertex pair for which $\pi(s,t)$ is added to H during spanner completion. Observe that after d-light initialization, the first ℓ missing edges and the last ℓ missing edges from $\pi(s,t)$ do not overlap; otherwise no remaining edges from $\pi(s,t)$ would have been added to H. Let $u_1v_1, \ldots, u_\ell v_\ell$ denote the first ℓ missing edges on $\pi(s,t)$ which are added after d-light initialization, and let $u'_1v'_1, \ldots, u'_\ell v'_\ell$ denote the last ℓ missing edges, where u_i (or u'_i) is closer to s than v_i (or v'_i); see Fig. 1 for illustration.

[3] A vertex pair (s,t) is *satisfied* if the spanner inequality (1) holds for that pair, and *unsatisfied* otherwise.

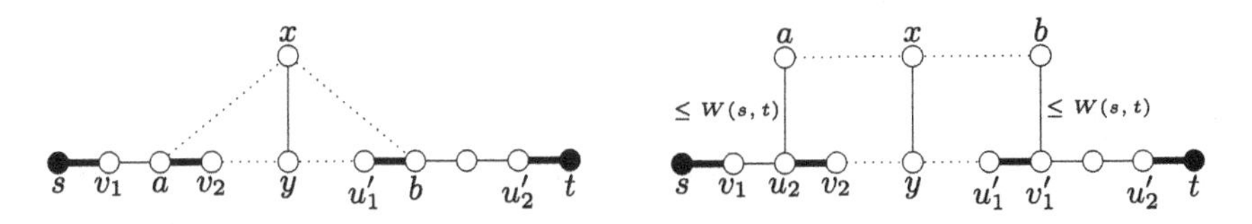

Fig. 1. Illustration of Lemma 2 (left) and Lemma 3 (right) with $\ell = 2$. Note that $s = u_1$ and $t = v_2'$ in this example. By adding $\pi(s,t)$, at least one of the pairs' (a,x) or (x,b) distance improves by at least $\frac{\varepsilon W(s,t)}{2}$. Note that a, x, b are not necessarily distinct.

We will refer to the set $\{u_1, \ldots, u_\ell\}$ as the *prefix* and the set $\{v_1', \ldots, v_\ell'\}$ as the *suffix*. Consider the shortest paths $\pi(s, v_\ell)$ and $\pi(u_1', t)$. By Lemma 1, there are $\Omega(d\ell)$ d-light neighbors which are adjacent to a vertex in the prefix and suffix, respectively, connected by an edge of weight at most $W(s,t)$.

Consider the subpath $\pi(v_\ell, u_1')$; suppose $z \geq 1$ edges of $\pi(s,t)$ are added during spanner completion. By Lemma 1 again, there are $\Omega(dz)$ d-light neighbors which are adjacent to a vertex on $\pi(v_\ell, u_1')$. In the following lemmas, denote by H_0 and H_1 the spanner immediately before and after $\pi(s,t)$ is added, respectively.

Lemma 2. *Let $(s,t) \in \mathcal{P}$ be such that $\pi(s,t)$ is added to H during $+(2+\varepsilon)W(\cdot,\cdot)$ spanner completion. Let a and b be vertices in the prefix and suffix respectively. Let x be a d-light neighbor of the path $\pi(v_\ell, u_1')$. Then both of the following hold:*

1. $d_{H_1}(a,x) \leq d_G(a,x) + 2W(s,t)$ *and* $d_{H_1}(b,x) \leq d_G(b,x) + 2W(s,t)$
2. $d_{H_0}(a,x) - d_{H_1}(a,x) > \frac{\varepsilon W(s,t)}{2}$ *or* $d_{H_0}(b,x) - d_{H_1}(b,x) > \frac{\varepsilon W(s,t)}{2}$.

For the pairwise $+(6+\varepsilon)W(\cdot,\cdot)$ spanner, we consider arbitrary d-light neighbors a and b of the prefix and suffix, and similarly consider vertex pairs (a,x), (b,x) whose distances sufficiently improve:

Lemma 3. *Let $(s,t) \in \mathcal{P}$ be such that $\pi(s,t)$ is added to H during $+(6+\varepsilon)W(\cdot,\cdot)$ spanner completion. Let a and b be d-light neighbors adjacent to vertices u_i and v_j' in the prefix and suffix, respectively. Let x be a d-light neighbor of the path $\pi(v_\ell, u_1')$. Then both of the following hold:*

1. $d_{H_1}(a,x) \leq d_G(a,x) + 4W(s,t)$ *and* $d_{H_1}(x,b) \leq d_G(x,b) + 4W(s,t)$
2. $d_{H_0}(a,x) - d_{H_1}(a,x) > \frac{\varepsilon W(s,t)}{2}$ *or* $d_{H_0}(x,b) - d_{H_1}(x,b) > \frac{\varepsilon W(s,t)}{2}$.

Proofs omitted due to space are provided in the arXiv version of this paper [3].

Lemma 4. *By setting $d = |\mathcal{P}|^{1/3}$ and $\ell = n/|\mathcal{P}|^{2/3}$, the pairwise $+(2+\varepsilon)W(\cdot,\cdot)$ construction outputs a subgraph H with $|E(H)| = O_\varepsilon\left(n|\mathcal{P}|^{1/3}\right)$.*

Proof. In the distance-oblivious phase, we add $O(nd + \ell|\mathcal{P}|) = O(n|\mathcal{P}|^{1/3})$ edges to H. A vertex pair (v,x) is *set-off* if it is the first time that $d_H(v,x) \leq d_G(v,x)+$

$(2+\varepsilon)W(s,t)$ and is *improved* if its distance in H decreases by at least $\frac{\varepsilon W(s,t)}{2}$. Suppose adding $\pi(s,t)$ during $+(2+\varepsilon)W(\cdot,\cdot)$ spanner completion adds $z \geq 1$ additional edges. Let x be a d-light neighbor of $\pi(v_\ell, u_1')$ and let a, b be vertices in the prefix and suffix. By Lemma 1, there are $\Omega(dz)$ vertices x adjacent to $\pi(v_\ell u_1')$. By Lemma 2, both of the pairs (a,x), (b,x) are set-off if not already, and at least one of the pairs is improved upon adding $\pi(s,t)$. Since there are $\Omega(\ell)$ choices for a or b, this gives $\Omega(dz \times \ell) = \Omega(nz/|\mathcal{P}|^{1/3})$ improvements upon adding z edges to H.

Once a pair (v,x) is set-off, it can only be improved $O(\frac{1}{\varepsilon})$ times; this follows since pairs are ordered by their maximum weight, so any improvement is by at least $\frac{\varepsilon W(s,t)}{2}$. If Z total edges are added during spanner completion, then the number of improvements is $\Omega(d\ell Z)$. There are $O(n^2)$ vertex pairs and once set-off, each vertex pair is improved $O\left(\frac{1}{\varepsilon}\right)$ times, in which we have $\Omega(d\ell Z) = O\left(\frac{n^2}{\varepsilon}\right)$. Since $d = |\mathcal{P}|^{1/3}$ and $\ell = n/|\mathcal{P}|^{2/3}$, we obtain $Z = O\left(\frac{1}{\varepsilon}n|\mathcal{P}|^{1/3}\right)$. Altogether we obtain $|E(H)| = O\left(\frac{1}{\varepsilon}n|\mathcal{P}|^{1/3}\right)$. □

Lemma 5. *By setting* $d = |\mathcal{P}|^{1/4}$ *and* $\ell = n/|\mathcal{P}|^{3/4}$, *the pairwise* $+(6+\varepsilon)W(\cdot,\cdot)$ *construction outputs a subgraph* H *with* $|E(H)| = O_\varepsilon(n|\mathcal{P}|^{1/4})$.

The proof is nearly identical to that of Lemma 4 except that a and b are d-light neighbors instead of prefix and suffix vertices. Lemmas 4 and 5 imply Theorem 1.1 and 1.2 respectively. Further, by setting $W = 1$ and $\varepsilon = 0.5$, these results imply pairwise +2 and +6 spanners of size $O(n|\mathcal{P}|^{1/3})$ and $O(n|\mathcal{P}|^{1/4})$ in *unweighted* graphs (as a +2.5 spanner of an unweighted graph is also a +2 spanner). These edge bounds match those of existing pairwise +2 spanners [11,25] and pairwise +6 spanners [24].

3.2 Pairwise $+2W(\cdot,\cdot)$ and $+4W(\cdot,\cdot)$ Spanners

We address Theorem 1.3 and 1.4: every weighted graph has a pairwise $+2W(\cdot,\cdot)$ spanner and $+4W(\cdot,\cdot)$ spanner on $O(n|\mathcal{P}|^{1/3})$ edges and $O(n|\mathcal{P}|^{2/7})$ edges respectively. This removes the ε dependence from Theorem 1.1 and uses local error instead of global W error as in [2]. First, we need the following simple lemma:

Lemma 6. *Let* H *be a* d*-light initialization, and let* $s,t \in V$. *Let* x *be a* d*-light neighbor of* $\pi(s,t)$ *connected to a vertex* $y \in \pi(s,t)$ *in* H. *Consider a shortest path tree in* G *rooted at* x. *Then the distance from* s *to* t *in this tree is at most* $d_G(s,t) + 2W(s,t)$.

This is proven by the triangle inequality and the fact that $w(xy) \leq W(s,t)$. The remainder of Theorem 1.3 and 1.4 is similar to the pairwise $+2W$ and $+4W$ spanners in [2] except we use the fact that d-light neighbors are connected by an edge of weight $\leq W(s,t)$. The $+4W(\cdot,\cdot)$ pairwise spanner can also be used to show existence of an all-pairs $+4W(\cdot,\cdot)$ spanner of size $\widetilde{O}(n^{7/5})$ (Theorem 1.5).

4 Lightweight Local Additive Spanners

In this section, we prove Theorem 2 by constructing *lightweight* additive spanners. The previous spanner algorithms based on d-light initialization can produce spanners with poor lightness: Let G be obtained by considering $K_{\frac{n}{2},\frac{n}{2}}$ with all edges of weight W, then adding two paths of weight 0 connecting the vertices within each bipartition. Then $w(\mathrm{MST}(G)) = W$, while d-light initialization already adds $\Omega(Wnd)$ weight to the spanner H, or $\Omega(nd)$ lightness.

In order to construct lightweight additive $+\beta$ spanners, we introduce a new initialization technique called *d-lightweight initialization*, which adds edges of total weight at most d for each vertex, starting from the MST of G. We first perform the following simple modifications to the input graph G in order:

1. Scale the edge weights of G linearly so that the weight of the MST is $\frac{n}{2}$.
2. Remove all edges of weight $\geq n$ from G.

Note that step 1 also scales $W(s,t)$ for vertex pairs (s,t), but the validity of an additive spanner or spanner path is invariant to scaling. Step 2 does not change the shortest path metric or the maximum edge weights $W(s,t)$; since the MST has weight $\frac{n}{2}$, this implies $W(s,t) \leq d_G(s,t) \leq \frac{n}{2}$ for all vertex pairs (s,t).

Given G and $d \geq 0$, the *d-lightweight initialization* is a subgraph H defined as follows: let H be an MST of G. If $d > 0$, then for each vertex $v \in V$, consider all incident edges to v which have not already been added to H, in nondecreasing order of weight. Add these incident edges one by one to H until the next edge causes the total edge weight added corresponding to v to be greater than d. This subgraph H has $O(n + nd)$ total edge weight, and $O(1 + d)$ lightness. Note that 0-lightweight initialization is simply the MST of G. After d-lightweight initialization, perform $+\beta$ spanner completion as before (Sect. 2).

4.1 Lightweight Initialization and Neighborhoods of Shortest Paths

In order to prove the desired lightness bounds, we consider a subdivided graph G' obtained as follows: for each MST edge e of $\mathrm{MST}(G)$, subdivide e into $\lceil w(e) \rceil$ edges of weight $\frac{w(e)}{\lceil w(e) \rceil}$. Recall that $w(\mathrm{MST}(G)) = \frac{n}{2}$, so this adds $O(n)$ vertices, all of which are on MST edges. We do not modify the maximum edge weights $W(s,t)$ for each pair $(s,t) \in V(G)$, even if such edges are subdivided in G'. We remark that the MST of G' also has weight $\frac{n}{2}$, and all MST edges have weight at most 1.

Lemma 7. *Let $e = uv$ be an edge in $E(G')$ which is not contained in the d-lightweight initialization H. Then there are $\Omega(d^{1/2})$ vertices x in G' such that $d_H(u,x) \leq w(e)$. Moreover, if $\varepsilon' \in (0,1]$, there are $\Omega(\varepsilon' w(e))$ vertices x such that $d_H(u,x) \leq \varepsilon' w(e)$.*

We now describe a lightweight analogue of Lemma 1:

Lemma 8. *Let H be a d-lightweight initialization, let $\pi(s,t)$ be a shortest s–t path in G', let z be the total weight of the edges in $\pi(s,t)$ which are absent in H, and let $\varepsilon' \in (0,1]$. Then there is a set N of at least $\frac{\varepsilon' z}{10} = \Omega(\varepsilon' z)$ vertices in G' which are of distance at most $\varepsilon' W(s,t)$ in H from some vertex in $\pi(s,t)$.*

4.2 Proof of Theorem 2

Using Lemmas 7 and 8, we can analyze the lightness of the above spanner constructions by considering the number of vertex pairs in the subdivided graph G' whose distances sufficiently improve upon adding $\pi(s,t)$, where $s,t \in V(G)$.

Lemma 9. *Let $(s,t) \in V(G)$ be a vertex pair such that $\pi(s,t)$ is added during $+\varepsilon W(\cdot,\cdot)$ spanner completion. Let $x \in N$ be a vertex in G' which is of distance $\leq \frac{\varepsilon}{4} W(s,t)$ from some vertex u in $\pi(s,t)$ in H. Let H_0 and H_1 denote the spanner before and after $\pi(s,t)$ is added. Then both of the following hold:*

1. $d_{H_1}(s,x) \leq d_{G'}(s,x) + \frac{\varepsilon}{2} W(s,t)$ *and* $d_{H_1}(t,x) \leq d_{G'}(t,x) + \frac{\varepsilon}{2} W(s,t)$
2. $d_{H_0}(s,x) - d_{H_1}(s,x) \geq \frac{\varepsilon}{4} W(s,t)$ *or* $d_{H_0}(t,x) - d_{H_1}(t,x) \geq \frac{\varepsilon}{4} W(s,t)$.

Similar to Lemma 2, statement 1. holds by the triangle inequality, and 2. can be proven by contradiction: if neither inequality was true, then $d_{H_0}(s,t) \leq d_G(s,t) + \varepsilon W(s,t)$, contradicting that $\pi(s,t)$ was added during spanner completion (Fig. 2).

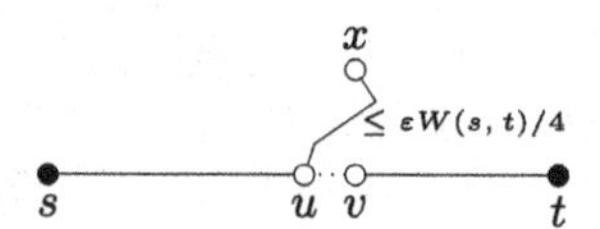

Fig. 2. Illustration of Lemma 9. By adding $\pi(s,t)$ to H during $+\varepsilon W(\cdot,\cdot)$ spanner completion, both pairs (s,x) and (t,x) are satisfied, and at least one of the pairs' distances improves by at least $\frac{\varepsilon}{4} W(s,t)$.

Proof (Theorem 2.1). By Lemma 8 with $\varepsilon' = \frac{\varepsilon}{4}$, there are $\Omega(\varepsilon z)$ vertices $x \in V(G')$ which are of distance $\leq \frac{\varepsilon}{4} W(s,t)$ from some vertex in $\pi(s,t)$, so adding path $\pi(s,t)$ of weight z improves $\Omega(\varepsilon z)$ vertex pairs. If Z is the total weight added during $+\varepsilon W(\cdot,\cdot)$ spanner completion, then there are $\Omega(\varepsilon Z)$ improvements.

Once a vertex pair is set-off, it is only improved a *constant* number of times (since any such pair (s,x) or (t,x), once set-off, has error $\frac{\varepsilon}{2} W(s,t)$, and any improvement is by $\Omega(\varepsilon W(s,t))$). Then by considering the number of improvements, we obtain $\Omega(\varepsilon Z) = O(n^2) \implies Z = O(\frac{1}{\varepsilon} n^2)$. This result does not depend on d-lightweight initialization, so set $d = 0$. The total weight of the spanner H is $O(\frac{1}{\varepsilon} n^2)$; since $w(\text{MST}(G)) = \frac{n}{2}$, we obtain lightness$(H) = O_\varepsilon(n)$ as desired. □

Lemma 10. *Let $(u, v) \in V(G)$ be a vertex pair such that $\pi(u, v)$ is added during $+(4+\varepsilon)W(\cdot,\cdot)$ spanner completion. Let a and b be d-lightweight neighbors of s and t in G', respectively, such that $d_{H_0}(x, a) \leq W(s,t)$ and $d_{H_0}(y, b) \leq W(s,t)$. Let $x \in N$ be a vertex in G' which is of distance $\leq \frac{\varepsilon}{4}W(s,t)$ from some vertex y in $\pi(s,t)$. Then both of the following hold:*

1. $d_{H_1}(a, x) \leq d_{G'}(a, x) + (2+\varepsilon)W(s,t)$ *and* $d_{H_1}(b, x) \leq d_{G'}(b, x) + (2+\varepsilon)W(s,t)$
2. $d_{H_0}(a, x) - d_{H_1}(a, x) \geq \frac{\varepsilon}{4}W(s,t)$ *or* $d_{H_0}(b, x) - d_{H_1}(b, x) \geq \frac{\varepsilon}{4}W(s,t)$.

Again, this is proved using the same methods as in Lemmas 2, 3, and 9; see Fig. 3.

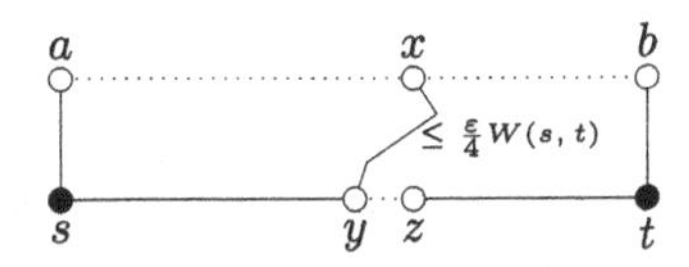

Fig. 3. Illustration of Lemma 10. By adding $\pi(s,t)$ to H during $+(4+\varepsilon)W(\cdot,\cdot)$ spanner completion, both pairs (a, x) and (b, x) are satisfied, and at least one of the pairs' distances improves by at least $\frac{\varepsilon}{4}W(s,t)$.

Proof (Theorem 2.2). Let (s,t) be a vertex pair for which $\pi(s,t)$ is added during $+(4+\varepsilon)W(\cdot,\cdot)$ spanner completion. Again, by Lemma 8 with $\varepsilon' = \frac{\varepsilon}{4}$, there are $\Omega(\varepsilon z)$ choices for x which are of distance $\leq \frac{\varepsilon}{4}W(s,t)$ from some vertex in $\pi(u,v)$.

Similar to [19], we observe that the first edge (say st_1) in $\pi(s,t)$ (starting from s) is absent in H_0 immediately before $\pi(s,t)$ is added. Suppose otherwise $st_1 \in E(H_0)$, then consider the pair (t_1, t). Since $+(4+\varepsilon)W(\cdot,\cdot)$ spanner completion considers all vertex pairs in nondecreasing $W(s,t)$ and then by distance $d_G(s,t)$, the pair (t_1, t) is already satisfied before considering (s,t). Then $d_{H_0}(s,t) \leq w(st_1) + d_{H_0}(t_1, t) \leq w(st_1) + [d_G(t_1,t) + (4+\varepsilon)W(s,t)] \leq d_G(s,t) + (4+\varepsilon)W(s,t)$, contradicting that $\pi(s,t)$ was added to H. Symmetrically, the last edge in $\pi(s,t)$ is absent in H_0.

By Lemma 7 and the above observation, we can establish there are $\Omega(d^{1/2})$ choices for a and b. Then for every choice of x, a, b, adding $\pi(s,t)$ sets off the pairs (a, x), (b, x) if not already, and improves at least one pair's distance by at least $\frac{\varepsilon}{4}W(s,t)$. By Lemma 8, this gives $\Omega(\varepsilon d^{1/2} z)$ improvements upon adding z total edge weight in $\pi(s,t)$.

If Z is the total weight added during $+(4+\varepsilon)W(\cdot,\cdot)$ spanner completion, then

$$\Omega(\varepsilon d^{1/2} Z) = O\left(\frac{n^2}{\varepsilon}\right) \implies Z = O_\varepsilon\left(\frac{n^2}{d^{1/2}}\right).$$

By setting $d := n^{2/3}$, we obtain that the total weight of the spanner H is $O_\varepsilon(n^{5/3})$; since $w(\text{MST}(G)) = \frac{n}{2}$, this implies lightness$(H) = O_\varepsilon(n^{2/3})$ as desired. □

Acknowledgements. The authors wish to thank Michael Elkin, Faryad Darabi Sahneh, and the anonymous reviewers for their discussion and comments.

References

1. Abboud, A., Bodwin, G.: The 4/3 additive spanner exponent is tight. J. ACM (JACM) **64**(4), 1–20 (2017)
2. Ahmed, R., Bodwin, G., Darabi Sahneh, F., Kobourov, S., Spence, R.: Weighted additive spanners. In: Adler, I., Müller, H. (eds.) WG 2020. LNCS, vol. 12301, pp. 401–413. Springer, Cham (2020). https://doi.org/10.1007/978-3-030-60440-0_32
3. Ahmed, R., Bodwin, G., Hamm, K., Kobourov, S., Spence, R.: Weighted sparse and lightweight spanners with local additive error. arXiv preprint arXiv:2103.09731 (2021)
4. Ahmed, R., et al.: Graph spanners: a tutorial review. Comput. Sci. Rev. **37**, 100253 (2020)
5. Ahmed, R., Bodwin, G., Sahneh, F.D., Hamm, K., Kobourov, S., Spence, R.: Multi-level weighted additive spanners. arXiv preprint arXiv:2102.05831 (2021)
6. Aingworth, D., Chekuri, C., Indyk, P., Motwani, R.: Fast estimation of diameter and shortest paths (without matrix multiplication). SIAM J. Comput. **28**, 1167–1181 (1999)
7. Althöfer, I., Das, G., Dobkin, D., Joseph, D.: Generating sparse spanners for weighted graphs. In: Gilbert, J.R., Karlsson, R. (eds.) SWAT 1990. LNCS, vol. 447, pp. 26–37. Springer, Heidelberg (1990). https://doi.org/10.1007/3-540-52846-6_75
8. Baswana, S., Kavitha, T., Mehlhorn, K., Pettie, S.: Additive spanners and (α, β)-spanners. ACM Trans. Algorithms (TALG) **7**(1), 5 (2010)
9. Bodwin, G.: A note on distance-preserving graph sparsification. arXiv preprint arXiv:2001.07741 (2020)
10. Cai, L., Keil, J.M.: Computing visibility information in an inaccurate simple polygon. Int. J. Comput. Geom. Appl. **7**(6), 515–538 (1997)
11. Censor-Hillel, K., Kavitha, T., Paz, A., Yehudayoff, A.: Distributed construction of purely additive spanners. In: Gavoille, C., Ilcinkas, D. (eds.) DISC 2016. LNCS, vol. 9888, pp. 129–142. Springer, Heidelberg (2016). https://doi.org/10.1007/978-3-662-53426-7_10
12. Chandra, B., Das, G., Narasimhan, G., Soares, J.: New sparseness results on graph spanners. In: Proceedings of the Eighth Annual Symposium on Computational Geometry, pp. 192–201. ACM (1992)
13. Chechik, S.: New additive spanners. In: Proceedings of the Twenty-Fourth Annual ACM-SIAM Symposium on Discrete Algorithms (SODA), pp. 498–512. Society for Industrial and Applied Mathematics (2013)
14. Chechik, S., Wulff-Nilsen, C.: Near-optimal light spanners. In: Proceedings of the Twenty-Seventh Annual ACM-SIAM Symposium on Discrete Algorithms (SODA), pp. 883–892. Society for Industrial and Applied Mathematics (2016)
15. Cygan, M., Grandoni, F., Kavitha, T.: On pairwise spanners. In: Proceedings of 30th International Symposium on Theoretical Aspects of Computer Science (STACS 2013), vol. 20, pp. 209–220 (2013)
16. Dobson, A., Bekris, K.E.: Sparse roadmap spanners for asymptotically near-optimal motion planning. Int. J. Robot. Res. **33**(1), 18–47 (2014)
17. Dor, D., Halperin, S., Zwick, U.: All-pairs almost shortest paths. SIAM J. Comput. **29**(5), 1740–1759 (2000)

18. Elkin, M., Gitlitz, Y., Neiman, O.: Almost shortest paths and PRAM distance oracles in weighted graphs. arXiv preprint arXiv:1907.11422 (2019)
19. Elkin, M., Gitlitz, Y., Neiman, O.: Improved weighted additive spanners. arXiv preprint arXiv:2008.09877 (2020)
20. Elkin, M., Neiman, O., Solomon, S.: Light spanners. In: Esparza, J., Fraigniaud, P., Husfeldt, T., Koutsoupias, E. (eds.) ICALP 2014. LNCS, vol. 8572, pp. 442–452. Springer, Heidelberg (2014). https://doi.org/10.1007/978-3-662-43948-7_37
21. Elkin, M., Peleg, D.: $(1+\epsilon, \beta)$-spanner constructions for general graphs. SIAM J. Comput. **33**(3), 608–631 (2004)
22. Erdős, P.: Extremal problems in graph theory. In: Proceedings of the Symposium on Theory of Graphs and its Applications, p. 2936 (1963)
23. Filtser, A., Solomon, S.: The greedy spanner is existentially optimal. In: Proceedings of the 2016 ACM Symposium on Principles of Distributed Computing, pp. 9–17. ACM (2016)
24. Kavitha, T.: New pairwise spanners. Theory Comput. Syst. **61**(4), 1011–1036 (2017)
25. Kavitha, T., Varma, N.M.: Small stretch pairwise spanners and approximate d-preservers. SIAM J. Discrete Math. **29**(4), 2239–2254 (2015)
26. Khuller, S., Raghavachari, B., Young, N.: Balancing minimum spanning trees and shortest-path trees. Algorithmica **14**(4), 305–321 (1995)
27. Knudsen, M.B.T.: Additive spanners: a simple construction. In: Ravi, R., Gørtz, I.L. (eds.) SWAT 2014. LNCS, vol. 8503, pp. 277–281. Springer, Cham (2014). https://doi.org/10.1007/978-3-319-08404-6_24
28. Le, H., Solomon, S.: A unified and fine-grained approach for light spanners. arXiv preprint arXiv:2008.10582 (2020)
29. Marble, J.D., Bekris, K.E.: Asymptotically near-optimal planning with probabilistic roadmap spanners. IEEE Trans. Robot. **29**(2), 432–444 (2013)
30. Peleg, D., Schäffer, A.A.: Graph spanners. J. Graph Theory **13**(1), 99–116 (1989)
31. Peleg, D., Upfal, E.: A trade-off between space and efficiency for routing tables. J. ACM (JACM) **36**(3), 510–530 (1989)
32. Pettie, S.: Low distortion spanners. ACM Trans. Algorithms (TALG) **6**(1), 7 (2009)
33. Salzman, O., Shaharabani, D., Agarwal, P.K., Halperin, D.: Sparsification of motion-planning roadmaps by edge contraction. Int. J. Robot. Res. **33**(14), 1711–1725 (2014)
34. Woodruff, D.P.: Additive spanners in nearly quadratic time. In: Abramsky, S., Gavoille, C., Kirchner, C., Meyer auf der Heide, F., Spirakis, P.G. (eds.) ICALP 2010. LNCS, vol. 6198, pp. 463–474. Springer, Heidelberg (2010). https://doi.org/10.1007/978-3-642-14165-2_40

Labeling Schemes for Deterministic Radio Multi-broadcast

Colin Krisko and Avery Miller(✉)

University of Manitoba, Winnipeg, MB, Canada
avery.miller@umanitoba.ca

Abstract. We consider the multi-broadcast problem in arbitrary connected radio networks consisting of n nodes. There are k designated *source nodes* for some fixed $k \in \{1, \dots, n\}$, and each source node has a piece of information that it wants to share with all nodes in the network. We set out to determine the shortest possible labels so that multi-broadcast can be solved deterministically in the labeled radio network by some deterministic distributed algorithm. First, we show that every radio network G with maximum degree Δ can be labeled using $O(\min\{\log k, \log \Delta\})$-bit labels in such a way that multi-broadcast with k sources can be accomplished. This bound is tight for certain network topologies (e.g., complete graphs), but there are networks where significantly shorter labels are sufficient, e.g., we show how to construct a tree with maximum degree $\Theta(\sqrt{n})$ in which gossiping (i.e., multi-broadcast with n sources) can be solved after labeling each node with $O(1)$ bits. For all trees, we provide a labeling scheme and algorithm that will solve gossiping, and, we prove an impossibility result that demonstrates that our labeling scheme is optimal for gossiping in each tree. In particular, we prove that $\Theta(\log D(G))$-bit labels are necessary and sufficient in each tree G, where $D(G)$ denotes the *distinguishing number of* G.

1 Introduction

Information dissemination is one of the fundamental goals for network algorithms. One important primitive is known as k-broadcast: in a network of n nodes, there are k *source nodes* that each have some initial piece of information that they wish to share with all other nodes in the network.

We consider k-broadcast in synchronous radio networks, which is a particular model of wireless networks. More specifically, in a synchronous radio network, time proceeds in rounds, and each node in the network makes a decision in each round whether it will listen, or, transmit a message. In any round, a node receives a message if it listens and exactly one of its neighbours transmits. Otherwise, the node receives nothing, for one of three reasons: it is not listening, or, none of its neighbours are transmitting, or, two or more of its neighbours are transmitting (this case is known as a *collision*, which models radio signal interference).

The possibility of collisions introduces an interesting challenge, as many simultaneous transmissions can prevent information from spreading in the network. In order to solve k-broadcast (as well as many other problems) there needs

Ł. Kowalik et al. (Eds.): WG 2021, LNCS 12911, pp. 374–387, 2021.
https://doi.org/10.1007/978-3-030-86838-3_29

to be some way of breaking symmetry in the behaviour of the nodes, and one way this can be accomplished is by having each node use an assigned label during its execution. In fact, using assigned labels is necessary: it is impossible to deterministically solve 1-broadcast in a 4-cycle with unlabeled nodes (or identical labels) since, after the source node transmits its source message, its two neighbours will behave identically in all future rounds (i.e., both transmit or both listen), which means that the remaining node will never receive the source message. For this reason, many deterministic solutions to communication tasks in radio networks are designed for networks where each node has a unique identifier. In such networks, k-broadcast is always solvable using a simple round-robin algorithm: each node uses its unique identifier to ensure that it transmits in a round by itself, which avoids all transmission collisions, and this is repeated until all information has reached all nodes. So, we see that at least one-bit labels are required, and $O(\log n)$-bit labels are sufficient. This leads us to ask: what is the shortest label size that allows us to solve k-broadcast in radio networks using a deterministic algorithm? A result by Ellen, Gorain, Miller, and Pelc [20] demonstrates that 2-bit labels are necessary and sufficient in the special case of 1-broadcast. In this paper, we set out to answer this question more generally.

1.1 The Model and Problem

We consider networks modeled as simple undirected connected graphs with an arbitrary number of nodes n. For any fixed integer $k \in \{1, \ldots, n\}$, there are k nodes $s_1, \ldots, s_k$ that are designated as *sources*. For each $i \in \{1, \ldots, k\}$, source node s_i initially has a *source message* μ_i.

Execution proceeds in synchronous rounds: each node has a local clock and all local clocks run at the same speed. Each node has a radio that it can use to send or receive transmissions. In each round, each node must choose one radio mode: *transmit* or *listen*. In transmit mode, a node sends an identical transmission to all of its neighbours in the network. We place no restrictions on the contents of the transmission, e.g., it can contain information other than source messages. In listen mode, a node is silent and may receive transmissions. More specifically, in each round t at each node v: (1) if v is in transmit mode in round t, then v does not hear anything in round t; (2) if v is in listen mode in round t, and v has no neighbours in transmit mode in round t, then v does not hear anything in round t; (3) if v is in listen mode in round t, and v has exactly one neighbour w in transmit mode in round t, then v receives the message contained in the transmission by w; (4) if v is in listen mode in round t, and v has two or more neighbours in transmit mode in round t, then v does not hear anything in round t. This final case is often referred to as a *collision*, and we assume that nodes have no way of detecting when a collision occurs.

The *k-broadcast problem* is solved when each node in the network possesses all of the source messages. Two well-known special cases of this problem are when $k = 1$ (called *broadcast*) and $k = n$ (called *gossiping*). A variant of this problem, called *acknowledged k-broadcast*, requires that, at termination, all source nodes know that all nodes possess all of the source messages.

A *labeling scheme* for a network $G = (V, E)$ is any function λ from the set V of nodes into the set of finite binary strings. This function λ has complete information about G: the node set, the edge set, and the set of k designated sources. The string $\lambda(v)$ is called the *label* of the node v. Labels assigned by a labeling scheme are not necessarily distinct. The *length* of a labeling scheme is the maximum label length taken over all network nodes.

Suppose that each network G has been labeled by some labeling scheme. We consider solving the k-broadcast task using a *deterministic distributed algorithm.* In particular, each node initially knows its own label, and, for each $i \in \{1, \ldots, k\}$, the source node s_i possesses its source message μ_i. In each round t, each node makes a decision whether it will transmit or listen in round t, and, if it decides to transmit, it decides on a finite binary string that it will send in its transmission during round t. These decisions are based *only* on the current history of the node, that is: the label of the node, the node's source message (if it has one), and the sequence of messages received by the node before round t.

Our goal is to answer the following question: what is the minimum possible length of a labeling scheme λ such that there exists a deterministic distributed algorithm that solves k-broadcast in networks labeled with λ?

1.2 Our Results

In Sect. 2, we show that every radio network G with maximum degree Δ can be labeled using $O(\min\{\log k, \log \Delta\})$-bit labels in such a way that k-broadcast can be solved by a deterministic distributed algorithm, and we explicitly provide such an algorithm. In Sect. 2.4, we demonstrate that this bound is tight, in the sense that there exist networks such that every labeling scheme sufficient for k-broadcast requires $\Omega(\min\{\log k, \log \Delta\})$-bit labels. However, in Sect. 2.5, we demonstrate that these bounds are not tight in every network: there exist trees on n nodes with maximum degree $\Theta(\sqrt{n})$ in which n-broadcast (gossiping) can be solved after labeling each node with $O(1)$ bits. This inspires the question: can we prove tight bounds that hold in *every* graph?

Restricting to trees, in Sect. 3, we provide a labeling scheme and an accompanying deterministic distributed algorithm that solves gossiping, and we prove that the length of the labeling scheme is optimal for every tree. In particular, we prove that $\Theta(\log D(G))$-bit labels are necessary and sufficient for solving gossiping in each tree G, where $D(G)$ denotes the distinguishing number of G, i.e., the smallest integer c such that G has a node labeling using $\{1, \ldots, c\}$ that is not preserved by any non-trivial graph automorphism. This result also applies more generally to k-broadcast in trees with $k \in \{2, \ldots, n\}$ sources in the case where the k sources are not known when the labeling scheme is applied. From previous work about the distinguishing number of trees [36], our bound can range anywhere from $\Theta(1)$ to $\Theta(\log n)$ depending on the tree, although it is known that the distinguishing number of any tree is bounded above by Δ.

Due to lack of space, some details and proofs have been omitted. They will appear in the full version of the paper.

1.3 Related Work

Information dissemination tasks are well-studied in radio network models in the case where each node has been pre-assigned a unique identifier. One set of results concerns centralized algorithms, i.e., each node has complete knowledge of the network. In this case, much is known about efficient deterministic solutions for broadcast [2,6,7,17,22,27,34], gossiping [28,29], and multi-broadcast [35]. Another direction of research concerns distributed algorithms, i.e., each node initially only knows its own identifier. Again, there has been much progress in devising efficient algorithms for broadcast [8,9,13,15], gossiping [12,13,23,25, 26], and multi-broadcast [10,11,14,24,35].

In contrast to pre-assigned identifiers, there is much previous work related to solving tasks more efficiently after choosing labels for the nodes of the network, or for mobile agents moving in the network (see related surveys [3,16,21,32]). We restrict attention to previous work concerning tasks in radio networks. In [30], the authors proved that $\Theta(\log\log\Delta)$-bit labels are necessary and sufficient for topology recognition in tree radio networks. When nodes have collision detectors, the authors of [31] proved that $\Theta(\log\log\Delta)$-bit labels are necessary and sufficient for computing the size of any radio network, while $O(1)$-bit labels are sufficient for computing the diameter. In [33], the authors considered the set of radio networks where broadcast is possible in $O(1)$ rounds when all nodes know the complete network topology, and they prove that broadcast is possible in such networks if and only if the sum of the lengths of all labels is $\Theta(n)$.

Most relevant to our work are results involving labeling schemes for information dissemination tasks. In [19], the authors showed that broadcast could be achieved in any radio network after applying a labeling scheme with length 2. The worst-case number of rounds used by their algorithm was $\Theta(n)$. In [18], the authors once again showed that $O(1)$-bit labels were sufficient for solving broadcast in any radio network, but they provided faster algorithms: a non-constructive proof that an $O(\varepsilon\log n + \log^2 n)$-round algorithm exists, and an explicit algorithm that completes within $O(\varepsilon\log^2 n)$ rounds (where ε denotes the source eccentricity). In [5], the authors considered broadcast in level-separable radio networks: they proved that 1-bit labels were sufficient, and provided an accompanying algorithm using at most 2ε rounds. In [4], the authors considered arbitrary radio networks, but instead studied the convergecast task: each node has an initial message, and all of these must eventually reach a designated sink node. They provide a labeling scheme using $O(\log n)$-bit labels, and an accompanying convergecast algorithm that uses $O(n)$ rounds. They prove matching lower bounds for certain network topologies.

2 k-Broadcast in Arbitrary Graphs

2.1 Labeling Schemes and Algorithms for Acknowledged Broadcast

We recall some results from [19,20] about solving acknowledged broadcast, and use them to define a new algorithm that will be used later in our work.

Acknowledged Broadcast [20]**.** Given an arbitrary network G with a designated start node s_G, there is a labeling scheme λ_{ack} that labels each node of G using three bits called *join*, *stay*, and *ack*. There is a deterministic distributed algorithm $\mathcal{B}_{ack}$ that executes on the labeled version of G that solves acknowledged broadcast: the designated start node s_G possesses a message μ, the message μ is eventually received by all other nodes in G, and after this occurs, the node s_G eventually receives a message containing the string "ack". In fact, the algorithm $\mathcal{B}_{ack}$ can be viewed as two algorithms performed consecutively. First, a subroutine $\mathcal{B}$ is initiated by s_G, and this algorithm performs the broadcast of μ that eventually reaches all nodes, and, in the process, establishes a global clock (i.e., all nodes have synchronized their local clock value to be equal to s_G's local clock at the end of $\mathcal{B}$). Then, a subroutine $\mathcal{ACK}$ is initiated by a node z that sends the "ack" message that is eventually received by s_G. The "ack" travels one hop per round along the same path that μ traveled from s_G to z, but in reverse order, so the time to complete $\mathcal{ACK}$ is bounded above by the time to complete $\mathcal{B}$. In [20], the unique node z that initiates $\mathcal{ACK}$ is designated by the labeling scheme using the label 001, and z was chosen due to it being the last node to receive μ during the execution of $\mathcal{B}$. This choice of z is important for the correctness, as it ensures that the execution of $\mathcal{B}$ is finished so that $\mathcal{ACK}$ can run on its own (which prevents transmissions from the two subroutines from interfering with one another). However, we note that any node could initiate $\mathcal{ACK}$, as long as it does so after the execution of $\mathcal{B}$ is finished. We will use this fact below to create a modified version of $\mathcal{B}_{ack}$ that will work in the case where we want to designate the initiator of the $\mathcal{ACK}$ subroutine. Another useful observation is that the labeling scheme λ_{ack} never sets the *join*, *stay*, and *ack* bits all to 1 at any node. We will use this fact later to designate a special node in the network as a "coordinator" by setting these three bits to 1, and this will not affect the original labeling or the behaviour of $\mathcal{B}_{ack}$.

Bounded Acknowledged Broadcast [19]**.** Using the same labeling scheme λ_{ack} as above, there is a modification of $\mathcal{B}_{ack}$ so that it satisfies the following property: all nodes know an upper bound m on how many rounds it took to complete the broadcast of μ, and, there is a common round $t_{done} = 3m$ in which all nodes know that the broadcast of the message μ has been completed. We denote this version of the algorithm by $\mathcal{B}_{bounded}$.

Acknowledged Broadcast with Designated Acknowledger. Using the same labeling scheme λ_{ack} as above, and assuming that all nodes know an upper bound m on the number of rounds that elapse during the execution of $\mathcal{B}$ (which could be learned by first executing $\mathcal{B}_{bounded}$, for example), we describe a modification of $\mathcal{B}_{ack}$ so that the acknowledgement process begins from a designated node z_{des} that knows that it must initiate the $\mathcal{ACK}$ algorithm (e.g., it could be given a special label to indicate this). The algorithm consists of first executing $\mathcal{B}$, which performs the broadcast of μ starting at s_G, then, in round $m+1$, the node z_{des} initiates the $\mathcal{ACK}$ algorithm. This algorithm works and completes within $2m$ rounds, since: the execution of $\mathcal{B}$ is finished by round m (so all nodes know μ by round m), the execution of $\mathcal{ACK}$ takes at most an additional m rounds, and, there is no interference between the $\mathcal{B}$ and $\mathcal{ACK}$ subroutines. We denote this version of the algorithm by $\mathcal{B}_{ack:des}$.

2.2 Labeling Scheme λ_{kB} for k-Broadcast

In this section, we provide a labeling scheme λ_{kB} that will be used by our k-broadcast algorithm $\mathcal{KB}$ (described in Sect. 2.3). At a high level, the algorithm $\mathcal{KB}$ works in two steps: first, the k source messages are collected at a coordinator node (arbitrarily chosen by the labeling scheme), then, the coordinator broadcasts all the source messages to the entire network.

The number of bits used by our scheme is $O(\min\{\log k, \log \Delta\})$. To achieve this upper bound, two different labeling strategies are used, depending on the relationship between k and Δ. The labeling scheme has complete information about the network and the k designated sources, which it uses when choosing which labeling strategy to employ, and it uses a single bit in the labels to signal to the k-broadcast algorithm which strategy was used. Suppose that we are provided with a network G with k designated source nodes $s_1, \ldots, s_k$. We assign a label to each node v in G, and the label at each node v consists of 5 components: a *strat* bit, a *join* bit, a *stay* bit, an *ack* bit, and a binary string *sched*. The labeling scheme assigns values to the components as follows. Choose an arbitrary node $r \in G$. This node will act as the *coordinator*. Apply the labeling scheme λ_{ack} (see Sect. 2.1) to G with designated start node $s_G = r$. This will set the *join*, *stay*, and *ack* bits at each node v. For the coordinator node r, set $join = stay = ack = 1$. There are two cases based on k and Δ. If $k \leq \Delta$, then: Set *strat* to 0 at each node v. For each source node s_i with $i \in \{1, \ldots, k\}$, set *sched* at s_i to be the binary representation of i. For each node $v \notin \{s_1, \ldots, s_k\}$, set *sched* at v to the value 0. Otherwise, if $k > \Delta$, then: Set *strat* to 1 at each node v. Compute a distance-two colouring of the graph G. Let c be the number of colours used. For each node v, set *sched* at v to be the $\lceil \log_2(c+1) \rceil$-bit binary representation of the colour assigned to v in the distance-two colouring of G.

2.3 Algorithm $\mathcal{KB}$ for k-Broadcast

In this section, we describe our k-broadcast algorithm $\mathcal{KB}$ that is executed after the network nodes have been labeled using λ_{kB} from Sect. 2.2. The algorithm's execution consists of three subroutines performed consecutively: **Initialize**, **Aggregate**, and **Inform**. These subroutines make use of the algorithms $\mathcal{B}_{ack}$, $\mathcal{B}_{bounded}$, and $\mathcal{B}_{ack:des}$ described in Sect. 2.1.

The **Initialize** subroutine consists of executing $\mathcal{B}_{bounded}$. The start node is the coordinator r (the unique node with *join*, *stay*, and *ack* bits all set to 1), and the broadcast message is "init". At the conclusion of the execution, all nodes know an upper bound m on the number of rounds that elapsed during the broadcast of the "init" message, and, they all received this value before round $t_{done} = 3m$. In round $3m$, all nodes terminate the subroutine.

The **Aggregate** subroutine is designed to collect all the source messages at the coordinator r. There are two possible algorithms, and the nodes will run one of these two algorithms depending on the value of the *strat* bit that was set by the labeling scheme λ_{kB} (and this value is the same at all nodes). We present these two algorithms separately below.

If the *strat* bit is 0, the nodes run INDIVIDUAL-COLLECT, in which the source messages are gathered at r one at a time. More specifically, the execution proceeds in phases, each consisting of exactly $2m$ rounds. For each $i \geq 1$, at the start of the ith phase, the coordinator r initiates the $\mathcal{B}_{ack:des}$ algorithm with a broadcast message containing the value of i. In the $(m+1)^{th}$ round of the ith phase, the unique node that has its *sched* bits set to the binary of representation of i initiates the acknowledgement process, i.e., it will act as z_{des}. By the definition of λ_{kB}, note that $z_{des} = s_i$. In its transmitted "ack" message, z_{des} includes its source message μ_i. Eventually, there will be a phase j in which the coordinator r does not receive an "ack" message, and, in phase $j+1$, the coordinator r initiates the $\mathcal{B}_{ack}$ algorithm with broadcast message "done". Upon receiving the "done" message, each node terminates the INDIVIDUAL-COLLECT subroutine at the end of the current phase.

If the *strat* bit is 1, the nodes run ROUNDROBIN-COLLECT, in which a round-robin schedule is repeated until all source messages have reached r. More specifically, each node computes `numColours` using the calculation $2^{|sched|} - 1$, where $|sched|$ represents the number of bits in the *sched* part of its label. Then, the execution consists of m phases, each consisting of exactly `numColours` rounds. In the ith round of each phase, a node transmits if and only if its *sched* bits are equal to the binary representation of i, and, if it transmits, its message is equal to the subset of source messages $\{\mu_1, \ldots, \mu_k\}$ that it knows. At the end of the mth phase, each node terminates the ROUNDROBIN-COLLECT subroutine.

The **Inform** subroutine consists of executing $\mathcal{B}_{bounded}$. The start node is the coordinator r, and the broadcast message is equal to the subset of source messages $\{\mu_1, \ldots, \mu_k\}$ that r knows. All nodes terminate this subroutine at the same time, and they all know that k-broadcast has been completed.

Theorem 1. *Consider any n-node unlabeled network G with maximum degree Δ, and, for any $k \in \{1, \ldots, n\}$, consider any designated source nodes $s_1, \ldots, s_k$ with source messages $\{\mu_1, \ldots, \mu_k\}$. By applying the labeling scheme λ_{kB} and then executing algorithm $\mathcal{KB}$, all nodes possess the complete set of source messages $\{\mu_1, \ldots, \mu_k\}$. The length of λ_{kB} is $O(\min\{\log k, \log \Delta\})$.*

2.4 Existential Lower Bound

In this section, we prove that in any complete graph K_n, any labeling scheme that is sufficient for solving k-broadcast has length at least $\Omega(\min\{\log k, \log \Delta\})$. This matches the $O(\min\{\log k, \log \Delta\})$ worst-case upper bound guaranteed by Theorem 1, which means that the upper bound cannot be improved in general.

The idea behind the proof is to show that each source must be labeled differently by any labeling scheme: otherwise, using an indistinguishability argument, we prove that two sources with the same label will behave the same way in every round, which prevents any other node from receiving their source messages (either due to both being silent, or both transmitting and causing collisions everywhere). Since any labeling scheme using at least k different labels has length at least $\Omega(\log k)$, and $k \leq n = \Delta + 1$ in the complete graph, the result follows.

Theorem 2. *Consider any integer $n > 1$, any $k \in \{1, \ldots, n\}$ and any labeling scheme λ. If there exists a deterministic distributed algorithm that solves k-broadcast on the complete graph K_n labeled by λ, then the length of λ is at least $\Omega(\min\{\log k, \log \Delta\})$.*

2.5 An Example of Better Labeling

For infinitely many values of n, we construct a tree T_n on n nodes with maximum degree $\Delta \in \Theta(\sqrt{n})$ such that, after labeling each node with $O(1)$ bits, n-broadcast can be solved by a deterministic distributed algorithm. The length of the labeling scheme is significantly smaller than the upper and lower bounds from Theorems 1 and 2, which for T_n would give $\Theta(\min\{\log k, \log \Delta\}) = \Theta(\log n)$.

Let n be any triangular number greater than 1, i.e., there exists a positive integer $x \geq 2$ such that $n = x(x+1)/2$. The tree T_n consists of: a node r, and, for each $i \in \{2, \ldots, x\}$, a node ℓ_i and a path of length i with endpoints ℓ_i and r. Note that node r has degree $x - 1 \in \Theta(\sqrt{n})$, each node in $\{\ell_2, \ldots, \ell_x\}$ has degree 1, and all other nodes have degree 2.

We label each node of T_n with 2 bits: node r is given the label 11, each node in $\{\ell_2, \ldots, \ell_{x-1}\}$ is given the label 01, the node ℓ_x is given the label 10, and all other nodes are given the label 00. To solve n-broadcast, the idea is to initiate a broadcast from node r to send an "init" message that gets forwarded along the paths towards each leaf $\ell_2, \ldots, \ell_x$, and, when each leaf receives the "init" message, it sends a "gather" message back towards r. Each time a node forwards a "gather" message, it appends its own source message. Since the paths have distinct lengths, the "gather" messages along each path arrive back at r at different times, which prevents collisions at r. When ℓ_x sends its "gather" message, it also includes the string "last". After r receives "last", it initiates a broadcast of a message containing all the source messages it knows. Of all the leaf-to-r paths, the one involving leaf ℓ_x is the longest, which means that the "gather" message containing "last" is the last one that r receives, and this guarantees that r has all of the source messages before initiating the final broadcast.

3 Gossiping in Trees

We first review some relevant definitions about arbitrary graphs $G = (V, E)$.

Definition 1. *A bijection $\phi : V \to V$ is called an* automorphism *of G if, for every $u, v \in V$, we have that $\{u, v\} \in E$ if and only if $\{\phi(u), \phi(v)\} \in E$. An automorphism ϕ is* non-trivial *if there exists $v \in V$ such that $\phi(v) \neq v$. For a fixed graph G, the set of all its automorphisms is denoted by* Aut(G).

Definition 2 (Albertson and Collins [1]). *A labeling $\rho : V \to \{1,\ldots,c\}$ is called* c-distinguishing *if, for every non-trivial $\phi \in \text{Aut}(G)$, there exists $v \in V$ such that $\rho(v) \neq \rho(\phi(v))$. The* distinguishing number *of G, denoted by $D(G)$, is the smallest integer c such that G has a labeling that is c-distinguishing.*

3.1 Lower Bound

We prove that any labeling scheme that is sufficient for gossiping to be solved in a tree G must use labels of size $\Omega(\log D(G))$. In fact, the proof of this lower bound also works in the case of k-broadcast for each $k \in \{2,\ldots,n-1\}$, however, only under the additional condition that the source nodes are not known when the labeling scheme is applied. First, we prove a structural result about any non-trivial automorphism ϕ of a tree.

Lemma 1. *For any tree G and any non-trivial automorphism $\phi \in \text{Aut}(G)$, consider any node x such that $x \neq \phi(x)$. Let $\ell \geq 1$ be the length of the path with endpoints x and $\phi(x)$. Let $v_1 = x$ and let $v_{\ell+1} = \phi(x)$, and denote by $(v_1,\ldots,v_{\ell+1})$ the sequence of nodes along the path. For each $i \in \{0,\ldots,\lfloor \ell/2 \rfloor\}$, we have that $v_{\ell+1-i} = \phi(v_{1+i})$.*

To prove the desired lower bound, we show that any labeling scheme λ that allows gossiping to be solved in a tree G is also a distinguishing labeling of G. The proof is by contradiction: assume there is a non-trivial automorphism ϕ that preserves λ, take any node x such that $x \neq \phi(x)$, and consider the path P between x and $\phi(x)$ in G. By Lemma 1, there are two nodes u, w in P that are adjacent or have a common neighbour in P such that $w = \phi(u)$. We prove that they perform the same action (both transmit or both listen) in each round in the execution of any algorithm. This prevents the source message at x from reaching $\phi(x)$, which contradicts that gossiping can be solved.

Theorem 3. *Consider any labeling scheme λ and any deterministic distributed algorithm $\mathcal{A}$. If $\mathcal{A}$ solves the gossiping task when executed by the nodes of $\lambda(G)$ for some tree G, then the length of λ is $\Omega(\log D(G))$.*

3.2 Upper Bound

We present a $O(\log D(G))$-bit labeling scheme and a deterministic distributed gossiping algorithm that runs on the labeled network.

The Labeling Scheme λ_{gossip}. The label at each node v consists of: a *join* bit, a *stay* bit, an *ack* bit, a *term* bit, and a binary string *sched*. These values are assigned as follows. Choose an arbitrary 'coordinator' node $r \in G$. Apply the labeling scheme λ_{ack} (see Sect. 2.1) to G with designated start node $s_G = r$. This will set the *join*, *stay*, and *ack* bits at each node v. For the coordinator node r, set $join = stay = ack = 1$. For some $D(G)$-distinguishing labeling f of the nodes of G, set the *sched* bits at each node v to be the binary representation of $f(v)$. The *term* bit is set to 1 at exactly one node: the node whose source message is last to arrive at the coordinator r during the **Aggregate** subroutine (as described at the end of Sect. 3.2). The *term* bit is 0 all other nodes.

The Gossiping Algorithm ($\mathcal{GOSSIP}$). Assume that the tree's nodes have been labeled using labeling scheme λ_{gossip}. The algorithm's execution consists of three subroutines performed consecutively: **Initialize**, **Aggregate**, and **Inform**.

The first stage of the algorithm, **Initialize**, consists of executing $\mathcal{B}_{bounded}$ starting at r with message "init". As G is a tree, no collisions occur during the execution of $\mathcal{B}_{bounded}$, so the round number in which a node v first receives "init" allows the node to compute its exact distance from r (which we will denote by $d(v)$). For the same reason, the value of m computed by $\mathcal{B}_{bounded}$ is equal to the exact height of the tree rooted at r, since this is exactly how many rounds it took for the initial broadcast to complete. The third stage of the algorithm, **Inform**, consists of executing $\mathcal{B}_{ack}$ starting at r, and the broadcast message is equal to the set of source messages that r knows. The second stage of the algorithm, mainly consisting of the **Aggregate** subroutine, is the most interesting. The remainder of this section is dedicated to its description.

The main idea is for each node w in the tree rooted at r to compute an encoding $\text{enc}(w)$ of the subtree rooted at itself, and then use this encoding to compute a transmission delay value that it will use to avoid transmission collisions when sending information to its parent. In rounds in which w decides to transmit, it always includes in its transmitted message: the source messages it knows, its $\text{enc}(w)$ value, and its exact distance $d(w)$ to the coordinator (so that any recipient of its message can distinguish whether it is a parent or a child of w). Suppose that a node w has received messages from some (possibly empty) subset of its children (if it has any). Denote these children as $v_1, \ldots, v_\ell$, ordered arbitrarily. Then w computes its own encoding as $enc(w) = 2^{f(w)} \cdot 3^{d(w)} \cdot \prod_{j=1}^{\ell} p_{j+2}^{\text{enc}(v_j)}$, where p_{j+2} denotes the $(j+2)^{\text{th}}$ smallest prime number. For any two nodes a, b with a common ancestor c, we are able to show that this encoding ensures that $\text{enc}(a), \text{enc}(b), \text{enc}(c)$ are all different. At a high level: $\text{enc}(c)$ is different since $d(c)$ is different from $d(a), d(b)$, and, we can prove that $\text{enc}(a) \neq \text{enc}(b)$ either because $f(a) \neq f(b)$, or, the subtrees rooted at a and b are non-isomorphic (which must be the case if $f(a) = f(b)$, since f is a distinguishing labeling).

Each node w computes a transmission delay value $\tau(w) = p_{\text{enc}(w)}$ (again, using p_i to denote the ith-smallest prime number) and it transmits every $\tau(w)$ rounds. Using the fact that siblings a, b and any common ancestor c will have different $\text{enc}(a), \text{enc}(b), \text{enc}(c)$, it follows that all siblings and their common ancestors have mutually co-prime transmission delay values. This implies that each sibling will eventually succeed in transmitting its knowledge to its parent in G.

One challenge is that a node w may not have complete information, since the values $\text{enc}(w)$ and $\tau(w)$ are computed in each round using only the information that w has received before the current round (and its own label). There might be a long delay before w has heard from a certain child x, or, w may hear from a child x multiple times with different encoding values. To deal with this issue: we overwrite a previously saved encoding e if a received $\text{enc}(x)$ is a multiple of e, and otherwise we append $\text{enc}(x)$ to our list of saved encodings from children.

We are able to carefully prove that each node w *eventually* receives a message from each of its children and has complete information about its entire subtree, and from that point on, the transmission delays will diverge and allow w and all of its siblings to successfully transmit a message to their parent without collision.

Another challenge is that nodes must know when to stop the second stage of the algorithm. There are two challenges: (1) How does the coordinator know when it possesses the complete set of source messages so that it can begin the **Inform** stage? (2) How do we make sure that all of the other nodes have stopped executing **Aggregate** so that they will be listening when the **Inform** stage begins? To address the first challenge, the labeling scheme simulates the **Aggregate** subroutine described above, and makes note of the last source message μ_{last} to arrive at the coordinator r. The labeling scheme uses one bit called *term*, which it sets to 1 at the node s_{last} that started with source message μ_{last}, and sets to 0 at all other nodes. Then, during the actual execution of the algorithm, the node with *term* bit set to 1 will include with its source message the string "last". When r receives a message containing "last", it knows that it possesses all of the source messages. To address the second challenge, we introduce a new set of rounds that are interleaved with the rounds of **Aggregate**. In particular, in odd-numbered rounds of the second stage, the **Aggregate** subroutine is executed as described above, and initially, in even-numbered rounds, all nodes listen. Eventually, the coordinator r receives the "last" message in some round t (which is an odd-numbered round, as this occurs during the execution of **Aggregate**). Then, in round $t + 1$, coordinator r transmits a message containing "finish". Whenever a node receives a "finish" message for the first time, it stops executing **Aggregate** immediately, and it transmits a "finish" message two rounds later (in the next even-numbered round). As the height of the tree is m, it follows that all nodes have terminated **Aggregate** by round $t + 2m - 1$ and the last "finish" message is sent by round $t + 2m + 1$. Thus, the coordinator can safely begin the **Inform** stage of the algorithm in round $t + 2m + 2$.

Theorem 4. *Consider any n-node unlabeled tree G, and suppose that each node has an initial source message. By applying the $O(\log D(G))$-bit labeling scheme λ_{gossip} and then executing algorithm $\mathcal{GOSSIP}$, all nodes possess the complete set of source messages.*

4 Future Work

We focused on the optimal length of labeling schemes that can be used for multi-broadcast. We would like to generalize our results about trees, and we wonder if the optimal label length is related to the distinguishing number for more general graphs. Interesting open problems remain about optimizing the round complexity and message size of the multi-broadcast algorithms, and determining tradeoffs between these quantities versus the length of the labeling scheme.

Acknowledgments. Avery Miller acknowledges the support of NSERC, Discovery Grant RGPIN-2017-05936.

References

1. Albertson, M.O., Collins, K.L.: Symmetry breaking in graphs. Electron. J. Comb. **3**(1) (1996). http://www.combinatorics.org/Volume_3/Abstracts/v3i1r18.html
2. Alon, N., Bar-Noy, A., Linial, N., Peleg, D.: A lower bound for radio broadcast. J. Comput. Syst. Sci. **43**(2), 290–298 (1991). https://doi.org/10.1016/0022-0000(91)90015-W
3. Boyar, J., Favrholdt, L.M., Kudahl, C., Larsen, K.S., Mikkelsen, J.W.: Online algorithms with advice: a survey. ACM Comput. Surv. **50**(2), 19:1–19:34 (2017). https://doi.org/10.1145/3056461
4. Bu, G., Lotker, Z., Potop-Butucaru, M., Rabie, M.: Lower and upper bounds for deterministic convergecast with labeling schemes. Research report, Sorbonne Université, May 2020. https://hal.archives-ouvertes.fr/hal-02650472
5. Bu, G., Potop-Butucaru, M., Rabie, M.: Wireless broadcast with short labels. In: Georgiou, C., Majumdar, R. (eds.) NETYS 2020. LNCS, vol. 12129, pp. 146–169. Springer, Cham (2021). https://doi.org/10.1007/978-3-030-67087-0_10
6. Chlamtac, I.: The wave expansion approach to broadcasting in multihop radio networks. IEEE Trans. Commun. **39**(3), 426–433 (1991). https://doi.org/10.1109/26.79285
7. Chlamtac, I., Kutten, S.: On broadcasting in radio networks-problem analysis and protocol design. IEEE Trans. Commun. **33**(12), 1240–1246 (1985). https://doi.org/10.1109/TCOM.1985.1096245
8. Chlebus, B.S., Gasieniec, L., Gibbons, A., Pelc, A., Rytter, W.: Deterministic broadcasting in ad hoc radio networks. Distrib. Comput. **15**(1), 27–38 (2002). https://doi.org/10.1007/s446-002-8028-1
9. Chlebus, B.S., Gasieniec, L., Östlin, A., Robson, J.M.: Deterministic radio broadcasting. In: Montanari, U., Rolim, J.D.P., Welzl, E. (eds.) ICALP 2000. LNCS, vol. 1853, pp. 717–729. Springer, Heidelberg (2000). https://doi.org/10.1007/3-540-45022-X_60
10. Chlebus, B.S., Kowalski, D.R., Pelc, A., Rokicki, M.A.: Efficient distributed communication in ad-hoc radio networks. In: Aceto, L., Henzinger, M., Sgall, J. (eds.) ICALP 2011. LNCS, vol. 6756, pp. 613–624. Springer, Heidelberg (2011). https://doi.org/10.1007/978-3-642-22012-8_49
11. Chlebus, B.S., Kowalski, D.R., Radzik, T.: Many-to-many communication in radio networks. Algorithmica **54**(1), 118–139 (2009). https://doi.org/10.1007/s00453-007-9123-5
12. Christersson, M., Gąsieniec, L., Lingas, A.: Gossiping with bounded size messages in *ad hoc* radio networks. In: Widmayer, P., Eidenbenz, S., Triguero, F., Morales, R., Conejo, R., Hennessy, M. (eds.) ICALP 2002. LNCS, vol. 2380, pp. 377–389. Springer, Heidelberg (2002). https://doi.org/10.1007/3-540-45465-9_33
13. Chrobak, M., Gasieniec, L., Rytter, W.: Fast broadcasting and gossiping in radio networks. J. Algorithms **43**(2), 177–189 (2002). https://doi.org/10.1016/S0196-6774(02)00004-4
14. Clementi, A.E.F., Monti, A., Silvestri, R.: Distributed multi-broadcast in unknown radio networks. In: Proceedings of the Twentieth Annual ACM Symposium on Principles of Distributed Computing, PODC 2001, pp. 255–264. Association for Computing Machinery, New York (2001). https://doi.org/10.1145/383962.384040
15. Clementi, A.E.F., Monti, A., Silvestri, R.: Distributed broadcast in radio networks of unknown topology. Theoret. Comput. Sci. **302**(1–3), 337–364 (2003). https://doi.org/10.1016/S0304-3975(02)00851-4

16. Dobrev, S., Kralovic, R., Kralovic, R., et al.: Computing with advice: when knowledge helps. Bull. EATCS **2**(110) (2013)
17. Elkin, M., Kortsarz, G.: An improved algorithm for radio broadcast. ACM Trans. Algorithms **3**(1), 8:1–8:21 (2007). https://doi.org/10.1145/1219944.1219954
18. Ellen, F., Gilbert, S.: Constant-length labelling schemes for faster deterministic radio broadcast. In: SPAA 2020: 32nd ACM Symposium on Parallelism in Algorithms and Architectures, pp. 213–222 (2020). https://doi.org/10.1145/3350755.3400238
19. Ellen, F., Gorain, B., Miller, A., Pelc, A.: Constant-length labeling schemes for deterministic radio broadcast. CoRR abs/1710.03178 (2017). http://arxiv.org/abs/1710.03178
20. Ellen, F., Gorain, B., Miller, A., Pelc, A.: Constant-length labeling schemes for deterministic radio broadcast. In: The 31st ACM on Symposium on Parallelism in Algorithms and Architectures, SPAA 2019, pp. 171–178 (2019). https://doi.org/10.1145/3323165.3323194
21. Feuilloley, L.: Introduction to local certification. CoRR abs/1910.12747 (2019). http://arxiv.org/abs/1910.12747
22. Gaber, I., Mansour, Y.: Centralized broadcast in multihop radio networks. J. Algorithms **46**(1), 1–20 (2003). https://doi.org/10.1016/S0196-6774(02)00292-4
23. Gąsieniec, L.: On efficient gossiping in radio networks. In: Kutten, S., Žerovnik, J. (eds.) SIROCCO 2009. LNCS, vol. 5869, pp. 2–14. Springer, Heidelberg (2010). https://doi.org/10.1007/978-3-642-11476-2_2
24. Gasieniec, L., Kranakis, E., Pelc, A., Xin, Q.: Deterministic M2M multicast in radio networks. Theoret. Comput. Sci. **362**(1–3), 196–206 (2006). https://doi.org/10.1016/j.tcs.2006.06.017
25. Gasieniec, L., Lingas, A.: On adaptive deterministic gossiping in ad hoc radio networks. Inf. Process. Lett. **83**(2), 89–93 (2002). https://doi.org/10.1016/S0020-0190(01)00312-X
26. Gasieniec, L., Pagourtzis, A., Potapov, I., Radzik, T.: Deterministic communication in radio networks with large labels. Algorithmica **47**(1), 97–117 (2007). https://doi.org/10.1007/s00453-006-1212-3
27. Gasieniec, L., Peleg, D., Xin, Q.: Faster communication in known topology radio networks. Distributed Comput. **19**(4), 289–300 (2007). https://doi.org/10.1007/s00446-006-0011-z
28. Gąsieniec, L., Potapov, I.: Gossiping with unit messages in known radio networks. In: Baeza-Yates, R., Montanari, U., Santoro, N. (eds.) Foundations of Information Technology in the Era of Network and Mobile Computing. ITIFIP, vol. 96, pp. 193–205. Springer, Boston, MA (2002). https://doi.org/10.1007/978-0-387-35608-2_17
29. Gasieniec, L., Potapov, I., Xin, Q.: Time efficient centralized gossiping in radio networks. Theoret. Comput. Sci. **383**(1), 45–58 (2007). https://doi.org/10.1016/j.tcs.2007.03.059
30. Gorain, B., Pelc, A.: Short labeling schemes for topology recognition in wireless tree networks. In: Das, S., Tixeuil, S. (eds.) SIROCCO 2017. LNCS, vol. 10641, pp. 37–52. Springer, Cham (2017). https://doi.org/10.1007/978-3-319-72050-0_3
31. Gorain, B., Pelc, A.: Finding the size and the diameter of a radio network using short labels. Theoret. Comput. Sci. (2021). https://doi.org/10.1016/j.tcs.2021.02.004
32. Ilcinkas, D.: Structural information in distributed computing. Ph.D. thesis, Université de Bordeaux (2019). https://tel.archives-ouvertes.fr/tel-03099722

33. Ilcinkas, D., Kowalski, D.R., Pelc, A.: Fast radio broadcasting with advice. Theoret. Comput. Sci. **411**(14–15), 1544–1557 (2010). https://doi.org/10.1016/j.tcs.2010.01.004
34. Kowalski, D.R., Pelc, A.: Optimal deterministic broadcasting in known topology radio networks. Distrib. Comput. **19**(3), 185–195 (2007). https://doi.org/10.1007/s00446-006-0007-8
35. Levin, L., Kowalski, D.R., Segal, M.: Message and time efficient multi-broadcast schemes. Theoret. Comput. Sci. **569**, 13–23 (2015). https://doi.org/10.1016/j.tcs.2014.12.006
36. Tymoczko, J.: Distinguishing numbers for graphs and groups. Electron. J. Comb. **11**(1) (2004). http://www.combinatorics.org/Volume_11/Abstracts/v11i1r63.html

On 3-Coloring of $(2P_4, C_5)$-Free Graphs

Vít Jelínek[1], Tereza Klimošová[1], Tomáš Masařík[1,2,3](✉), Jana Novotná[1,2], and Aneta Pokorná[1]

[1] Faculty of Mathematics and Physics, Charles University, Prague, Czech Republic
{jelinek,pokorna}@iuuk.mff.cuni.cz
{tereza,masarik,janca}@kam.mff.cuni.cz
[2] Faculty of Mathematics, Informatics and Mechanics, University of Warsaw, Warsaw, Poland
[3] Simon Fraser University, Burnaby, BC, Canada

Abstract. The 3-coloring of hereditary graph classes has been a deeply-researched problem in the last decade. A hereditary graph class is characterized by a (possibly infinite) list of minimal forbidden induced subgraphs $H_1, H_2, \ldots$; the graphs in the class are called $(H_1, H_2, \ldots)$-free. The complexity of 3-coloring is far from being understood, even for classes defined by a few small forbidden induced subgraphs. For H-free graphs, the complexity is settled for any H on up to seven vertices. There are only two unsolved cases on eight vertices, namely $2P_4$ and P_8. For P_8-free graphs, some partial results are known, but to the best of our knowledge, $2P_4$-free graphs have not been explored yet. In this paper, we show that the 3-coloring problem is polynomial-time solvable on $(2P_4, C_5)$-free graphs.

Keywords: 3-coloring · Hereditary classes · $2P_4$-free graphs · Cographs

---

V. Jelínek was supported by project 18-19158S of the Czech Science Foundation. T. Klimošová is supported by the grant no. 19-04113Y of the Czech Science Foundation (GAČR) and the Center for Foundations of Modern Computer Science (Charles Univ. project UNCE/SCI/004). T. Masařík received funding from the European Research Council (ERC) under the European Union's Horizon 2020 research and innovation programme Grant Agreement 714704. He completed a part of this work while he was a postdoc at Simon Fraser University in Canada. J. Novotná and A. Pokorná were supported by SVV-2017-260452 and GAUK 1277018.
A preprint of the full version of this paper is available from arXiv [26].

Ł. Kowalik et al. (Eds.): WG 2021, LNCS 12911, pp. 388–401, 2021.
https://doi.org/10.1007/978-3-030-86838-3_30

1 Introduction

Graph coloring is a notoriously known and well-studied concept in both graph theory and theoretical computer science. A *k-coloring* of a graph $G = (V, E)$ is defined as a mapping $c : V \to \{1, \ldots, k\}$ which is *proper*, i.e., it assigns distinct colors to $u, v \in V$ if $uv \in E$. The k-COLORING problem asks whether a given graph admits a k-coloring. For any $k \geq 3$, the k-coloring is a well-known NP-complete problem [27]. We also define a more general *list-k-coloring* where each vertex v has a list $P(v)$ of allowed colors such that $P(v) \subseteq \{1, \ldots, k\}$. In that case, the coloring function c, in addition to being proper, has to respect the lists, that is, $c(v) \in P(v)$ for every vertex v.

A graph class is *hereditary* if it is closed under vertex deletion. It follows that a graph class $\mathcal{G}$ is hereditary if and only if $\mathcal{G}$ can be characterized by a unique (not necessarily finite) set $\mathcal{H}_\mathcal{G}$ of minimal forbidden induced subgraphs. Special attention was given to hereditary graph classes where $\mathcal{H}_\mathcal{G}$ contains only one or only a very few elements. In such cases, when $\{H\} = \mathcal{H}_\mathcal{G}$, or $\{H_1, H_2, \ldots\} = \mathcal{H}_\mathcal{G}$, we say that $G \in \mathcal{G}$ is *H-free*, or *$(H_1, H_2, \ldots)$-free*, respectively. We let P_t denote the path on t vertices, and C_ℓ the cycle on ℓ vertices. We let $\overline{H}$ denote the complement of a graph H. For two graphs H_1 and H_2, we let $H_1 + H_2$ denote their disjoint union. We write kH for the disjoint union of k copies of a graph H.

In recent years, a lot of attention has been paid to determining the complexity of k-coloring of H-free graphs. Classical results imply that for every $k \geq 3$, k-coloring of H-free graphs is NP-complete if H contains a cycle [14] or an induced claw [23,30]. Hence, it remains to consider the cases where H is a *linear forest*, i.e., a disjoint union of paths. The situation around complexity of (list) k-coloring on P_t-free graphs where $k \geq 4$ has been resolved completely. The cases $k = 4, t \geq 7$ and $k \geq 5, t \geq 6$ are NP-complete [24] while cases for $k \geq 1, t = 5$ are polynomial-time solvable [22]. In fact, k-coloring is polynomial-time solvable on $sP_1 + P_5$-free graphs for any $s \geq 0$ [13]. The borderline case where $k = 4, t = 6$ has been settled recently. There the 4-coloring problem (even the precoloring extension problem with 4 colors) is polynomial-time solvable [33] while the list 4-coloring problem is NP-complete [18].

Now, we move our focus towards the complexity of the 3-coloring problem, which was less well understood, in spite of the amount of the research interest it received in the past years. However, a considerable progress has been made in 2020; a quasi-polynomial algorithm running in time $n^{O(\log^2(n))}$ on n-vertex P_t-free graphs (t is a constant) was shown by Pilipczuk et al. [31] (extending results in [15]). In the realms of polynomiality, Bonomo et al. [1] found a polynomial-time algorithm for P_7-free graphs. Klimošová et al. [28] completed the classification of 3-coloring of H-free graphs, for any H on up to 7 vertices. These results were subsequently extended to $P_6 + rP_3$-free graphs, for any $r \geq 0$ [5]. There are only two remaining graphs on at most 8 vertices, namely P_8 and $2P_4$, for which the complexity of 3-coloring is still unknown. Algorithms for subclasses of P_t-free graphs which avoid one or more additional induced subgraphs, usually cycles, have been studied. They might be a first step in the attempt to settle the case of P_t-free graphs. This turned out to be the case for 3-coloring of P_7-free graphs (as can be seen from preprints [2,7,8] leading to [1]) and 4-coloring of P_6-free graphs [6]. Note that the problem of 4-coloring is NP-complete even when some (P_t, C_ℓ)-free graphs are considered when $t \geq 7$. Hell and Huang [21] and Huang et al. [25] settled many NP-complete cases of this type. These results, in combination with the polynomiality of P_6-free case, leave open only the following cases: (P_7, C_7)-, (P_8, C_7)-, and (P_t, C_3)-free graphs, for $7 \leq t \leq 21$.

Chudnovsky and Stacho [11] studied the problem of 3-coloring of P_8-free graphs which additionally avoid induced cycles of two distinct lengths; specifically, they consider graphs that are (P_8, C_3, C_4)-, (P_8, C_3, C_5)-, and (P_8, C_4, C_5)-free. For the first two cases, they show that all such graphs are 3-colorable. For the last one, they provide a complete list of 3-*critical graphs*, i.e., the graphs with no 3-coloring such that all their proper induced subgraphs are 3-colorable. Independently, using a computer search, Goedgebeur and Schaudt [16] showed that there are only finitely many 3-critical (P_8, C_4)-free graphs. In fact, 3-coloring is polynomial-time solvable on (P_t, C_4)-free graphs for any $t \geq 1$ [19].

The situation concerning $2P_4$ or P_8 is still far from being determined when two forbidden induced subgraphs are considered; in particular, it is not known whether (P_8, C_3)-, (P_8, C_5)-, $(2P_4, C_3)$-, or $(2P_4, C_5)$-free graphs can be 3-colored in polynomial time[1]. This is in contrast with the algorithm for (P_7, C_3)-free graphs [3] which is considerably simpler than the one for P_7-free graphs [1]. Recently, Rojas and Stein [32] approached the problem by showing that for any odd $t \geq 9$, there exists a polynomial-time algorithm that solves the 3-coloring problem in P_t-free graphs of odd girth at least $t - 2$. In particular, their result implies that 3-coloring is polynomial-time solvable for (P_9, C_3, C_5)-free graphs.

Freshly, a similar question was resolved in the case where, instead of a cycle, a 1-subdivision of $K_{1,s}$ (a star with s leaves), denoted as $SDK_{1,s}$, is considered. Chudnovsky, Spirkl, and Zhong have shown that the class of $(SDK_{1,s}, P_t)$-free graphs is list-3-colorable in polynomial time for any $s, t \geq 1$ [10]. For other related results and history of the problem, please consult a recent survey [17].

In this paper, we resolve one of the remaining open problems mentioned above, which considers $2P_4$-free graphs, as we will describe a polynomial-time algorithm for 3-coloring of $(2P_4, C_5)$-free graphs. To the best of our knowledge, this is a first attempt to attack the 3-coloring of $2P_4$-free graphs.

Theorem 1. *The 3-coloring problem is polynomial-time solvable on* $(2P_4, C_5)$*-free graphs.*

To prove our result, we will make use of some relatively standard techniques. Let $\omega(G)$ be the size of the largest clique of graph G. We use a seminal result of Grötschel, Lovász, and Schrijver [20] that shows the k-coloring problem on *perfect graphs*, i.e., graphs where each induced subgraph G' is $\omega(G')$-colorable, can be solved in polynomial time. Perfect graphs are characterized by the strong perfect graph theorem [9] as the graphs that have neither odd-length induced cycles nor complement of odd-length induced cycles on at least five vertices.

[1] First two cases were explicitly mentioned as open in [17] and [32], the latter two cases are open to the best of our knowledge.

As K_4 and $\overline{C_7}$ graphs are not 3-colorable, we can assume that our graph is $(2P_4, C_5, \overline{C_7}, K_4)$-free. As $K_4 \subseteq \overline{C_\ell}$ whenever $\ell \geq 8$ and $2P_4 \subseteq C_\ell$ whenever $\ell \geq 10$, it follows that either the graph is perfect, or it contains C_7 or C_9. In the first case, we are done by the aforementioned polynomial-time algorithm. For the latter cases, we divide the analysis into two further subcases. First, we suppose that the graph is $(2P_4, C_5, C_7, \overline{C_7}, K_4)$-free and therefore it must contain C_9. We first analyze this case then we suppose that graph contains C_7 for the rest of the proof. We will exploit the fact that once we find an induced P_4, the vertices that are not adjacent to it must induce a P_4-free graph (also known as *cograph*). Such graphs were among the first H-free graphs studied, and have many nice properties, e.g., any greedy coloring gives a proper coloring using the least number of colors [4]. We will make use of a slightly stronger statement that handles the list-3-coloring problem.

Theorem 2 ([17]). *The list-3-coloring problem on P_4-free graphs can be solved in polynomial time.*

The 3-coloring algorithm that we develop to prove Theorem 1 cannot be directly extended to solve the more general list-3-coloring problem, since it uses the 3-coloring algorithm for perfect graphs to deal with graphs avoiding C_7 and C_9. However, apart from this one case, the algorithm works with the more general setting of list-3-coloring. In fact, we use reductions of lists as one of our base techniques. After several branching steps with polynomially many branches and suitable structural reductions of the original graph G, the algorithm will transform a 3-coloring instance of a $(2P_4, C_5)$-free graph G to a set of polynomially many heavily structured list-3-coloring instances. These structured instances can then be encoded by a 2-SAT formula, whose satisfiability is solvable in linear time [29].

2 Proof of Theorem 1

We are given a $(2P_4, C_5)$-free graph $G = (V, E)$, and our goal is to determine whether it is 3-colorable. We will present an algorithm that solves this problem in polynomial time. The algorithm begins by checking that the graph is $\overline{C_7}$-free, and that the neighborhood of each vertex induces a bipartite graph, rejecting the instance if the check fails. Note that this check ensures, in particular, that G is K_4-free.

The algorithm then partitions the graph into connected components, solving the 3-coloring problem for each component separately. From now on, we assume that the graph $G = (V, E)$ is connected, $\overline{C_7}$-free, and each of its vertices has a bipartite neighborhood. The basic idea of the algorithm is to choose an initial subgraph N_0 of bounded size, try all possible proper 3-colorings of N_0, and analyze how the precoloring of N_0 affects the possible colorings of the remaining vertices. We let N_1 denote the vertices in $V \setminus N_0$ which are adjacent to at least one vertex of N_0, and we let N_2 be the set $V \setminus (N_0 \cup N_1)$. We will use the notation $N(x)$ for the set of neighbors of x in G, and $N_i(x)$ for $N_i \cap N(x)$.

Our algorithm will iteratively color the vertices of G. We will assume that throughout the algorithm, each vertex v has a list $P(v) \subseteq \{1, 2, 3\}$ of *available colors*. We call $P(v)$ the *palette of* v. The goal is then to find a proper coloring of G in which each vertex is colored by one of its available colors. The problem of deciding the existence of such coloring is known as the *list-3-coloring problem*, and is a generalization of the 3-coloring problem. Whenever a vertex x of G is colored by a color c in the course of the algorithm, we immediately remove c from the palette of x's neighbors. Additionally, if the vertex x is not in N_0, it is then deleted. The vertices in N_0 are kept in G even after they are colored. We then update the list-3-coloring instance using the following *basic reductions*:

- If a vertex y has only one color c' left in $P(y)$, we color it by the color c' and remove c' from the palettes of its neighbors. If $y \notin N_0$, we then delete y.
- If $P(y)$ is empty for a vertex y, the instance of list-3-coloring is rejected.
- If, for a vertex $y \notin N_0$, the size of $P(y)$ is greater than the degree of y, we delete y.
- *Diamond consistency rule*: If y and y' are a pair of nonadjacent vertices such that $P(y) \neq P(y')$, and if $N(y) \cap N(y')$ is not an independent set, then any valid 3-coloring of G must assign the same color to y and y'; we therefore replace both $P(y)$ and $P(y')$ with $P(y) \cap P(y')$.
- *Neighborhood domination rule*: If y and y' are a pair of nonadjacent vertices such that $N(y) \subseteq N(y')$ and $P(y') \subseteq P(y)$, and if y is not in N_0, we delete y.
- If G has a connected component in which every vertex has at most two available colors, we determine whether the component is colorable by reducing the problem to an instance of 2-SAT. If the component can be colored, we remove it from G and continue, otherwise we reject the whole instance.
- If a connected component of G is P_4-free, we solve the list-3-coloring problem for this component by Theorem 2. If the component is colorable we remove it, otherwise we reject the whole instance G.

It is clear that the rules are correct in the sense that the instance of list-3-coloring produced by a basic reduction is list-3-colorable if and only if the original instance was list-3-colorable. It is also clear that we may determine in polynomial time whether an instance of list-3-coloring (with fixed N_0) permits an application of a basic reduction, and perform the basic reduction, if available. Throughout the algorithm, we apply the basic reductions greedily as long as possible, until we reach a situation where none of them is applicable.

The basic reductions by themselves are not sufficient to solve the 3-coloring problem for G. Our algorithm will sometimes also need to perform branching, i.e., explore several alternative ways to color a vertex or a set of vertices. Formally, this means that the algorithm reduces a given instance G of list-3-coloring to an equivalent set of instances $\{G_1, \ldots, G_k\}$; here saying that a list-3-coloring instance G is *equivalent* to a set $\{G_1, \ldots, G_k\}$ of instances means that G has a solution if and only if at least one of $G_1, \ldots, G_k$ has a solution.

In the beginning of the algorithm, we attach to each vertex v of G the list $P(v) = \{1, 2, 3\}$ of available colors, thereby formally transforming it to an instance of list-3-coloring. The algorithm will then try all possible proper

3-colorings of N_0, and for each such coloring, apply basic reductions as long as any basic reduction is applicable. If this fails to color all the vertices, more complicated reduction steps and further branching will be performed, to be described later. Overall, the algorithm will ensure that the initial instance G is eventually reduced to a set of at most polynomially many smaller instances, each of which can be transformed to an equivalent instance of 2-SAT, which then can be solved efficiently.

The case when G is C_7-free can be handled in a simple way.

Proposition 1 *(♣)*[2]. *The 3-coloring problem for a $(2P_4, C_5, C_7)$-free graph G can be solved in polynomial time.*

From now on, we assume that the graph G contains an induced C_7. We choose one such C_7 as N_0, and define N_1 and N_2 accordingly.

More Complicated Reductions. Apart from the basic reductions described previously, which we will apply whenever opportunity arises, we will also use more complicated reductions, to be applied in specific situations.

Cut Reduction. Suppose $G = (V, E)$ is a connected instance of list-3-coloring. Let $X \subseteq V$ be a vertex cut of G, let C be a union of one or more connected components of $G - X$, and let C_X be the subgraph of G induced by $C \cup X$. Suppose further that the following conditions hold.

- C has at least two vertices.
- X is an independent set in G.
- All the vertices in X have the same palette, which has size 2.
- For any two vertices x, x' in X, we have $N(x) \cap C = N(x') \cap C$.
- The graph C_X is P_4-free.

Assume without loss of generality that all the vertices of X have palette equal to $\{1, 2\}$. Let us say that a coloring $c \colon X \to \{1, 2\}$ of X is *feasible for* C, if it can be extended into a proper 3-coloring of the list-3-coloring instance C_X. Note that the feasibility of a given coloring can be determined in polynomial time by Theorem 2, because C_X is a cograph.

We distinguish three types of possible colorings of X: the *all-1* coloring colors all the vertices of X by the color 1, the *all-2* coloring colors all the vertices of X by color 2, and a *mixed* coloring is a coloring that uses both available colors on X. Observe that if X admits at least one mixed coloring feasible for C, then every (not necessarily mixed) coloring of X by colors 1 and 2 is feasible for C. This is because when we extend a mixed coloring of X to a coloring of C_X, all the vertices $y \in C$ must receive the color 3. If such a coloring of C exists, we can combine it with any coloring of X by colors 1 and 2.

The *cut reduction* of X and C is an operation which reduces G to a smaller, equivalent list-3-coloring instance, determined as follows. We choose an arbitrary mixed coloring c of X, and check whether it is feasible for C. If it is feasible, we

[2] Due to space limitations, we defer some proofs to the full version of our paper [26]. We mark the respective statements by (♣).

reduce the instance G to $G - C$, leaving the palettes of the remaining vertices unchanged. The new instance is equivalent to the original one, since any proper list-3-coloring of $G - C$ can be extended to a coloring of G, because all the colorings of X are feasible for C.

If the mixed coloring c is not feasible for C, we know that no mixed coloring is feasible. We then test the all-1 and the all-2 coloring for feasibility. If both are feasible, we reduce the instance G by replacing C with a single new vertex v, with palette $P(v) = \{1, 2\}$, and connecting v to all the vertices of X. Note that the reduced instance is an induced subgraph of the original one. It is easy to see that the reduced instance is equivalent to the original one.

If only one coloring of X is feasible for C, we delete C, color the vertices of X using the unique feasible coloring, and delete the corresponding color from the palettes of the neighbors of X in $G - C$. If no coloring of X is feasible for C, we declare that G is not list-3-colorable.

Neighborhood Collapse. Let G be an instance of list-3-coloring, and let v be a vertex of G. Suppose that $N(v)$ induces in G a connected bipartite graph with nonempty partite classes X and Y. Suppose furthermore that all the vertices of X have the same palette P_X, and all the vertices in Y have the same palette P_Y. The *neighborhood collapse* of the vertex v is the operation that replaces X and Y by a pair of new vertices x and y, adjacent to each other and to v, with the property that any vertex of $G - Y$ adjacent to at least one vertex in X will be made adjacent to x, and similarly every vertex adjacent to Y in $G - X$ will be adjacent to y. We then set $P(x) = P_X$ and $P(y) = P_Y$.

It is clear that the collapsed instance is equivalent to the original one. However, since the new instance is not necessarily an induced subgraph of the original one, it might happen, e.g., that a collapse performed in a C_5-free graph will introduce a copy of C_5 in the collapsed instance. In our algorithm, we will only perform collapses at a stage when C_5-freeness will no longer be needed.

On the other hand, $2P_4$-freeness is preserved by collapses, as we now show.

Lemma 2 (♣). *Let G be a $2P_4$-free instance of list-3-coloring in which a neighborhood collapse of a vertex v may be performed, and let G^* be the graph obtained by the collapse. Then G^* is $2P_4$-free.*

Graphs Containing C_7. We now turn to the most complicated part of our coloring algorithm, which solves the 3-coloring problem for a $(2P_4, C_5)$-free graph G that contains an induced C_7. We let N_0 be an induced copy of C_7 in this graph, and define N_1 and N_2 accordingly.

We let $v_1, v_2, \ldots, v_7$ denote the vertices of N_0, in the order in which they appear on the cycle N_0. We evaluate their indices modulo 7, so that, e.g., $v_8 = v_1$.

Fix a proper coloring of N_0, and apply the basic reductions to G until no basic reduction is applicable. We now analyze the structure of G at this stage of the algorithm. We again let $N_0(x)$ denote the set of neighbors of x in N_0.

Lemma 3 (♣). *After fixing the coloring of N_0 and applying all available basic reductions, the graph G has the following properties.*

- *Each vertex x of N_1 satisfies either $N_0(x) = \{v_i\}$ for some i, or $N_0(x) = \{v_i, v_{i+2}\}$ for some i.*
- *Each induced copy of P_4 in G has at most two vertices in N_2.*
- *G is connected.*

Lemma 3 is the last part of the proof that makes use of the C_5-freeness of G. From now on, we will not need to use the fact that G is C_5-free. In particular, we will allow ourselves reduction operations, such as the neighborhood collapse, which do not preserve C_5-freeness.

We will assume, without mentioning explicitly, that after performing any modification of the list-3-coloring instance G, we always apply basic reductions until no more basic reductions are available.

In the rest of the proof, we use the term *top component* to refer to a connected component of N_2. Observe that every top component is P_4-free and therefore has a dominating set of size at most 2 [12]. We say that a top component is *relevant*, if it contains a vertex z with $|P(z)| = 3$. Note that if G has no relevant top component, then all its vertices have at most two available colors, and the coloring problem can be solved by a single basic reduction. We will say that a vertex x of N_1 is *relevant* if x is adjacent to a vertex belonging to a relevant top component. Let $x \in N_1$ be a vertex, and let C be a top component. We say that x is a *partial neighbor* of C, if x is adjacent to at least one but not all the vertices of C. We say that x is a *full neighbor* of C, if it is adjacent to every vertex of C.

Lemma 4 *(♣). Suppose $x \in N_1$ is a partial neighbor of a top component C. Then x is not a neighbor of any other top component. Moreover, $|N_0(x)| = 2$.*

We will now reduce G to a set of polynomially many instances in which the set of relevant vertices has special form. We first eliminate the relevant vertices that have only one neighbor in N_0. Let R_i be the set of relevant vertices that are adjacent to v_i and not adjacent to any other vertex of N_0.

Lemma 5 *(♣). For any $i \in \{1, \dots, 7\}$, we can reduce G to an equivalent set of at most two instances, both of which satisfy $R_i = \emptyset$.*

From now on, we deal with instances of G where every relevant vertex has exactly two neighbors in N_0. Let S_i be the set of relevant vertices adjacent to v_i.

Lemma 6 *(♣). For any $i \in \{1, \dots, 7\}$, we can reduce G to an equivalent set of polynomially many instances, each of which satisfies $S_i = \emptyset$ or $S_{i+3} = \emptyset$.*

From now on, assume that we deal with an instance G in which for every i, one of the two sets S_i and S_{i+3} is empty. Unless the instance is already completely solved, there must be at least one relevant vertex. Assume without loss of generality that G has a relevant vertex adjacent to v_1 and v_3. It follows that S_1 and S_3 are nonempty, and hence S_4, S_5, S_6 and S_7 are empty. Moreover, as any relevant vertex is adjacent to a pair of vertices of the form $\{v_i, v_{i+2}\}$, it follows that S_2 is empty as well. In particular, every relevant vertex x satisfies

$N_0(x) = \{v_1, v_3\}$. It follows that all the relevant vertices have the same palette of size 2; assume without loss of generality that this palette is $\{1, 2\}$.

We will now focus on describing the structure of the subgraph of G induced by the relevant vertices and the relevant top components adjacent to them. Let R denote the set of relevant vertices. Note that the subgraph of G induced by $R \cup N_2$ does not contain P_4, otherwise we could use the path $v_4v_5v_6v_7$ to get a $2P_4$ in G. Note also that if two relevant vertices x and y are adjacent, then any common neighbor of x and y must be colored by color 3, thanks to the diamond consistency rule. We thus know that adjacent relevant vertices have no common neighbors outside N_0. We may also assume that the graph induced by the relevant vertices is bipartite, otherwise G would clearly not be 3-colorable.

Lemma 7 (♣). *Suppose that x and y are two adjacent relevant vertices. Let us write $X' = N_2(x)$ and $Y' = N_2(y)$. Then there are disjoint sets $X, Y \subseteq R$, with $x \in X$ and $y \in Y$, satisfying these properties:*

1. *Every vertex in $X \cup Y'$ is adjacent to every vertex in $Y \cup X'$.*
2. *X and Y are independent sets of G.*
3. *The vertices in $X' \cup Y'$ are only adjacent to vertices in $X \cup Y \cup X' \cup Y'$; in particular, $X' \cup Y'$ induce a top component.*

Suppose $G[R]$ contains at least one edge xy, and let X, Y, X', Y' be as in the previous lemma. Note that there are only two possible ways to color $G[X \cup Y]$ – either X is colored 1 and Y is colored 2, or vice versa. We can check in polynomial time which of these two colorings can be extended to a valid coloring of $G[X \cup Y \cup X' \cup Y']$. If neither of the two colorings extends, we reject G, if only one of the two coloring extends, we color $X \cup Y$ accordingly, and if both colorings extend, we remove the vertices $X' \cup Y'$ from G, resulting in a smaller equivalent instance, in which $X \cup Y$ are no longer relevant. Repeating this for every component of $G[R]$ that contains at least one edge, we eventually reduce the problem to an instance in which the relevant vertices form an independent set. From now on, we assume R is independent in G. For a vertex $x \in R$, let $C_2(x)$ denote the set of top components that contain at least one neighbor of x.

Lemma 8 (♣). *For any two relevant vertices x and y, we either have $C_2(x) = C_2(y)$, or $C_2(x)$ and $C_2(y)$ are disjoint.*

Let us say that two relevant vertices x and y are *equivalent* if $C_2(x) = C_2(y)$. As the next step in our algorithm, we will process the equivalence classes one by one, with the aim to reduce the instance G to an equivalent instance in which each relevant vertex is adjacent to a single top component.

Let $x \in R$ be a vertex such that $|C_2(x)| \geq 2$, and let R_x be the equivalence class containing x. By Lemma 4, each vertex in R_x is a full neighbor of any component in $C_2(x)$, and by Lemma 8, no vertex outside of R_x may be adjacent to a relevant top component in $C_2(x)$. Thus, R_x is a vertex cut separating the relevant top components in $C_2(x)$ from the rest of G. We may therefore apply the cut reduction through the vertex cut R_x to reduce G to a smaller instance in which the vertices of R_x are no longer relevant.

We repeat the cut reductions until there is no relevant vertex adjacent to more than one top component. From now on, we deal with instances in which each relevant vertex is adjacent to a unique top component; note that this top component is necessarily relevant.

Lemma 9 (♣). *Let x be a relevant vertex, let C be the top component adjacent to x, let R_x be the equivalence class of x, and let $y \in R_x \cup C$ be a vertex not adjacent to x. Then y is adjacent to at least one vertex in $N_2(x)$. Moreover, if $N_2(x)$ induces a disconnected subgraph of G, then y is adjacent to all the vertices of $N_2(x)$.*

Fix now a relevant top component C and let R be set of relevant vertices in N_1 adjacent to C. Fix a vertex $x \in R$ so that $N_2(x)$ is as large as possible. Let R_x be the equivalence class containing x. We distinguish several possibilities, based on the structure of $N_2(x)$.

$N_2(x)$ is Disconnected. Suppose first that $N_2(x)$ induces in G a disconnected subgraph. By Lemma 9, any vertex in R_x is adjacent to all vertices in $N_2(x)$. By our choice of x, this implies that for any $x' \in R_x$ we have $N_2(x') = N_2(x)$. We may therefore apply the cut reduction for the cut R_x that separates C from the rest of G, to obtain a smaller instance in which the vertices of R_x are no longer relevant.

$N_2(x)$ is Connected, with ≥3 Vertices. Now suppose that $N_2(x)$ induces a connected graph, and that $N_2(x)$ has at least three vertices. We now verify that $N_2(x)$ induces a complete bipartite graph, otherwise C contains P_4 or G is not 3-colorable. Let Y and Z be the two partite classes of $N_2(x)$. Note that any two vertices y, y' in Y have the same neighbors in G: indeed if u were a vertex adjacent to y but not to y', then $uyxy'$ would induce a copy of P_4. By the same argument, all the vertices in Z have the same neighbors in G as well. Diamond consistency enforces that all the vertices in Y have the same palette, and similarly for Z. We may then invoke neighborhood domination to delete from Y all vertices except a single vertex y, and do the same with Z, reducing G to an equivalent instance in which $N_2(x)$ consists of a single edge.

$N_2(x)$ is a Single Vertex. Suppose that $N_2(x)$ consists of a single vertex y. If y is the only vertex of C, then y must have the palette $\{1, 2, 3\}$, otherwise C would not be a relevant component. In such case, we may simply color y with color 3 and delete it, as this does not restrict the possible colorings of $G - y$ in any way. If, on the other hand, C has more than one vertex, it follows from Lemma 9 that all the vertices of R_x are adjacent to y, and by the choice of x, every vertex in R_x is adjacent to y as its only neighbor in C. We may then apply cut reduction for the cut R_x. In all cases, we obtain a smaller equivalent instance, in which the vertices in R_x are no longer relevant.

$N_2(x)$ is a Single Edge. The last case to consider deals with the situation when $N_2(x)$ contains exactly two adjacent vertices u and v. Assume that $\deg_G(u) \geq \deg_G(v)$. Recall that the set R of relevant vertices is independent. Note that for any vertex $x' \in R_x$, $N_2(x')$ is connected, otherwise Lemma 9 implies that $N_2(x')$ is contained in $N_2(x)$, contradicting $N_2(x)$ being a single edge.

We first claim that any vertex $y \in R_x \cup (C - u)$ adjacent to v is also adjacent to u. Suppose this is not the case. Then, since $\deg_G(u) \geq \deg_G(v)$, there must also be a vertex $z \in R_x \cup (C - v)$ adjacent to u but not to v. If yz is an edge, then $zyvx$ is a copy of P_4, and if yz is not an edge, then $zuvy$ is a copy of P_4. In both cases we have a contradiction, establishing the claim. Note that the claim, together with Lemma 9, implies that u is adjacent to all the other vertices of $R_x \cup C$.

Next, we show that if C contains a vertex adjacent to both u and v, then we may reduce G to a smaller equivalent instance. Suppose $y \in C$ is adjacent to u and v. Then $P(y) = P(x) = \{1, 2\}$ by diamond consistency. We now claim that y has no other neighbors in G beyond u and v. Suppose that $z \notin \{u, v\}$ is a neighbor of y. Then z cannot be adjacent to v, since $uvyz$ would form a K_4. Therefore $zyvx$ is a copy of P_4, a contradiction. We conclude that $N(y) = \{u, v\} \subseteq N(x)$, and since $P(y) = P(x)$, we may delete y due to neighborhood domination.

From now on, we assume that u and v have no common neighbor in C. Recall that u is adjacent to all the other vertices in $C \cup R_x$. We now reduce G to an instance where $C - u$ is an independent set. We already know that v is isolated in $C - u$ by the previous paragraph. Suppose that $C - u$ has a component D with more than one vertex. If D has a vertex v' adjacent to a vertex $x' \in R_x$, we can repeat the reasoning of the previous paragraph with x' and v' in the place of x and v, showing that u and v' cannot have any common neighbor in C, contradicting the assumption that D has more than one vertex. We can thus conclude that D is not adjacent to any vertex in R_x. Then u is a cut-vertex separating D from the rest of G. We may test which colorings of u can be extended into D (since D is P_4-free, this can be done efficiently), then restrict the palette of u to only the feasible colors, and then delete D.

We are now left with a situation when C is a star with center u, and every vertex of R_x is adjacent to u and to at most one vertex of $C - u$. If there is a vertex $w \in C - u$ adjacent to more than one vertex in R_x, it means that the neighborhood of w is a connected bipartite graph (a star with center u) to which we may apply neighborhood collapse.

Suppose now that every vertex $w \in C - u$ has only one neighbor in R_x (if w had no neighbor in R_x, it would have degree 1 and we could remove it). If w's palette has 3 colors, we can remove it, so we may assume that every vertex in $C - u$ has a palette of size 2. Then u's palette must have 3 colors, otherwise C would not be a relevant component. If a vertex in $C - u$ has palette $\{1, 2\}$, then u must be colored 3 and then R_x is no longer relevant.

It remains to consider the case when each vertex of $C - u$ has the palette $\{1, 3\}$ or $\{2, 3\}$. Let W_1 and W_2 be the sets of vertices of $C - u$ having palette $\{1, 3\}$ and $\{2, 3\}$, respectively. Let X_1 and X_2 be the sets of vertices of R_x that are adjacent to a vertex in W_1 and W_2, respectively. Let X_0 be the set of vertices in R_x that have no neighbor in $C - u$. Let us consider the possible colorings of $C \cup R_x$. If u is colored by 3, then the whole set W_1 is colored by 1, W_2 is colored by 2, hence X_1 is colored by 2 and X_2 by 1, while the vertices in X_0 can be colored arbitrarily by 1 or 2. On the other hand, if u receives a color $\alpha \neq 3$, then all the vertices in R_x receive the color $\beta \in \{1,2\} \setminus \{\alpha\}$, and the vertices in $C - u$ can be colored by 3. The set R_x therefore admits three types of feasible colorings: the all-1 coloring, the all-2 coloring, and any coloring where the set X_1 is colored by 2 and X_2 by 1. This set of colorings can be equivalently characterized by the following properties:

- If a vertex in X_1 is colored by 1, then the whole R_x receives 1.
- If a vertex in X_2 is colored by 2, then the whole R_x is colored by 2.
- All the colors in X_1 are equal and all the colors in X_2 are equal.

The above properties can be encoded by a 2-SAT formula whose variables correspond to vertices of R_x. □

References

1. Bonomo, F., Chudnovsky, M., Maceli, P., Schaudt, O., Steinz, M., Zhong, M.: Three-coloring and list three-coloring of graphs without induced paths on seven vertices. Combinatorica **38**(4), 779–801 (2018). https://doi.org/10.1007/s00493-017-3553-8
2. Bonomo, F., Schaudt, O., Stein, M.: 3-colouring graphs without triangles or induced paths on seven vertices. CoRR (2014). https://arxiv.org/abs/1410.0040v1
3. Bonomo-Braberman, F., et al.: Better 3-coloring algorithms: excluding a triangle and a seven vertex path. Theoret. Comput. Sci. **850**, 98–115 (2020). https://doi.org/10.1016/j.tcs.2020.10.032
4. Christen, C.A., Selkow, S.M.: Some perfect coloring properties of graphs. J. Comb. Theory Ser. B **27**(1), 49–59 (1979). https://doi.org/10.1016/0095-8956(79)90067-4
5. Chudnovsky, M., Huang, S., Spirkl, S., Zhong, M.: List 3-coloring graphs with no induced $P_6 + rP_3$. Algorithmica **83**(1), 216–251 (2020). https://doi.org/10.1007/s00453-020-00754-y
6. Chudnovsky, M., Maceli, P., Stacho, J., Zhong, M.: 4-coloring P_6-free graphs with no induced 5-cycles. J. Graph Theory **84**(3), 262–285 (2017). https://doi.org/10.1002/jgt.22025
7. Chudnovsky, M., Maceli, P., Zhong, M.: Three-coloring graphs with no induced seven-vertex path I: the triangle-free case. CoRR (2014). https://arxiv.org/abs/1409.5164
8. Chudnovsky, M., Maceli, P., Zhong, M.: Three-coloring graphs with no induced seven-vertex path II: using a triangle. CoRR (2015). https://arxiv.org/abs/1503.03573
9. Chudnovsky, M., Robertson, N., Seymour, P., Thomas, R.: The strong perfect graph theorem. Ann. Math. **164**(1), 51–229 (2006). http://www.jstor.org/stable/20159988
10. Chudnovsky, M., Spirkl, S., Zhong, M.: List-three-coloring P_t-free graphs with no induced 1-subdivision of $K_{1,s}$. Discrete Math. **343**(11), 112086 (2020). https://doi.org/10.1016/j.disc.2020.112086
11. Chudnovsky, M., Stacho, J.: 3-colorable subclasses of P_8-free graphs. SIAM J. Discrete Math. **32**(2), 1111–1138 (2018). https://doi.org/10.1137/16m1104858
12. Corneil, D.G., Perl, Y.: Clustering and domination in perfect graphs. Discrete Appl. Math. **9**(1), 27–39 (1984). https://doi.org/10.1016/0166-218x(84)90088-x

13. Couturier, J.F., Golovach, P.A., Kratsch, D., Paulusma, D.: List coloring in the absence of a linear forest. Algorithmica **71**(1), 21–35 (2013). https://doi.org/10.1007/s00453-013-9777-0
14. Emden-Weinert, T., Hougardy, S., Kreuter, B.: Uniquely colourable graphs and the hardness of colouring graphs of large girth. Comb. Probab. Comput. **7**(4), 375–386 (1998). https://doi.org/10.1017/S0963548398003678
15. Gartland, P., Lokshtanov, D.: Independent set on P_k-free graphs in quasi-polynomial time. In: 61st IEEE Annual Symposium on Foundations of Computer Science, FOCS 2020, Durham, NC, USA, 16–19 November 2020, pp. 613–624. IEEE (2020). https://doi.org/10.1109/FOCS46700.2020.00063
16. Goedgebeur, J., Schaudt, O.: Exhaustive generation of k-critical H-free graphs. J. Graph Theory **87**(2), 188–207 (2017). https://doi.org/10.1002/jgt.22151
17. Golovach, P.A., Johnson, M., Paulusma, D., Song, J.: A survey on the computational complexity of coloring graphs with forbidden subgraphs. J. Graph Theory **84**(4), 331–363 (2016). https://doi.org/10.1002/jgt.22028
18. Golovach, P.A., Paulusma, D., Song, J.: Closing complexity gaps for coloring problems on H-free graphs. Inf. Comput. **237**, 204–214 (2014). https://doi.org/10.1016/j.ic.2014.02.004
19. Golovach, P.A., Paulusma, D., Song, J.: Coloring graphs without short cycles and long induced paths. Discrete Appl. Math. **167**, 107–120 (2014). https://doi.org/10.1016/j.dam.2013.12.008
20. Grötschel, M., Lovász, L., Schrijver, A.: Polynomial algorithms for perfect graphs. In: Topics on Perfect Graphs, pp. 325–356. Elsevier (1984). https://doi.org/10.1016/s0304-0208(08)72943-8
21. Hell, P., Huang, S.: Complexity of coloring graphs without paths and cycles. Discrete Appl. Math. **216**, 211–232 (2017). https://doi.org/10.1016/j.dam.2015.10.024
22. Hoàng, C.T., Kamiński, M., Lozin, V., Sawada, J., Shu, X.: Deciding k-colorability of P_5-free graphs in polynomial time. Algorithmica **57**(1), 74–81 (2008). https://doi.org/10.1007/s00453-008-9197-8
23. Holyer, I.: The NP-completeness of edge-coloring. SIAM J. Comput. **10**(4), 718–720 (1981). https://doi.org/10.1137/0210055
24. Huang, S.: Improved complexity results on k-coloring P_t-free graphs. Eur. J. Comb. **51**, 336–346 (2016). https://doi.org/10.1016/j.ejc.2015.06.005
25. Huang, S., Johnson, M., Paulusma, D.: Narrowing the complexity gap for colouring (C_s, P_t)-free graphs. Comput. J. **58**(11), 3074–3088 (2015). https://doi.org/10.1093/comjnl/bxv039
26. Jelínek, V., Klimošová, T., Masařík, T., Novotná, J., Pokorná, A.: Note on 3-coloring of $(2P_4, C_5)$-free graphs. CoRR (2020). https://arxiv.org/abs/2011.06173
27. Karp, R.M.: Reducibility among Combinatorial Problems, pp. 85–103. Springer, Boston (1972). https://doi.org/10.1007/978-1-4684-2001-2_9
28. Klimošová, T., Malík, J., Masařík, T., Novotná, J., Paulusma, D., Slívová, V.: Colouring $(P_r + P_s)$-free graphs. Algorithmica **82**(7), 1833–1858 (2020). https://doi.org/10.1007/s00453-020-00675-w
29. Krom, M.R.: The decision problem for a class of first-order formulas in which all disjunctions are binary. Zeitschrift für Mathematische Logik und Grundlagen der Mathematik **13**(1–2), 15–20 (1967). https://doi.org/10.1002/malq.19670130104
30. Leven, D., Galil, Z.: NP completeness of finding the chromatic index of regular graphs. J. Algorithms **4**(1), 35–44 (1983). https://doi.org/10.1016/0196-6774(83)90032-9

31. Pilipczuk, M., Pilipczuk, M., Rzążewski, P.: Quasi-polynomial-time algorithm for independent set in P_t-free graphs via shrinking the space of induced paths. In: Symposium on Simplicity in Algorithms (SOSA), pp. 204–209. Society for Industrial and Applied Mathematics, January 2021. https://doi.org/10.1137/1.9781611976496.23
32. Rojas, A., Stein, M.: 3-colouring P_t-free graphs without short odd cycles. CoRR (2020). https://arxiv.org/abs/2008.04845
33. Spirkl, S., Chudnovsky, M., Zhong, M.: Four-coloring P_6-free graphs. In: Proceedings of the Thirtieth Annual ACM-SIAM Symposium on Discrete Algorithms, pp. 1239–1256. Society for Industrial and Applied Mathematics, January 2019. https://doi.org/10.1137/1.9781611975482.76

Correction to: Bears with Hats and Independence Polynomials

Václav Blažej, Pavel Dvořák, and Michal Opler

Correction to:
Chapter "Bears with Hats and Independence Polynomials" in: Ł. Kowalik et al. (Eds.): *Graph-Theoretic Concepts in Computer Science*, LNCS 12911, https://doi.org/10.1007/978-3-030-86838-3_22

Chapter "Bears with Hats and Independence Polynomials" was previously published non-open access. It has now been changed to open access under a CC BY 4.0 license and the copyright holder updated to 'The Author(s)'. The book has also been updated with this change.

The updated version of this chapter can be found at
https://doi.org/10.1007/978-3-030-86838-3_22

Ł. Kowalik et al. (Eds.): WG 2021, LNCS 12911, p. C1, 2022.
https://doi.org/10.1007/978-3-030-86838-3_31

Correction to: Graph-Theoretic Concepts in Computer Science

Łukasz Kowalik, Michał Pilipczuk, and Paweł Rzążewski

Correction to:
Chapters 1 and 6 in Ł. Kowalik et al. (Eds.):
***Graph-Theoretic Concepts in Computer Science*, LNCS 12911,**
https://doi.org/10.1007/978-3-030-86838-3

"Preprocessing to Reduce the Search Space: Antler Structures for Feedback Vertex Set", written by Huib Donkers, Bart M. P. Jansen;

"FPT Algorithms to Compute the Elimination Distance to Bipartite Graphs and More", written by Bart M. P. Jansen, Jari J. H. de Kroon.

With the authors' decision to opt for Open Choice the copyright of the chapter changed on 16 August 2022 to © Author(s) 2023 and the chapter is forthwith distributed under a Creative Commons Attribution.

Funded by the European Research Council (ERC) under the European Union's Horizon 2020 research and innovation programme (grant agreement No 803421, ReduceSearch).

The updated versions of these chapters can be found at
https://doi.org/10.1007/978-3-030-86838-3_1
https://doi.org/10.1007/978-3-030-86838-3_6

Ł . Kowalik et al. (Eds.): WG 2021, LNCS 12911, p. C2, 2023.
https://doi.org/10.1007/978-3-030-86838-3_32

Author Index

Printed in the United States
by Baker & Taylor Publisher Services